FOUNDATIONS OF POINT SET THEORY

AMERICAN MATHEMATICAL SOCIETY
COLLOQUIUM PUBLICATIONS, VOLUME XIII

FOUNDATIONS OF POINT SET THEORY

R. L. MOORE

Revised Edition

AMERICAN MATHEMATICAL SOCIETY
PROVIDENCE, RHODE ISLAND

First edition published 1932
Completely revised and enlarged edition 1962
Reprinted with corrections 1970
Reprinted 1987

International Standard Book Number 0-8218-1013-8
Library of Congress Catalog Card Number 62-8325

Printed in the United States of America

The paper used in this book is acid-free and falls within the guidelines established to ensure permanence and durability.

TABLE OF CONTENTS

PREFACE

In this revised edition there is again presented what may be roughly termed a largely self-contained treatment of the foundations of continuity, or point set-theoretic, analysis situs (topology).

All the numbered propositions of Chapter I are proved on the basis of two axioms (0 and 1) that hold true in a very large class of spaces including many spaces that are not locally connected. If these axioms hold true in a space Σ and S' is *any* inner limiting set in Σ then, under a suitable interpretation of the word *region*, they hold true also in a space in which *point* means point of S'.

The numbered propositions of Chapter II hold true in all locally connected spaces that satisfy the first three axioms. This class of spaces includes Euclidean spaces of any finite number of dimensions, Hilbert space, and the spaces of infinitely many dimensions which Fréchet designates by the symbols D_ω and E_ω. If Axioms 0, 1 and 2 hold true in a space Σ and S' is a locally connected inner limiting set in Σ then, under a suitable interpretation of the word region, they hold true also in a space in which the word *point* is restricted to mean point of S'.

It is shown, in Chapters III and IV, that, on the basis of Axioms 0, 1, 2, 3, 4 and 5, it is possible to prove a very considerable portion of the well known topological propositions of the plane. Nevertheless there exist spaces which satisfy these axioms, and therefore in which all the numbered theorems of these chapters hold true, but which are neither metric, locally compact nor separable and in which, moreover, it is not even true that if P is a point of a domain D then there exists a domain lying in D, containing P, and bounded by a simple closed curve.

Chapter V is concerned particularly with upper semi-continuous collections.

In Chapter VI there is formulated a set of axioms having the property that every compact space that satisfies all of them is topologically equivalent to a sphere, while every noncompact one is topologically equivalent to a plane.

In his paper *Concerning R. L. Moore's Axiom* 5, F. Burton Jones showed that if Axiom 5′ denotes the axiom obtained by replacing *simple closed curve* by *compact continuum* in the statement of Axiom 5 of the first edition of this book then the original Axiom 5 is a consequence of Axioms 0–4 and 5′. This Axiom 5′ is the Axiom 5 of the present edition. The author considers this improvement, due to Jones, to be a major one. No single axiom in the set thus revised implies the existence of a simple closed curve or even of an arc.

In each of a comparatively small number of instances, the name of a mathematician has been written in conjunction with a theorem. There are probably many cases in which it could have been done just as appropriately as in the cases where it has been done. The author did not however care to assume the responsibility of fixing the credit for more than a small proportion of the propositions proved. Thus the fact that no name is attached to a particular theorem and that no reference is made in connection with it does not, by any means, necessarily imply that no one other than the author has stated the theorem in question and proved it by an argument that would apply, with little or no change, on the basis of the axioms employed in the treatment given here. On the other hand, the fact that the name of a mathematician is given in connection with a theorem does not necessarily imply that the proof given in this book is due to that mathematician or even that any argument that has been given by him would be at all sufficient to prove the theorem in question on the basis of the axioms here employed.

There is appended a bibliography which is *much* more extensive than the one in the first edition but which is far from being complete either as to authors listed or (with possibly a few exceptions) as to the publications of those who are listed. In particular, much of the recent work on topological groups, special types of mappings, homotopy, orbit theory, periodicity, etc., has been omitted.

The author is not at all satisfied with this bibliography. But he feels that he would not be satisfied with any selection which, under existing self-imposed limitations of space and of time, he would be able to make (for this listing) from the great number of publications that now constitute the literature in this field.

Among the subjects which have been either omitted entirely from the book or only slightly treated are (1) Dimension Theory, (2) certain branches of what may be roughly termed Metrical Point Set Theory, including the Theory of Measure, with applications, in particular, to both Real and Complex Variable Theory, (3) a large body of propositions relating to continuous transformations and their Fixed Points, (4) the relationship between the Foundations of Geometry and Topology and, to the author's regret, (5) a certain (now considerable) group of propositions concerning paracompactness and related subjects.

I sincerely thank the Colloquium Editorial Committee of The American Mathematical Society for giving its approval to the preparation of this second edition.

AUSTIN, TEXAS R. L. MOORE

INTRODUCTION

We shall begin at the foundations of the subject, basing the treatment on a set of axioms. The undefined notions peculiar to the subject will be *point* and *region*. Other special notions will be defined in terms of these. General logical notions and propositions including the logic of classes, the notion *sensed* pair, and fundamental propositions concerning natural numbers will be taken for granted and used freely, ordinarily without explicit formulation.

The words set, collection and family are used synonymously. Thus "a family of collections of sets" is synonymous with "a set of sets of sets" but the former phrase would seem preferable for rhetorical reasons.

Among the fundamental propositions of the logic of classes the author includes the Zermelo Axiom, which may be stated as follows.

ZERMELO AXIOM. *If G is a collection of mutually exclusive sets there exists a set K such that* (1) *each set of the collection G contains one and only one element of K and* (2) *each element of K belongs to some set of the collection G.*

With the aid of this *axiom* of logic it is possible to prove the more general Zermelo *Proposition* which relates to collections of sets which are not necessarily mutually exclusive.

ZERMELO PROPOSITION. *If G is a collection of sets there exists a collection H of sensed pairs such that* (1) *if (x,y) is a sensed pair of the collection H then x, the first term of (x,y) is a set of the collection G and y, the second term of this pair, is an element of the set x,* (2) *each set of the collection G is the first term of some sensed pair of the collection H,* (3) *no two sensed pairs of the collection H have the same first term.*

PROOF. For each set g of the collection G let g' denote the set of all sensed pairs (g,x) where x is an element of g. Let G' denote the collection of all such sets g' for all sets g of the collection G. Suppose g_1' and g_2' are two different sets of this collection. Then there exist two different sets g_1 and g_2 of the collection G such that g_1 is the first term of every sensed pair which is an element of g_1' and g_2 is the first term of every sensed pair which is an element of g_2'. Hence no element of g_1' is also an element of g_2'. Thus no two sets of the collection G' have an element in common. Therefore, by the Zermelo Axiom, there exists a set H such that (1) each set of the collection G' contains one and only one element of H and (2) each element of H belongs to some set of the collection G'. The set H satisfies all the requirements of the Zermelo Proposition.

DEFINITION. Suppose M is a set and G is a collection of sensed pairs such that (1) if (x,y) is an element of G, x and y are different elements of M, (2) if

(x,y) is an element of G, (y,x) is not an element of G, (3) if x and y are different elements of M, either (x,y) or (y,x) is an element of G, (4) if (x,y) and (y,x) are elements of G then (x,z) is an element of G. Then the collection G is called a *sequence*, the elements of M are called the *terms* of this sequence and x is said to *precede* y in the sequence G if, and only if, (x,y) is an element of G.

The sequence α is said to be a sequence of elements of the set M if the terms of α are elements of M. The subsequence β of the sequence α is said to be an *initial segment* of α if, and only if, it is true that if a term of α *precedes*, in α, a term of β then it *is* a term of β.

It is clear that if x and y are two terms of a sequence α then (1) either x precedes y or y precedes x, (2) if x precedes y, y does not precede x, (3) if x precedes y and y precedes z then x precedes z.

DEFINITION. The sequence α is said to be *well ordered* if, and only if, it is true that if N is any set of terms of α then there is a first term of α in N, that is to say there is an element of N which precedes, in the sequence α, every other element of N.

THEOREM. *If M is any set there exists a well ordered sequence α whose terms are the elements of M, that is to say there exists a well ordered sequence α such that each term of α is an element of M and each element of M is a term of α.*

INDICATION OF PROOF. Let G denote the collection of all subsets of M. Let H denote a collection of sensed pairs satisfying, with respect to G, all the conditions of the Zermelo Proposition. For each element g of G let P_g denote the second term of the sensed pair of the collection H of which g is the first term. There exists* a sequence α such that (1) P_M is the first term of α, that is to say it precedes every other term of α, (2) if K is any set of terms of α which contains every term of α that precedes, in α, some term of K and $M - K$ exists, then P_{M-K} is the first term of α which is preceded, in α, by every element of K. The sequence α is well ordered and every element of M is a term of α.

If K is a set and (1) for each natural number n, P_n is an element of K and (2) for each element x of K there exists only one natural number n such that x is P_n, then by the infinite sequence $P_1, P_2, P_3, \cdots$ is meant a sequence α such that (1) for each n, P_n is a term of α, (2) if X is a term of α, there exists a natural number n such that X is P_n, (3) if i and j are natural numbers then P_i precedes P_j if, and only if, $i < j$.

If m is a natural number and K is a set of m elements and, for each n less than $m + 1$, P_n is an element of K and there do not exist two natural numbers i and j, less than $m + 1$, such that P_i is identical with P_j, then by the finite sequence $P_1, P_2, P_3, \cdots, P_m$ is meant a finite sequence α such that (1) for

* For a detailed proof of the existence of such a sequence see E. Zermelo, *Beweis, dass jede Menge wohlgeordnet werden kann*, Math. Ann. **59** (1904), 514, and *Neuer Beweis für die Möglichkeit der Wohlordnung*, ibid. **65** (1908), 107.

each n less than $m + 1$, P_n is a term of α, (2) if X is a term of α there exists a natural number n less than $m + 1$ such that X is P_n, (3) if i and j are natural numbers, each less than $n + 1$, then P_i precedes P_j in α if, and only if, $i < j$.

By a simple sequence $P_1, P_2, P_3, \cdots$ is meant either an infinite sequence $P_1, P_2, P_3, \cdots$ or a finite sequence $P_1, P_2, P_3, \cdots, P_m$. Hereafter, in this book, the word *sequence* will be regarded as synonymous with *simple sequence* except when the contrary is indicated as for example where it is preceded by the qualifying phrase *well ordered*.

A set is said to be *countable* if there exists a simple sequence whose terms are the elements of that set.

According to these definitions if $P_1, P_2, P_3, \cdots$ is a simple infinite sequence and i is distinct from j then P_i is distinct from P_j. However if, for example, A and B are two elements of a set and, for every even natural number n, P_n is B and, for every odd one, P_n is A, then the phrase *the sequence* $P_1, P_2, P_3, \cdots$ may be used as an abbreviation for *the sequence* $(A,1)$, $(B,1)$, $(A,2)$, $(B,2)$, $(A,3)$, $(B,3), \cdots$. In this case, the statement that every odd term of the sequence $P_1, P_2, P_3, \cdots$ is A and every even one is B is to be regarded as an abbreviation for the statement that if n is any natural number P_{2n} is B and P_{2n-1} is A. This example illustrates an abbreviated mode of expression that will be adopted for convenience.

CHAPTER I

CONSEQUENCES OF AXIOMS 0 AND 1

AXIOM 0. *Every region is a point set.*

DEFINITIONS. A point P is said to be a *limit point* of a point set M if and only if it is true that if a region contains P it contains a point of M distinct from P. A point set is said to be *closed* if there is no limit point of it that does not belong to it. The *boundary* of a point set M is a set such that X belongs to it if and only if every region that contains X contains at least one point of M and at least one point that does not belong to M. A point of the boundary of M is called a boundary point of M. If G is a collection of sets, by the *sum* of all the sets of the collection G is meant a set such that X belongs to it if and only if X belongs to some set of the collection G. The sum of all the point sets of the collection G will be denoted by the symbol G^*. The sum of the sets M and N is denoted by $M + N$. By a *proper* subset of the set M is meant a subset of M that is not identical with M. If N is a proper subset of M then $M - N$ denotes a set such that X belongs to it if and only if X belongs to M but not to N. If there exists at least one element common to all the sets of the collection G, the *common part* of the sets of the collection G is the set to which X belongs if and only if X belongs to every set of the collection G. If the set M has an element in common with the set N it is said to *intersect* N and the common part of M and N is denoted by $M \cdot N$. Two point sets are said to be *mutually separated* if they are mutually exclusive and neither of them contains a limit point of the other one. A point set is said to be *compact* if and only if there is no infinite subset of it that has no limit point. The collection G of sets is said to *cover* the set M if and only if each element of M belongs to some set of the collection G.

NOTATION. Hereafter, throughout this book, the letter S will be used to denote the set to which X belongs if and only if X is a point. If M is a point set, the notation $\bar{M}$ will be used to designate a point set such that X belongs to it if and only if X either belongs to M or is a limit point of M. The point set $\bar{M}$ will sometimes be called the *closure* of M.

AXIOM 1. *There exists a sequence $G_1, G_2, G_3, \cdots$ such that* (1) *for each n, G_n is a collection covering S such that each element of G_n is a region,* (2) *for each n, G_{n+1} is a subcollection of G_n,* (3) *if R is a region, X is a point of R and Y is a point of R, either identical with X or not, then there exists a natural number m such that if g is any region belonging to the collection G_m and containing X then $\bar{g}$ is a subset of R and, unless Y is X, $\bar{g}$ does not contain Y,* (4) *if M_1, M_2,*

$M_3, \cdots$ is a sequence of closed point sets such that, for each n, M_n contains M_{n+1} and, for each n, there exists a region g_n of the collection G_n such that M_n is a subset of $\bar{g}_n$, then there is at least one point common to all the point sets of the sequence $M_1, M_2, M_3, \cdots$.

THEOREM 1. *No point of a region is a boundary point of that region.*

PROOF. If the region R contained a point P of its boundary then R itself would be a region containing P and therefore containing at least one point not belonging to R.

THEOREM 2. *If M is a point set, $\bar{M}$ is closed.*

PROOF. Suppose P is a limit point of $\bar{M}$. Then if R is a region containing P, R contains a point X of $\bar{M}$ distinct from P and either X belongs to M or X is a limit point of M in which case R contains a point of M distinct from X.

DEFINITION. The collection G of sets is said to be a *monotonic* collection if and only if, for every two sets of the collection G, one of them is a subset of the other one.

DEFINITION. The point set M is said to be *perfectly compact* if and only if it is true that if G is a monotonic collection of subsets of M then either the elements of G have a common point or they have a common limit point.

THEOREM 3. *If M is a closed and perfectly compact point set and β is a well ordered sequence of regions such that M is covered by the collection whose elements are the terms of β then there exists a term g of β such that M is covered by the collection whose elements are g and the terms of β that precede g.*

PROOF. Suppose there exists no such region g. For each term x of β let M_x denote the set of all points of M that are not in x or in any term of β that precedes x. Suppose P is a limit point of M_x. If P were not in M_x then it would be either in x or in some region that precedes x in β and this particular region would contain a point of M_x, contrary to the definition of M_x. Hence M_x is closed. So the collection of all M_x's for all terms x of the sequence β is a monotonic collection of closed subsets of M. Hence there exists a point C which belongs to M_x for every term x of β. But C belongs to some region y of the sequence β. Hence C is not in M_y. This involves a contradiction.

DEFINITION. The point set M is said to have the *Borel-Lebesgue* property if and only if it is true that if G is a collection of regions covering M then some finite subcollection of G covers M.

THEOREM 4. *Every closed and perfectly compact point set has the Borel-Lebesgue property.*

PROOF. Suppose G is a collection of regions covering the closed and perfectly compact point set M. There exists a well ordered sequence β whose terms are the regions of G. By Theorem 3, there exists at least one term g of β such that M is covered by the collection composed of g and all terms of β that precede g, together with not more than a finite number of terms of β that follow g. Let g_1 denote the first such g and let $g_2, g_3, g_4, \cdots, g_n$ denote a finite set of terms of β following g_1 such that M is covered by $g_1, g_2, g_3, \cdots, g_n$ together with all terms of β that precede g_1. Suppose g_1 is not the first term of β. Then if g_1 and all the succeeding terms of β are removed, there remains a well ordered sequence β_1 whose terms are the remaining terms of β arranged in exactly the same order as in the original sequence β. Let β_2 denote a third well ordered sequence, having for its *first* n terms the regions $g_1, g_2, g_3, \cdots, g_n$ in the order indicated and having, as its subsequent terms, the terms of β_1, in the same order, *with respect to each other*, as in β_1. The sequence β_2 has no last term. The point set M is covered by the collection whose elements are the terms of β_2 but there does not exist a term g of β_2 such that M is covered by the collection whose elements are g and the regions that precede g in the sequence β_2. But this is contrary to Theorem 3. It follows that g_1 is the first term of β and that M is covered by the finite set of regions $g_1, g_2, g_3, \cdots, g_n$.

THEOREM 5. *If a point set has the Borel-Lebesgue property it is perfectly compact.*

PROOF. Suppose the point set M has the Borel-Lebesgue property and G is a monotonic collection of subsets of M. Let H denote a collection such that h belongs to it if and only if h is $S - \bar{g}$ for some element g of G such that $\bar{g}$ does not contain M. Suppose the closures of the elements of G have no point in common. Then the collection H covers M. But every element of H is a domain. Therefore M is covered by some finite subcollection F of H. Since F is monotonic and finite there exists an element f of F which contains all the others. Since $M - f \cdot M$ exists and is not covered by F, this involves a contradiction.

THEOREM 6. *Suppose M is an uncountable point set such that every uncountable subset of M has a limit point. Suppose the word "precedes" has such a meaning that* (1) *if X and Y are points of M either X precedes Y or Y precedes X,* (2) *if X precedes Y, X and Y are points of M and Y does not precede X. Suppose furthermore that X is said to follow Y if, and only if, Y precedes X. Then some point is a limit point of the set of all points that precede it and also of the set of all those that follow it.*

PROOF. Suppose H is an uncountable subset of M such that if P is a point of H there exists a region which contains P but no point of H preceding P. Then for each point P of H there exists a positive integer n_P such that no region of G_{n_P} contains P and a point of H preceding P. Since H is uncountable there exist a number k and an uncountable subset K of H such that, for each point P of K, $n_P = k$. Let A denote a limit point of K and let R denote a region of G_k containing A. The region R contains two points B and C of K. Since one of these points precedes the other one, this involves a contradiction. So every uncountable subset H of M contains a point which is a limit point of the set of all points of H that precede it. Hence if M_1 denotes the set of all points P of M such that P is a limit point of the set of all points of M that precede it then either M_1 is M or $M - M_1$ is countable. Similarly if M_2 denotes the set of all points P of M_1 such that P is a limit point of the set of all points of M_1 that follow P then either M_2 is M_1 or $M_1 - M_2$ is countable. Every point of M_2 is a limit point both of the set of all points of M that precede it and of the set of all those that follow it and either M_2 is M or $M - M_2$ is countable.

THEOREM 7. *If P is a point there exists an infinite sequence $R_1, R_2, R_3, \cdots$ such that* (1) *for each n, R_n is a region,* (2) *P is the only point that belongs to every region of this sequence,* (3) *for every n, $\bar{R}_{n+1}$ is a subset of R_n,* (4) *if R is a region containing P then there exists a positive integer n such that $\bar{R}_n$ is a subset of R.*

PROOF. Let R_1 denote some region of G_1 that contains P. Let n_1 denote the smallest positive integer n such that if g is any region of G_n that contains P then $\bar{g}$ is a subset of R_1 and let R_2 denote some region containing P and belonging to G_{n_1}. Continue this process, obtaining an infinite increasing sequence of positive integers $n_1, n_2, n_3, \cdots$ and a sequence $R_1, R_2, R_3, \cdots$ such that, for each j, (1) n_j is the smallest positive integer n such that if g is a region of G_n that contains P then $\bar{g}$ is a subset of R_j and (2) R_{j+1} is a region belonging to G_{n_j} and containing P. It is clear that the sequence $R_1, R_2, R_3, \cdots$ satisfies all of the conditions of Theorem 7.

DEFINITION. The sequence $R_1, R_2, R_3, \cdots$ is said to *close down* on the point P if it satisfies, with respect to P, all the conditions of Theorem 7. If, for each n, P_n is a point, the point P is said to be a *sequential limit point* of the sequence $P_1, P_2, P_3, \cdots$ if, for every region R containing P there exists a positive integer δ such that if $n > \delta$ then P_n lies in R. If a sequence α has a sequential limit point it is said to *converge* to that point and to be a *convergent* sequence.

THEOREM 8. *If the point P is common to all the regions of the finite set G then there exists a region which contains P and which is a subset of every region of the set G.*

PROOF. Let $K_1, K_2, K_3, \cdots, K_m$ denote the regions of the set G. Let $R_1, R_2, R_3, \cdots$ denote a sequence of regions closing down on P. For each i $(1 \leqq i \leqq m)$ there exists a number n_i such that R_{n_i} is a subset of K_i. Let j denote the greatest number in the finite set $n_1, n_2, n_3, \cdots, n_m$. The region R_j is a subset of every region of the set G.

THEOREM 9. *If a point is not a limit point of any member of a finite collection of point sets it is not a limit point of their sum.*

PROOF. Suppose the point P is not a limit point of any point set of the finite collection $M_1, M_2, M_3, \cdots, M_m$. Then, for each i $(1 \leqq i \leqq m)$ there exists a region K_i containing P but no point of M_i. By Theorem 8, there exists a region K which contains P and which is a subset of the common part of all the regions $K_1, K_2, K_3, \cdots, K_m$. The region K contains no point of $M_1 + M_2 + M_3 + \cdots + M_m$. Hence P is not a limit point of this point set.

THEOREM 9(a). *If a point set has the Borel-Lebesgue property it is closed.*

PROOF. Suppose the point set M has the Borel-Lebesgue property and P is a limit point of M that does not belong to it. By Theorem 7, for each point X of M there exists a region R_x containing X such that $\bar{R}_x$ does not contain P. By hypothesis, M is covered by some finite subcollection G of the set of all such regions R_x. Since P is a limit point of M and G^* contains M therefore P is a limit point of G^*. This contradicts Theorem 9.

THEOREM 10. *If P is a limit point of the point set M then every region that contains P contains infinitely many points of M and, indeed, there exists an infinite sequence of points belonging to M and all distinct from each other and from P such that P is a sequential limit point of this sequence.*

PROOF. Suppose R is a region containing P. Let $R_1, R_2, R_3, \cdots$ denote a sequence of regions closing down on P. There exists a positive integer n_1 such that R_{n_1} is a subset of R. Since P is a limit point of M, the region R_{n_1} contains a point P_1 belonging to M and distinct from P. There exists a positive integer n_2 such that R_{n_2} does not contain P_1. The region R_{n_2} contains a point P_2 belonging to M and distinct from P. This process may be continued. Thus there exist a monotonically increasing sequence of positive integers $n_1, n_2, n_3, \cdots$ and a sequence of points $P_1, P_2, P_3, \cdots$ belonging to M such that, for each i, P_i belongs to R_{n_i} but not to $R_{n_{i+1}}$. The point P is a sequential limit point of the sequence $P_1, P_2, P_3, \cdots$.

THEOREM 11. *If, for each n, P_n is a point and the point set $P_1 + P_2 + P_3 + \cdots$ is compact then there exists an increasing sequence of positive integers $n_1, n_2, n_3, \cdots$ such that the sequence $P_{n_1}, P_{n_2}, P_{n_3}, \cdots$ is convergent.*

PROOF. If there are infinitely many distinct points in the sequence $P_1, P_2, P_3, \cdots$ then, by hypothesis, $P_1 + P_2 + P_3 + \cdots$ has a limit point and, by Theorem 10, there exists a sequence of distinct numbers $m_1, m_2, m_3, \cdots$ such that $P_{m_1}, P_{m_2}, P_{m_3}, \cdots$ is a convergent sequence of distinct points. If there are not infinitely many distinct points in the sequence $P_1, P_2, P_3, \cdots$ then there exists a sequence of distinct numbers $m_1, m_2, m_3, \cdots$ such that $P_{m_1}, P_{m_2}, P_{m_3}, \cdots$ all denote the same point P. The sequence $P_{m_1}, P_{m_2}, P_{m_3}, \cdots$ converges to P. In either case, let n_1 denote the smallest of the numbers $m_1, m_2, m_3, \cdots$, let n_2 denote the next smallest one and so on. The sequence $P_{n_1}, P_{n_2}, P_{n_3}, \cdots$ is convergent.

THEOREM 12. *No point is a limit point of a finite point set.*

PROOF. Suppose P is a limit point of a finite point set M. By Axiom 1 there exists a region R containing P. By Theorem 10, R contains infinitely many points of M, contrary to the hypothesis that M is finite.

THEOREM 13. *If A and B are two points there exist two regions R_A and R_B containing A and B respectively such that $\bar{R}_A$ and $\bar{R}_B$ have no point in common.*

PROOF. By Theorem 12, there exists a region R containing A but not B. By Theorem 7, there exists a region R_A containing A such that $\bar{R}_A$ is a subset of R. By Theorem 2, there exists a region H containing B but no point of $\bar{R}_A$, and, by Theorem 7, there exists a region R_B containing B such that $\bar{R}_B$ is a subset of H and therefore contains no point of $\bar{R}_A$.

THEOREM 14. *No sequence has more than one sequential limit point and if P is a sequential limit point of the sequence of points $P_1, P_2, P_3, \cdots$ then no other point is a limit point of the point set $P_1 + P_2 + P_3 + \cdots$.*

PROOF. Suppose that the point P is a sequential limit point of the sequence $P_1, P_2, P_3, \cdots$ and that this sequence has a sequential limit point X distinct from P. By Theorem 12, there exists a region R containing P but not X. By Theorem 7, there exists a region K, containing P, such that $\bar{K}$ is a subset of R. There exists a number δ such that if $n > \delta$ then P_n lies in K. But X does not belong to $\bar{K}$. Hence there exists a region L containing X but no point of $\bar{K}$. If there were a number δ' such that, for every n greater than δ', P_n belongs to L then $P_{\delta+\delta'}$ would belong to L whereas, as a matter of fact, it belongs to K and therefore to $S - L$. Hence X is not a sequential limit point of the sequence $P_1, P_2, P_3, \cdots$. Therefore $P_1, P_2, P_3, \cdots$ does not have two sequential limit points. It follows, by Theorem 10, that the point set $P_1 + P_2 + P_3 + \cdots$ has no limit point except P.

THEOREM 15. *If the point set whose elements are the terms of the infinite sequence α is compact and there do not exist two points such that each of them is a sequential limit point of some subsequence of α then α has a sequential limit point.*

PROOF. By Theorem 11, there exist a point O and an infinite subsequence β of α such that β converges to O. Suppose R is a region containing O. Suppose there exists an infinite subsequence γ of α such that no term of γ is a point of R. Then, by Theorem 11, there exists an infinite subsequence γ' of γ converging to some point O'. Since no term of γ' belongs to R, therefore

O' is distinct from O. But this is contrary to hypothesis. Hence there exists a number δ such that, for every n greater than δ, P_n belongs to R. Thus O is the sequential limit point of α.

DEFINITIONS. The point set D is said to be a *domain* if for each point P of D there exists a region containing P and lying in D. A domain is also called an *open* point set. A subset K of a point set M is said to be a *domain with respect to M* or a *domain relatively to M* if, for each point P of K, there exists a region R containing P such that $R \cdot M$ is a subset of K. Such a subset K of M is also said to be *open with respect to M* or *open relatively to M*.

Every domain with respect to the point set M is the common part of M and some domain.

THEOREM 16. *If M and N are two mutually exclusive closed point sets and one of them is perfectly compact then there exist two domains D_M and D_N, containing M and N respectively, such that $\bar{D}_M$ and $\bar{D}_N$ are mutually exclusive.*

PROOF. Suppose M is perfectly compact. For each point P of M there exists a region R_P, containing P, such that $\bar{R}_P$ contains no point of N. Let G denote the set of all such regions R_P for all points P of M. By Theorem 4, M is covered by some finite subset G_1 of G. Let D_1 denote the sum of all the regions of the set G_1. The domain D_1 contains M. Since no point of N either belongs to, or is a limit point of, any region of the finite set G_1 therefore, by Theorem 9, N contains no point of $\bar{D}_1$. Let D_N denote the domain $S - \bar{D}_1$. The sets M and $\bar{D}_N$ are closed and mutually exclusive and M is perfectly compact. Hence there exists a domain D_M, containing M, such that $\bar{D}_M$ and $\bar{D}_N$ have no point in common.

THEOREM 17. *Suppose H and K are two mutually exclusive closed point sets and G is a monotonic collection of closed and perfectly compact point sets such that each of them intersects both H and K but no one of them is the sum of two mutually exclusive closed point sets one intersecting H and the other intersecting K. Then if M denotes the common part of all the point sets of the collection G, the closed point set M is not the sum of two mutually exclusive closed point sets one intersecting H and the other intersecting K.*

PROOF. Suppose, on the contrary, that M is the sum of two such point sets M_H and M_K intersecting H and K respectively. By Theorem 16, there exists a domain D, containing M_H, such that $\bar{D}$ contains no point of M_K. Suppose g is a point set of the collection G. If g contained no point of β, the boundary of D, then g would be the sum of two mutually separated point sets $D \cdot g$ and $(S - \bar{D}) \cdot g$ intersecting M_H and M_K respectively. Since β and g are closed, $\beta \cdot g$ is closed. Let Z denote a collection such that z belongs to it if and only if, for some point set g of the collection G, z is $\beta \cdot g$. The collection Z is monotonic and its elements are closed and perfectly compact point sets. Hence the point sets of this collection have a point O in common. The point O belongs to $M \cdot \beta$. But M and β have no point in common. This involves a contradiction.

THEOREM 18. *If M is a point set and G is a set of regions covering $\bar{M}$, then M contains a point set K, with no limit point, such that* (1) *no region of G contains two points of K,* (2) *if P is a point of M, some region of G contains both P and some point of K.*

PROOF. There exists a well ordered sequence α whose terms are the points of M. There exists a well ordered sequence β such that (1) the first term of β is the first term of α, (2) the initial segment γ of β is distinct from β if, and only if, there is a point of α belonging to no region of G which intersects γ, (3) if the initial segment γ of β is distinct from β, the first point of α that belongs to no region of G that intersects γ is the first point of β that follows every point of γ. Let K denote the set of all points of β. Every point of M belongs to some region of G that intersects K and no region of G contains two points of K. Suppose K has a limit point P. The point P belongs to $\bar{M}$ and therefore to some region g of G, and g contains two points of K. This involves a contradiction.

THEOREM 19. *If H is a collection of regions covering S, and G is a collection of regions and, for each point P of M, there exists a region of the collection G that contains every region of H that contains P then* (1) *G covers $\bar{M}$,* (2) *there exist a subset K of M with no limit point and a subcollection U of G covering M but not more numerous than K, and* (3) *if M is not covered by any subcollection of G less numerous than G then there exists a subset of M equally numerous with G but with no limit point.*

PROOF. Suppose A is a limit point of M. There exists a region h belonging to H and containing A. The region h contains a point P of M. There exists a region g of the collection G containing every region of H that contains P. The region g contains h and therefore A. Thus G covers $\bar{M}$. By Theorem 18, M contains a point set K, with no limit point, such that (1) no region of H contains two points of K, (2) if P is a point of M some region of H contains both P and some point of K. For each point P of K there exists a region belonging to G and containing every region of H that contains P. For each point P of K, let g_P denote one definite region of G that does this, and let U denote the set of all such regions g_P for all points P of K. The collection U covers M and it is not more numerous than K. Suppose M is not covered by any subcollection of G less numerous than G. Then U is not less numerous than G. Hence it is equally numerous with G. So K is not less numerous than G. Hence K contains a point set L which is equally numerous with G. Since K has no limit point, neither does L.

THEOREM 20. *If H is a collection of regions covering S, M is a point set, Z is a subset of M, and G is a collection of regions and, for each point P of M,* *there exists a region of the collection G that contains every region of the collection H that contains P then* (1) *if every subset of M which is more numerous than Z has a limit point, M is covered by a subcollection of G not more numerous than Z,* (2) *if every infinite subset of M has a limit point, M is covered by a finite subcollection of G, and* (3) *if every uncountable subset of M has a limit point then M is covered by a countable subcollection of G.*

PROOF. By Theorem 19 there exist a subset K of M with no limit point and a subcollection U of G covering M, but not more numerous than K. Suppose every subset of M more numerous than Z has a limit point. Then K is not more numerous than Z. Hence U is not more numerous than Z. But U covers M.

Suppose every infinite subset of M has a limit point. Then K is finite. Hence, since it is not more numerous than K, U is finite. But it covers M.

Suppose every uncountable subset of M has a limit point. Let Z denote a countably infinite subset of M. Then every subset of M more numerous than Z has a limit point. Hence M is covered by a subcollection of G which is not more numerous than Z and is therefore countable.

THEOREM 21. *If every uncountable subset of the point set M has a limit point and G is a collection of regions covering M then some countable subcollection of G covers M.*

PROOF. For each point P of M there exists a positive integer k such that some region of G contains every region of G_k that contains P. For each point P of M let k_P denote the smallest such number k. For each k which is equal to k_P for some point P of M, let M_k denote the set of all points P of M such that $k_P = k$. For each such k, by Theorem 20, there exists a countable subcollection G'_k of G that covers M_k. Furthermore each point of M belongs to M_k for some k. Thus M is covered by the countable collection of all point sets g such that, for some k, g belongs to the countable collection G'_k.

DEFINITIONS. The point set M is said to be *separable* if, and only if, it contains a countable point set K such that every point of M either belongs to K or is a limit point of K. The point set M is said to be *completely separable* if, and only if, there exists a countable set H such that each element of H is a domain with respect to M and if D is any domain whatsoever with respect to M and P is a point of D then there exists a domain with respect to M belonging to H, containing P, and lying in D.

THEOREM 22. *If every uncountable subset of the point set M has a limit point, M is completely separable.*

PROOF. By Theorem 21, for each n there exists a countable subcollection T_n of G_n that covers M. Let T denote the collection of all regions t such that, for some n, t belongs to T_n. The collection T is countable. But, for each point P of M and region R containing P, there exists a region belonging to T, lying in R and containing P. Hence M is completely separable.

THEOREM 23. *If, for every collection G of regions covering the point set M, some countable subcollection of G covers M then every uncountable subset of M has a limit point.*

PROOF. Suppose the uncountable subset K of M has no limit point. Then every subset of K is closed. For each point P of M let R_P denote a definite region containing P but no point of the closed point set consisting of all points of K other than P. Let G denote a collection such that g belongs to it if and only if g is R_P for some point P of M. The collection G covers M but no countable subcollection of it does so since each point of the uncountable point set K belongs to only one region of the collection G.

THEOREM 24. *If G is a collection of domains covering the completely separable point set M some countable subcollection of G does so.*

PROOF. By hypothesis there exists a countable collection H of domains with respect to M such that if R is a domain of the collection G containing the point P of M there exists a point set of the collection H containing P and lying in R. For each point set h of the collection H which is a subset of some domain of the collection G, let R_h denote a *definite* domain of G of which it is a subset. The set of all such domains for all point sets h of the collection H is a countable subcollection of G that covers M.

THEOREM 25. *If every uncountable subset of the uncountable point set M has a limit point there exists a subset T of M such that $M - T$ is countable and every region that intersects T contains uncountably many points of T.*

PROOF. By Theorem 22 there exists a countable collection H of domains with respect to M such that if a region contains the point P of M it contains a point set of the collection H containing P. Suppose for every point P of M there is a region R_P containing P but not containing uncountably many points of M. For each point P of M there is a point set D_P of the collection H containing P and lying in R_P. Let K denote a collection such that k belongs to it if, and only if, k is D_P for some point P of M. The point set K^* is the sum of countably many countable point sets and is therefore countable. But this is impossible since K^* is M. Hence there exists a point P of M such that every region that contains P contains uncountably many points of M. Let T' denote the set of all such points P. Suppose $M - T'$ is uncountable. Then there exists a point X of $M - T'$ such that every region that contains X contains uncountably many points of $M - T'$, and therefore of M, and therefore X belongs to T'. This involves a contradiction.

THEOREM 26. *If every point set is separable, S is completely separable.*

PROOF. Suppose S is not completely separable. By Theorem 22, there exists an uncountable point set T with no limit point. Suppose K is a countable subset of T. There exists a point P belonging to $T - K$. The point P neither belongs to K nor is a limit point of it. Since T is separable, this involves a contradiction.

DEFINITIONS. A point set is said to be *connected* if it is not the sum of two mutually separated point sets. A point set which is both closed and connected is called a *continuum*. A *maximal connected* subset of a point set M is a connected subset of M which is not a proper subset of any other connected subset of M. A maximal connected subset of a point set is also called a *component* of that point set. A set is said to be *degenerate* if it has only one element.

THEOREM 27. *If M is a nondegenerate connected point set, every point of M is a limit point of M.*

PROOF. If P is a point of M, by hypothesis, $M - P$ exists and either P is a limit point of $M - P$ or $M - P$ contains a limit point of the set whose only element is P. The latter is impossible by Theorem 12.

THEOREM 28. *If H and K are two mutually separated point sets, every connected subset of $H + K$ is a subset either of H or of K.*

PROOF. Suppose, on the contrary, that T is a connected subset of $H + K$ containing both a point of H and a point of K. Then T is the sum of the two mutually separated point sets $H \cdot T$ and $K \cdot T$ and thus T is not connected, contrary to hypothesis.

THEOREM 29. *If M is a connected point set and L is a point set consisting of M together with some or all of its limit points, then L is connected.*

PROOF. Suppose, on the contrary, that L is the sum of two mutually separated point sets H and K, where H contains a point of M. By Theorem 28, M is a subset of H. But every point of $L - M$ is a limit point of M. Therefore K contains no point of $L - M$. This involves a contradiction.

THEOREM 30. *If G is a collection of connected point sets and one of them contains a point or limit point of each of the others then the sum of all the point sets of the collection G is connected.*

PROOF. Suppose h is a point set belonging to the collection G and containing at least one point or limit point of each of the remaining point sets of this collection. Let M denote the sum of all the point sets of the collection G. Suppose M is not connected. Then it is the sum of two mutually separated point sets H and K such that H contains a point of h. By Theorem 28, h is a subset of H, and K contains a point set k of the collection

G. By hypothesis, some point of h, and therefore of H, is a point or limit point of k, and therefore of K. But this involves a contradiction.

THEOREM 31. *If a closed point set is not connected it is the sum of two mutually exclusive closed point sets.*

PROOF. Suppose the closed point set M is the sum of two mutually separated point sets H and K. Suppose P is a limit point of H. Since M is closed, P belongs to $H + K$. But K contains no limit point of H. Hence P belongs to H. Therefore H is closed. Similarly, K is closed.

THEOREM 32. *If the point P is a boundary point of the common part of the point sets M and N then P is a boundary point either of M or of N.*

PROOF. Suppose P is a boundary point of $M \cdot N$. Then either (1) P belongs to $M \cdot N$ and is a limit point of $S - M \cdot N$ or (2) P belongs to $S - M \cdot N$ and is a limit point of $M \cdot N$. In case (1), since $S - M \cdot N = (S - M) + (S - N)$ therefore P is a limit point either of $S - M$ or of $S - N$ and therefore, since it belongs both to M and to N, it is a boundary point either of M or of N. In case (2), P belongs either to $S - M$ or to $S - N$ and it is a limit point both of M and of N. Therefore, in this case also, it is a boundary point either of M or of N.

THEOREM 33. *If the connected point set I_2 does not intersect the boundary of the point set I_1 then either I_1 and I_2 are mutually exclusive or I_1 contains I_2.*

PROOF. Suppose I_1 intersects I_2 but does not contain it. Then, since I_2 is connected and intersects both I_1 and $S - I_1$, it intersects the boundary of I_1, contrary to hypothesis.

THEOREM 34. *Each point of a point set M belongs to some component of M.*

PROOF. If P is a point of M and G is the set of all connected subsets of M that contain P, then, by Theorem 30, G^* is connected. It is not a subset of any other connected subset of M.

THEOREM 35 (F. B. JONES). *If the point set M has at least n components, it is the sum of n mutually separated point sets.*

PROOF. Since it has two components, M is the sum of two mutually separated point sets. Suppose it is not the sum of n mutually separated point sets. Let k denote a number such that M is the sum of k mutually separated sets $M_1, M_2, \cdots, M_k$ but not of $k + 1$ mutually separated sets. For each j less than $k + 1$, M_j contains a component C_j of M, and for some i less than $k + 1$, M_i contains two components of M and is therefore the sum of two mutually separated point sets L_1 and L_2. The collection whose elements are L_1, L_2 and the elements of $M_1, M_2, \cdots, M_k$ other than M_i is a collection of $k + 1$ mutually separated subsets of M. This involves a contradiction.

THEOREM 36. *Every perfectly compact point set is compact.*

PROOF. Suppose the perfectly compact point set M contains an infinite point set K with no limit point. There exists a simple countably infinite sequence of distinct points of K. For each n, let K_n denote the set of all points of this sequence following the nth one. Since K_n has no limit point it is closed. Let G denote a collection such that g belongs to it if, and only if, g is K_n for some n. Since G is a monotonic collection of closed subsets of the perfectly compact point set M, therefore the point sets $K_1, K_2, K_3, \cdots$ have a point in common. But this involves a contradiction.

THEOREM 37. *Every compact point set is perfectly compact.*

PROOF. Suppose M is a compact point set which is not perfectly compact. Then there exists a monotonic collection G of subsets of M whose closures are distinct and have no point in common. There exists a subcollection H of G such that if the word "precedes" means "contains" then, with respect to this meaning, H is well ordered and every point set of the collection G contains one of the collection H.

Suppose some countable sequence W of elements of H runs through H in the sense that every element of H contains some element of W. Then there exists a simple countably infinite subsequence $M_1, M_2, M_3, \cdots$ of W running through H in that sense. For each n, let P_n denote a point of M_n. By hypothesis and Theorem 11, there exists an increasing sequence of positive integers $n_1, n_2, n_3, \cdots$ such that the sequence $P_{n_1}, P_{n_2}, P_{n_3}, \cdots$ converges to some point P. For each k, every point of this sequence following the kth one belongs to M_k and therefore, since $\bar{M}_k$ is closed, P belongs to $\bar{M}_k$. Thus P is common to all the point sets of the sequence $\bar{M}_1, \bar{M}_2, \bar{M}_3, \cdots$ and therefore to all the closures of the sets of the collection G. But this involves a contradiction.

For each point set h of the collection H, let P_h denote some definite point belonging to h but not to the closure of t_h, the first point set of the collection H which follows, and therefore is a subset of, h and let n_h denote the smallest positive integer n such that no region of G_n contains both P_h and a point of the closed point set $\bar{t}_h$. Let Q denote a set such that q belongs to it if, and only if, q is P_h for some element h of H. Since H is uncountable and the set of all positive integers is countable, therefore there exist a positive integer m and an uncountable subcollection H_1 of H such that $m = n_h$ for every point set h of the collection H_1. Let Q_1 denote the set of all points q of Q such that q is P_h for some element h of H_1. Since M is compact and Q_1 is infinite, Q_1 has a limit point A. There exists a region R of G_m containing A. The region R contains two points X and Y of Q_1. There exist two point sets x and y of the collection H_1 such that X is P_x and Y is P_y. One of these point sets precedes, and therefore contains, the other one. Suppose x precedes y. Then no region of G_{n_x} contains both P_x and a point of y. But m is n_x and R

is a region of G_m containing P_x and P_y. This involves a contradiction. So M is perfectly compact.

DEFINITION. If a point set is both connected and closed it is called a *continuum*.

THEOREM 38. *If G is a monotonic collection of perfectly compact continua the common part of all the continua of G is a continuum.*

PROOF. Suppose M, their common part, is the sum of two mutually exclusive closed point sets H and K. Each continuum of G intersects (in fact contains) both H and K and is not the sum of two mutually exclusive closed point sets one intersecting H and the other intersecting K. Hence, by Theorem 17, M is not the sum of two mutually exclusive closed point sets one intersecting H and the other intersecting K. But M *is* $H+K$ and H and K are mutually exclusive.

THEOREM 39. *If G is a collection of sets such that if H is a monotonic subcollection of G there is a set of the collection G which is a subset of every set of the collection H then there exists a set of the collection G which contains no other set of the collection G.*

PROOF. There exists a well ordered sequence α of distinct sets such that every element of α is a set of the collection G and every set of the collection G belongs to the sequence α. There exists a sequence β such that (1) the first term of β is the first term of α, (2) if ω is any subsequence of β such that every term of β that precedes a term of ω belongs to ω and there exists a term of α which is a proper subset of every term of ω then the first such term of α is the first term of β that follows all the terms of ω. Suppose that the sequence β has no last term. Since the terms of β form a monotonic collection, there exists a term g of α which is a subset of every term of β. Let h denote any term of β. Since there is no last term of β there exists a term q of β which is a proper subset of h. But g is a subset of q. Therefore g is a proper subset of h. Thus there exists a term of α which is a proper subset of every term of β. Let W denote the set of all such terms of α. Since α is a well ordered sequence there exists a term w of W which, in the sequence α, precedes every other term of W. By the definition of β, the term w of α belongs to β. But it is a proper subset of every term of β. Thus the supposition that β has no last term leads to a contradiction. Let e denote the last term of β. The set e belongs to G but no proper subset of it does so.

DEFINITION. If H and K are two mutually exclusive closed point sets, the continuum M is said to be an *irreducible continuum from* H to K if M intersects both H and K but no proper subcontinuum of M intersects both of them.

DEFINITION (W. A. WILSON). The continuum M is said to be an *irreducible continuum about* the point set H if M contains H but no proper subcontinuum of M contains H.

THEOREM 40. *If the closed point set M intersects each of the two mutually exclusive closed point sets H and K and is not the sum of two mutually exclusive closed point sets one intersecting H and the other intersecting K but every closed proper subset of M that intersects H and K is the sum of two such point sets then M is an irreducible continuum from H to K.*

PROOF. Suppose M is not a continuum. Then it is the sum of two mutually exclusive closed point sets M_{H+K} and T where M_{H+K} contains both $H \cdot M$ and $K \cdot M$. By hypothesis, M_{H+K} is the sum of two mutually exclusive closed point sets M_H and M_K intersecting H and K respectively. The closed point sets M_H and $M_K + T$ are mutually exclusive, one intersects H and the other intersects K and their sum is M. Thus the supposition that M is not a continuum leads to a contradiction. By hypothesis, no proper subcontinuum of M intersects both H and K.

THEOREM 41. *If the closed and perfectly compact point set M intersects each of the two mutually exclusive closed point sets H and K and is not the sum of two mutually exclusive closed point sets one intersecting H and the other intersecting K, then M contains a continuum which is irreducible from H to K.*

PROOF. Let G denote a collection such that g belongs to it if, and only if, g is a closed subset of M which intersects both H and K but is not the sum of two mutually exclusive closed point sets one intersecting H and the other intersecting K. By Theorems 17 and 39, there exists a point set g_0 belonging to G such that no proper closed subset of g_0 belongs to G. By Theorem 40, g_0 is an irreducible continuum from H to K.

THEOREM 42. *If the closed and perfectly compact point set M intersects each of the two mutually exclusive closed point sets H and K but M contains no continuum intersecting both H and K, then M is the sum of two mutually exclusive closed point sets, one intersecting H and the other intersecting K.*

THEOREM 43. *If H and K are mutually exclusive closed subsets of the perfectly compact continuum M then there is a subcontinuum of M that is irreducible from H to K.*

Theorems 42 and 43 are direct corollaries of Theorem 41.

THEOREM 44. *If H and K are mutually exclusive closed subsets of the perfectly compact closed point set M but M contains no continuum intersecting both H and K, then M is the sum of two mutually exclusive closed point sets one containing H and the other containing K.*

PROOF. Let P denote some point of H. If X is any point whatsoever of K then, by Theorem 42, M is the sum of two mutually exclusive closed point sets containing P and X respectively. Therefore, by Theorem 16, there exists a domain E_{PX} with respect to M containing X, but not P, such that E_{PX} has no boundary with respect to M. Let G_P denote the collection of all such point sets E_{PX}, for all points X of K. By Theorem 4, K is covered by some finite subcollection H_P of G_P. Let E_P denote the sum of all the point sets of the collection H_P. The point set E_P is a domain with respect to M. It contains K. By Theorem 9, it has no boundary with respect to M. Let D_P denote the point set $M - E_P$. This set contains P but no point of K. It has no boundary with respect to M. Let U denote the collection of all point sets D_P for all points P of H. There exists a finite subcollection W of U such that H is covered by W. Let D denote the sum of all the point sets of the collection W and let E denote $M - D$. The point sets D and E are closed and mutually exclusive; D contains H, E contains K, and $D + E = M$.

THEOREM 45. *If H and K are mutually exclusive closed subsets of the closed point set M and $H + K$ contains a domain D with respect to M such that $M - D$ is perfectly compact and intersects both H and K but M contains no continuum intersecting both $(H + D) - D$ and $(K + D) - D$ then M is the sum of two mutually separated point sets one containing H and the other containing K.*

PROOF. Let H' and K' denote $(H + D) - D$ and $(K + D) - D$ respectively. The point sets H' and K' are mutually exclusive closed subsets of the closed and perfectly compact point set $M - D$. Hence, by Theorem 44, $M - D$ is the sum of two mutually exclusive closed point sets H'' and K'' containing H' and K' respectively. If a point of K belongs to D it is not a limit point of H'' since H'' contains no point of D and D is a domain with respect to M. If a point of K does not belong to D it is a point of K' and therefore of K'' and therefore it is not a limit point of H''. So no point of K is a limit point of H''. Since H'' and K'' are mutually separated and K'' contains $(K + D) - D$ therefore no point of H'' is a limit point of $(K + D) - D$. Suppose H'' contains a point P which is a limit point of $K \cdot D$. Then P belongs to K but not to D and therefore it belongs to K' and thus to K'', which is impossible since H'' and K'' are mutually separated. It follows that K and H'' are mutually separated. Similarly H and K'' are mutually separated. Hence $H'' + H$ and $K'' + K$ are mutually separated. But their sum is M.

THEOREM 46. *If K is a closed subset of the perfectly compact continuum M, some subset of M is an irreducible continuum about K.*

PROOF. Let G denote the collection of all subcontinua of M that contain K. If H is a monotonic subcollection of G the common part of all the continua of H exists and, by Theorem 38, it belongs to G. Hence, by Theorem 39, there exists a continuum g of the collection G such that no other continuum of G is a subset of g.

THEOREM 47. *If H and K are two mutually exclusive closed point sets and the perfectly compact continuum M is an irreducible continuum from H to K*

then $M - M\cdot(H + K)$ *and* $M - M\cdot H$ *are connected and every point of* $M\cdot H$ *is a limit point of* $M - M\cdot(H + K)$.

PROOF. Let h and k denote $M\cdot H$ and $M\cdot K$ respectively. Suppose that $M - (h + k)$ is the sum of two mutually separated point sets T and N. Then $T + h + k$ and $N + h + k$ are both closed and perfectly compact and neither of them contains a continuum intersecting both h and k. Hence, by Theorem 44, $T + h + k$ is the sum of two mutually exclusive closed point sets T_h and T_k, and $N + h + k$ is the sum of two mutually exclusive closed point sets N_h and N_k such that T_h and N_h contain h while T_k and N_k contain k. The point sets $T_h + N_h$ and $T_k + N_k$ are closed and mutually exclusive and their sum is M, contrary to the hypothesis that M is connected. Thus the supposition that $M - (h + k)$ is not connected has led to a contradiction. Since h and k are mutually exclusive closed subsets of the continuum M, each of them contains at least one limit point of $M - (h + k)$. Let T denote the set of all points of $h + k$ that are limit points of $M - (h + k)$. The point set $T + [M - (h + k)]$ is a continuum intersecting both H and K. But M is irreducible from H to K. Hence $T + [M - (h + k)]$ is identical with M and therefore T is identical with $h + k$. Thus every point of k is a limit point of the connected point set $M - (h + k)$. Hence $k + M - (h + k)$, that is to say $M - h$, is connected.

THEOREM 48 (W. A. WILSON). *If* H *and* K *are two mutually exclusive continua and the compact continuum* M *is irreducible about* $H + K$ *then* $M - H$, $M - K$ *and* $M - (H + K)$ *are connected.*

PROOF. By Theorem 43, M contains a continuum N which is irreducible from H to K. Let L denote the point set $N - (H + K)\cdot N$. By Theorem 47, L is connected and each of the continua H and K contains a limit point of L. Therefore, by Theorem 30, $H + L$ and $K + L$ are connected and the closed point set $H + K + L$ is a continuum. Therefore, by hypothesis, $H + K + L$ is identical with M. In other words, $M - (H + K)$ is the connected point set L, and $M - K$ and $M - H$ are identical with the connected point sets $H + L$ and $K + L$.

Theorem 48 does not remain true even in a Euclidean plane if the supposition that H and K are mutually exclusive continua is replaced by the supposition that they are mutually exclusive closed point sets. In a Euclidean plane, let A, B and C denote three colinear points such that B is between A and C. Let M denote the straight line interval AC, let H denote the point set whose only element is A, and let K denote the set $B + C$. Here the continuum M is irreducible about $H + K$. But $M - (H + K)$ is the sum of the two mutually separated straight line segments AB and BC.

THEOREM 49. *If* K *is a component of the closed and perfectly compact point set* M *and* H *is a closed subset of* $M - K$ *then* M *is the sum of two mutually separated point sets one containing* K *and the other containing* H.

PROOF. The point set K is closed and no subcontinuum of M intersects both H and K. The conclusion follows by Theorem 44.

THEOREM 50. *If T is a component of the domain D relative to the continuum M, D is a proper subset of M and $\bar{D}$ is perfectly compact, then the boundary, with respect to M, of D contains at least one limit point of T.*

PROOF. Let β denote the boundary of D with respect to M. Let O denote a point of T. The closed point set $\bar{D}$ is not the sum of two mutually exclusive closed point sets M_β and M_O containing β and O respectively. For if it were then M would be the sum of the two mutually exclusive closed point sets M_O and $M_\beta + (M - \bar{D})$, contrary to the hypothesis that M is a continuum. But $\bar{D}$ is perfectly compact. Hence, by Theorem 44, $\bar{D}$ contains a continuum containing both O and a point of β. Hence, by Theorem 43, $\bar{D}$ contains an irreducible continuum H from O to β. By Theorem 47, $H - H\cdot\beta$ is connected and every point of $H\cdot\beta$ is a limit point of $H - H\cdot\beta$. Hence $H - H\cdot\beta$ is a subset of T and every point of $H\cdot\beta$ is a limit point of T.

THEOREM 51. *If K is a closed proper subset of the continuum M, and $\overline{(M - K)}$ is perfectly compact and H is a component of $M - K$ then K contains a limit point of H.*

PROOF. Let β denote the boundary of $M - K$ with respect to M. Since K is closed, $M - K$ is a domain with respect to M. Hence, by Theorem 50, β contains a limit point of H. But β is a subset of K. Therefore K contains a limit point of H.

That it is not true even in the Euclidean plane that every component of a domain with respect to a compact continuum is a domain with respect to that continuum may be seen from the following example.

EXAMPLE 1. Interpret the word "point" as meaning a point of the Cartesian plane and the word "region" as meaning the interior of a circle lying in that plane. It is easy to see that, for this interpretation of "point" and "region", Axioms 0 and 1 hold true. Let A denote the origin of coordinates, let B denote the point (2,0) and, for each n, let B_n denote the point $(2,1/n)$. Let AB, AB_1, AB_2, $AB_3, \cdots$ denote straight line intervals with end points as indicated. Let M denote the compact continuum (Figure 1) obtained by adding together all of the point sets AB, AB_1, $AB_2, \cdots$. Let D denote the set of all points of M at a distance from A greater than 1 and let H denote the common part of D and AB. The point set D is a domain with respect to M and H is a component of D. But every point of H is a limit point of $M - H$. Hence H is not a domain with respect to M.

Theorem 17 does not remain true in the plane on the removal of the condition requiring that every element of G be compact. Consider the following example.

EXAMPLE 2. Let "region" and "point" be interpreted as in Example 1. Let B denote the origin of coordinates, let A denote the point (2,0) and, for each n, let A_n denote the point between B and A whose abscissa is $(2n + 1)/(n + 1)$, and let C_n denote the point $(1,n)$. For each n, let BC_n

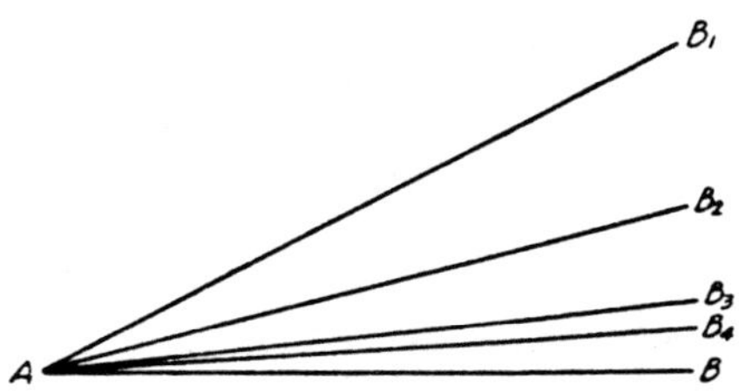

FIGURE 1

and C_nA_n denote straight line intervals with endpoints as indicated and let t_n denote the point set $BC_n + C_nA_n$. Let H denote the set of all points $(2,y)$ such that y is not negative and let K denote the set of all points $(0,y)$ with non-negative y. For each n, let M_n (Figure 2) denote $H + K + t_n + t_{n+1} + t_{n+2} + \cdots$. Let G denote a collection such that g belongs to it

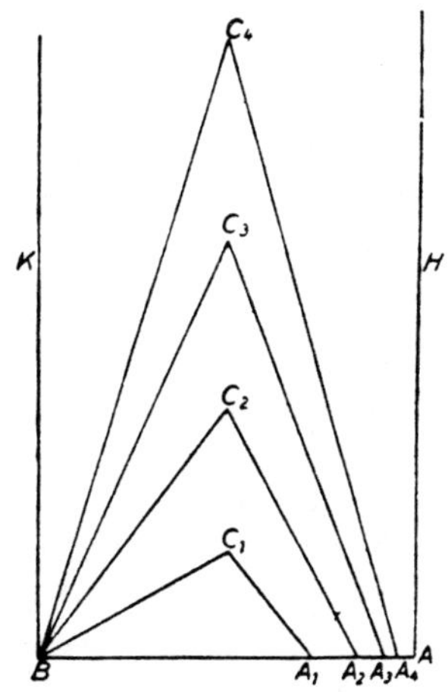

FIGURE 2

if and only if, for some n, g is M_n. Here all the conditions of the hypothesis of Theorem 17 are satisfied except that the elements of G are not compact. But the common part of all the point sets of the collection G is the sum of the two point sets H and K.

Theorem 17 does not remain true in the plane on the removal of the condition requiring that every element of G be closed. Consider the following example.

EXAMPLE 3. Let "point" and "region" be interpreted as in Example 2 and let B, A and $A_1, A_2, A_3, \cdots$ have the same significance as in that example; but let H and K denote sets whose only elements are A and B respectively and, for each n, let C_n denote the point whose abscissa is 1 and whose ordinate is $n/(n+1)$. For each n, let t_n denote the sum of the intervals BC_n and C_nA and let M_n denote the point set $H + K + t_n + t_{n+1} + t_{n+2} + \cdots$. Let G denote a collection such that g belongs to it if, and only if, for some n, g is M_n.

Theorem 42 does not remain true in the plane on the omission of the condition requiring that M be perfectly compact. Consider the following example.

EXAMPLE 4. Let "point" and "region" be interpreted as in Example 2 and let H, K, B, A and $A_1, A_2, A_3, \cdots$ and $C_1, C_2, C_3, \cdots$ have the same significance as in that example. For each n, let B_n denote the point between B and A whose abscissa is $1/(n+1)$. For each n, let t_n denote the sum of the intervals B_nC_n and C_nA_n. Let M denote $H + K + t_1 + t_2 + t_3 + \cdots$. The closed point set M contains both of the two mutually exclusive closed point sets H and K and it contains no continuum intersecting both H and K. But M is not the sum of two mutually exclusive closed point sets one intersecting H and the other intersecting K.

To see that Theorem 43 does not remain true in the plane on the removal of the condition requiring that M be perfectly compact, let M denote the point set M_1 of Example 2.

Theorem 44 does not remain true in the plane if the phrase "intersecting both H and K" is replaced by "containing both H and K". Consider the following example.

EXAMPLE 5. Let H denote a point set consisting of two points A and B, let t denote a straight line interval, or any other continuum, containing B and another point C but not containing A. Let M denote $t + A$ and let K denote the set whose only element is C.

Theorem 47 does not remain true in the plane if the requirement that M be irreducible from H to K is replaced by the requirement that it be irreducible about $H + K$. This may be seen from an example in which M denotes a straight line interval AB, H denotes the set whose only element is A, and K denotes the set whose elements are B and a point C lying between A and B.

That Theorem 51 does not remain true in the plane on the omission of the requirement that $(\overline{M - K})$ be compact may be seen with the aid of Example 2. Let M denote the continuum M_1 of that example. Here H is a component of $M - K$. But K contains no limit point of H. If, on the other hand, K' denotes the point set consisting of all points of M on or without a circle with center at A and radius equal to one half the distance from A to B, then $\overline{(M - K')}$ is compact and the component of $M - K'$ that contains A is a subset of H of which K' contains a limit point.

Example 2 may also be used to show that Theorem 50 does not remain true in the plane on the removal of the requirement that $\bar{D}$ be compact.

THEOREM 52. *If K is a component of the closed and perfectly compact point set M and T is a closed point set containing no point of K then* (1) *there exists a domain D containing K but no point of T and such that there is no point of $M + T$ on the boundary of D and* (2) *K is a component of $M + T$.*

PROOF. If M and T intersect then the point sets K and $M \cdot T$ are mutually exclusive closed subsets of the perfectly compact point set M and there is no connected subset of M containing both a point of K and a point of $M \cdot T$. Hence, by Theorem 44, M is the sum of two mutually exclusive closed point sets M_K and M_T containing K and $M \cdot T$ respectively. Let N denote $M_T + T$ or T according as $M \cdot T$ does or does not exist; and if $M \cdot T$ does not exist let M_K denote M. The point sets M_K and N are closed and mutually exclusive and M_K is compact. Hence, by Theorem 16, there exists a domain D, containing M_K but no point of N, such that neither M_K nor N contains a boundary point of D.

Suppose now that K is not a component of $M + T$. Then $M + T$ contains a continuum H containing K and a point of T. Since the connected set H intersects both D and $S - D$ it contains a boundary point of D. This involves a contradiction. Hence K is a component of $M + T$.

If the requirement that M be perfectly compact is removed from the hypothesis of Theorem 52 then neither part (1) nor part (2) of the conclusion follows from these axioms.

EXAMPLE 6. Let M be as described in Example 4. Let T denote the straight line interval whose endpoints are the points A and A_1. It is interesting to note that, in this example, although (a) K is a component of M and (b) T contains no point of K, nevertheless K is not a component of $M + T$. Indeed $M + T$ is a continuum.

THEOREM 52(a). *If K is a component of the closed and compact proper subset H of the continuum M then K contains a limit point of $M - H$.*

PROOF. Suppose K contains no limit point of $M - H$. Let T denote $\overline{(M - H)}$. The closed point set T contains no point of K. Hence, by Theorem 52, K is a component of $H + T$. Since $H + T$ is the connected set M and K is a proper subset of M, this involves a contradiction.

DEFINITION. The point set M is *locally compact* at the point P if some compact domain with respect to M contains P. If M is locally compact at each point of M it is said to be *locally compact.*

THEOREM 53. *No locally compact closed point set M is the sum of countably many closed point sets such that if g is any one of them every point of g is a limit point of $M - g$.*

PROOF. Suppose the locally compact closed point set M is the sum of countably many closed sets $M_1, M_2, \cdots$ such that, for each n, every point of M_n is a limit point of $M - M_n$. Let n_1 denote the smallest n such that M_n is not a subset of M_1. Let P_1 denote a point belonging to M_{n_1} but not to M_1. There exists a region R_1 containing P_1 and such that (1) $\bar{R}_1$ contains no point of M_1, (2) $M \cdot \bar{R}_1$ is compact. Since P_1 is a limit point of $M - M_{n_1}$ there exists a positive integer n such that M_n contains a point of R_1 not belonging to M_{n_1}. Let n_2 denote the smallest such number and let P_2 denote a point of $R_1 \cdot M_{n_2} \cdot (M - M_{n_1})$. There exists a region R_2, containing P_2, such that $\bar{R}_2$ is a subset of R_1 containing no point of M_{n_1}. This process may be continued indefinitely. Thus there exists a sequence of regions $R_1, R_2, R_3, \cdots$ such that (1) $M \cdot \bar{R}_1$ is compact, (2) for each n, $\bar{R}_{n+1}$ is a subset of R_n, (3) for each n, R_n contains a point of M but no point of any one of the point sets $M_1, M_2, M_3, \cdots, M_n$. By Theorem 37, there exists a point P common to all the point sets $M \cdot \bar{R}_1$, $M \cdot \bar{R}_2$, $M \cdot \bar{R}_3, \cdots$. But P does not belong to any one of the point sets $M_1, M_2, M_3, \cdots$. Thus the supposition that Theorem 53 is false leads to a contradiction.

THEOREM 54. *If the closed and compact point set K is a subset of the locally compact closed point set M and N is a domain containing K then there exists a domain D with respect to M containing K and such that $\bar{D}$ is a compact subset of N.*

PROOF. For each point P of M there exists a domain D_P with respect to M, containing P, such that $\bar{D}_P$ is a compact subset of N. By Theorems 37 and 4, there exists a finite subcollection H of point sets covering M such that each element of H is D_P for some point P of K. The point set H^* is a domain with respect to M and it is a compact subset of N.

THEOREM 55. *If H is a countable collection of closed and compact point sets such that no nondegenerate component of an element of H intersects another element of H, then the sum of all the elements of H is not a locally compact continuum.*

PROOF. Let $M_1, M_2, M_3, \cdots$ denote the elements of the collection H. Suppose H^* is a locally compact continuum M. Let n_1 denote the smallest integer n such that M_n is not a subset of M_1, let P_1 denote a point of M_{n_1} not belonging to M_1 and let K_1 denote the component of M_{n_1} that contains P_1. If K_1 is degenerate, its only element is P_1 and therefore it does not intersect M_1. If it is nondegenerate then, by *hypothesis*, it does not intersect M_1. Hence, in either case, by Theorem 54, there exists a domain D with respect to M, containing K_1, such that $\bar{D}$ is a compact subset of $M - M_1$. By Theorem 50, the boundary with respect to M of D intersects L_1, the closure of the component of D that contains P_1. Let H_1 denote a collection such that h belongs to it if and only if h is the common part of L_1

and some element of H. The collection H_1 is nondegenerate, its elements are closed, their sum is compact and no one of them intersects a nondegenerate component of any other one of them. It follows that there exists an infinite sequence $L_1, L_2, L_3, \cdots$ of subcontinua of M such that, for each n, L_{n+1} is a subcontinuum of L_n that intersects two point sets of the sequence $M_1, M_2, M_3, \cdots$, but none of subscript less than n. The continua of the sequence $L_1, L_2, L_3, \cdots$ have a point O in common. The point O belongs to M but to no point set of the collection H. This involves a contradiction.

DEFINITION. The point set M is said to be *totally disconnected* if no nondegenerate subset of it is connected.

THEOREM 56 (SIERPIŃSKI). *No locally compact continuum is the sum of countably many mutually exclusive closed and compact point sets.*

THEOREM 57. *No nondegenerate locally compact continuum is the sum of countably many closed and totally disconnected point sets.*

PROOF. Suppose, on the contrary, that there exists a locally compact continuum M which is the sum of the elements of a countable collection G of closed and totally disconnected point sets. Suppose g is an element of G, P is a point of g and R is a region containing P. There exists a domain D, relatively to M, containing P such that $\bar{D}$ is a compact proper subset of M lying in R. By Theorems 50 and 12, there exists a nondegenerate connected subset of M lying in D and containing P. Hence D, and therefore R, contains a point of $M - g$. Thus every point of g is a limit point of $M - g$. This contradicts Theorem 53.

DEFINITION. If, for each positive integer n, M_n is a point set then by the *limiting set* of the sequence $M_1, M_2, M_3, \cdots$ is meant a point set such that the point P belongs to it if and only if it is true that if R is a region containing P there exist infinitely many integers n such that M_n contains a point of R.

THEOREM 58 (JANISZEWSKI). *If $M_1, M_2, M_3, \cdots$ is a sequence of connected point sets such that the closure of $M_1 + M_2 + M_3 + \cdots$ is compact and there exists a convergent sequence $A_1, A_2, A_3, \cdots$ such that, for each n, A_n is a point of M_n then the limiting set of the sequence $M_1, M_2, M_3, \cdots$ is a continuum.*

PROOF. Let A denote the sequential limit point of the sequence $A_1, A_2, A_3, \cdots$ and let M denote the limiting set of the sequence $M_1, M_2, M_3, \cdots$. Suppose that M is not a continuum. Then, since it is closed, it is the sum of two mutually exclusive closed point sets H and K where H contains A. By Theorem 16, there exists a domain D containing H such that $\bar{D}$ contains no point of K. Let B denote some point of K. The domains D and $S - \bar{D}$ contain A and B respectively. There exists an ascending sequence of numbers $n_1, n_2, n_3, \cdots$ such that, for each m, M_{n_m} contains a point B_m of $S - \bar{D}$. There exists a number δ such that, for every m greater than δ,

A_m belongs to D. For each i, the connected point set $M_{n_{\delta+i}}$ contains both a point of D and a point of $S - \bar{D}$ and therefore a point X_i of the boundary of D. There exists an ascending sequence $m_1, m_2, m_3, \cdots$ such that the sequence $X_{m_1}, X_{m_2}, X_{m_3}, \cdots$ converges to some point O. The point O belongs both to M and to the boundary of D. Thus the supposition that M is not connected leads to a contradiction.

That Theorem 58 does not remain true on the omission of the requirement that the closure of $M_1 + M_2 + M_3 + \cdots$ is compact, may be seen with the aid of Example 4. For each n, let M_n denote the continuum t_n of that example. The limiting set of the sequence $M_1, M_2, M_3, \cdots$ is the point set $H + K$. But this point set is not connected.

DEFINITION. The point set M is said to be the *sequential limiting set* of the sequence α of point sets if M is the limiting set of every infinite subsequence of α. Under these conditions α is said to *converge* to M.

THEOREM 59. *If the limiting set of the sequence α is compact then some subsequence of α has a sequential limiting set.*

PROOF. Let O denote some point of M, the limiting set of α. There exist a subsequence $M_1, M_2, M_3, \cdots$ of α and a sequence $P_1, P_2, P_3, \cdots$ having O as sequential limit point such that, for each n, P_n is a point of M_n. Let β_O denote the sequence $M_1, M_2, M_3, \cdots$. There exists a sequence $H_1, H_2, H_3, \cdots$ such that, for each n, (1) H_n is a finite subcollection of G_n covering M, (2) if g is a region of H_{n+1}, $\bar{g}$ is a subset of some region of H_n. There exist a subcollection U_1 of H_1 and an infinite subsequence β_1 of β_O such that if x is any point set of the sequence β_1 then U_1 is the set of all regions of the collection H_1 that intersects x. There exist a subcollection U_2 of H_2 and an infinite subsequence β_2 of β_1 such that if x is any point set of the sequence β_2 then U_2 is the set of all regions of H_2 that intersect x. This process may be continued. Thus there exist two sequences $\beta_1, \beta_2, \beta_3, \cdots$ and $U_1, U_2, U_3, \cdots$ such that (1) for each n, β_n is an infinite subsequence of β_{n-1} and (2) if x is any point set of the sequence β_n then U_n is the collection of all regions of the collection H_n that intersect x. For each n, let T_n denote the sum of all the regions of the collection U_n. Let T denote the common part of all the point sets of the sequence $\bar{T}_1, \bar{T}_2, \bar{T}_3, \cdots$. Let n_1 denote the smallest n such that M_n belongs to the sequence β_1. Let n_2 denote the smallest n greater than n_1 such that M_n belongs to β_2. This process may be continued. Thus there exists an infinite sequence $n_1, n_2, n_3, \cdots$ such that, for each i, n_{i+1} is the smallest n greater than n_i such that M_n belongs to β_{i+1}. The point set T is the sequential limiting set of the sequence $M_{n_1}, M_{n_2}, M_{n_3}, \cdots$.

DEFINITION. The point set T is said to *separate the point set H from the point set K in the connected point set M* if and only if T is a subset of M and $M - T$ is the sum of two mutually separated point sets one containing H and the other containing K. The statement that T *separates* H *from* K means that it separates H from K in S. The point O will be called a *cut point* of the connected point set M if $M - O$ is not connected.

THEOREM 60. *If T is a connected subset of the connected point set M and $M - T$ is the sum of two mutually separated point sets H and K then $H + T$ is connected.*

PROOF. Suppose, on the contrary, that $H + T$ is the sum of two mutually separated point sets L and N where L intersects T. Since T is connected therefore, by Theorem 28, it is a subset of L. Therefore N is a subset of H. But H and K are mutually separated. Therefore N and K are mutually separated. But so are N and L. Therefore $K + L$ and N are mutually separated and therefore M is not connected, contrary to hypothesis.

THEOREM 61. *If T is a subset of the connected point set M, A and B are points of M and $M - T$ is the sum of two mutually separated point sets H and K containing A and B respectively, but no proper subset of T separates A from B in M, then $H + T$ is connected and every point of T is a limit point of H.*

PROOF. Suppose that $H + T$ is the sum of two mutually separated point sets L and N where N contains A. Suppose L contains T. Then $L + K$ and N are mutually separated and their sum is M, contrary to the supposition that M is connected. A similar contradiction would be reached in case N contained T. Furthermore T does not contain both L and N. Suppose it does not contain N. The point set $M - N \cdot T$ is the sum of two mutually separated point sets $N - N \cdot T$ and $(K + L)$. But these point sets contain A and B respectively and $N \cdot T$ is a proper subset of T, contrary to hypothesis. The supposition that T does not contain L leads to a contradiction in a similar manner.

Suppose the point P of T is not a limit point of H. Then there exists a region R_P, containing P, such that $\bar{R}_P$ contains no point of $H + A + B$. The point set $T - T \cdot R$ is a proper subset of T and $M - (T - T \cdot R)$ is the sum of two mutually separated point sets, H and $K + T \cdot R$, containing A and B respectively. This involves a contradiction.

THEOREM 62. *Under the hypothesis of Theorem* 61, *no connected subset of $M - T$ separates two points of T from each other in M.*

PROOF. Suppose the connected subset Z of $M - T$ separates the points C and F of T from each other in M. By Theorem 61, $H + T$ and $K + T$ are connected. But both of them contain C and F. Therefore they intersect Z. Hence, since it does not intersect T, Z intersects both H and K. But Z is a subset of $H + K$ and H and K are mutually separated. This is contrary to Theorem 28.

It is not true in the plane that, under the hypothesis of Theorem 61, if N

is a component of $M - T$ then $N + T$ is connected. Consider an example indicated in Figure 3. Here T consists of two points A' and B', $B'O - B'$ is a component of $M - T$ and neither A' nor B' is a cut point of M. But $(B'O - B') + B' + A'$ is not connected.

FIGURE 3

THEOREM 63. *If the points A and B of the connected point set M are separated in M by the subset T_1 of M, and the point O of T_1 is separated from B in M by the subset T_2 of M, then there exists a subset of $T_1 + T_2$ which separates $A + O$ from B in M and which is closed in case M, T_1 and T_2 are closed.*

PROOF. By hypothesis, the point set $M - T_1$ is the sum of two mutually separated point sets H_1 and K_1 containing A and B respectively and $M - T_2$ is the sum of two mutually separated point sets H_2 and K_2 containing O and B respectively. The point sets $H_1 + H_2$ and $K_1 \cdot K_2$ are mutually separated and contain $A + O$ and B respectively. Let T denote $M - (H_1 + H_2 + K_1 \cdot K_2)$. Suppose P is a point of T. Then P does not belong to H_1 or to H_2. Therefore it belongs either to T_1 or to K_1 and it also belongs either to T_2 or to K_2. But it does not belong to K_1 and to K_2. Hence it belongs either to T_1 or to T_2. Therefore T is a subset of $T_1 + T_2$. Thus $A + O$ is separated from B in M by a subset of $T_1 + T_2$. It is easy to see that T is closed provided M, T_1 and T_2 are closed.

THEOREM 64. *If M is a compact irreducible continuum from the point A to the point B, and a point separates M into two sets, one of them contains A and the other one contains B.*

PROOF. Suppose, on the contrary, that, for some point P of M, $M - P$ is the sum of two mutually separated point sets M_A and M_B such that M_B contains neither A nor B. By Theorem 47, P is neither A nor B. Therefore M_A contains both A and B. But, by Theorem 60, $M_A + P$ is a continuum. But it contains no point of M_B. Thus it is a proper subcontinuum of M containing both A and B, contrary to the hypothesis that M is an irreducible continuum from A to B.

DEFINITION. If T is a proper subset of the point set M, $M - T$ is called the complement of T *relative to* (or *with respect to*) M. The complement with respect to S of a point set is called the *complement* of that point set.

THEOREM 65. *If S is connected, K is a closed point set that does not have infinitely many components and $S - K$ is the sum of two mutually separated point sets S_1 and S_2 then every component of $K + S_1$ contains a component of K.*

PROOF. Suppose this is not true. Then $K + S_1$ is not a component of itself. Hence it is the sum of two mutually exclusive closed point sets T_{11} and T_{12}. If one of these point sets contained no point of K then it and its complement would be mutually separated contrary to the hypothesis that S is connected. So both of them intersect K. But each of them contains every component of K that it intersects and, if j is 1 or 2, $S - K \cdot T_{1j}$ is the sum of the two mutually separated point sets $T_{1j} - K \cdot T_{1j}$ and $T_{1k} + S_2$ (where k is $3 - j$) and each component of $S_1 + K$ is a component either of T_{11} or T_{12}. Hence either if $j = 1$ or if $j = 2$ there is some component of T_{1j} that does not contain a component of K and therefore T_{1j} is the sum of two mutually exclusive closed point sets T_{1j1} and T_{1j2}. Each of these point sets intersects K and contains every component of K that it intersects. But every component of T_{1j} is either a component of T_{1j1} or a component of T_{1j2}. It follows that either some component of T_{1j1} or some component of T_{1j2} fails to intersect K. This process may be continued indefinitely. Hence K has infinitely many components, contrary to hypothesis.

THEOREM 66. *If M is a continuum, H and K are mutually exclusive compact subcontinua of M and each point of H is separated in M from each point of K by some point of M then H is separated from K in M by some one point.*

PROOF. Let Y denote a definite point of K. For each point X of H there exists a point P of M such that $M - P$ is the sum of two mutually separated point sets D_{XY} and T_{XY} containing X and Y respectively. The point set D_{XY} is a domain with respect to M whose boundary with respect to M is the single point P. Thus for each point X of H there exists a domain D_{XY} with respect to M, containing X, such that (1) $\bar{D}_{XY}$ does not contain Y, (2) the boundary with respect to M of D_{XY} is a single point that separates X from Y in M. Let G denote the collection of all such point sets D_{XY} for all points X of H. With the help of Theorem 4, it may be seen that there exists a finite subcollection G' of G such that each point of H belongs to some point set of the collection G'. Let D_Y denote the sum of all the point sets of the collection G'. The point set D_Y is a domain with respect to M. Since M is connected and D_Y is a proper subset of M therefore the boundary of D_Y exists. Let Z denote a point belonging to it. The point Z belongs to the boundary of some point set $D_{X'Y}$ of the collection G'. The point X' is separated from Y in M by Z. But, since H is a subset of D_Y, therefore Z does not belong to H. Furthermore H is a connected subset of M containing X'. Hence Z separates H from Y in M. In other words, $M - Z$ is the sum of two mutually separated point sets L_Y and N_Y containing H and Y

respectively. Thus, for each point Y of K, there exists a domain N_Y with respect to M, containing Y, such that (1) $\bar{N}_Y$ contains no point of H, (2) the boundary of N_Y with respect to M is a single point that separates Y from H. Let W denote the collection of all such point sets N_Y for all points Y of K. There exists a finite subcollection W' of W such that each point of K belongs to some point set of the collection W'. Let E denote the sum of all the point sets of the collection W'. The boundary of E with respect to M exists. Let F denote a point belonging to it. The point F does not belong to K and it separates H, in M, from some point of K. Since K is connected, it follows that F separates H from K in M.

THEOREM 67. *If H and K are two mutually exclusive closed and compact subsets of the continuum M and W is a collection of closed subsets of M and each two points that belong to H and K respectively are separated from each other in M by some element of W then H and K are separated from each other in M by some closed point set which is a subset of the sum of a finite number of elements of W.*

PROOF. Let X denote a point of H. For each point Y of K there exists an element T_{XY} of W such that $M - T_{XY}$ is the sum of two mutually separated point sets D_{XY} and E_{XY} containing X and Y respectively. Let G_X denote a collection such that g belongs to it if and only if g is E_{XY} for some point Y of K. Each point set of the collection G_X is a domain with respect to M bounded, with respect to M, by a subset of some element of W, G_X covers K and K is closed and compact. Therefore K is covered by some finite subcollection G'_X of G_X. Let E'_X denote the sum of all the domains (with respect to M) of the collection G'_X. The point set E'_X is a domain with respect to M, it contains K and its boundary with respect to M is a closed subset of the sum of a finite number of elements of W. Let D'_X denote the domain $M - \bar{E}'_X$ with respect to M. Let Q denote the collection of all point sets D'_X, for all points X of H. The closed and compact point set H is covered by a finite subcollection Q' of Q. Let D denote the sum of all the point sets of the collection Q' and let β denote the boundary of D with respect to M. The point set β separates H from K in M and it is a closed subset of the sum of a finite number of elements of W.

If the condition requiring that K be closed is omitted from the hypothesis of Theorem 67, the resulting proposition does not hold true in the plane even for the case where H contains only one point and W has only one element, that element being a degenerate point set. Consider the following example.

EXAMPLE. Let M, A, B and B_1, B_2, $B_3, \cdots$ be as described in Example 1. Let H denote the set whose only element is B, let W denote the collection whose only element is the point set whose only element is A and let K denote the point set $B_1 + B_2 + B_3 + \cdots$.

THEOREM 68. *If the closed subsets H and K of the continuum M are separated from each other in M by the closed point set T, and K contains the* *boundary with respect to M of a domain D with respect to M that contains no point of H then H is separated from $K + D$ in M by some closed subset of T.*

PROOF. By hypothesis, $M - T$ is the sum of two mutually separated point sets M_H and M_K containing H and K respectively. Suppose D contains T. Then, since $D + M_K$ does not intersect H, $M - (D + M_K)$ exists and is a subset of M_H. But M is connected and $D + M_K$ is a domain with respect to M and therefore contains no limit point of $M - (D + M_K)$. Hence $M - (D + M_K)$ contains a point P which is a limit point of $D + M_K$ and therefore of D. The point P belongs to β, the boundary of D with respect to M, contrary to the hypothesis that β is a subset of K. So either D contains no point of T, in which case T separates H from $K + D$ in M, or $T - D \cdot T$ exists. Suppose $D \cdot T$ exists. The point sets $M_H - D \cdot M_H$ and $M_K + D \cdot M_H + D \cdot T$ are mutually separated and they contain H and $K + D$ respectively. But their sum is $M - (T - D \cdot T)$. Thus $T - D \cdot T$ separates H from $K + D$ in M. Since T is closed and D is a domain with respect to M therefore $T - D \cdot T$ is closed.

THEOREM 69. *If the point B separates the point A from the point C in the connected point set M then A does not separate B from C in M and no point separates B from $A + C$ in M.*

PROOF. By hypothesis, $M - B$ is the sum of two mutually separated point sets M_A and M_C containing A and C respectively. By Theorem 60, $M_C + B$ is connected. Hence no point not belonging to $M_C + B$ separates B from C in M. In particular A does not do so. Similarly no point not belonging to the connected point set $M_A + B$ separates B from A in M. But no point except B belongs both to $M_A + B$ and to $M_C + B$. Therefore no point separates B from $A + C$ in M.

THEOREM 70. *If A, B and O are three points of a connected point set M and no point separates O either from A or from B in M then no point other than O separates A from B in M.*

PROOF. Suppose, on the contrary, that there exists a point X belonging to $M - O$ such that $M - X$ is the sum of two mutually separated point sets M_A and M_B containing A and B respectively. Then O belongs either to M_A or to M_B. In the first case X separates O from B in M and in the second case X separates O from A in M.

NOTE. One might be inclined to argue that since O is not separated either from A or from B, in M, by any one point, therefore if X is a point of $M - O$, $M - X$ contains a connected point set containing O and A and another one containing O and B and since these two connected sets have the point O in common their sum is a connected subset of $M - X$ containing both A and B and therefore A and B are not separated from each other in M by X. Such an argument would however be fallacious since it can not be proved, on the basis of Axioms 0 and 1, even for the case where M is a compact continuum, that if X does not separate A from O in M then there is a connected subset of M containing A and O but not X.

NOTATION. If M, X, Y and Z are point sets, "XYZ in M", used as a sentence, means Y separates X from Z in M. If X, Y and Z are point sets, "XYZ", used as a sentence, means XYZ in S.

THEOREM 71. *If S is connected, M_1, M_2 and M_3 are closed and $M_1M_2M_3$ is true then $M_1M_3M_2$ is not true.*

PROOF. By hypothesis $S - M_2$ is the sum of two mutually separated point sets D_{21} and D_{23} containing M_1 and M_3 respectively. Suppose that $S - M_3 = D_{31} + D_{32}$ where D_{31} and D_{32} are mutually separated point sets containing M_1 and M_2 respectively. Then $D_{21} \cdot D_{31}$ and $D_{23} + D_{32}$ are mutually separated and their sum is S. Since S is connected this involves a contradiction.

THEOREM 72. *If S is connected, M_1, M_2, M_3 and M_4 are closed and $M_1M_2M_3$ and $M_2M_3M_4$ then $M_1M_2(M_3 + M_4)$.*

PROOF. By hypothesis there exist two mutually separated point sets D_{21} and D_{23} containing M_1 and M_3 respectively and two mutually separated point sets D_{32} and D_{34} containing M_2 and M_4 respectively such that $S - M_2 = D_{21} + D_{23}$ and $S - M_3 = D_{32} + D_{34}$. Suppose D_{32} and D_{21} are mutually exclusive. Then, since they are domains, they are mutually separated. Since D_{23} and D_{21} are mutually separated it follows that $D_{32} + D_{23}$ and D_{21} are mutually separated. But since D_{32} contains M_2, $(D_{32} + D_{23}) + D_{21} = S$. This involves a contradiction. So $D_{32} \cdot D_{21}$ exists. Since D_{32} contains M_2, D_{34} does not intersect M_2.

Suppose now that $D_{34} \cdot D_{21}$ exists. Then $D_{34} \cdot (M_2 + D_{21}) = D_{34} \cdot D_{21}$. But M_3 contains no point of $M_2 + D_{21}$. Hence $M_2 + D_{21}$ is the sum of two mutually separated point sets $D_{34} \cdot D_{21}$ and $D_{32} \cdot (M_2 + D_{21})$. Hence S is the sum of the point sets $M_2 + D_{32} \cdot D_{21} + D_{23}$ and $D_{34} \cdot D_{21}$. But these point sets are mutually separated. Thus the supposition that D_{34} intersects D_{21} involves a contradiction. It follows that D_{34} does not intersect $M_2 + D_{21}$. Hence it is a subset of D_{23} and therefore D_{23} contains M_4, and M_2 separates M_1 from $M_3 + M_4$.

THEOREM 73. *If S is connected, M_1, M_2, M_3 and M_4 are closed and mutually exclusive and $M_1M_2M_3$ and $M_1M_3M_4$ then $M_1M_2(M_3 + M_4)$.*

PROOF. By hypothesis there exist two mutually separated sets D_{21} and D_{23} containing M_1 and M_3 respectively and two mutually separated sets D_{31} and D_{34} containing M_1 and M_4 respectively such that $S - M_2 = D_{21} + D_{23}$ and $S - M_3 = D_{31} + D_{34}$. The point sets $D_{31} \cdot D_{21}$ and $M_3 + D_{23} + (D_{34} + M_2) - M_2$ are mutually separated, their sum is $S - M_2$ and they contain M_1 and $M_3 + M_4$ respectively.

THEOREM 74 (J. B. BATCHA, R. D. DAVIS, C. L. KITE, L. P. MAHER, E. D. ROACH AND S. W. YOUNG). *If S is connected, M_1, M_2, M_3 and M_4 are closed and mutually exclusive and M_2 separates M_1 from M_4 and M_2 does not separate two points of M_3 from each other then either $M_1M_2(M_3 + M_4)$ or $M_4M_2(M_3 + M_1)$.*

PROOF. By hypothesis the point set $S - M_2$ is the sum of two mutually separated point sets D_{21} and D_{24} such that D_{21} contains M_1 and D_{24} contains M_4. Since M_2 does not separate any two points of M_3 from each other therefore M_3 does not intersect both D_{21} and D_{24}. But since, by hypothesis, it contains no point of M_2 therefore it is a subset of the sum of D_{21} and D_{24}. It follows that it is a subset of one of them. Suppose it is a subset of D_{21}. Then, since M_4 is a subset of D_{24} and D_{21} contains both M_1 and M_3, M_2 separates $M_1 + M_3$ from M_4. In case M_3 is a subset of D_{24} then, since M_1 is a subset of D_{21} and D_{24} contains both M_4 and M_3, M_2 separates $M_3 + M_4$ from M_1.

Theorem 74 does not remain true, even on a circle, if, in its hypothesis, the requirement that M_2 do not separate two points of M_3 from each other is replaced by the requirement that no proper subset of M_2 or M_3 separate M_1 from M_4. To see this consider the following example.

EXAMPLE. Let S denote a circle, let A_1, A_2, B_3, A_4, B_2 and A_3 denote points lying on S in the order $A_1A_2B_3A_4B_2A_3A_1$, let M_1 denote the set whose only element is A_1, let M_4 denote the one whose only element is A_4, let M_2 denote $A_2 + B_2$ and let M_3 denote $A_3 + B_3$.

THEOREM 75. *If S is connected, M_1, M_2, M_3 and M_4 are closed and $M_1M_2M_3$ and $M_1M_3M_4$ and M_3 does not separate two points of M_2 from each other then $M_2M_3M_4$.*

PROOF. There exist two mutually separated sets D_{31} and D_{34} containing M_1 and M_4 respectively such that $S - M_3 = D_{31} + D_{34}$. Since M_3 does not separate two points of M_2 from each other, M_2 is a subset either of D_{31} or of D_{34}. Since $M_1M_2M_3$, therefore, by Theorem 71, $M_1M_3M_2$ is not true. Hence M_2 is not in D_{34}. Therefore it is in D_{31}. Hence $M_2M_3M_4$.

Even on a straight line, Theorem 75 does not remain true if the condition in its hypothesis requiring that M_3 do not separate two points of M_2 from each other is replaced by one requiring that M_2 do not separate any two points of M_3 from each other. Consider the following example.

EXAMPLE. Let A, B, C, D and E denote points in the order indicated on some straight line S, let M_1, M_3 and M_4 denote degenerate sets whose elements are A, C and E respectively, and let M_2 denote the set whose elements are B and D.

THEOREM 76. *If M_1, M_2, M_3 and M_4 are closed and mutually exclusive, S is connected and $M_1M_2M_3$ and $M_1M_3M_4$, and M_2 does not separate two points of M_3 from each other and no closed proper subset of M_2 separates M_1 from M_4 then $M_2M_3M_4$.*

Proof. By hypothesis and Theorem 73, there exist two mutually separated sets D_{21} and D_{24} containing M_1 and M_4 respectively, and two mutually separated sets D_{31} and D_{34} containing M_1 and M_4 respectively, such that $S - M_2 = D_{21} + D_{24}$ and $S - M_3 = D_{31} + D_{34}$. Suppose M_3 does not separate M_2 from M_4. Then M_2 is not a subset of D_{31}. Suppose it is a subset of D_{34}. Then $M_1M_3M_2$. But, by Theorem 71, this is impossible since, by hypothesis, $M_1M_2M_3$. So M_2 is not a subset either of D_{31} or of D_{34}. But it is a subset of their sum. Hence it intersects both of them. Since M_2 does not separate two points of M_3 from each other, M_3 does not intersect both D_{21} and D_{24}. But it is a subset of their sum. Hence it is a subset of one of them. If M_3 is a subset of D_{24} then $S - M_2 \cdot D_{31}$ is the sum of the two mutually separated sets $D_{31} \cdot D_{21}$ and $D_{24} + D_{34}$ which contain M_1 and M_4 respectively and if M_3 is a subset of D_{21} then $S - M_2 \cdot D_{34}$ is the sum of the mutually separated sets $D_{34} \cdot D_{24}$ and $D_{21} + D_{31}$ which contain M_4 and M_1 respectively. In the first case $M_2 \cdot D_{31}$, and in the second case $M_2 \cdot D_{34}$, is a closed proper subset of M_2 that separates M_1 from M_4. Thus the supposition that M_3 does not separate M_2 from M_4 has led to a contradiction.

If the condition requiring that no closed proper subset of M_2 separate M_1 from M_4 be omitted from the hypothesis of Theorem 76, the resulting proposition is not a consequence of Axioms 0 and 1. See the example given above concerning Theorem 75.

Notation. Used as a sentence, "$XYZW$" means XYZ, XYW, XZW and YZW.

Theorem 77. *If S is connected and M_1, M_2, M_3 and M_4 are closed, and $M_1M_2M_3$ and $M_2M_3M_4$ then $M_1M_2M_3M_4$.*

Theorem 77 is a consequence of Theorem 72.

Theorem 78. *If S is connected and M_1, M_2, M_3 and M_4 are closed and mutually exclusive and $M_1M_2M_3$ and $M_1M_3M_4$, and M_3 does not separate any two subsets of M_2 from each other, then $M_1M_2M_3M_4$.*

Theorem 78 is a consequence of Theorems 73 and 75.

Theorem 79. *If S is connected and M_1, M_2, M_3 and M_4 are closed and mutually exclusive and $M_1M_2M_3$ and $M_1M_3M_4$, and M_2 does not separate any two subsets of M_3 from each other, and no closed proper subset of M_2 separates M_1 from M_4, then $M_1M_2M_3M_4$.*

Theorem 79 is a consequence of Theorems 73 and 76.

Theorem 80. *If S is connected and M_1, M_2, M_3 and M_4 are closed and mutually exclusive and $M_1M_2M_4$ and $M_1M_3M_4$ and M_2 does not separate two points of M_3 from each other and either M_3 does not not separate two points of M_2 from each other or no closed proper subset of M_2 separates M_1 from M_4 then either $M_1M_2M_3M_4$ or $M_1M_3M_2M_4$.*

Theorem 80 is a consequence of Theorems 74, 75 and 76.

DEFINITION. If A and B are two points of a connected point set M and there exists at least one point that separates A from B in M then the point set such that X belongs to it if and only if it separates A from B in M is called the *segment* AB of M and the segment AB of M plus the pair of points A and B is called the *interval* AB of M. The points A and B are called the *endpoints* of the segment AB, and of the interval AB, of M.

DEFINITION. If AB is an interval of the connected point set M, the point X is said to *precede* the point Y, *in* M, *on* AB (or *in the order from* A *to* B, *in* M, *on* AB) if X and Y are points of M and either (1) Y is B or (2) Y separates $A + X$ from B in M. The point X is said to *follow* the point Y, *in* M, *on* AB if Y precedes X, in M, on AB. A point is said to be between the points X and Y *in* M *on* AB if it precedes one of those points, and follows the other one, in M, on AB, in the order from A to B.

With the help of Theorems 71–80, the following theorem may be established.

THEOREM 81. *If X and Y are two points of the interval AB of the connected point set M then either X precedes Y, or Y precedes X, in M, on AB and if X precedes Y, in M, on AB then Y does not precede X, in M, on AB. If X, Y and Z are three points of AB one of them is between the two others in M on AB and if X precedes Y, in M, on AB and Y precedes Z, in M, on AB then X precedes Z, in M, on AB. If X precedes Y, in M, on AB, in the order from A to B then Y precedes X, on that interval, in the order from B to A. If X and Y are two points of AB, and the interval XY of M exists then the point P belongs to that interval if and only if P either is X or Y or lies between X and Y, in M, on AB.*

THEOREM 82. *If X is a point of the interval AB of the connected point set M and $M - X = H_1 + K_1 = H_2 + K_2$, where H_1 and K_1 are mutually separated, H_2 and K_2 are mutually separated, A belongs to H_1 and to H_2 and B belongs to K_1 and to K_2, then $H_1 \cdot (AB) = H_2 \cdot (AB)$ and $K_1 \cdot (AB) = K_2 \cdot (AB)$.*

PROOF. Each of the point sets $H_1 \cdot (AB)$ and $H_2 \cdot (AB)$ is the set of all points that precede X on AB. Hence $H_1 \cdot (AB)$ is identical with $H_2 \cdot (AB)$. Similarly $K_1 \cdot (AB)$ is identical with $K_2 \cdot (AB)$.

THEOREM 83. *If the point X of the interval AB of the connected point set M is a limit point of the subset T of AB then every point that follows X on AB follows infinitely many points of T on AB.*

PROOF. Suppose Y is a point following X on AB. If Y is B it follows every point of T, other than B, on AB. If Y is not B then $M - Y$ is the sum of two mutually separated point sets H and K such that H contains $A + X$, and K contains B. The point X is not a limit point of $K + Y$. But X is a limit point of T and every point of T belongs either to $K + Y$ or to H. Hence X is a limit point of $H \cdot T$ and therefore there are infinitely many points of T in H. But Y follows every point of H.

Definition. If for the point O of the connected point set M, the set $M - O$ is the sum of two mutually separated point sets H and K then H and K will be called *sects* of M from O, and M will be said to be separated by the omission of O into the two sets H and K. If P is a point of M distinct from O then in case there is only one sect of M from O that contains P that sect will be called *the sect* OP.

Definition. If A and B are cut points of the connected point set M, the *section* AB of M is a subset of M such that X belongs to it if and only if (1) X is A or B or (2) X belongs to M but to no sect of M from A or B which contains neither A nor B. Unless A and B are cut points of M the section AB of M does not exist.

Clearly a point X of M belongs to the section AB of M according to this definition if and only if A and B are cut points of M and either X is one of the points A and B or X belongs to every sect of M from A which contains B and to every sect of M from B which contains A.

Theorem 84. *If A is a cut point of the connected point set M and the sect H of M from A is the sum of two mutually separated sets H_1 and H_2 then H_1 and H_2 are sects of M from A.*

Proof. By hypothesis H and $M - (A + H)$ are mutually separated. Since H_1 and H_2 are mutually separated then so are H_1 and $H_2 + M - (A + H)$. But their sum is $M - A$. So H_1 is a sect of M from A.

Theorem 85. *If A is a cut point of the connected point set M and H is a connected sect of M from A then H is a component of $M - A$.*

Proof. By hypothesis H and $M - (A + H)$ are mutually separated. If H is a proper subset of the subset T of $M - A$ then $T - H$ is a subset of $M - (A + H)$ and therefore H and $T - H$ are mutually separated and therefore T is not connected. Hence H is a component of $M - A$.

Theorem 86. *If A is a cut point of the connected point set M, and the point B of $M - A$ does not belong to infinitely many sects of M from A then B belongs to only one connected sect of M from A and every sect of M from A to B contains that one.*

Proof. Suppose no connected sect of M from A contains B. The point set $M - A$ is the sum of two mutually separated point sets E and H_1 where H_1 contains B. By definition, H_1 is a sect of M from A. Since it is not connected, H_1 is the sum of two mutually separated sets. Let H_2 denote the one of these two that contains B. By Theorem 84, it is a sect of M from A containing B. Hence H_2 is the sum of two mutually separated sets of which one, H_3, contains B. The set H_3 is a sect of M from A containing B. This process may be continued indefinitely. So B belongs to infinitely many sects $H_1, H_2, H_3, \cdots$ of M from A. But this is contrary to hypothesis. So there exists a connected sect L of M from A containing B. By Theorem 85,

L is a component of $M - A$. Since there is only one component of $M - A$ containing B, L is the only connected sect of M from A to B. Suppose K is a sect of M from A to B not containing L. Then it is not L. Therefore it is the sum of two mutually separated sects of M from A. If this process is continued a contradiction may be reached as in the above proof of the existence of L.

THEOREM 87. *If A and B are two points of the connected point set M and K is the set of all points that separate A from B in M and there do not exist uncountably many mutually exclusive domains with respect to M then there exists a countable subset H of K such that, if X is a point of K not belonging to H, $M - X$ is the sum of two mutually separated connected point sets one containing A and the other containing B.*

PROOF. Suppose there exists an uncountable subset K' of K such that, for each point X of K', $M - X$ is the sum of two mutually separated point sets, one containing A and the other containing B, such that one of them is not connected. Suppose X is a point of K' such that $M - X$ is the sum of two mutually separated point sets M_{XA} and M_{XB} containing A and B respectively such that M_{XA} is not connected. Then M_{XA} is the sum of two mutually separated point sets M_{XAA} and D_X such that M_{XAA} contains A. It follows that $M - X$ is the sum of two mutually separated point sets D_X and $M_{XAA} + M_{XB}$. Let E_X denote $M_{XAA} + M_{XB}$. The point set E_X contains both A and B, and D_X contains neither of them. Similarly, if M_{XA} is connected and M_{XB} is not connected, $M - X$ is the sum of two mutually separated point sets D_X and E_X such that D_X contains neither A nor B.

Suppose now that X and Y are two points of K'. The point set $E_X + X$ is connected and it contains A and B but no point of D_X. Hence Y does not belong to D_X. Therefore it belongs to E_X. Similarly X does not belong to $D_Y + Y$. Thus the connected point set $D_Y + Y$ is a subset of the sum of the two mutually separated point sets D_X and E_X and it contains the point Y of E_X. Hence $D_Y + Y$ is a subset of E_X. Therefore D_X and D_Y are mutually exclusive domains with respect to M. Since there are uncountably many such domains with respect to M this involves a contradiction.

THEOREM 88. *If A is a point of the connected point set M then $M - A$ is the sum of two mutually separated connected point sets if and only if there exists only one pair of mutually separated point sets whose sum is $M - A$. There exists only one such pair if and only if there are only two sects of M from A.*

It is not true in the plane that if A and B are cut points of the connected point set M then the segment AB of M or the section AB of M minus its endpoints is either a domain with respect to M or a connected point set.

EXAMPLE. In a Cartesian plane, let O denote the origin, let E denote the point $(0, -1)$, let F denote the point $(0,1)$, let A and B denote the points $(0,2)$ and $(0, -2)$ respectively and, for each positive integer n, let A_n and C_n

denote the points $(-n, 1 + 1/n)$ and $(0, 1 + 1/n)$ respectively and let B_n and D_n denote the points $(-n, -1 - 1/n)$ and $(0, -1 - 1/n)$ respectively. Let J denote the circle with center at O that contains the points E and F. Let α_n denote the straight line interval AA_n plus the straight ray A_nC_n and let β_n denote the straight line interval BB_n plus the straight ray B_nD_n. Let M denote the point set $J + E + F + A + B + \alpha_1 + \alpha_2 + \cdots + \beta_1 + \beta_2 + \cdots$. The point set M is connected. The segment AB of M is the point set whose elements are E and F, and the point set $E + F + A + B$ in both the interval and the section AB of M. Neither the segment AB, nor the section AB minus $A + B$, is a domain with respect to M.

However, the following theorem holds true.

THEOREM 89. *If A and B are points of the connected point set M and there are only two sects of M from A and only two from B then the section AB of M is a connected point set T and $T - (A + B)$ is a domain with respect to M.*

PROOF. By hypothesis, $M - A$ is the sum of two mutually separated point sets H_{AB} and K, and $M - B$ is the sum of two mutually separated point sets H_{BA} and N such that H_{AB} and H_{BA} contain B and A respectively. Since $K + A$ is a connected subset of the sum of the two mutually separated point sets H_{BA} and N, and A belongs to H_{BA} therefore $K + A$ is a subset of H_{BA}. Hence $K + A$ and $N + B$ are mutually separated. Since H_{AB} is a domain with respect to M containing the limit point B of H_{BA} therefore $H_{AB} \cdot H_{BA}$ exists. Suppose $H_{AB} \cdot H_{BA} + A + B$ is the sum of two mutually separated point sets T_{A+B} and Q such that T_{A+B} contains A and B. Then Q is a subset of $H_{AB} \cdot H_{BA}$, and $K + N + T_{A+B}$ and Q are mutually separated. But their sum is M. Suppose, on the other hand, that $H_{AB} \cdot H_{BA} + A + B$ is the sum of two mutually separated sets T_A and T_B containing A and B respectively. Then M is the sum of the two mutually separated sets $K + T_A$ and $N + T_B$. This also involves a contradiction. So $H_{AB} \cdot H_{BA} + A + B$ is connected. But it is T, the section AB of M. The point sets H_{AB} and H_{BA} are domains with respect to M therefore so is their common part, that is to say $T - (A + B)$.

THEOREM 90. *If T is an uncountable subset of the interval AB of the connected point set M and there do not exist uncountably many mutually exclusive domains with respect to M then there exists a countable subset K of T such that each point of $T - K$ is separated from A in M by uncountably many points of $T - K$ and is also separated from B in M by uncountably many points of $T - K$.*

PROOF. Suppose no point of T is separated from A in M by uncountably many points of T. By Theorem 87, there exists an uncountable subset T' of T such that, if X is a point of T', $M - X$ is the sum of two mutually separated *connected* point sets, one containing A and the other containing B. There exists a well ordered sequence α whose terms are the points of T'. Let

P_1 denote the first term of this sequence. Uncountably many points of T' follow P_1 in the order from A to B in M on the interval AB. Let P_2 denote the first such point that follows P_1 in the sequence α. Let P_3 denote the first point of T' in the α sense that follows P_2 in the order from A to B, in M, on the interval AB. There exists a well ordered subsequence β of α such that the first, second and third terms of α are P_1, P_2 and P_3 respectively and if γ is an initial segment of β then (1) γ is identical with β if and only if there exists no term of α following every term of γ both in the sequence α and in the order from A to B on the interval AB, (2) if γ is not identical with β then the first term following every term of γ in β is the first term of α which follows every term of γ both in the sequence α and in the order from A to B, in M, on the interval AB. Since T' is uncountable and no point of T' is preceded, in the order from A to B on the interval AB, by uncountably many points of T', therefore β is uncountable. There exists an uncountable set Z of points of β such that (1) each point X of Z has an immediate predecessor L_X and an immediate successor R_X in α, (2) if the point Y of Z follows the point X of Z in α then L_Y follows R_X in α. For each point X of Z, let D_X denote the set of all points of the section L_XR_X of M except L_X and R_X. For each X in Z, D_X is a domain with respect to M. There are uncountably many such domains, with respect to M, and they are mutually exclusive, contrary to hypothesis.

So there is at least one point of T which is separated from A in M by uncountably many points of T. Let T_1 denote a point set such that X belongs to it if and only if X is such a point. Suppose $T - T_1$ exists and is uncountable. Then $T - T_1$ contains a point which is separated from A in M by uncountably many points of T. This involves a contradiction. So $T - T_1$ either does not exist or is countable. Hence T_1 is uncountable. It follows that there exists a subset T_2 of T_1 such that either T_2 is T_1 or $T_1 - T_2$ is countable and such that every point of T_2 is separated from B in M by uncountably many points of T_1. Every point of T_2 is separated from A, in M, by uncountably many points of T and is also separated from B, in M, by uncountably many points of T. But T_2 is either T or T minus a countable point set. Hence every point of T_2 is separated from A in M by uncountably many points of T_2 and is also separated from B in M by uncountably many points of T_2.

THEOREM 91. *Suppose M is an uncountable completely separable point set and "precedes" has such a meaning that* (1) *if X precedes Y, X and Y are points of M and Y does not precede X,* (2) *if X and Y are points of M, either X precedes Y or Y precedes X,* (3) *if X precedes Y, there exist a region R_X containing X and a region R_Y containing Y such that every point of $M \cdot R_X$ precedes Y and every point of $M \cdot R_Y$ is preceded by X. Then there exists a countable subset T of M such that, if P is a point of $M - T$, every region that contains P contains uncountably many points that precede P and uncountably many points that are preceded by P.*

PROOF. Suppose K is an uncountable subset of M. By Theorems 23, 24 and 25 there exists an uncountable subset K_1 of M such that every region that intersects K_1 contains uncountably many points of K_1. By Theorem 6 there exists a point P of K_1 which is a limit point both of the set of all points of K_1 that precede it and of the set of all those that follow it. Suppose R is a region containing P. The region R contains a point A of K_1 preceding P and a point B of K_1 following P. There exist two regions R_A and R_B lying in R and containing A and B respectively such that every point of K_1 in R_A precedes P and every one in R_B follows P. Each of these regions contains uncountably many points of K_1.

THEOREM 92. *If H is an uncountable set of domains and M is a completely separable point set and, for each domain h of H, P_h is a definite point of M belonging to h, then there exist a region R, a point A of R, and an uncountable subset H' of H such that* (1) *if h is any domain of H', h contains R and* (2) *if Z is any region containing A, there exist uncountably many domains h of the set H' such that P_h belongs to Z.*

PROOF. For each domain h of H, let P_h denote some definite point of $h \cdot M$. For each h there exists a number n_h such that if g is a region belonging to G_{n_h} and containing P_h then g is a subset of h. There exist a number m and an uncountable subset W of H such that if h is any domain of W then $n_h = m$. Since M is completely separable therefore, by Theorems 23, 24 and 25, there exists a domain h_0 belonging to the set W such that if D is any domain containing P_{h_0} then there exists an uncountable subset Q_D of W such that if q is any domain of the set Q_D then P_q belongs to D. There exists a domain g_0 belonging to G_m and containing P_{h_0}. If q is any domain of the uncountable set Q_{g_0} then g_0 contains P_q and therefore lies in q. Let R denote g_0 and let H' denote Q_{g_0}. There exists a domain q_0 belonging to the set H' such that if Z is any region containing P_{q_0} then there exist uncountably many domains h of the set H' such that P_h belongs to Z. Let A denote the point P_{q_0}.

THEOREM 93. *Every nondegenerate compact continuum has at least two non-cut points.*

PROOF. Suppose M is a nondegenerate compact continuum. Suppose E is a point of M such that every other point of M is a cut point of M. Let G denote a collection such that g belongs to it if and only if g is the closure of a sect of M that neither contains E nor has E as a starting point. Suppose H is a monotonic subcollection of G. By Theorem 37, the continua of the collection H have a point P in common. The point set $M - P$ is the sum of two mutually separated sets M_{PE} and K such that M_{PE} contains E. Suppose h is an element of the collection H. There exists a point X such that $M - X$ is the sum of two mutually separated point sets M_{XE} and $h - X$

where M_{XE} contains E. Since h belongs to H it contains P. So P does not belong to the connected point set $M_{XE} + X$. Hence $M_{XE} + X$ is a subset either of M_{PE} or of K. But it contains E. Hence it is a subset of M_{PE}. Therefore h contains $K + P$. But $K + P$ belongs to G. So the element $K + P$ of G is a subset of every element of H. Hence, by Theorem 39, there exists an element g_0 of G which contains no other element of G. There exists a point E_0 such that $M - E_0$ is the sum of two mutually separated point sets such that one of them is $g_0 - E_0$ and the other one contains E. Let Y denote some point of g_0. The point set $M - Y$ is the sum of two mutually separated point sets M_{YE} and T such that T does not contain E. The point set $T + Y$ is a subset of $g_0 - E_0$. This involves a contradiction.

So, for every point E of M, there is a non-cut point of M distinct from E. It follows that M has at least two non-cut points.

THEOREM 94. *If H is a connected proper subset of the compact continuum K, then $K - H$ contains a non-cut point of K.*

PROOF. Suppose, on the contrary, that every point of $K - H$ is a cut point of K. Let P denote some definite point of $K - H$. Then $K - P$ is the sum of two mutually separated point sets. Since H is a connected subset of $K - P$ it must lie wholly in one of these sets. Let K_1 denote the one that contains H and let K_2 denote the other one. By Theorem 93, K_2 contains a point O whose omission does not disconnect $K_2 + P$. Since the connected point sets $(K_2 + P) - O$ and $K_1 + P$ have the point O in common, their sum is connected. But their sum is $K - O$. Thus K is not disconnected by the omission of O. But since O belongs to K_2 it belongs to $K - H$.

DEFINITION. An *arc* is a compact nondegenerate continuum that does not have more than two non-cut points.

From this definition and Theorem 93 it follows that every arc has only two non-cut points. They will be called its *endpoints* and if A and B are the endpoints of the arc M, M will be referred to as an arc from A to B, and the statement that AB is an arc is to be interpreted to mean that AB is an arc from A to B.

THEOREM 95. *If O is a point of the arc M from A to B, and $M - O$ is the sum of two mutually separated point sets H and K, then one of these sets contains A and the other one contains B. Thus the interval AB of the arc AB is the arc AB.*

PROOF. Suppose K contains neither A nor B. By Theorem 93, the compact continuum $K + O$ contains a point X distinct from O such that $K + O - X$ is connected. The point set $H + O$ is also connected and the sum of these two connected point sets is $M - X$. Hence $M - X$ is connected. This involves a contradiction.

By Theorem 95, if M is an arc from A to B then M is the interval AB of M and therefore Theorems 81 and 82 apply to it. Since Theorem 82 applies to it, it is permissible to speak of *the* two sets into which the arc AB is separated by the omission of any one of its points except A and B.

EXAMPLE 7. Let point and region be interpreted as in Example 1. Let K denote a set of four points A, B, X and Y. It is easy to see that there exist two connected point sets M_1 and M_2 containing these four points and such that (1) K is the interval AB of M_1 and also the interval AB of M_2, and (2) X precedes Y, in the order from A to B, in M_1, on AB but Y precedes X, in the order from A to B, in M_2, on AB.

From this example, it is clear that it might not, in all cases, be desirable to omit "in M" from the statement "X precedes Y, in M, on AB", since, for a given AB, X might precede Y, in the order from A to B, in *one* M, on AB and Y might precede X, in the order from A to B, in *another* M, on the *same* AB. If, however, AB is an *arc* from A to B, then it may be easily seen that if M is any connected point set such that AB is an interval of M, the point X precedes the point Y, in the order from A to B, in M, on AB if, and only if, X precedes Y, in the order from A to B, in AB, on AB and therefore if M_1 and M_2 are two different connected point sets such that the arc AB is an interval of each of them then X precedes Y, in the order from A to B, in M_1 on AB if and only if X precedes Y, in the order from A to B, in M_2 on AB. If AB is an arc it is then permissible to substitute "X precedes Y, in the order from A to B, on AB" in place of "X precedes Y, in the order from A to B, in AB, on AB".

THEOREM 96. *If M is an arc from A to B, M is irreducible with respect to the property of being a connected point set that contains A and B.*

PROOF. Suppose there exists a connected proper subset K of M containing A and B. There exists a point P belonging to M but not to K. The point set $M - P$ is the sum of two mutually separated point sets H and L containing A and B respectively. The sets $H \cdot K$ and $L \cdot K$ are mutually separated. But their sum is the connected point set K. This involves a contradiction.

THEOREM 97. *If M is an arc from A to B and P is a point of M distinct from A and B, and $M - P$ is the sum of two mutually separated point sets H and K, and H contains A, then $H + P$ is an arc from A to P and H is connected.*

PROOF. Suppose that X is a point of $H + P$ distinct from A and from P. Then, by Theorem 95, $M - X$ is the sum of two mutually separated sets T and L, where T contains A and L contains B. The connected set $K + P$ is a subset of L. Let U denote the common part of $(H + P) - X$ and T and let V denote the common part of $(H + P) - X$ and L. Since T and L are mutually separated and U and V are subsets of T and L respectively, therefore U and V are mutually separated. But $(H + P) - X = U + V$. Thus the continuum $H + P$ is disconnected by the omission of any one of its points except A and P. Therefore it is an arc from A to P. It follows that $(H + P) - P$, that is to say H, is connected.

THEOREM 98. *If M is an arc from A to B, then $M - (A + B)$ is connected.*

Theorem 98 is a consequence of Theorems 96 and 47.

THEOREM 99. *Every interval of an arc is an arc.*

PROOF. Suppose XY is an interval of the arc AB.

Case 1. Suppose the points X and Y are the points A and B. Then, by Theorem 95, the interval XY of AB is the arc AB.

Case 2. Suppose X is A and Y is not B. Of the two sets into which the omission of Y separates AB, let H denote the one that contains A. By Theorem 97, $H + Y$ is an arc from A to Y. But $H - A$ is the set of all points that lie between A and Y on AB. Hence, by Theorem 81, $H + Y$ is the interval XY.

Case 3. Suppose X and Y are both distinct from A and from B. One of them precedes the other one on AB in the order from A to B. Suppose X precedes Y in that order. Let AY denote the interval of AB from A to Y. By Case 2, AY is an arc from A to Y. By Theorem 81, the set of all points that separate X from Y in AB is the same as the set of all those that separate X from Y in AY. Therefore the interval XY of AB is the interval XY of AY. But, by Case 2, the latter is an arc from X to Y.

THEOREM 100. *If M is an arc from A to B and α is a sequence of points $P_1, P_2, P_3, \cdots$, belonging to M, such that, for every n, P_n precedes P_{n+1} in the order from A to B, on M, then this sequence has a sequential limit point P and every term of it precedes P in that order on M.*

PROOF. Since M is compact therefore, by Theorem 11, there exists an infinite increasing sequence $n_1, n_2, n_3, \cdots$ such that the sequence $P_{n_1}, P_{n_2}, P_{n_3}, \cdots$ has a sequential limit point P. Unless every P_n precedes P then there exists a number j such that P precedes P_{n_j}. The point set $P_{n_j} + P_{n_{j+1}} + P_{n_{j+2}} + \cdots$ lies wholly on the interval $P_{n_j}B$ of AB. Hence P is not a limit point of this point set. But neither is it a limit point of the finite point set $P_{n_1} + P_{n_2} + \cdots + P_{n_{j-1}}$. Hence, by Theorem 9, P is not a limit point of $P_{n_1} + P_{n_2} + \cdots$. Thus the supposition that not every P_n precedes P leads to a contradiction. Suppose that P is not a sequential limit point of α. Then, by Theorems 14 and 15, there exists an infinite increasing sequence $m_1, m_2, m_3, \cdots$ such that the sequence $P_{m_1}, P_{m_2}, P_{m_3}, \cdots$ has, as sequential limit point, some point O distinct from P. One of the points O and P precedes the other one. Suppose that O precedes P. Then every P_n belongs to the segment AO and P belongs to the segment BO. Hence P is not a limit point of the point set $P_1 + P_2 + P_3 + \cdots$. Thus the supposition that P is not the sequential limit point of the sequence α has led to a contradiction.

THEOREM 101. *If M is a subset of the arc AB, then in order that the point P of AB be a limit point of M it is necessary and sufficient that every segment of AB*

that contains P contain a point of M distinct from P, or that every interval of AB that contains P contain a point of M distinct from P, according as P is, or is not, distinct from A and from B.

PROOF. Suppose first that P is neither A nor B and that every segment of AB that contains P contains a point of M distinct from P. Then either every such segment contains a point of M preceding P or every one contains a point of M following P. Suppose that the former of these alternatives holds true. The point P is a limit point of the segment AP. It follows, with the help of Theorems 10, 11, 14 and 100, that P is the sequential limit point of some sequence α of points $P_1, P_2, P_3, \cdots$, all lying on AP, such that, for each n, P_n precedes P_{n+1}. Let H denote the set of all points of M that precede P. The segment P_1P contains a point X_1 of H. Let n_1 denote the smallest n such that P_n is between X_1 and P. The segment PP_{n_1} contains a point X_2 of H. This process may be continued indefinitely. Thus there exists a sequence β of points $X_1, P_{n_1}, X_2, P_{n_2}, X_3, P_{n_3}, \cdots$ such that if one point precedes another one in this sequence, it also precedes it in the order from A to B on AB. It follows, by Theorem 100, that the sequence β has a sequential limit point Z. But the sequence α and the sequence β have an infinite subsequence in common. Hence Z is P. Therefore P is the sequential limit point of the sequence of distinct points $X_1, X_2, X_3, \cdots$. But these points all belong to M. Therefore P is a limit point of M.

Suppose, secondly, that P is distinct from A and B and that it is a limit point of M, and that X is a point of AB preceding P and Y is one following P, in the order from A to B, on AB. If the segment XY of AB contains no point of M distinct from P, then $(M + P) - P$ is a subset of the sum of the two intervals AX and YB of AB. But these intervals are closed point sets and neither of them contains P. Hence P is not a limit point of their sum. Therefore it is not a limit point of $(M + P) - P$. Hence it is not a limit point of M. Thus the supposition that there is a segment of AB containing P but no point of M leads to a contradiction.

The case where P is A or B may be treated by similar methods.

THEOREM 102 (DEDEKIND-CUT PROPOSITION). *If H and K are two subsets of the arc AB such that each point of AB belongs either to H or to K and each point of H precedes each point of K in the order from A to B on AB, then there is either a last point of H or a first point of K in that order.*

PROOF. Since AB is a connected point set, one of the sets H and K contains a limit point of the other one. Suppose H contains a point P which is a limit point of K. Suppose X is a point of AB following P. Then, by Theorem 101, the segment AX of AB contains a point Y which belongs to K. Since Y precedes X therefore X belongs to K. Thus every point that follows P belongs to K. Therefore P is the last point of H. Similarly, if K contains a point P which is a limit point of H, then P is the first point of K.

THEOREM 103. *If K is a closed point set lying on the arc AB, there is a first point of K in the order from A to B on AB.*

PROOF. If A belongs to K it is the first point of K. Suppose A does not belong to K. Let H denote the set of all points X of AB such that X precedes every point of K, and let T denote the point set $AB - H$. It is clear that every point of H precedes every point of T. Suppose X is a point of H. Since K is closed and does not contain X therefore, by Theorem 101, there exists a point Y between X and B such that the interval XY of AB contains no point of K. Hence Y precedes every point of K. Therefore it belongs to H. Thus for every point X of H there is a point Y of H following X. In other words there is no last point of H. Therefore, by Theorem 102, there exists a point P which is the first point of T. If P did not belong to K it would precede every point of K and therefore it would belong to H which is impossible since it belongs to T. Hence P belongs to K. Therefore it is the first point of K.

THEOREM 104. *If M is a connected point set there do not exist a domain D and an uncountable collection H of mutually exclusive closed subsets of M such that $D \cdot M$ is a separable point set containing H^* and if h is any element of H then $M - h$ is the sum of two mutually separated point sets of which one contains $H^* - h$.*

PROOF. Suppose, on the contrary, that there exist such a domain D and collection H. For each point set h of the uncountable collection H, $M - h$ is the sum of two mutually separated point sets T_h and K_h where K_h contains $H^* - h$ and therefore T_h contains no point of H^*. Suppose now that x and y are two elements of H. The point set x separates T_x from y in M and y separates x from T_y in M. Therefore, by Theorem 77, x separates T_x from T_y in M. Hence T_x and T_y are mutually exclusive. Now let G denote the collection of all point sets T_h for all sets h of H and let N denote a countable subset of $D \cdot M$ such that $D \cdot M$ is a subset of $\bar{N}$. Since each point set of the collection G is a domain with respect to M therefore, for each element g of G, $g \cdot D$ contains a point of N. But the point sets of the collection G have been shown to be mutually exclusive, there are uncountably many of them and N is countable. This involves a contradiction.

THEOREM 105. *If M is a separable and connected point set, there do not exist a connected subset K of M and an uncountable collection H of mutually exclusive closed subsets of K such that each point set of the collection H disconnects M but not K.*

PROOF. Suppose, on the contrary, that there do exist such a set K and collection H. Then, by Theorem 28, if h is an element of H, $M - h$ is the sum of two mutually separated point sets of which one contains $K - h$. But this is contrary to Theorem 104.

THEOREM 106. *No separable and connected point set M contains infinitely many mutually exclusive connected point sets $M_1, M_2, M_3, \cdots$ and a point set K such that* (1) *the sequence $M_1, M_2, M_3, \cdots$ has a sequential limiting set that contains K and* (2) *there exists an uncountable collection H of mutually exclusive closed subsets of K each of which separates M but intersects no set of this sequence.*

PROOF. Suppose, on the contrary, that there exists such a point set M. Suppose h is an element of H. The point set $M - h$ is the sum of two mutually separated point sets T_h and K_h where K_h contains a point P of K. Since K_h contains P and P belongs to the sequential limiting set of the sequence $M_1, M_2, M_3, \cdots$ therefore there are infinitely many point sets of this sequence each intersecting K_h. But each M_n is connected and if a connected subset of $M - h$ contains a point of K_h then it is a subset of K_h. Hence there exists an infinite subsequence α of the sequence $M_1, M_2, M_3, \cdots$ such that every point set of the sequence α is a subset of K_h. But K is a subset of $\bar{E} \cdot M$, where E is the sum of all the point sets of the sequence α, and $\bar{E} \cdot M$ is a subset of $K_h + h$. Hence $K - h$ is a subset of K_h. Thus if h is any element of the uncountable collection H, then $M - h$ is the sum of two mutually separated point sets of which one, K_h, contains $K - h$. This contradicts Theorem 104.

DEFINITION. A *simple closed curve* is a nondegenerate compact continuum which is disconnected by the omission of any two of its points.

THEOREM 107. *If A and B are two points of the simple closed curve M then M is the sum of two arcs which have in common only their endpoints A and B.*

PROOF. By definition, $M - (A + B)$ is the sum of two mutually separated point sets H and K. Suppose the closed point set $H + A + B$ is not connected. Then it is the sum of two mutually exclusive closed point sets H_A and H_B where H_A contains A. If H_B did not contain B then M would be the sum of two mutually separated point sets $K + H_A$ and H_B. Hence H_B contains B. The point set H_A is connected. Otherwise it would be the sum of two mutually separated point sets L and N such that L contains A, and M would be the sum of two mutually separated sets $K + H_B + L$ and N. The point set $K + A + B$ is connected since, if it were not, it would be the sum of two mutually separated point sets K_A and K_B, containing A and B respectively, and thus M would be the sum of two mutually separated point sets $H_A + K_A$ and $H_B + K_B$. By Theorem 93, the continuum H_A has a non-cut point X distinct from A, and H_B has a non-cut point Y distinct from B. The point set $M - (X + Y)$ is the sum of the connected point sets $K + A + B, H_A - X$ and $H_B - Y$. But $H_A - X$ has A, and $H_B - Y$ has B, in common with $K + A + B$. Therefore this sum is connected. Thus M is not disconnected by the omission of X and Y and the supposition that $H + A + B$ is not connected leads to a contradiction. By a similar argument, $K + A + B$ is not connected. Suppose now that P is a point of H, and $(H + A + B) - P$ is connected. Then if Q is any point whatsoever of

K, $(K + A + B) - Q$ is connected. Otherwise $M - (P + Q)$ would be the sum of two connected point sets having the points A and B in common and therefore it would be connected. Therefore, by definition, $K + A + B$ is an arc from A to B. Let W denote some definite point of K. Since $K + A + B$ is an arc from A to B, therefore, by Theorems 95 and 97, $(K + A + B) - W$ is the sum of two mutually separated connected sets Z_A and Z_B containing A and B respectively. Therefore $M - (P + W)$ is the sum of the connected point sets $(H + A + B) - P$, Z_A and Z_B. But Z_A has A, and Z_B has B, in common with $(H + A + B) - P$. Therefore $M - (P + W)$ is connected. Thus the supposition that $H + A + B$ is not disconnected by the omission of P has led to a contradiction. Therefore $H + A + B$ is an arc from A to B. Similarly $K + A + B$ is an arc from A to B. These arcs have only A and B in common.

THEOREM 108. *If P is a point of a simple closed curve M, then $M - P$ is connected.*

PROOF. There exists a point O belonging to M but distinct from P. By Theorem 107, M is the sum of two arcs PXO and PYO having only their endpoints in common. The point sets $PXO - P$ and $PYO - P$ are connected. Since they have O in common therefore their sum is connected. But their sum is $M - P$.

THEOREM 109. *If the simple closed curve J is a subset of the separable and connected point set M then J contains a countable point set T such that no point of $J - T$ is a cut point of M.*

PROOF. By Theorem 108 no point of J separates J. Hence, by Theorems 22 and 105, there do not exist uncountably many points of J which are cut points of M. But, by Theorems 27 and 53, J contains uncountably many points. Hence uncountably many of them are non-cut points of M.

THEOREM 110. *If t is an arc and H and K are mutually exclusive closed point sets, there do not exist infinitely many mutually exclusive segments of t each having one endpoint in H and the other one in K.*

PROOF. Suppose there exists an infinite sequence α of such mutually exclusive segments of t. By Theorem 59, there exists a convergent subsequence β of α. By Theorem 58, the sequential limiting set of β is a continuum L. The continuum L contains a point of H and a point of K and is therefore nondegenerate. By Theorem 53, L is uncountable. But, by Theorem 106, L does not contain uncountably many cut points of t. Hence it contains three non-cut points of t, contrary to the definition of an arc.

THEOREM 111. *If K is an uncountable set of cut points of the separable and connected point set M then there exist two points of K that are separated from each other in M by uncountably many points of K.*

PROOF. Let O denote some definite point of K. By hypothesis, there exists a countable subset H of $M - O$ such that every point of M either belongs to H or is a limit point of it. Suppose X is any point of K distinct from O. Then $M - X$ is the sum of two mutually separated point sets L and N such that L contains O. The point set N contains a point Y_X belonging to H. Thus, for every point X of the set $K - O$, there exists a point Y_X of the set H such that X separates O from Y_X in M. But $K - O$ is uncountable and H is countable. Hence there exist a point A of H and an uncountable subset K' of K such that if X is any point of K' then Y_X is A. Every point of the uncountable point set K' separates O from A in M. Since M is separable, there do not exist uncountably many mutually exclusive domains with respect to M. Hence, by Theorems 90 and 81, there exists a point of K' which is separated from O in M by uncountably many points of K'.

THEOREM 112. *If M is an uncountable subset of an arc AB there exists a point O of M distinct from A and from B such that every segment of AB that contains O contains uncountably many points of M preceding O on AB and uncountably many following O on AB.*

PROOF. By Theorems 22, 91 and 95, there exists a point O of M distinct from A and B such that every region containing O contains both uncountably many points of M that precede O on AB and uncountably many points of M that follow O on AB. But, by Theorem 101, for every segment t of AB that contains O there exists a region R containing O such that $R \cdot (AB)$ is a subset of t. Hence Theorem 112 holds true.

DEFINITION. A collection G of point sets is said to be *coherent* if there do not exist two subcollections U and V of G such that (1) every element of G is an element either of U or of V, (2) no element of U has a point in common with any element of V.

THEOREM 113. *If M is a closed and completely separable point set there exists a sequence $Q_1, Q_2, Q_3, \cdots$ such that* (1) *for each n, Q_n is a subcollection of G_n covering M and Q_{n+1} is a subcollection of Q_n,* (2) *if H is a closed and compact subset of M, K is a closed subset of M having no point in common with H, and m is a positive integer, then there exists a positive integer δ such that if g is a region of the collection Q_δ then $\bar{g}$ does not intersect both H and K and, indeed, if U is a coherent collection of m regions all belonging to Q_δ and U^*, the sum of all the regions of U, contains a point of H, then U^* contains no point of K.*

PROOF. By hypothesis there exists a countable set Q such that each element of Q is a domain with respect to M and such that if P is a point belonging to a domain D with respect to M then there exists an element of Q containing P and lying wholly in D. Let W denote the set of all pairs (x,y) such that x and y both belong to Q and $\bar{x}$ is a subset of y. The pairs of the collection W form a simple infinite sequence (x_1,y_1), (x_2,y_2), $(x_3,y_3), \cdots$.

Let Q_1 denote the set of all regions of G_1 that do not intersect both $\bar{x}_1$ and $M - y_1$. It is easy to see that if P is a point belonging to a domain D with respect to M then there is a region R belonging to Q_1 and containing P such that $R \cdot M$ is a subset of D. Similarly, let Q_2 denote the set of all elements of Q_1 which belong to G_2 and do not intersect both $\bar{x}_2$ and $M - y_2$. This process may be continued. Thus there exists a sequence $Q_1, Q_2, Q_3, \cdots$ such that, for each integer n greater than 1, Q_n is the set of all elements g of Q_{n-1} which belong to G_n and which do not intersect both $\bar{x}_n$ and $M - y_n$. Suppose that H and K are two mutually exclusive closed subsets of M and that H is compact. If P is a point of H there exists a number n_P such that x_{n_P} contains P and y_{n_P} contains no point of K. By Theorems 37 and 4, there exists a finite set of points $P_1, P_2, P_3, \cdots, P_k$, belonging to H, such that H is covered by the set of point sets $x_{n_{P_1}}, x_{n_{P_2}}, x_{n_{P_3}}, \cdots, x_{n_{P_k}}$. Let δ_1 denote the largest number in the set $n_{P_1}, n_{P_2}, n_{P_3}, \cdots, n_{P_k}$ and let U_1 and V_1 denote the domains, with respect to M, $x_{n_{P_1}} + x_{n_{P_2}} + x_{n_{P_3}} + \cdots + x_{n_{P_k}}$ and $y_{n_{P_1}} + y_{n_{P_2}} + y_{n_{P_3}} + \cdots + y_{n_{P_k}}$ respectively. The domain, with respect to M, V_1 is a subset of $S - K$ and if $n > \delta_1$ and g is a region of Q_n containing a point of U_1 then the point set $g \cdot M$ is a subset of V_1. Similarly there exist a number δ_2, greater than δ_1, and two domains U_2 and V_2, with respect to M, such that (1) U_2 contains H, and V_2 is a subset of U_1, (2) if $n > \delta_2$ and g is a region of Q_n containing a point of U_2 then $g \cdot M$ is a subset of V_2. This process may be continued. It follows that there exist an infinite sequence of positive integers $\delta_1, \delta_2, \delta_3, \cdots$ and two sequences $U_1, U_2, U_3, \cdots$ and $V_1, V_2, V_3, \cdots$ of domains with respect to M, such that V_1 is a subset of $M - K$ and, for each m, (1) U_m contains H, (2) V_{m+1} is a subset of U_m and (3) if $n > \delta_m$ and g is a region of Q_n containing a point of U_m then $g \cdot M$ is a subset of V_m. It is easy to see that, for each m, δ_m satisfies the conditions required in the statement of Theorem 113.

THEOREM 114. *If Q is a property and, for each positive integer n, H_n is a finite set and, for each n, each element of H_{n+1} has property Q with respect to some element of H_n then there exists a simple infinite sequence $P_1, P_2, P_3, \cdots$ such that, for each n, P_n is an element of H_n, and P_{n+1} has the property Q with respect to P_n.*

PROOF. With the help of the fact that H_1 is finite, it may be seen that there exists an element P_1 of H such that, for each n, there exists at least one finite sequence W such that (1) W has only n terms, the first one is P_1 and, for each i less than $n + 1$, the ith one is an element of H_i, (2) for each i less than $n + 1$, the $(i + 1)$th term of W has property Q with respect to the ith one. For each n, let Z_{1n} denote the set of all sequences W satisfying these conditions with respect to P_1 and n. With the help of the fact that H_2 is finite, it may be seen that there exists an element P_2 of H_2 such that, for every n greater than 1, P_2 is the second term of some sequence of the set

Z_{1n}. For each n, let Z_{2n} denote the set of all sequences belonging to Z_{1n} and having P_2 as their second term. There exists an element P_3 of H_3 such that, for every n greater than 2, P_3 is the third term of some sequence of the set Z_{2n}. Let Z_{3n} denote the set of all sequences belonging to Z_{2n} and having P_3 as their third term. There exists an element P_4 of H_4 such that, for every n greater than 3, P_4 is the fourth term of some sequence of the set Z_{3n}. This process may be continued indefinitely. The sequence $P_1, P_2, P_3, \cdots$ so obtained satisfies all the requirements of Theorem 114.

THEOREM 115. *Suppose $M_1, M_2, M_3, \cdots$ is a sequence of point sets and $H_1, H_2, H_3, \cdots$ is a sequence such that, for each n,* (1) *H_n is a finite collection of point sets covering M_n,* (2) *each point set of the collection H_n is a subset of some region of G_n,* (3) *if h is a set of H_{n+1} then $\bar{h}$ is a subset of some set of H_n,* (4) *$\bar{M}_{n+1}$ is a subset of M_n. Then there exists at least one point common to all the point sets of the sequence $M_1, M_2, M_3, \cdots$ and T, the set of all their common points, is compact and if, for each n, P_n is a point of M_n, some subsequence of the sequence $P_1, P_2, P_3, \cdots$ converges to a point of T.*

PROOF. If, for some n, M_n is finite, the common part of the sets of the sequence $M_1, M_2, M_3, \cdots$ clearly exists. Suppose that, for every n, M_n is infinite. Then there exists an infinite sequence α of distinct points $P_1, P_2, P_3, \cdots$ such that, for each n, P_n belongs to M_n. For each n, let H_n' denote the collection of all point sets of H_n that contain infinitely many points of the sequence α. By Theorem 114, there exists a sequence $h_1, h_2, h_3, \cdots$ such that, for each n, h_n belongs to H_n' and $\bar{h}_{n+1}$ is a subset of h_n. The point sets $h_1, h_2, h_3, \cdots$ have a point P in common. Suppose R is a region containing P. There exists a number δ such that R contains every region of the collection G_δ which contains P. But h_δ contains P and it is a subset of some region of G_δ. Hence R contains infinitely many points of the set $P_1 + P_2 + P_3 + \cdots$. Hence P belongs to every point set of the sequence $\bar{M}_1, \bar{M}_2, \bar{M}_3, \cdots$ and therefore to every one of the sequence $M_1, M_2, M_3, \cdots$. So T, the common part of all the point sets of this sequence, exists. The same argument shows that if $P_1, P_2, P_3, \cdots$ is *any* infinite sequence of distinct points belonging to T, or indeed such that, for each n, P_n belongs to M_n, then T contains a limit point of the point set $P_1 + P_2 + P_3 + \cdots$. Hence T is compact and contains every point which is a limit point of such a point set $P_1 + P_2 + P_3 + \cdots$.

THEOREM 116. *If, under the same hypothesis as in Theorem* 115, *each point set of the sequence $M_1, M_2, M_3, \cdots$ is connected then their common part is a continuum.*

PROOF. Let T denote the common part of the point sets of this sequence. Suppose it is not a continuum. Then, since it is closed, it is the sum of two mutually exclusive closed and compact point sets K and L. By Theorem 16, there exists a domain D containing K such that $\bar{D}$ contains no point of L.

Let β denote the boundary of D. For each n, the connected point set M_n intersects both D and $S - D$ and therefore contains a point of β. By Theorem 115, the point sets $\beta \cdot M_1$, $\beta \cdot M_2$, $\beta \cdot M_3, \cdots$ have a point P in common. The point P belongs to T but neither to K nor to L. Thus the supposition that T is not a continuum leads to a contradiction.

THEOREM 117. *If M is a closed and compact point set there exists an infinite sequence W_1, W_2, $W_3, \cdots$ such that* (1) *for each n, W_n covers M and is a finite subcollection of G_n,* (2) *if g is a region of W_{n+1}, $\bar{g}$ is a subset of some region of W_n,* (3) *if H and K are two mutually exclusive closed subsets of M, and k is a positive integer there exists a positive integer m such that if U is a coherent collection of k regions each belonging to some collection of the sequence W_m, W_{m+1}, $W_{m+2}, \cdots$ and U^*, the sum of all the regions of the collection U, contains a point of H, then U^* contains no point of K,* (4) *if, for each n, the points A_n and B_n lie together in some region of W_n and A_1, A_2, $A_3, \cdots$ converges to the point P then so does B_1, B_2, $B_3, \cdots$.*

PROOF. Let Q_1, Q_2, $Q_3, \cdots$ denote a sequence of collections satisfying all the conditions of Theorem 113. There exists a finite subcollection W_1 of Q_1 covering M. For each point P of M there exists a region g_P belonging to Q_2 and containing P and such that $\bar{g}_P$ is a subset of some region of W_1. Let U_2 denote the set of all such regions g_P for all points P of M. There exists a finite subcollection W_2 of U_2 covering M. This process may be continued indefinitely. With the help of Theorem 115, it may be seen that the sequence W_1, W_2, $W_3, \cdots$ satisfies all the requirements of Theorem 117.

DEFINITION. The point set M is said to be *properly* covered by the collection G of point sets if G covers M and each set of the collection G contains at least one point of M.

THEOREM 118. *If G is an uncountable collection of closed subsets of the closed and compact point set M there exists an uncountable subcollection G' of G such that if g' is an element of G' and W is a finite collection of domains properly covering g', then there exist uncountably many point sets of the collection G' each properly covered by W.*

PROOF. Let W_1, W_2, $W_3, \cdots$ denote a sequence satisfying, with respect to M, all the conditions of Theorem 117. For each n, let H_n denote the collection of all sets x such that x is a finite set of regions belonging to W_n. Let H denote the collection of all sets y such that, for some n, y belongs to H_n. Suppose that for every set g of the collection G there exists a set h_g of the collection H such that (1) g is properly covered by h_g, (2) there do not exist uncountably many point sets each belonging to G and each properly covered by h_g. Let H' denote the collection of all such sets h_g for all sets g of the collection G. Every point set of the collection G is properly covered by some set of the countable collection H' and for each set of the collection H' there are only countably many elements of G each properly covered by that set. Therefore G is countable. But this involves a contradiction. Therefore there exists an element g of G such that if a set of the collection H properly covers g then it properly covers each of uncountably many elements of G. Let G' denote the set of all such elements g. The set G' is uncountable and indeed it contains all but countably many of the elements of G. For if $G - G'$ were uncountable then, as has been proved, it would itself have an element q such that if h is an element of H properly covering q then there are uncountably many elements of $G - G'$, and therefore of G, properly covered by h, contrary to the supposition that q does not belong to G'.

Suppose now that W is any finite set of domains properly covering an element g' of G'. There exists a set h belonging to H, properly covering g' and such that (1) every region of H is a subset of some domain of W and (2) every domain of W contains some region of h. Every point set that is properly covered by h is also properly covered by W. But there are uncountably many elements of G each properly covered by h and all but countably many of the elements of G belong to G'. Hence uncountably many elements of G' are properly covered by W.

THEOREM 119. *If M is a closed and compact point set, m is a positive integer and G is an uncountable collection of subsets of M each containing m points then there exist m regions $R_1, R_2, R_3, \cdots, R_m$ such that their closures are mutually exclusive and such that there are uncountably many elements of G each intersecting all of the regions $R_1, R_2, R_3, \cdots, R_m$.*

PROOF. For each point set g of the collection G, let q_g denote a definite set of m points of g. Let G' denote the set of all such points q_g for all sets g of G. By Theorem 118, there exists a point set g_0 of the collection G' such that if H is any finite collection of domains properly covering g_0 then there exist uncountably many elements g of G such that q_g is properly covered by H. Let $A_1, A_2, A_3, \cdots, A_m$ denote the points of g_0. There exist regions $R_1, R_2, R_3, \cdots$ such that, for each n less than $m + 1$, R_n contains A_n and such that $\bar{R}_1, \bar{R}_2, \bar{R}_3, \cdots, \bar{R}_m$ are mutually exclusive. The point set g_0 is properly covered by $R_1, R_2, R_3, \cdots, R_m$. Hence there exist uncountably many point sets of the collection G each intersecting all of these regions.

THEOREM 119(a). *If G is an uncountable collection of mutually exclusive nondegenerate subsets of the closed and compact point set M, there exist two points A and B of M such that, if R_A and R_B are regions containing A and B respectively, uncountably many elements of G intersect both R_A and R_B.*

THEOREM 120. *Suppose H and K are two mutually exclusive closed subsets of the compact continuum M, and G is a collection of closed subsets of M such that if x and y are any two elements of G then every closed subset of the point set $x + y$ is an element of G. Suppose furthermore that if X is a point of H and Y is a point of K then X is separated from Y in M by some point set of the collection G. Then H is separated from K in M by some point set of the collection G.*

PROOF. Suppose X is a point of H. Then for each point Y of K there exists a point set g_{XY} belonging to G such that $M - g_{XY}$ is the sum of two mutually separated point sets D_{XY} and E_{XY} containing X and Y respectively. For each point Y of the closed and compact point set K, E_{XY} is a domain with respect to M containing Y. It follows, with the help of Theorems 4 and 37, that there exists a finite set of points $Y_1, Y_2, Y_3, \cdots, Y_k$ belonging to K such that K is a subset of the point set $E_{XY_1} + E_{XY_2} + E_{XY_3} + \cdots + E_{XY_k}$. Since each of the point sets $E_{XY_1}, E_{XY_2}, \cdots, E_{XY_k}$ is a domain with respect to M therefore so is E_X, their sum, and since the boundary of the sum of a finite number of point sets is a subset of the sum of their boundaries, therefore the boundary of E_X, with respect to M, is a point set β_X belonging to the collection G. Let D_X denote $M - (E_X + \beta_X)$. Since $E_X + \beta_X$ is closed, D_X is a domain with respect to M and its boundary with respect to M is a subset of $E_X + \beta_X$. But E_X is a domain with respect to M and therefore contains no limit point of D_X. Hence γ_X, the boundary of D_X with respect to M, is a subset of β_X and therefore belongs to the collection G. Thus, for each point X of H, the domain D_X with respect to M contains X, and its boundary with respect to M is a point set γ_X of the collection G. There exists a finite set of points $X_1, X_2, X_3, \cdots, X_h$ belonging to H such that H is a subset of D, the sum of the point sets $D_{X_1}, D_{X_2}, \cdots, D_{X_h}$. Let γ denote the boundary of D. Since D is a domain with respect to M, it contains no limit point of $M - (D + \gamma)$ and since every limit point of D which belongs to $M - D$ is a point of γ, therefore $M - (D + \gamma)$ contains no limit point of D. Thus D and $M - (D + \gamma)$ are mutually separated point sets. But they contain H and K respectively and their sum is $M - \gamma$. Thus γ separates H from K in M. But γ is a subset of the sum of the boundaries, with respect to M, of the point sets $D_{X_1}, D_{X_2}, \cdots, D_{X_h}$ and therefore it belongs to the collection G.

NOTATION. If G is a collection of subsets of the connected point set M and P is a point of M, the notation M_{GP} will be used to designate the set of all points X of M such that X is not separated from P in M by any point set of the collection G.

DEFINITION. If G is a collection of subsets of the connected point set M and P is a point of M and there do not exist two points A and B such that (1) A and B are separated from each other in M by some point set of the collection G which contains P and (2) P belongs to every point set of the collection G which separates A from B in M, then M_{GP} will be called a *link of M with respect to the collection G*, and P will be called a *proper point of M with respect to G*. A point of M which is not a proper point of M with respect to G will be called an *improper* point of M with respect to G.

THEOREM 121. *If M is a compact continuum and G is a collection of closed subsets of M and no element of G contains uncountably many improper points of*

M with respect to G, then there do not exist uncountably many improper points of M with respect to G.

PROOF. Suppose there exists an uncountable collection N of such points. For each point P of N, select just one pair of points of M such that there is a point set belonging to G and separating them from each other in M and such that every point set that belongs to G and separates them from each other in M contains P. For each P, let q_P denote the pair so selected and let h_P denote the set whose elements are the point P and the points of the set q_P. Let H denote the collection of all such sets h_P for all points P of N. By Theorem 119, there exist an uncountable subset H' of H and three regions R_1, R_2, R_3 such that $\bar{R}_1$, $\bar{R}_2$ and $\bar{R}_3$ are mutually exclusive and such that every point set of the collection H' intersects each of these regions. There exist an uncountable subset N' of N and a region K_2 of the set R_1, R_2, R_3 such that if K_1 and K_3 denote the remaining regions of this set then (1) if P is any point of N', P belongs to K_2, and one of the points of q_P belongs to K_1 and the other one belongs to K_3. Let H'' denote the collection of all point sets h_P for all points P of N'. By Theorem 118, there exists a point set h of the collection H'' such that if a finite collection of domains covers h properly then it properly covers uncountably many point sets of the collection H''. Let A, O and B denote the points of h belonging to K_1, K_2 and K_3 respectively. There exists a point set L belonging to G, containing O, and such that $M - L$ is the sum of two mutually separated point sets M_A and M_B containing A and B respectively. Since M_A and M_B are domains with respect to M containing A and B respectively, there exists an uncountable subset N'' of N' such that if P is any point of N'' then one point A_P of q_P is in $K_1 \cdot M_A$ and the other one, B_P, is in $K_3 \cdot M_B$. There exists a point X belonging to N'' but not to L. Since A_X and B_X belong to M_A and M_B respectively, they are separated from each other in M by L. But every point set belonging to G and separating A_X from B_X in M contains the point X, and X does not belong to L. Thus the supposition that Theorem 121 is false has led to a contradiction.

THEOREM 122. *Suppose A, O and B are three points of the continuum M and G is a collection of closed subsets of M such that if x is a closed proper subset of some element of G and y is a closed subset of an element of G then $x + y$ is an element of G. Suppose furthermore that A and B are separated from each other in M by some point set of the collection G but not by any one that does not contain O. Then no point set of the collection G separates O either from A or from B, in M.*

PROOF. Suppose, on the contrary, that there exists a point set T_2 belonging to G and separating O from say B in M. By hypothesis there exists a point set T_1 belonging to G, separating A from B in M and containing O. By Theorem 63, $A + O$ is separated from B in M by some point set L which is a closed subset of $T_1 + T_2$. Since T_1 contains O, and L does not, therefore L is the sum of two point sets of which one is a closed proper subset of T_1 and the other is a closed subset of T_2. Therefore L belongs to G. Thus A is separated from B in M by a point set belonging to the collection G and not containing O. But this is contrary to hypothesis.

THEOREM 123. *If M is a compact continuum and G is either the collection of all degenerate subsets of M, or a collection of closed subsets of M such that every closed subset of the sum of two point sets that belong to G is itself a point set belonging to G, then, for every point P of M, M_{GP} is a continuum.*

PROOF. The point set M_{GP} is closed. For suppose, on the contrary, that there exists a point O which is a limit point of M_{GP} but which does not belong to M_{GP}. Then there exists a point set T belonging to G such that $M - T$ is the sum of two mutually separated point sets H and K containing P and O respectively. Since O is a limit point of M_{GP} and K is a domain with respect to M containing O therefore K contains points of M_{GP}. Every point of $K \cdot M_{GP}$ is separated from P in M by the point set T. But no point of M_{GP} is separated from P in M by a point set of the collection G. Thus the supposition that M_{GP} is not closed leads to a contradiction.

It will be shown that M_{GP} is also connected. Two cases will be considered.

Case 1. Suppose G is the collection of all degenerate subsets of M. Suppose A is a point of M_{GP} distinct from P. By Theorem 43, there exists a subcontinuum T of M which is irreducible with respect to the property of being a continuum containing A and B. Suppose that T contains a point Y not belonging to M_{GP}. Then there exists a point O such that $M - O$ is the sum of two mutually separated point sets H and K containing P and Y respectively. The point O must belong to T, and the point sets $H \cdot T$ and $K \cdot T$ are mutually separated. By Theorem 64, one of them contains P and the other one contains A. But $H \cdot T$ contains P. Therefore $K \cdot T$ contains A. Thus K contains A and therefore O separates A from P in M, contrary to the supposition that A belongs to M_{GP}. Thus the supposition that T is not a subset of M_{GP} leads to a contradiction. Therefore not only is M_{GP} connected, but it contains the whole of every subset of M which is an irreducible continuum from P to another point of M_{GP}.

Case 2. Suppose G is a collection of closed subsets of M such that if x and y are two point sets of the collection G then every closed subset of $x + y$ is itself a point set of the collection G. Suppose that M_{GP} is not connected. Then it is the sum of two mutually exclusive closed point sets H and K such that H contains P. There exists a domain D with respect to M containing K such that $\bar{D}$ contains no point of H. Let β denote the boundary of D with respect to M. Each point of β is separated from P in M by a point set of the collection G. Therefore, by Theorem 120, β is separated from P, in M, by a point set T belonging to the collection G. By Theorem 68, P is separated from D, in M, by some closed subset of T and therefore by a point set of the collection G. But D contains at least one point of M_{GP} and no such point is separated from P in M by any point set of the collection G. Thus the supposition that M_{GP} is not connected leads to a contradiction.

THEOREM 124. *If G is a collection of subsets of the continuum M, A and B are proper points of M with respect to the collection G, and B is a point of M_{GA} then M_{GB} is identical with M_{GA}.*

PROOF. Suppose there exists a point P belonging to M_{GA} but not to M_{GB}. There exists a point set belonging to the collection G and separating P from B in M. Hence, by hypothesis, there exists a point set T belonging to G and not containing A, but such that $M - T$ is the sum of two mutually separated point sets H and K containing P and B respectively. The point A belongs either to H or to K. If it belongs to H it is separated from B in M by the point set T and therefore B does not belong to M_{GA}, contrary to hypothesis. If it belongs to K then it is separated from P by T, contrary to the supposition that P belongs to M_{GA}. Thus the supposition that not every point of M_{GA} belongs to M_{GB} leads to a contradiction. Hence M_{GA} is a subset of M_{GB}. Since A clearly belongs to M_{GB} it follows, by the same argument, that M_{GB} is a subset of M_{GA}. Hence M_{GB} is M_{GA}.

DEFINITION. If G is a collection of subsets of the continuum M, the point P is said to be a *regular point of M with respect to the collection G* if and only if it is true that if X is any point of M distinct from P then there exists a point set belonging to G and separating X from P in M.

THEOREM 125. *If M is a continuum and G is either the collection of all degenerate subsets of M or a collection of closed subsets of M such that if x is a closed proper subset of an element of G and y is a closed subset of an element of G then every closed subset of $x + y$ is itself a point set of the collection G, then every regular point of M with respect to G is a proper point of M with respect to G and if P is a regular point of M with respect to G then M_{GP} is a degenerate link of M with respect to G.*

PROOF. Suppose P is not a proper point of M with respect to G. Then there exist two points A and B of M which are separated from each other in M by a point set T_1 belonging to G and containing P but not by any point set belonging to G and not containing P. By hypothesis there exists a point set T_2 belonging to G and separating P from B in M. By Theorem 63, there exists a closed subset T of $T_1 + T_2$ that separates $A + P$ from B in M. The point set T belongs to G, it separates A from B in M and it does not contain P. But this is contrary to supposition. Hence P is a proper point of M. It is clear that M_{GP} is the set whose only element is P. Thus it is a degenerate link of M with respect to G.

THEOREM 126. *Suppose M is a compact continuum and G is either the collection of all degenerate subsets of M or a collection of closed subsets of M such that every closed subset of the sum of two sets that belong to G is itself an element* *of G. Suppose that no element of G contains uncountably many points that are not proper points of M with respect to the collection G. Then every nonregular point of M with respect to G belongs to a nondegenerate link of M with respect to G.*

PROOF. Suppose O is a nonregular point of M with respect to G. Let B denote a point of $M - O$ which is not separated from O in M by any point set of the collection G. By Theorem 123, M_{GO} is a continuum containing O and B. Hence, by Theorem 56, it is uncountable. Therefore, by Theorem 121, there exists a point P of M_{GO} such that if two points of M are separated from each other in M by a point set of the collection G then they are separated from each other in M by a subset of $M - P$ belonging to G. By definition, M_{GP} is a link of M with respect to G. But M_{GP} contains O. If P is O, M_{GP} contains the two points O and B. If P is distinct from O, M_{GP} contains the points P and O. Hence, in any case, it is nondegenerate.

THEOREM 127. *If M is a compact continuum and G is either the collection of all degenerate subsets of M or a collection of closed subsets of M such that every closed subset of the sum of two elements of G is itself an element of G and no element of G contains uncountably many improper points of M with respect to G, then every point of M belongs to some link of M with respect to G.*

Theorem 127 is an immediate consequence of Theorems 125 and 126.

THEOREM 128. *If G is a collection of closed subsets of the continuum M and A and B are points belonging to the same link of M with respect to G then A and B are not separated from each other in M by any point set of the collection G.*

PROOF. By hypothesis there exists a point P such that P is a proper point of M with respect to G and A and B belong to M_{GP}. Suppose A and B are separated from each other in M by some point set of the collection G. Then, since P is a proper point of M with respect to G, there exists a point set g belonging to G, not containing P, but such that $M - g$ is the sum of two mutually separated point sets K_A and K_B containing A and B respectively. Since it does not belong to g, the point P belongs either to K_A or to K_B. In the first case it is separated from B, in M, by g and, in the second case, it is separated from A, in M, by g. This involves a contradiction.

THEOREM 129. *Suppose M is a compact continuum and G is either the collection of all degenerate subsets of M or a collection of closed subsets of M such that every closed subset of the sum of two elements of G is an element of G and no element of G contains uncountably many improper points of M with respect to G. Then every nonregular point of M with respect to the collection G belongs to a nondegenerate continuum of such points.*

Theorem 129 is a consequence of Theorems 126 and 128.

THEOREM 130. *If G is a collection of closed subsets of a compact continuum M such that no element of G contains uncountably many improper points of M*

with respect to G, and x and y are two links of M with respect to G having at least one point in common, then $x \cdot y$ is a subset of some element of G and no point of $x \cdot y$ is a proper point of M with respect to G and thus $x \cdot y$ is a countable point set.

PROOF. Let P denote a point of $x \cdot y$. If P were a proper point of M with respect to G then, by Theorem 124, M_{GP} would be identical with x and with y and thus x would be identical with y, contrary to hypothesis. It follows, by Theorem 121, that $x \cdot y$ is countable. Let A and B denote two proper points of M with respect to G belonging to x and y respectively. By Theorem 124, M_{GA} is identical with x. But B does not belong to x. Hence there exists a point set g, belonging to G, such that $M - g$ is the sum of two mutually separated point sets H_A and H_B containing A and B respectively. If C is a point of x not belonging to g, C belongs either to H_A or to H_B. But if it belonged to H_B then C would be separated from A, in M, by g which is impossible since C belongs to M_{GA}. Thus $x - x \cdot g$ is a subset of H_A. Similarly $y - y \cdot g$ is a subset of H_B. Therefore $x - x \cdot g$ and $y - y \cdot g$ are mutually exclusive. Hence $x \cdot y$ is a subset of g.

DEFINITION. If G is the collection of all degenerate subsets of the continuum M, then every link of M with respect to G will be called a *simple* link of M.

THEOREM 131. *If G is a collection of closed subsets of a compact continuum M, no link of M with respect to G is a proper subset of any other link of M with respect to G.*

PROOF. By Theorem 124, if the link H of M with respect to G were a proper subset of the link K of M with respect to G then H would contain no proper point of M with respect to G, contrary to definition.

THEOREM 132. *In order that the points A and B of the compact continuum M lie together in some simple link of M it is necessary and sufficient that A and B be not separated from each other, in M, by any one point.*

PROOF. By Theorem 128, this condition is necessary. It will be shown that it is also sufficient. Suppose A and B are not separated from each other, in M, by any one point. By Theorem 43, there exists a subset L of M which is an irreducible continuum from A to B. By Theorems 56 and 121, L contains a point P which is a proper point of M with respect to G, the collection of all degenerate subsets of M. If there exists a point X such that $M - X$ is the sum of two mutually separated point sets H_A and H_P containing A and P respectively, then $L - X$ is the sum of two mutually separated point sets $L \cdot H_A$ and $L \cdot H_P$ and, by Theorem 64, the point B belongs to $L \cdot H_P$ and thus X separates A from B in M, contrary to hypothesis. Therefore no point separates A from P in M. In other words, A belongs to M_{GP}. Similarly B belongs to M_{GP}. But, since P is a proper point of M with respect to G, M_{GP} is a simple link of M.

THEOREM 133. *If G is the collection of all degenerate subsets of the compact continuum M and the point P of M is not a proper point of M with respect to G then P belongs to two simple links of M.*

PROOF. By hypothesis, there exist two points A and B such that P is the only point that separates A from B in M. By Theorem 73, no point separates P either from A or from B in M. Therefore, by Theorem 132, there exists a simple link a of M containing A and P and a simple link b of M containing B and P. Since P separates A from B in M therefore, by Theorem 132, a and b are distinct.

THEOREM 134. *In order that a point set be a simple link of the compact continuum M it is necessary and sufficient that it be either* (1) *a cut point K of M such that if X is any other point of M then K is separated from X, in M, by some one point or* (2) *a point set K such that, for some non-cut point P of M, K is a set such that X belongs to it if and only if X belongs to M and is not separated from P in M by any one point.*

PROOF. This condition is necessary. For suppose K is a simple link of M. If K is a set whose only element is the cut point P of M then each point of $M - K$ is separated from K in M by a point. If K is a set whose only element is the non-cut point P of M then, by definition, K is a set such that X belongs to it if and only if X belongs to M and is not separated from P in M by any one point. If K contains more than one point then, by Theorems 123, 56 and 121, it contains an uncountable point set L such that every point of L is a proper point of M with respect to G. By Theorem 132, no point of L separates two points of K from each other in M. Therefore, by Theorem 104, not every point of L is a cut point of M. Let P denote one which is not. By Theorem 124, M_{GP} is identical with K.

This condition is also sufficient. For suppose it is satisfied. Suppose first that K is a set whose only element is a cut point P of M such that each point of $M - K$ is separated from P, in M, by some point. Then, by Theorem 73, P is a proper point of M with respect to G, the collection of all degenerate subsets of M with respect to G, the collection of all degenerate subsets of M. Hence M_{GP} is a link of M. But M_{GP} is identical with K. Suppose secondly that, for some non-cut point P of M, K is a set such that X belongs to it if and only if X is a point of M which is not separated from P in M by any one point. Then P is a proper point of M with respect to G, the collection of all degenerate subsets of M, and K is the link M_{GP}.

DEFINITION. If M is a continuum, a *composant* of M is a point set K such that, for some point P of M, the point X belongs to K if and only if there is a proper subcontinuum of M containing both P and X.

Even in the plane the composants of a continuum are not, in all cases, mutually exclusive. For example if AB is a straight line interval, its composants are AB, $AB - A$, $AB - B$ and $AB - (A + B)$, and $AB - (A + B)$ is common to every two of them.

THEOREM 135. *If K is a composant of a nondegenerate compact continuum M, every point of M is a limit point of K.*

PROOF. By definition, there exists a point P of M such that the point X belongs to K if and only if it belongs to M and M is not an irreducible continuum from P to X. Let Y denote any point of M distinct from P. Suppose R is a region containing Y. Then there exists a region R_1 containing Y such that $\bar{R}_1$ is a subset of R not containing P. Let T denote the component of $M - M \cdot \bar{R}_1$ that contains P. By Theorem 51, $\bar{R}_1$ contains a point X of $\bar{T}$. But X belongs to R and $\bar{T}$ is a proper subcontinuum of M containing P and X. Thus every region that contains Y contains a point of K. Hence Y is a limit point of K. Since K is a nondegenerate connected point set containing P therefore P is a limit point of K. Hence every point of M is a limit point of K.

THEOREM 136. *Every composant of a compact continuum M is the sum of countably many proper subcontinua of M.*

PROOF. Suppose A is a point of the compact continuum M. Let L denote the composant of M consisting of all points X of M such that M is not irreducible from A to X. By Theorem 22, there exists a countable set G of domains with respect to M such that if P is any point of M and D is any domain with respect to M containing P then there exists an element of G containing P and lying wholly in D. For each element R of the set G that does not contain A let M_R denote the component of $M \cdot (S - R)$ that contains A. Let H denote the collection of all sets M_R for all such elements R of G and let T denote the sum of all the continua of the set H. The set H is countable. If P is a point of $M - T$, M is irreducible from A to P. For if there exists a proper subcontinuum N of M containing A and P there exists an element g of the set G such that $\bar{g}$ contains no point of N and therefore P belongs to M_g, contrary to hypothesis. Thus L is a subset of T. But clearly T is a subset of L. Hence L is identical with T. But T is the sum of countably many continua.

DEFINITION. The continuum M is said to be *indecomposable* if it is nondegenerate but it is not the sum of two continua both distinct from it.

DEFINITION. The continuum K is said to be a *continuum of condensation* of the point set M if K is a nondegenerate subset of M and every point of K is a limit point of $M - K$.

THEOREM 137 (JANISZEWSKI AND KURATOWSKI). *In order that the continuum M be indecomposable it is necessary and sufficient that every nondegenerate proper subcontinuum of M be a continuum of condensation of M.*

PROOF. This condition is sufficient. For suppose the continuum M is not indecomposable. Then it is the sum of two continua H and K both distinct from it. The continuum H contains a point P not belonging to K.

The point P is not a limit point of K. Hence it is not a limit point of $M - H$ and therefore H is not a continuum of condensation of M.

This condition is also necessary. For suppose, on the contrary, that there exists an indecomposable continuum M containing a continuum K such that $M - K$ exists but not every point of K is a limit point of $M - K$. Let T denote $M - K$. The point set T is connected. For if it were the sum of two mutually separated point sets H and L then M would be the sum of two continua $H + K$ and $L + K$, both different from M, contrary to hypothesis. Since T is connected $\overline{T}$ is a continuum. Both K and $\overline{T}$ are proper subcontinua of M, and $K + \overline{T}$ is M. Thus the supposition that the condition of Theorem 137 is not necessary leads to a contradiction.

THEOREM 138. *No two composants of an indecomposable continuum have a point in common.*

PROOF. For each point P of the indecomposable continuum M, let M_P denote the set of all points X such that M is not irreducible from P to X. If B is a point of M_A then M_B is identical with M_A. For suppose that X is a point of M_A and Y is a point of M_B. There exist three proper subcontinua of M, K_1, K_2 and K_3 such that K_1 contains A and B, K_2 contains A and X, and K_3 contains B and Y. Since M is indecomposable, the continuum $K_1 + K_2$ is a proper subset of M and therefore, for the same reason, so is the continuum $K_1 + K_2 + K_3$. But $K_1 + K_2 + K_3$ is a subset both of M_A and of M_B. Thus X belongs to M_B and Y belongs to M_A. Hence M_B is identical with M_A. Suppose now that H and K are two composants of M having a point O in common. Then H is identical with M_O and so is K, and thus H is K.

THEOREM 139. *Every compact indecomposable continuum has uncountably many composants.*

PROOF. Suppose M is a compact indecomposable continuum. By Theorem 136, every composant of M is the sum of countably many proper subcontinua of M. Hence if M had only countably many composants it would itself be the sum of countably many continua all distinct from it. Since, by Theorem 137, every proper subcontinuum of M is a continuum of condensation of M, this would involve a contradiction of Theorem 53.

THEOREM 140 (MAZURKIEWICZ). *In order that the compact continuum M be indecomposable it is necessary and sufficient that there exist three points such that M is irreducible between every two of them.*

PROOF. This condition is sufficient. For suppose that the compact continuum M is irreducible between each two of the points A, B and C. Suppose M is the sum of the two continua H and K. Either H or K contains two of the points A, B and C and is therefore identical with M. Hence M is indecomposable.

This condition is also necessary. For suppose M is an indecomposable compact continuum. By Theorem 139, M has uncountably many composants. There exists a point set T such that (1) each point of T belongs to some composant of M and (2) each composant of M contains only one point of T. The continuum M is irreducible between every two points of the uncountable set T.

THEOREM 141. *If A is a point of a compact indecomposable continuum M and K is the set of all points X such that M is irreducible from A to X then every point of M is a limit point of K.*

PROOF. By Theorem 139, there exist uncountably many composants of M distinct from the one that contains A. Let K' denote one of them. By Theorem 135, every point of M is a limit point of K'. But M is irreducible from A to each point of K'. That is to say, K' is a subset of K. Hence every point of M is a limit point of K.

THEOREM 142. *If T is the sum of countably many proper subcontinua of the compact indecomposable continuum M, then $M - T$ is a connected point set.*

PROOF. If G is a countable set of proper subcontinua of the compact indecomposable continuum M each continuum of the set G is a subset of some composant of M and therefore, by Theorems 138 and 139, there exists a composant K of M that contains no point of T, the sum of all the continua of the set G. By Theorem 135, every point of M is a limit point of K. In particular, every point of $M - T$ is a limit point of K. But K is connected. Hence $M - T$ is connected.

THEOREM 143. *If A is a point of a decomposable continuum M, there exists a domain D containing A such that M is not irreducible from A to any point of D.*

PROOF. Let T denote the set of all points X if there exist any such that M is irreducible from A to X. Suppose that A is a limit point of T. The continuum M is the sum of two continua H and K, both distinct from M, such that H contains A. Since H is a proper subcontinuum of M containing A, it contains no point of T. Hence T is a subset of K. But A is a limit point of T. Therefore K contains A. Thus K is a proper subcontinuum of M containing both A and T, contrary to the fact that M is irreducible from A to every point of T.

THEOREM 144. *If A and B are two points, M is a continuum which is irreducible from A to B, and T is a proper subcontinuum of M containing B, then $M - T$ is connected.*

PROOF. Suppose, on the contrary, that $M - T$ is the sum of two mutually separated point sets H and K such that K does not contain A. Then $H + T$ contains both A and B. But $H + T$ is a proper subset of M and, by Theorem 60, it is connected. Hence it is a proper subcontinuum of M containing A and B, contrary to the hypothesis that M is irreducible from A to B.

THEOREM 145. *If M is a compact irreducible continuum from the point A to some point B that belongs to no indecomposable subcontinuum of M then the set of all points X such that M is an irreducible continuum from A to X is a continuum.*

PROOF. Suppose C is a point of $M - (A + B)$ such that M is an irreducible continuum from A to C. There exists a subcontinuum K of M irreducible from B to C. By hypothesis and Theorem 140, K does not contain A. Suppose there exists a proper subcontinuum T of M containing A and some point X of K. By hypothesis K is the sum of two continua K_B and K_C, both distinct from K, such that K_B contains B. If K_B contained C, K would not be an irreducible continuum from B to C. Hence K_C contains C. But it does not contain B. The point X belongs either to K_B or to K_C. Since T is a proper subcontinuum of M containing A it contains neither B nor C. It follows that if X belongs to K_B, $T + K_B$ is a subcontinuum of M containing A and B but not C. If X belongs to K_C then $T + K_C$ is a subcontinuum of M containing A and C but not B. In either case there is a proper subcontinuum of M containing A and one of the points B and C. Thus the supposition that M is not irreducible from A to each point of K has led to a contradiction.

Let L denote the set of all points X of M such that M is irreducible from A to X. If X is a point of L distinct from B, every subcontinuum of M which is irreducible from B to X is a subset of L. Hence L is connected.

Suppose $M - L$ contains a point O which is a limit point of L. There exists a proper subcontinuum N of M containing A and O and there exists a subcontinuum Q of M which is irreducible from B to O. Since the continuum $N + Q$ contains both A and B it is identical with M. But N is a proper subcontinuum of M containing A and therefore no point of L. Hence L is a subset of Q. Since Q is a subcontinuum of M containing B it is the sum of two continua both distinct from it. One of them, Q_B, contains B but not O and the other one, Q_O, contains O but not B. The point O is not a limit point of Q_B. But O is a limit point of L and L is a subset of $Q_O + Q_B$. Hence Q_O contains a point E of L. The continuum $N + Q_O$ is a proper subcontinuum of M containing A and E. This involves a contradiction. Hence L is closed.

DEFINITION. The point set M is said to be *normal* if and only if it is true that if the mutually exclusive subsets H and K of M are closed relatively to M then there exist two mutually exclusive domains with respect to M one containing H and the other containing K.

THEOREM 146. *If S is connected it is normal if and only if it is true that if two mutually exclusive closed point sets are separated from each other by a domain then they are separated from each other by a closed subset of that domain.*

PROOF. Suppose, first, that S is normal and H and K are two mutually exclusive closed point sets that are separated from each other by the domain

D. Then $S - D$ is the sum of two mutually separated point sets M_H and M_K containing H and K respectively. Since D is a *domain* containing no point of M_H it contains no limit point of M_H. Neither does M_K. So M_H is closed. Similarly, so is M_K. Hence there exist two mutually exclusive domains D_H and D_K containing M_H and M_K respectively. Since D_H is not S and S is connected, D_H has a boundary β_H. The point set β_H is a closed subset of D that separates H from K.

Suppose, secondly, that it is true that if a domain separates two mutually exclusive closed point sets from each other then some closed subset of it does so. Suppose H and K are two mutually exclusive closed point sets. The point set $S - (H + K)$ is a domain which separates H from K. Hence it contains a closed point set T such that $S - T$ is the sum of two mutually separated point sets D_H and D_K containing H and K respectively. Since T is closed, these point sets are domains. So S is normal.

It cannot be proved from Axioms 0 and 1 that S is normal. Indeed it does not follow from these axioms that if D is a domain and $S - D$ is the sum of two mutually separated point sets containing the points A and B respectively then there exists a closed subset of D whose complement is the sum of two mutually separated point sets containing A and B respectively.

Consider the following example.

Example 8. Let Σ' denote a space whose points are the points of a Cartesian space of three dimensions and whose regions are the interiors of spheres in that space. If Z is a straight line interval and I is a segment of Z whose length is one third that of Z and whose midpoint is that of Z then I will be called the middle third segment of Z.

Let Z_1 denote some definite interval on the X-axis. Let H_1 denote the collection whose only element is I_1, the middle third segment of Z_1. Let H_2 denote the collection whose elements are the middle third segments of the components of $Z_1 - H_1^*$, that is to say of $Z_1 - I_1$. Let H_3 denote the collection whose elements are the middle third segments of the components of $Z_1 - (H_1^* + H_2^*)$. There exists a sequence $H_1, H_2, H_3, \cdots$ such that (1) H_1 and H_2 are as described, (2) for each n greater than 1, H_{n+1} is a collection whose elements are the middle third segments of the components of $Z_1 - (H_1^* + H_2^* + \cdots + H_n^*)$. The set $Z_1 - (H_1^* + H_2^* + H_3^* + \cdots)$ is a totally disconnected perfect point set K called a Cantor set. For each positive integer n, let T_n denote a point set such that the point P belongs to it if and only if P is a point in the XY plane whose y coordinate is $1/n$ and whose projection onto the X-axis is the midpoint of some segment of the collection H_n. The point set T_1 is degenerate. For each n greater than 1, T_n is a set of 2^{n-1} points. Let T denote $T_1 + T_2 + \cdots$. Let L_1 denote a set such that x belongs to it if and only if x is a straight line interval having, as one endpoint, F, the only element of T_1 and, as its other endpoint, one of the two points of the set T_2. For each n greater than 1, let L_n denote a set such that x belongs to it if and only if x is a straight interval having, as one of its endpoints, a point P of T_n and, as its other endpoint, one of the only two points of T_{n+1} such that P is equidistant from them. Let L denote a set such that x belongs to it if and only if, for some n, x is an interval of the set L_n. Let M denote the closure of the set L^*. It may be seen that M is $L^* + K$.

The point set K is the sum of two mutually exclusive point sets K_1 and K_2 such that a point belongs to K_1 if and only if it is an endpoint of Z_1 or of a component of $Z_1 - K$. Let F_1 denote some definite point not in the XY plane and let F_2 denote some definite point which is neither in the XY plane nor in the plane that contains F_1 and the X-axis. Let S denote a point set such that P belongs to it if and only if P belongs to T or to some straight line interval whose endpoints are either F_1 and some point of K_1, or F_2 and some point of K_2.

If two points of M are denoted by the letters X and Y, there is only one arc in M having X and Y as its endpoints and, in the description of this example, this arc will be denoted by the symbol XY.

Let Σ denote a space obtained by reinterpreting the word *region* to mean a subset R of S such that either (1) R is a degenerate subset of T or (2) R is a domain with respect to $S - T$ or (3) either for $i = 1$ or for $i = 2$, there exists a point P of K_i such that R is the sum of P and two point sets, the first one being the common part of T and some segment of PF with one endpoint at P, and the second one being a segment of PF_i that has P as one of its endpoints, or finally (4) either for $i = 1$ or for $i = 2$, R is a segment of PF_i. If a point of K belongs to a region of Σ it will be called the K-point of that region.

If e is a positive number and G is a collection of regions in Σ each containing a point of K and a point of M at a distance more than e from the X-axis, and the point A is a limit point, in the Σ' sense, of the set of all K-points of regions of G then every region in Σ that has A as a K-point intersects some region of the collection G.

Suppose Q_1 and Q_2 are two collections of regions in Σ such that, if j is 1 or 2, each region of Q_j has some point of K_j as its K-point and each point of K_j is the K-point of only one region of Q_j. There exists a positive number d such that some regions of Q_1 contain points of T at a distance more than d from the X-axis and so do some regions of Q_2. If j is 1 or 2, then for each positive integer n let Q_{jn} denote the collection of all regions of Q_j that have points of M at a distance more than d/n from the X-axis. Suppose no region of Q_1 intersects any region of Q_2. For each n, let Z_{jn} denote the set of all K-points of regions of Q_{jn}. If, for some n, some point of K_1 were a limit point, in the Σ' sense, of Z_{2n} then each region of Q_1 that contains that point would intersect some region of Q_2. So every limit point of Z_{2n} belongs to K_2. But K_2 is the sum of all the point sets $Z_{21}, Z_{22}, Z_{23}, \cdots$. Hence it is the sum of countably many closed point sets $\bar{Z}_{21}, \bar{Z}_{22}, \bar{Z}_{23}, \cdots$. Let U_2 denote the collection whose elements are these closed point sets, and let U_1 denote a collection such that x

belongs to it if, and only if, in Σ', x is the closure of some component of $Z_1 - K$. Let U denote the collection $U_1 + U_2$. In Σ', U is a collection of closed point sets whose sum is the compact continuum Z_1. Furthermore every nondegenerate component of an element of U *is* that element and belongs to U_1 and therefore is a subinterval of Z_1 having, in common with K, only its endpoints, and both of them belong to K_1. So no component of a nondegenerate element of U intersects another element of U. This contradicts Theorem 55. It follows that Q_1^* intersects Q_2^*. Hence, in the space Σ, every domain that contains K_1 intersects every one that contains K_2. But, in the space Σ, no subset of K has a limit point. So every subset of K is closed. Hence, in this space, K_1 and K_2 are mutually exclusive closed point sets. Therefore Σ is not normal.

For each n less than 3, let E_n denote the sum of all straight line intervals with one endpoint at F_n and the other one at some point of K_n. In Σ, since E_1 and E_2 are both closed, the point set $S - (E_1 + E_2)$ is a domain D. If A is a point of E_1 and B is a point of E_2, and N is a closed subset of D, $S - D$ is the sum of two mutually separated point sets containing A and B respectively but $S - N$ is not the sum of two such point sets. For suppose N is a closed subset of D such that $S - N$ is the sum of two mutually separated point sets S_{NA} and S_{NB} containing A and B respectively. In Σ, S_{NA} and S_{NB} are mutually exclusive domains. Hence, as has been shown, it is not true that S_{NA} contains K_1 and S_{NB} contains K_2. But E_1 and E_2 are connected subsets of $S_{NA} + S_{NB}$ containing A and B respectively. Hence E_1, and therefore K_1, is a subset of S_{NA}, and E_2, and therefore K_2, is a subset of S_{NB}. This involves a contradiction.

Definition. The point set M is said to be an *inner limiting set* if and only if either M is a domain or there exists a countable collection of domains such that M is the common part of all of them.

Theorem 147. *If M is an inner limiting set either M is a domain or there exists a sequence of domains $D_1, D_2, D_3, \cdots$ such that M is their common part and, for each n, D_n contains D_{n+1}.*

Proof. By hypothesis, if M is not a domain, there exists a simple countable sequence α such that M is the common part of all the domains of α. Let D_1 denote the first term of α and, for each n greater than 1, let D_n denote the common part of the first n terms of α.

Theorem 148. *A proper subset of S is an inner limiting set if and only if its complement either is degenerate or is the sum of countably many closed point sets.*

Theorem 149. *Every closed point set is an inner limiting set.*

Proof. Suppose M is a closed point set. For each n, let H_n denote the set of all regions of G_n that intersect M. Suppose there exists a point P belonging to each of the domains $H_1^*, H_2^*, \cdots$ but not to M. There exists a region R containing P but no point of M. There exists a positive integer m

such that every region of G_m that contains P is a subset of R and therefore contains no point of M. But some region of H_m contains P and every such region belongs to G_m and intersects M. This involves a contradiction.

THEOREM 150. *If S is normal and M is a closed point set which is not a domain then there exists a sequence of domains $D_1, D_2, D_3, \cdots$ such that* (1) *each of them contains the closure of the next one,* (2) *their common part is M, and* (3) *if, for each n, P_n is a point of D_n and the sequence $P_1, P_2, P_3, \cdots$ converges then it converges to a point of M.*

PROOF. By hypothesis there exists a point X_1 not belonging to M. There exists a positive integer n_1 such that no region of G_{n_1} contains both X_1 and a point of M. Let D_1 denote the sum of all regions of G_{n_1} that intersect M. The point set $S - D_1$ is closed. Since S is normal there exists a domain E_1 containing $S - D_1$ such that $\bar{E}_1$ does not intersect M. There exists a collection H_2 of regions of G_2, all intersecting M, such that D_2, their sum, contains no point of $\bar{E}_1$. The point set $\bar{D}_2$ contains no point of E_1. But E_1 contains $S - D_1$. So $\bar{D}_2$ is a subset of D_1. There exists a domain E_2 containing $S - D_2$ such that $\bar{E}_2$ contains no point of M. This process may be continued indefinitely. Thus there exists an infinite sequence $D_1, D_2, D_3, \cdots$ such that D_1 and D_2 are as indicated and (1) for each n, D_n is the sum of some regions of G_n that intersect M, (2) for each n, D_n contains $\bar{D}_{n+1}$. As shown in the above proof of Theorem 149, M is the common part of the domains $D_1, D_2, D_3, \cdots$. Suppose now that, for each n, P_n is a point of D_n and X is the sequential limit point of the sequence $P_1, P_2, P_3, \cdots$. Then, for each n, $P_{n+1} + P_{n+2} + \cdots$ is a subset of $\bar{D}_{n+1}$ and therefore X belongs to $\bar{D}_{n+1}$ and therefore to D_n. Hence X belongs to M.

It does not however follow from Axioms 0 and 1 and the supposition that S is normal that if M is a closed point set there exists a sequence $D_1, D_2, D_3, \cdots$ of domains such that all of the conditions of the conclusion of Theorem 150 hold true and furthermore every domain that contains M contains some domain of this sequence. Indeed this is not true even in the plane. But the following theorem holds true.

THEOREM 151. *If M is a point set then in order that there exist a sequence of domains $D_1, D_2, D_3, \cdots$ such that each of them contains the next one and their common part is M and every domain that contains M contains some one of them, it is necessary and sufficient that there exist a sequence of domains $D_1, D_2, D_3, \cdots$, with M as their common part, such that each of them contains the next one and if, for each n, P_n is a point of $D_n - M$ then the point set $P_1 + P_2 + \cdots$ has a limit point belonging to M.*

PROOF. Suppose $D_1, D_2, D_3, \cdots$ is a sequence of domains with M as their common part and $P_1, P_2, P_3, \cdots$ is a sequence of points such that, for each n, P_n belongs to $D_n - M$ and $P_1 + P_2 + \cdots$ has no limit point belonging to M. Then the closure of $P_1 + P_2 + \cdots$ is a closed point set K

containing no point of M and $S - K$ is a domain containing M but no domain of the sequence $D_1, D_2, \cdots$. So this condition is necessary.

Suppose now that there exists a sequence of domains $D_1, D_2, \cdots$, with M as their common part, such that if, for each n, P_n is a point of $D_n - M$ then the point set $P_1 + P_2 + \cdots$ has a limit point P belonging to M. Suppose D is a domain containing M and not containing any domain of this sequence. For each n let P_n denote a point of $D_n - D \cdot D_n$. The point set $P_1 + P_2 + \cdots$ has a limit point P belonging to M. But it is a subset of $S - D$ and P is a point of D. This involves a contradiction. So the condition of this theorem is also sufficient.

If the phrase "belonging to M" is omitted from the statement of Theorem 151, the resulting proposition is not a consequence of Axioms 0 and 1. Consider the following example.

EXAMPLE 9. Let Δ denote a space obtained from the space of Example 8 by adding certain points and regions as follows.

Using the notation of Example 8, let Q denote a collection such that q belongs to it if and only if q is a sequence of points of T converging in Σ' (not in Σ) to some point of K such that (1) each point of q is higher than every one that follows it in q, (2) no arc in M contains infinitely many points of q. There is a meaning for the word "precede" with respect to which the collection Q is well ordered. There exists a subcollection Z of Q such that (1) the first element of Q belongs to Z, (2) if Z' is a proper subcollection of Z such that if x is an element of Z' which is preceded by some element of Z, every element of Z that precedes x belongs to Z', then the first element of Z that follows every element of Z' is the first sequence q of the collection Q such that no infinite subsequence of q is an infinite subsequence of any sequence belonging to the collection Z'. Let Δ denote a space such that (1) x is a point of it if and only if either x is a point of $K + T$ or x is a sequence of the collection Z, (2) R is a region in Δ if and only if either R is a region in Σ or, for some sequence z of the collection Z and some point y of z, R is a set such that r belongs to it if and only if r is either the point z or a point following y in z. The space Δ satisfies Axioms 0 and 1 and, in this space, the set of all points of T is compact, but $\overline{T} - T$ (which consists of K plus all points of Δ that are not points of Σ) has no limit point, though it is uncountable. So, in Δ, the closure of the compact point set T is not compact. In the space Δ there exists a sequence $D_1, D_2, D_3, \cdots$ of domains, with K as their common part, such that if, for each n, P_n is a point of $D_n - K$ then the point set $P_1 + P_2 + P_3 + \cdots$ has a limit point. But this statement does not remain true if "limit point" is replaced by "limit point belonging to K".

THEOREM 152. *If S is normal, D is a domain and M is a closed subset of $\overline{D}$ intersecting D and the boundary of D then there exists a domain D', lying in D and containing $M \cdot D$, such that the boundary of D contains no point of $\overline{D}'$ that does not belong to M.*

PROOF. Let $D_1, D_2, D_3, \cdots$ denote a sequence of domains satisfying all the requirements of Theorem 150 except for the substitution of $M\cdot(\bar{D}-D)$ in place of M. Since S is normal there exists a domain E_1 such that (1) $\bar{E}_1$ contains no point of the closed point set $S-D$ and therefore lies wholly in D, and (2) E_1 contains the closed point set $M\cdot(S-D_2)$ if that point set exists. There exists a domain E_2 such that (1) $\bar{E}_2$ is a subset of $D_1\cdot D$ and (2) E_2 contains $M\cdot(\bar{D}_2-D_3)$ if M intersects $\bar{D}_2-D_3$. Similarly, for each n, there exists a domain E_{n+1} such that (1) $\bar{E}_{n+1}$ lies wholly in $D\cdot D_n$ and (2) E_{n+1} covers $M\cdot(\bar{D}_{n+1}-D_{n+2})$ if M intersects $\bar{D}_{n+1}-D_{n+2}$. Let D' denote $E_1+E_2+\cdots$. This domain satisfies all the requirements of the conclusion of Theorem 152.

THEOREM 153. *If S is normal and H and K are two mutually separated point sets there exist two mutually exclusive domains D_H and D_K containing H and K respectively such that either $\bar{D}_H$ and $\bar{D}_K$ are mutually exclusive or $\bar{D}_H\cdot\bar{D}_K$ is $\bar{H}\cdot\bar{K}$.*

PROOF. Let D denote the domain $S-\bar{K}$. If $\bar{H}\cdot\bar{K}$ exists the closed point set $\bar{H}$ is a subset of $\bar{D}$ intersecting both D and the boundary of D. Hence, by Theorem 152, there exists a domain D_H, lying in D and containing H, such that the boundary of D contains no point of $\bar{D}_H$ that does not belong to $\bar{H}$. Let E denote the domain $S-\bar{D}_H$. The point set $\bar{K}$ is a subset of $\bar{E}$. By Theorem 152, there exists a domain D_K, lying in E and containing K, such that the boundary of E contains no point of $\bar{D}_K$ that does not belong to $\bar{K}$. The domains D_H and D_K satisfy the requirements of the conclusion of Theorem 153.

THEOREM 154. *If S is normal and connected and the point set M separates the point set H from the point set K, some closed subset of M does so.*

PROOF. By hypothesis, $S-M$ is the sum of two mutually separated point sets H' and K' containing H and K respectively. By Theorem 153 there exist two mutually exclusive domains D_H and D_K containing H' and K' respectively. Since S is connected, $S-(D_H+D_K)$ exists. It is a closed subset of M and it separates H from K.

It is not true, even in the plane, that if A and B are two points of the compact continuum M and the closed subset K of M separates A from B in M then some closed subset of K is irreducible with respect to the property of being a closed point set that separates A from B in M.

EXAMPLE. Let M, A and B be as described in Example 1. Here *every* point set that separates A from B in M has a closed proper subset that does so.

THEOREM 155. *If M is a closed and compact point set there exists a sequence $D_1, D_2, D_3, \cdots$ such that* (1) *for each n, D_n contains $\bar{D}_{n+1}$,* (2) *M is the common part of all the domains of this sequence,* (3) *every domain that contains M contains some domain of this sequence.*

PROOF. Since M is closed and compact, it is covered by some finite subcollection H_1 of G_1. There exists a finite subcollection H_2 of G, covering M, such that, if h belongs to H_2, $\bar{h}$ is a subset of some element of H_1. There exists an infinite sequence $H_1, H_2, H_3, \cdots$ such that H_1 and H_2 are as indicated and if n is a positive integer, H_n is a finite subcollection of G_n covering M and if the region h belongs to H_{n+1} then $\bar{h}$ is a subset of some region of H_n. Suppose D is a domain containing M. Suppose, for each n, there exists a region of H_n containing a point of the closed point set $S - D$. By Theorem 114 there exists a sequence $h_1, h_2, h_3, \cdots$ such that, for each n, h_n belongs to H_n and intersects $S - D$ and $\bar{h}_{n+1}$ is a subset of h_n. Since they are closed and compact, the point sets $\bar{h}_1 \cdot M, \bar{h}_2 \cdot M, \bar{h}_3 \cdot M, \cdots$ have a point A in common. There exists a number m such that if a region of G_m contains A it is a subset of D. But h_m belongs to G_m and contains both A and a point of $S - D$. This involves a contradiction.

If P is a point not belonging to M, $D - P$ is a domain containing M and therefore there exists a positive integer k such that H_k^* is a subset of $D - P$. So no point not belonging to M belongs to every domain of the sequence $H_1^*, H_2^*, H_3^*, \cdots$.

THEOREM 156. *If the point set M contains no point set K such that every point of K is a limit point of K then M is an inner limiting set.*

PROOF. Let M_1 denote the set of all points of M that are not limit points of M. If $M - M_1$ exists, let M_3 denote the set of all points of it that are not limit points of it. There exists a well ordered sequence α such that (1) M_1 is the first term of α, (2) if β is an initial segment of α such that β^* is not M then the first term of α following all the terms of β is the set of all points of $M - \beta^*$ which are not limit points of $M - \beta^*$. The point set α^* is identical with M.

If P is a point of M and n is a positive integer there exists a region g_{nP} of G_n containing P but no other point belonging either to the member of α that contains P or to any member of α following the one containing P. For each n let D_n denote the sum of the regions g_{nP} for all points P of M. Suppose there exists a point O not belonging to M but belonging to all the domains of the sequence $D_1, D_2, D_3, \cdots$. There exists a point P_1 of M such that g_{1P_1} contains O. If n_1 denotes 1 there exist an ascending sequence of positive integers $n_1, n_2, n_3, \cdots$ and an infinite sequence of points $P_1, P_2, P_3, \cdots$ such that (1) for each i, $g_{n_iP_i}$ contains O, and $\bar{g}_{n_{i+1}P_{i+1}}$ is a subset of $g_{n_iP_i}$ not containing P_i. Since, for each i, $g_{n_iP_i}$ contains P_{i+1} therefore, for each i, the point set of the sequence α that contains P_{i+1} precedes, in α, the one that contains P_i. Thus if, for each i, T_i is the point set of the well ordered sequence α that contains P_i then $T_1, T_2, T_3, \cdots$ is an *infinite* sequence such that, for each i, T_{i+1} precedes T_i in the *well ordered* sequence α. But this is impossible. Hence M is an inner limiting set.

THEOREM 157. *If H is a countable collection of mutually exclusive closed and compact point sets such that no one of them contains a limit point of the sum of all the others, then there exists a collection G of domains such that their closures are mutually exclusive and each point set of the collection H is a subset of only one domain of the collection G and each domain of the collection G contains only one point set of the collection H.*

PROOF. Let $M_1, M_2, M_3, \cdots$ denote the point sets of the collection H. By hypothesis and either Theorem 155 or 16, M_1 is a subset of some domain D_1 whose closure is a subset of the domain obtained by subtracting the closure of $M_2 + M_3 + \cdots$ from S. The point set M_2 is a subset of some domain D_2 such that $\bar{D}_2$ is a subset of the complement of the sum of $\bar{D}_1$ and the closure of $M_3 + M_4 + \cdots$. This process may be continued indefinitely. It follows that there exists a sequence of domains $D_1, D_2, D_3, \cdots$ such that their closures are mutually exclusive and, for each n, D_n contains M_n.

THEOREM 158. *If $D_1, D_2, D_3, \cdots$ is a sequence of domains such that, for each n, D_n contains $\bar{D}_{n+1}$ and the common part of all of them is the point set M and $D_1 - M$ is compact and every point of M is a limit point of $S - M$, then every domain that contains M contains a domain of this sequence and, if S is normal, M is compact.*

PROOF. Suppose D is a domain containing M such that, for each n greater than 1, D_n contains a point not belonging to D. For each such n, the point set $\bar{D}_n \cdot (S - D)$ is compact and $\bar{D}_n \cdot (S - D)$ contains $\bar{D}_{n+1} \cdot (S - D)$. Hence there is a point which, for every n greater than 1, belongs to $\bar{D}_n \cdot (S - D)$ and therefore to M which is impossible.

Suppose S is normal and M is not compact. Then there exists a countably infinite subset K of M with no limit point. Let $A_1, A_2, A_3, \cdots$ denote the points of K. Let $Z_1, Z_2, Z_3, \cdots$ denote a sequence of domains satisfying, with respect to K, all of the conditions stated, in Theorem 150, as being satisfied by $D_1, D_2, D_3, \cdots$ with respect to M. By Theorem 157, there exists a collection of mutually exclusive regions $R_1, R_2, R_3, \cdots$ such that, for each n, R_n contains A_n. For each n there exists a point B_n common to Z_n, R_n and $D_1 - M$. Since $D_1 - M$ is compact, $B_1 + B_2 + B_3 + \cdots$ has a limit point B. By (3) of Theorem 150, for some n, B is A_n. Thus A_n is a limit point of $B_1 + B_2 + B_3 + \cdots$. But R_n is a region containing A_n and no point of this set except B_n. This involves a contradiction. Therefore M is compact.

DEFINITION. The statement that S has *property D* means that if M is a countably infinite point set with no limit point there exists a collection G of mutually exclusive regions such that (1) G properly covers M and each region of G contains only one point of M, and (2) if K is a point set properly covered by G such that each region of G contains only one point of K then K has no limit point.

THEOREM 159. *If S has property D and M is a closed point set with a boundary which is not compact then there exists no countable collection of domains such that M is their common part and every domain that contains M contains one of them.*

PROOF. By hypothesis β, the boundary of M, contains a countably infinite point set K with no limit point. Since S has property D, there exist a sequence of mutually exclusive regions $Z_1, Z_2, Z_3, \cdots$ and a sequence $P_1, P_2, P_3, \cdots$ of points such that (1) $P_1, P_2, P_3, \cdots$ are the points of K and, for each n, Z_n contains P_n but no other point of K, (2) if, for each n, X_n is a point of Z_n, the point set $X_1 + X_2 + X_3 + \cdots$ has no limit point.

Suppose there exists a sequence of domains $D_1, D_2, D_3, \cdots$ such that M is their common part and every domain that contains M contains one of them. For each n, D_n contains the point P_n of the boundary of M and therefore contains some point X_n belonging to Z_n but not to M. Since $X_1 + X_2 + X_3 + \cdots$ has no limit point, it is closed and therefore $S - (X_1 + X_2 + X_3 + \cdots)$ is a domain D. But $X_1 + X_2 + X_3 + \cdots$ contains no point of M. Hence D contains M. But if n is a positive integer D does not contain X_n and therefore does not contain D_n.

THEOREM 160. *If S is normal it has property D.*

PROOF. Suppose M is a countably infinite point set with no limit point. The point set M is closed. Hence, by Theorem 150, there exists a sequence $D_1, D_2, D_3, \cdots$ such that (1) for each n, D_n is a domain and, for each n, D_n contains $\bar{D}_{n+1}$, (2) if the point P is the sequential limit point of a sequence $X_1, X_2, X_3, \cdots$ such that, for each n, X_n is a point of D_n, then P belongs to M, (3) M is the common part of $D_1, D_2, D_3, \cdots$. Let $P_1, P_2, P_3, \cdots$ denote the points of M. There exists a collection H of mutually exclusive regions $R_1, R_2, R_3, \cdots$ such that (1) for each n, R_n contains P_n and is a subset of D_n. Suppose K is a set of points $X_1, X_2, X_3, \cdots$ such that, for each n, X_n is a point of R_n. Suppose K has a limit point X. Since, for each n, X_n belongs to D_n, X belongs to M. Hence, for some m, X is P_m. But R_m is a region containing X_m and no other point of K. It follows that K has no limit point. Hence S has property D.

THEOREM 161. *If S has property D and M is a compact point set then $\bar{M}$ is compact.*

PROOF. Suppose the point set M is compact and $\bar{M}$ is not. There exists a countably infinite subset K of $\bar{M} - M$ with no limit point. Suppose G is a collection of mutually exclusive domains such that each of them contains only one point of K and each point of K belongs to only one of them. For each domain g of G let P_g denote some definite point of $g \cdot M$. Let T denote a point set such that P belongs to it if and only if P is P_g for some domain g of G. Since M is compact, T has a limit point. This is contrary to the hypothesis that S has property D. So $\bar{M}$ is compact.

THEOREM 162. *No point set which contains a domain is a subset of the sum of countably many closed point sets no one of which contains a domain.*

PROOF. Suppose M is a point set which contains a domain D and which is a subset of the sum of countably many point sets $M_1, M_2, M_3, \cdots$ no one of which contains a domain. There exists a point P_1 of D not belonging to M_1. There exists a region R_1 of the collection G_1 containing P_1 and lying in D but containing no point of M_1. There exists a point P_2 of R_1 not belonging to M_2. There exists a region R_2 of the collection G_2, containing P_2 but no point of M_2, such that $\bar{R}_2$ is a subset of R_1. This process may be continued indefinitely. Thus there exists an infinite sequence of regions $R_1, R_2, R_3, \cdots$ such that, for each n, R_n belongs to G_n, lies in D, and contains no point of M_n, and $\bar{R}_{n+1}$ is a subset of R_n. The regions of this sequence have a point P in common. The point P belongs to no point set of the sequence $M_1, M_2, M_3, \cdots$. But it belongs to D and therefore to M. This involves a contradiction.

NOTATION. The notation "Axiom 1_3" will be used to denote the axiom obtained by deleting stipulation (4) from the statement of Axiom 1.

Theorem 162 is the third numbered theorem in this chapter which is not a consequence of Axioms 0 and 1_3. Consider the following example.

EXAMPLE 10. In a Cartesian plane, let S denote the set of all points whose coordinates are both rational and interpret the word region to mean the common part of S and the interior of a circle. The space so obtained satisfies Axiom 1_3. In this space no degenerate point set contains a domain, there are only countably many degenerate point sets and the sum of all of them is S which is a domain.

THEOREM 163. *If G is a collection of closed point sets no one of which contains a domain and every point is a limit point of M, the sum of all the point sets of the collection G, then M is not an inner limiting set.*

PROOF. Suppose M is an inner limiting set. Then $S - M$ is the sum of countably many closed point sets $K_1, K_2, K_3, \cdots$. Hence S is the sum of countably many closed point sets each of them being either a point set of this sequence or a point set of the collection G. But no point set of this sequence contains a domain and neither does any point set of the collection G. This contradicts Theorem 162.

THEOREM 164. *If there do not exist uncountably many mutually exclusive domains then S is separable.*

PROOF. There exists a well ordered sequence α whose terms are the regions of G_1. There exists a well ordered subsequence α_1 of α such that (1) the first term of α_1 is the first term of α, (2) if β is a well ordered subsequence of α_1, distinct from α_1, such that every term of α_1 which precedes a term of β belongs to β then (a) there exists a region R belonging to α and intersecting no region of β and (b) the first such region R in α is the first term of α_1 following all the terms of β. Let H_1 denote the collection of all regions of the sequence α_1. The regions of H_1 are mutually exclusive and every point either belongs to H_1^* or is a limit point of it. Similarly there exists a collection H_2 of mutually exclusive regions such that (1) H_2 is a subcollection of G_2, (2) if x is a region of H_2, $\bar{x}$ is a subset of some region of H_1, (3) if x is a region of H_1 every point of x belongs to, or is a limit point of, the sum of all the regions of H_2 which are subsets of x. This process may be continued. Thus there exists a sequence $H_1, H_2, H_3, \cdots$ such that, for each n, (1) H_n is a subcollection of G_n, (2) if x is a region of H_{n+1}, $\bar{x}$ is a subset of some region of H_n, (3) the regions of H_n are mutually exclusive, (4) if x is a region of H_n every region that intersects x intersects the sum of all the regions of H_{n+1} which are subsets of x.

Let H denote the set of all regions h such that, for some n, h belongs to H_n. Since, for each n, the regions of H_n are mutually exclusive, the collection H is countable. Suppose R is a region. There exists a region R_1 such that $\bar{R}_1$ is a subset of R. The region R_1 intersects some region x_1 of H_1. It also intersects some region x_2 of H_2 whose closure is a subset of x_1. This process may be continued. Thus there exists a sequence of regions $x_1, x_2, x_3, \cdots$ such that, for each n, x_n belongs to H_n, intersects R_1 and contains $\bar{x}_{n+1}$. But, for each n, x_n belongs to G_n. Hence the regions of this sequence have a point O in common. Suppose O does not belong to $\bar{R}_1$. There exists a number n such that if g is any region of G_n containing O then $\bar{g}$ is a subset of the domain $S - \bar{R}_1$, contrary to the fact that x_n contains O, belongs to G_n and intersects $\bar{R}_1$. Hence O belongs to $\bar{R}_1$ and therefore to R. It follows that, for some n, x_n is a subset of R. Thus every region contains some region of H. Since H is countable it follows that S is separable.

Let Q denote the axiom obtained from Axiom 1 by replacing stipulation (4) concerning $G_1, G_2, G_3, \cdots$ in the statement of that axiom by the stipulation that if R is a region, R contains a region R_1 such that if $g_1, g_2, g_3, \cdots$ is a sequence of regions such that, for each n, g_n belongs to G_n, contains $\bar{g}_{n+1}$ and intersects R_1 then, for some n, g_n is a subset of R. By an easily made modification of the above proof, it may be shown that Theorem 164 holds true in every space that satisfies Axioms 0 and Q.

DEFINITION. If A and B are points, a *simple chain from A to B* is a finite sequence of domains $R_1, R_2, R_3, \cdots, R_n$ such that (1) R_i contains A if and only if $i = 1$, (2) R_i contains B if and only if $i = n$, (3) if j is less than or equal to n and i is less than j, then R_i and R_j have a point in common if and only if $j = i + 1$. The domain R_k $(1 \leqq k \leqq n)$ is said to be the kth link of the chain $R_1R_2R_3 \cdots R_n$.

THEOREM 165. *If M is a connected point set, A and B are two points of M and G is a collection of domains covering M then there exists a simple chain from A to B such that every link of this chain is a domain of the collection G.*

PROOF. If there is no such chain from A to B then M is the sum of two sets S_A and S_B where S_A is a set such that P belongs to it if and only if either

P is A or P is a point of M and there is a simple chain from A to B every link of which is a domain of the collection G, and S_B is $M - S_A$. Since M is connected there exists a point X which belongs to one of the sets S_A and S_B and is a limit point of the other one. The point X lies in some domain R of the collection G. The domain R contains a point A_1 belonging to S_A, and a point B_1 belonging to S_B, where A_1 is X or B_1 is X according as X belongs to S_A or to S_B. The point A can be joined to A_1 by a simple chain $R_1R_2R_3 \cdots R_n$ every link of which is a domain of the set G. Let R_k denote the first link of this chain that has a point in common with R. Then $R_1R_2R_3 \cdots R_kR$ is a simple chain from A to B and every one of its links belongs to G. This involves a contradiction.

THEOREM 166. *If M is a completely separable closed point set and G is an uncountable collection of nondegenerate subsets of M, there exist two domains H and K, with respect to M, and an uncountable subcollection G' of G such that $\bar{H}$ and $\bar{K}$ are mutually exclusive and every element of G' intersects both H and K.*

PROOF. For each point set g of the collection G, let A_g and B_g denote two definite points of g. There exists a sequence $D_1, D_2, D_3, \cdots$ of domains with respect to M such that if P is a point of M every domain with respect to M that contains P contains a member of this sequence containing P. By Theorem 13, for each g there exist two definite elements x_g and y_g of this sequence, containing A_g and B_g respectively, such that $\bar{x}_g$ and $\bar{y}_g$ are mutually exclusive. Let U denote the set of all sensed pairs (x_g, y_g) for all point sets g of the collection G. The set of all sensed pairs of terms of the sequence $D_1, D_2, D_3, \cdots$ is countable. Hence U is countable. But G is uncountable. Hence there exist two domains H and K, with respect to M, of the sequence $D_1, D_2, D_3, \cdots$ and an uncountable subcollection G' of G such that, if g is any element of G', x_g is H and y_g is K. Each point set of the collection G' intersects both H and K.

THEOREM 167. *If G is an uncountable collection of nondegenerate subsets of the closed and compact point set M, there exist two points A and B of M such that, if R_A and R_B are two regions containing A and B respectively, then uncountably many elements of G intersect both R_A and R_B.*

Theorem 167 may be established with the help of Theorem 166.

THEOREM 167(a). *If, for each point P of the completely separable point set M, D_P is a domain containing P, and G is the collection of all domains D_P for all points P of M then there exists a domain D containing uncountably many points P such that D is a subset of D_P.*

PROOF. By hypothesis there exists a countable set H of domains such that if P is a point of M then there exists a domain belonging to H, containing P, and lying in D_P. For each P, let h_P denote one such domain of the set H. Since M is uncountable and H is countable there exist a domain

D, belonging to H, and an uncountable subset M' of M such that if P is any point of M' then $h_P = D$. If P is a point of M', P belongs to D, and D is a subset of D_P.

DEFINITION. Σ is a (topological) *space* if and only if, for some meaning of *point* and some meaning of *limit point*, either (1) Σ is a two termed sequence such that (a) the first term of Σ is a set such that P is an element of it if, and only if, P is a poi.., and (b) the second term of Σ is a collection G such that g belongs to G if and only if g is a two termed sequence such that its first term is a point set M and its second term is a point which is a limit point of M, or (2) no point is a limit point of any point set and Σ is a set such that P belongs to it if and only if P is a point.

In accordance with this definition, a space Σ is determined if and only if (1) for each object P it is determined whether or not P is a *point* in the Σ sense and (2) for each point P, and point set M, in the Σ sense, it is determined whether or not P is a *limit point* of M in the Σ sense.

DEFINITION. The space Σ' is called a *subspace* of the space Σ if and only if it is true that if S denotes a set such that P belongs to it if and only if P is a point in the Σ sense, and S' is a set such that P belongs to it if and only if P is a point in the Σ' sense, then (1) S' is a subset of S and (2) if P is a point of S' and M is a subset of S' then P is a limit point of M in the Σ' sense if and only if it is a limit point of M in the Σ sense.

Consider a space Σ satisfying Axioms 0 and 1_3.

Suppose M is a point set in Σ. Suppose the word "point" is interpreted to mean "point of M" and the word "region" is interpreted to mean "domain with respect to M". It is easy to see that if "point" and "region", wherever they occur in the statement of Axiom 0 or Axiom 1_3, are interpreted in this manner, these axioms remain true. But all of the first 161 theorems of this chapter except Theorems 115 and 116 were proved on the basis of these two axioms. It follows that all of these theorems hold true in Σ', the subspace of Σ whose points are the points of M.

The statement that a certain proposition (call it Proposition A) holds true under this interpretation implies that Proposition A holds true if (1) point and region wherever they occur *explicitly* in the statement of Proposition A, are interpreted as indicated and (2) any terms, used in the statement of Proposition A, which have been defined in terms of the original point and region, continue to be defined in exactly the same words, it being understood, of course, that now "point" and "region", wherever they occur, are to be given the interpretation in question.

It is of particular importance to note that *if P is a point of M and K is a subset of M* then every region in the original sense that contains P contains a point of K distinct from P if and only if every region in the new sense (that is to say every domain with respect to M) that contains P contains a point of K distinct from P and that *therefore P is a limit point of K, in the new sense, if and only if it is a limit point of K in the old sense.*

It is not, however, necessarily true that every subset of M which is closed in the new sense is also closed in the original sense. For suppose that M is not closed in the original sense. It is, nevertheless, closed in the new sense. For suppose P is a limit point of M in the new sense of the term "limit point of M". Then P belongs to M. Otherwise it would not be a point at all in the new sense of the word "point".

The point set M is compact in the new sense if and only if it is closed and compact in the old sense. The subset K of M is compact in the new sense if and only if K is compact in the old sense, and $\bar{K}$, in the old sense, is a subset of M.

It is clear, however, that a subset of M is *connected* in the new sense if and only if it is connected in the old sense.

Suppose now that Σ is a space satisfying Axioms 0 and 1. Then the following theorem holds true.

THEOREM 168. *Suppose M is a point set in the space Σ. Then it is necessary and sufficient that M be an inner limiting set in order that it be true that if the word "point" is reinterpreted to mean "point of M" then there exists a reinterpretation of the word "region" such that* (1) *for these new interpretations Axioms* 0 *and* 1 *hold true and* (2) *the notion of limit point is preserved in the sense that the point P of M is a limit point of the subset K of M under the new interpretation of "region" if and only if it is a limit point of K under the original one.*

PROOF. This condition is necessary. For suppose there exists a sequence $G_1', G_2', G_3', \cdots$ satisfying the conditions of Axiom 1 except that S is replaced by M, region is interpreted to mean "region in the new sense" and certain other terms employed are interpreted accordingly as explained above. Every region in the new sense is necessarily a domain with respect to M in the original sense. For each n, let D_n denote the domain in the original sense obtained by adding together all regions g in the original sense, belonging to G_n and intersecting M such that $\bar{g} \cdot M$ is a subset of some point set of the collection G_n'. Let K denote the common part of all the domains D_1, D_2, $D_3, \cdots$. The point set M is a subset of K.

Suppose now that K contains a point X not belonging to M. There exist an element g_1 of the collection G_1 and an element q_1 of the collection G_1' such that g_1 contains X and $M \cdot \bar{g}_1$ is a subset of q_1. Let n_1 denote 1. There exists a number n_2 such that if g is an element of G_{n_2} containing X then $\bar{g}$ is a subset of g_1. There exist an element q_2 of the collection G_{n_2}' and an element g_2 of G_{n_2} such that g_2 contains X and $M \cdot \bar{g}_2$ is a subset of q_2. This process may be continued. Thus there exist two infinite sequences $g_1, g_2, g_3, \cdots$ and $q_1, q_2, q_3, \cdots$ such that, for each n, g_n belongs to G_n, q_n belongs to G_n', $\bar{g}_{n+1}$ is a subset of g_n and $M \cdot \bar{g}_n$ is a subset of q_n. For each n, let M_n denote $M \cdot \bar{g}_n$. The point sets of the sequence $M_1, M_2, M_3, \cdots$ are all closed in the new sense (regardless of whether they are closed in the original sense), for each n, M_{n+1} is a subset of M_n and M_n is a subset of a region of the collection G'_n. Hence there is a point P common to all the point sets of the sequence $M_1, M_2, M_3, \cdots$. The point P is common to all the regions, in the original sense, of the sequence $g_1, g_2, g_3, \cdots$. But P belongs to M and therefore is distinct from X, and X is the only point common to all of these regions. Thus the supposition that K is not a subset of M has led to a contradiction. Hence K is identical with M and therefore M is an inner limiting set.

This condition is also sufficient. For suppose M is the common part of the domains of a sequence $D_1, D_2, D_3, \cdots$ such that, for each n, D_n contains D_{n+1}. For each n, let G'_n denote the collection of all point sets h such that h is the common part of M and some region g such that $\bar{g}$ is a subset of D_n and of some region of the collection G_n. The sequence $G'_1, G'_2, G'_3, \cdots$ satisfies all the conditions required of $G_1, G_2, G_3, \cdots$ in the statement of Axiom 1 except that S is replaced by M, region is interpreted to mean domain with respect to M and certain other terms employed are interpreted accordingly.

THEOREM 169. *If the closed point set M intersects the connected point set K and $M \cdot K$ is the sum of two mutually separated point sets H and L then H intersects the boundary of M.*

PROOF. The closed point set H is a proper subset of the connected point set K. Hence H contains a point P which is a limit point of $K - H$. There exists a region R_1 containing P but no point of L. Suppose P is not a boundary point of M. Then there exists a region R_2 containing P and lying wholly in M. The point set $R_1 \cdot R_2$ is a domain lying in M and containing P but no point of $M \cdot K - H$. Since $K \cdot R_2$ is a subset of $M \cdot K$, $R_1 \cdot R_2$ contains no point of $K - H$. This involves a contradiction.

DEFINITION. The point set M is said to be *compactly connected* if and only if every two points of M lie together in some compact continuum which is a subset of M.

THEOREM 170. *If β is the boundary of the compactly connected continuum M and, for every component D of $M - \beta$, T_D is a connected point set lying in M and containing the common part of β and the boundary of D, and T is the sum all the point sets T_D for all such D's, then $\beta + T$ is connected.*

PROOF. Suppose $\beta + T$ is the sum of two mutually separated point sets H and K. The point sets $H \cdot \beta$ and $K \cdot \beta$ exist and are mutually exclusive and closed. There exists a subcontinuum N of M which is irreducible from $H \cdot \beta$ to $K \cdot \beta$. The connected point set $N - N \cdot (H \cdot \beta + K \cdot \beta)$ is a subset of some component L of $M - \beta$. Each of the point sets $H \cdot \beta$ and $K \cdot \beta$ contains a limit point of $N - N \cdot (H \cdot \beta + K \cdot \beta)$ and therefore of L. Hence T_L intersects both $H \cdot \beta$ and $K \cdot \beta$. This involves a contradiction.

THEOREM 171. *If β is the boundary of the compact continuum M and, for every component D of $M - \beta$, the common part of β and the boundary of D is connected, then β is connected.*

Theorem 171 is a corollary of Theorem 170, but it does not remain true, in every space satisfying Axioms 0 and 1, if, in the statement of its hypothesis, "the common part of β and the boundary of D" is replaced by "the boundary of D", even though the resulting hypothesis is strengthened by the addition of the stipulation that space is compact. Consider the following example.

EXAMPLE 11. In a Cartesian plane E let Q denote a compact totally disconnected perfect point set lying on the Y-axis, let A and B denote the points $(-1,0)$ and $(-2,0)$ respectively, let S denote the sum of the straight line interval AB and all straight line intervals with one endpoint at A and midpoint at a point of Q, let M denote the sum of all straight line intervals with one endpoint at A and the other one at a point of Q and let β denote the point set $A + Q$. Let Σ denote the subspace of E whose points are the points of S. In the compact space Σ, the boundary of the continuum M is β and if D is a component of $M - \beta$ then the boundary of D is the continuum $\bar{D}$. But β is not connected.

That Theorem 171 does not remain a consequence of Axioms 0 and 1 if the stipulation requiring that M be compact is replaced by the stipulation that β is compact may be seen from the following example.

EXAMPLE 12. In a Cartesian plane E, let O and A denote the points $(0,0)$ and $(0,2)$ respectively and, for each n, let B_n denote $(-1/n,0)$. Let M denote the sum of O and the straight intervals AB_1, AB_2, $AB_3, \cdots$, and let K denote a semi-circle with extremities at A and O and containing the point $(1,1)$. Let S denote $M + K$. Let Σ denote the subspace of E whose points are the points of S. Here β is $A + O$ and, for each component D of $M - \beta$, $\beta \cdot \bar{D}$ is A.

THEOREM 172. *If β is the boundary of the continuum M, D is a component of $M - \beta$ and K is a continuum intersecting D but not lying wholly in it and $M \cdot K$ is compact then every component of $K \cdot \bar{D}$ contains a point of β.*

PROOF. Suppose there exists a component N of $K \cdot \bar{D}$ containing no point of β. Let Z denote the domain $M - \beta$. The continuum N is a subset of $D \cdot Z$. Let A denote some point of N. Since D is a component of Z and K is a continuum containing the point A of D but not lying wholly in D therefore K is not a subset of Z. But $K \cdot M$, and therefore $K \cdot Z$, is compact. Therefore, by Theorem 50, β contains a limit point of T, the component of $K \cdot Z$ that contains A. The point set $N + T$ is a connected subset of Z containing N. Hence N contains T. But N is closed. Hence N is T and therefore intersects β.

Theorem 172 does not remain true in every space satisfying Axioms 0 and 1 if the supposition that $M \cdot K$ is compact is replaced by the supposition that $\bar{D}$ is compact and M is identical with K. Consider the following example.

EXAMPLE 13. In a Cartesian plane E, let A and B denote the points $(0,2)$ and $(0,1)$ respectively and, for each n, let B_n denote the point $(-1/n,0)$ and let AB_n denote the straight interval whose endpoints are A and B_n.

Let both K and M denote the point set $B + AB_2 + AB_3 + \cdots$ and let S denote $M + AB_1$. Let Σ denote the subspace of E whose points are the points of S. Here β is the set whose only element is A and D, the set whose only element is B, is a component of $M - \beta$. Though $\bar{D}$ is compact, the component D of $K \cdot \bar{D}$ does not contain A.

That Theorem 172 does not remain true even in the plane if the supposition that $M \cdot K$ is compact is replaced by the supposition that space is the plane and D is the only component of $M - \beta$ may be seen from the following example.

EXAMPLE 14. In a Cartesian plane E, let M denote the set of all points (x,y) such that either $-1/x^2 \leqq y \leqq 1/x^2$ or $x = 0$ and let K denote the set of all points (x,y) such that either $x = 0$ or $y = (2 \sin 1/x^2)/x^2$.

THEOREM 173. *If β is the boundary of the continuum M, D is a component of $M - \beta$ and K is a continuum intersecting $\bar{D}$ such that $M \cdot K$ is compact and either $\bar{D}$ does not intersect β or there exists a subcontinuum of K lying in $\bar{D}$ and containing $\beta \cdot \bar{D}$, then $K \cdot \bar{D}$ is a continuum.*

Theorem 173 may be easily established with the help of Theorem 172. It does not remain true in every space satisfying Axioms 0 and 1 if $M \cdot K$ is replaced by $\bar{D}$ in the statement of its hypothesis. Consider the following example.

EXAMPLE 15. Let space be that of Example 12, but now let K denote the point set $O + AB_2 + AB_3 + \cdots$, let M denote $A + (S - AB_1)$ and let D denote $S - (AB_1 + AB_2 + \cdots)$. The point set $M \cdot K$ is now K, $\bar{D}$ is compact and β, the boundary of M, is the point set whose only element is A and β is a subcontinuum of K lying in $\bar{D}$ and coinciding with $\beta \cdot \bar{D}$. But $K \cdot \bar{D}$ is $A + O$ which is not connected.

Theorem 173 does not remain true in every space satisfying these axioms if, in the statement of its hypothesis, the supposition that $M \cdot K$ is compact is replaced by the supposition that $\bar{D}$ is compact, even though the resulting hypothesis is strengthened by the elimination of the word "either" and the phrase "or there exists a subcontinuum of K lying in $\bar{D}$ and containing $\beta \cdot \bar{D}$". Consider the following example.

EXAMPLE 16. In a Cartesian plane E, let A, A_1 and A_2 denote the points (0,4), (0,1) and (0,3) respectively and, for each n, let B_n denote $(-1/n,0)$ and let AB_n denote the straight interval whose extremities are A and B_n. Let D denote a semi-circle passing through (1,2) and with its extremities at A_1 and A_2. Let K denote the point set $A_1 + A_2 + AB_2 + AB_3 + \cdots$, let M denote $K + D$ and let S denote $M + AB_1$. Let Σ denote the subspace of E whose points are the points of S. Here A is the only point of β and $\bar{D}$ is identical with D and does not intersect β. But $K \cdot \bar{D}$ is $A_1 + A_2$ which is not connected.

THEOREM 174. *If M is a compact continuum with a boundary β such that, if D is a component of $M - \beta$, $\beta \cdot \bar{D}$ is connected, and K is a continuum intersecting β, then $M \cdot K + \beta$ is a continuum.*

PROOF. By Theorem 171, β is connected. Hence so is $K + \beta$. If $M \cdot K$ is K or a subset of β then $M \cdot K + \beta$ is $K + \beta$ or β. If $M \cdot K$ is neither K nor a subset of β then $(K + \beta) - \beta$ is the sum of two mutually separated point sets $M \cdot K - M \cdot K \cdot \beta$ and $K - M \cdot K$. Hence $(M \cdot K - M \cdot K \cdot \beta) + \beta$ is connected. But this point set is identical with $M \cdot K + \beta$.

Theorem 174 does not remain a consequence of these axioms if, in the statement of its hypothesis, the supposition that M is compact is replaced by the supposition that K is compact even though there is added, to this hypothesis, the additional supposition that the entire boundary of every component of $M - \beta$ is a connected subset of β and the conclusion is so weakened as to require only that there exist a continuum containing $M \cdot K$ and lying in $M \cdot K + \beta$. This may be seen with the help of Example 12.

THEOREM 175. *If the continuum K contains β, the boundary of the compact continuum M and, for every component D of $M - \beta$, $K \cdot \bar{D}$ is a continuum, then $M \cdot K$ is a continuum.*

Theorem 175 may be established with the help of Theorem 170.

With the help of Example 12, it may be seen that Theorem 175 does not remain a consequence of these axioms if K instead of M is required to be compact.

THEOREM 176. *If M is a continuum, K is a compact continuum intersecting M and, for every component E of $S - M$ that intersects K, T_E is a connected point set intersecting $M \cdot K$ such that, for every component Q of $K \cdot E$, T_E intersect every component of $\bar{Q} \cdot M$, and T is the sum of the point sets T_E for all E's then $M \cdot K + T$ is connected.*

PROOF. Suppose $M \cdot K + T$ is the sum of two mutually separated point sets F_1 and F_2. Every component of T intersects $M \cdot K$ and lies either in F_1 or in F_2. Hence $F_1 \cdot M \cdot K$ and $F_2 \cdot M \cdot K$ exist. Since these point sets are mutually separated and their sum is the closed set $M \cdot K$, therefore they are closed. Hence the compact continuum K contains an irreducible continuum C from $F_1 \cdot M \cdot K$ to $F_2 \cdot M \cdot K$. Let L denote $C - C \cdot (F_1 \cdot M \cdot K + F_2 \cdot M \cdot K)$. By Theorem 47, L is a connected point set such that $\bar{L}$ intersects $F_1 \cdot M \cdot K$ and $F_2 \cdot M \cdot K$. Let W denote the component of $S - M$ which contains L. The point set L lies in a component Z of $K \cdot W$ whose closure intersects M. So T_W intersects every component of $\bar{Z} \cdot M$. There exists a component H of $\bar{Z} \cdot M$ that intersects $\bar{L} \cdot F_1 \cdot M \cdot K$. Since Z is a subset of K, H is a subset of a component of $K \cdot M$. So T_W intersects a component N of $K \cdot M$ that intersects $\bar{L} \cdot F_1 \cdot M \cdot K$. But $F_1 \cdot M \cdot K$ *contains* every component of $K \cdot M$ that it intersects. Hence N is a subset of $F_1 \cdot M \cdot K$. Therefore T_W intersects $F_1 \cdot M \cdot K$. Similarly T_W intersects $F_2 \cdot M \cdot K$. But every connected subset of T is a subset either of F_1 or of F_2. This involves a contradiction.

Theorem 176 does not remain a consequence of these axioms if the supposition that K is compact is replaced by the supposition that M is compact. Consider the example obtained by interchanging M and K in Example 12.

THEOREM 177. *If M is a continuum and K is a compact continuum intersecting M and every component of $S - M$ that intersects K has a connected boundary then $M \cdot K$ plus the boundaries of all components of $S - M$ that intersect K is connected.*

Theorem 177 may be established with the help of Theorem 176. It does not remain a consequence of these axioms if its conclusion is replaced by "$M \cdot K$ is a subset of a component of $M \cdot K + \beta$ where β is the boundary of M". Indeed it does not remain a consequence of them if its hypothesis is strengthened by the addition of the supposition that the boundary of every component of $S - M$ is connected, and its conclusion is replaced by "$M \cdot K$ is a subset of some compact subcontinuum of M". Consider the following example.

EXAMPLE 17. In a Cartesian plane E let T denote a totally disconnected compact and perfect point set lying on OY such that the ordinate of its highest point is -10. Let A and B denote the points $(-10,0)$ and $(10,0)$ respectively. Let H denote the set of all points of the graph of $y = \sin(1/x)$ whose abscissae are neither less than $-1/\pi$ nor greater than $1/\pi$. Let N denote the set of all points of OX that lie between A and $(-1/\pi,0)$ or between B and $(1/\pi,0)$. Let M denote the point set $O + H + N + A + B$ and let K denote the sum of all straight intervals with one endpoint at one of the points A and B and the other one at a point of T. Let S denote $M + K$. Let Σ denote the subspace of E whose points are the points of S. Here β, the boundary of M, consists of two points, A and B.

However, the following theorem holds true.

THEOREM 178. *If β is the boundary of the continuum M and K is a compact continuum intersecting M and, for each component D of $S - M$ that intersects K, the common part of M and the boundary of D is connected, then $M \cdot K$ is a subset of a component of $M \cdot K + \beta$.*

THEOREM 179. *If the compact continuum K intersects the continuum M and, for every component E of $S - M$ that intersects K, $K \cdot \bar{E} \cdot M$ is a subset of some component of $M \cdot K$ then $M \cdot K$ is connected.*

PROOF. For each component E of $S - M$ that intersects K, let T_E denote the component of $M \cdot K$ that contains $K \cdot \bar{E} \cdot M$ and let T denote the sum of all T_E's. By Theorem 176, $M \cdot K + T$ is connected. But it is identical with $M \cdot K$.

It may be seen from the following example that Theorem 179 does not remain a consequence of these axioms if the supposition that K is compact is omitted even though the resulting hypothesis is strengthened by the addition of the supposition that (1) K contains β, the boundary of M, and (2) every component of $S - M$ is bounded by a connected subset of M and so is every component of $M - \beta$ and, furthermore, if D is a component of $M - \beta$, every component of $S - \bar{D}$ is so bounded.

EXAMPLE 18. In a Cartesian plane E, let A and D denote the points (0,2) and (0,1) and, for each n, let A_n and B_n denote the points $(1/n,0)$ and $(-1/n,0)$ respectively and let AA_n and AB_n denote straight line intervals with endpoints as indicated. Let M and K denote $D + AB_1 + AB_2 + \cdots$ and $D + AA_1 + AA_2 + \cdots$ respectively. Let S denote $M + K$. Let Σ denote the subspace of E whose points are the points of S.

Theorem 179 does not remain a consequence of these axioms if M instead of K is required to be compact. To see this, interchange M and K in the description of Example 12.

THEOREM 180. *If β is the boundary of the continuum M, D is a component of $M - \beta$, K is a compact continuum intersecting $\bar{D}$ but not lying wholly in it and Q is a component of $K \cdot (S - \bar{D})$ then every component of $\bar{Q} \cdot \bar{D}$ contains a point of β.*

PROOF. By Theorem 51, some point P of $K \cdot \bar{D}$ is a limit point of Q.

Case I. Suppose P does not belong to D. Then it belongs to $\bar{D} - D$ and therefore to β. But it also belongs to $\bar{Q} \cdot \bar{D}$.

Case II. Suppose P belongs to D. Then $D + Q$ is connected. But it has D as a proper subset. Hence $D + Q$ is not a subset of $M - \beta$. It follows that $D + Q$ contains a boundary point of $M - \beta$ and therefore of β. But D contains no point of β. Hence Q intersects β. But it contains the point P of D. Therefore there exists a connected subset L of $\bar{Q} - (P + \bar{Q} \cdot \beta)$ whose closure contains P and intersects β. The point set $D + L$ is a connected subset of $S - \beta$ containing D and having some point of β on its boundary. If $D + L$ is not a subset of M then, since it intersects both M and $S - M$, it contains a point of β. This involves a contradiction. So $D + L$ is a connected subset of $M - \beta$ having the component D of $M - \beta$ as a subset. Hence L is a subset of D. But it is also a subset of $\bar{Q}$, and $\bar{L}$ intersects β. Hence $\bar{Q} \cdot \bar{D}$ intersects β.

Suppose the component H of $\bar{Q} \cdot \bar{D}$ contains no point of β. Let O denote some point of $\bar{Q} - H$. By Theorem 52 there exists a domain W containing H, but no point of $O + \beta$, such that its boundary γ contains no point of $\beta + \bar{Q} \cdot \bar{D}$. The set $\bar{Q} \cdot W$ contains a connected set T such that $\bar{T}$ intersects γ and contains a point P of $\bar{D}$. Since P does not belong to β, it belongs to D. But $T + P$ is a connected subset of $S - \beta$. Hence it is a subset of D. But T is a subset of $\bar{Q}$. So it is a subset of $\bar{Q} \cdot \bar{D}$. Hence $\bar{Q} \cdot \bar{D}$ intersects γ. This involves a contradiction.

THEOREM 181. *If β is the boundary of the compact continuum M and D is a component of $M - \beta$ and K is a compact continuum intersecting $\bar{D}$ and, for every component E of $S - \bar{D}$ that intersects K, the common part of β and the boundary of E is connected, then $K \cdot \bar{D}$ is a subset of some component of $K \cdot \bar{D} + \beta$.*

PROOF. For every component E of $S - \bar{D}$ that intersects K, let T_E denote the common part of β and the boundary of E. With the help of

Theorem 180 it may be seen that if Q is a component of $K \cdot E$ every component of $\bar{Q} \cdot \bar{D}$ intersects β and therefore T_E. Hence, by Theorem 176, if T is the sum of the continua T_E for all E's, $K \cdot \bar{D} + T$ is connected. But T is a subset of β.

THEOREM 182. *If β is the boundary of the compact continuum M and, for every component D of $M - \beta$ and component E of $S - \bar{D}$, the common part of β and the boundary of E is connected and K is a compact continuum containing β then $M \cdot K$ is a continuum.*

PROOF. If D is a component of $M - \beta$, the continuum K intersects $\bar{D}$ and therefore, by hypothesis and Theorem 181, $K \cdot \bar{D}$ is a subset of a component T_D of $K \cdot \bar{D} + \beta$. Let T denote the sum of all the point sets T_D for all D's. By Theorem 170, $\beta + T$ is connected. But this point set is identical with $M \cdot K$.

Theorem 182 remains true if the supposition that the common part of β and the boundary of E is connected is replaced by the supposition that it is a subset of a component of $M \cdot K$. But it does not remain a consequence of these axioms if "the common part of β and the boundary of E" is replaced by "the common part of $\bar{D}$ and the boundary of E" or by "the boundary of E". Consider the following example.

EXAMPLE 19. In a Euclidean plane E let α denote a definite square and let γ denote a definite square enclosed by α. Let O denote the midpoint of one side of γ and let Q denote a totally disconnected perfect set lying on the opposite side of γ. Let M denote the point set obtained by adding together all straight line intervals with one endpoint at O and the other one at a point of Q. Let K denote the sum of α and γ and the set of all points that lie between them. Let S denote $M + K$ and let Σ denote the subspace of E whose points are the points of S. Here S is compact, β is $O + Q$, K contains β and, for every component D of $M - \beta$, $S - \bar{D}$ is connected and its boundary is $\bar{D}$. But $M \cdot K$ is the uncountable totally disconnected point set β.

Theorem 182 does not remain true in every space satisfying these axioms if the supposition that K is compact is omitted from its hypothesis. To see this, interchange K and M in the description of Example 12. To see that it does not remain true in every such space if the supposition that M is compact is omitted from its hypothesis, consider the following example.

EXAMPLE 20. In a Cartesian plane E, let O denote the origin and, for each positive integer n, let A_n, B_n, C_n, D_n and E_n denote the points $(0,1/n)$, $(1/n,1/n)$, $(1/n,-1/n)$, $(-1/n,-1/n)$ and $(-1/n,1)$ and let F_n denote a point lying midway between B_n and A_{n+1}. If X and Y are two points let XY denote the straight line interval whose extremities are X and Y. Let t_n denote the point set $A_nB_n + B_nC_n + C_nD_n + D_nE_n$. Let M and K denote the point sets $O + t_1 + t_2 + t_3 + \cdots$ and $O + A_1F_1 + F_1A_2 + A_2F_2 + F_2A_3 + \cdots$ respectively. Let S denote the point set $M + K$ and let Σ denote the subspace of E whose points are the points of S. Here β is the point set $O + A_1 + A_2 + \cdots$ and K is a compact continuum containing β. If D is a component of $M - \beta$ then, for some n, D is $t_n - A_n$ and if E is a component of $S - \bar{D}$ then E is $S - t_n$ and its boundary is the set whose only element is A_n, which is a connected subset of $\beta \cdot \bar{D}$. But $M \cdot K$ is not a continuum.

THEOREM 183. *If K is a closed point set lying in the compactly connected continuum M and containing β, the boundary of M, and, for every component D of $M - \beta$, $K \cdot \bar{D}$ is a continuum containing the boundary of D then K is a continuum.*

PROOF. For each component D of $M - \beta$, let T_D denote $K \cdot \bar{D}$. Let T denote the sum of all the point sets T_D. By Theorem 170, $\beta + T$ is connected. But it is identical with K.

That Theorem 183 does not remain true, even in the plane, on the removal, from its hypothesis, of the supposition that M is compactly connected, may be seen from the following example.

EXAMPLE 21. In a Cartesian plane E let A denote the origin, let B denote (0,1) and, for each n, let A_n denote the point $(-1/n,0)$. Let M denote A plus the sum of the straight line intervals BA_1, BA_2, $BA_3, \cdots$, let K denote $A + B$ and let S denote M plus the set of all points whose abscissae are positive and whose ordinates are either 0 or 1.

CHAPTER II

CONSEQUENCES OF AXIOMS 0, 1 AND 2

AXIOM 2. *If P is a point of a region R there exists a nondegenerate connected domain containing P and lying wholly in R.*

THEOREM 1. *If A and B are two points of a connected domain D there exists an arc from A to B that lies wholly in D.*

PROOF. Let H_1 denote the collection of all connected domains g such that $\bar{g}$ is a subset of D and of some region of G_1. By Axiom 1 and Theorem 164 of Chapter I, there exists a simple chain $R_{11}, R_{12}, \cdots, R_{1m_1}$ from A to B such that every link of this chain belongs to H_1. Call this chain C_1. For each i less than m_1, let P_{1i} denote a definite point common to R_{1i} and $R_{1(i+1)}$. Let P_{10} and P_{1m_1} denote A and B respectively. The point P_{10} can be joined to P_{11} by a simple chain C_{11} each link of which is a connected domain lying, with all of its limit points, in R_{11} and in some region of the collection G_2. Let C'_{11} denote the chain remaining after the removal from C_{11} of all of its links following the first one that contains a point of R_{12}. Some point common to R_{12} and the last link of C'_{11} is joined to P_{12} by a simple chain C_{12} each link of which is a connected domain lying, with all of its limit points, in R_{12} and in some region of G_2. Let C'_{12} denote the chain remaining after the removal from C_{12} of every one of its links that either precedes the last one that has a point in common with the last link of C'_{11} or follows the first one that has a point in common with R_{13}. This process may be continued. It follows that there exists a finite set of simple chains $C'_{11}, C'_{12}, \cdots, C'_{1m_1}$ such that (1) for each n $(1 \leqq n \leqq m_1)$ each link of C'_{1n} is a connected domain lying, with all of its limit points, in R_{1n} and in some region of G_2, (2) for each n less than m_1, the last link of C'_{1n} is the only one that intersects $R_{1(n+1)}$, (3) for each n $(1 < n \leqq m_1)$, the first link of C'_{1n} is the only one that intersects the last link of $C'_{1(n-1)}$, (4) the last link of C'_{1m_1} contains B. The links of the chains $C'_{11}, C'_{12}, \cdots, C'_{1m_1}$ form a simple chain C_2 from A to B. The chain C_2 has the important property that each of its links lies wholly in some single link of the chain C_1 and if a link x of C_2 lies in a link y of C_1 then every link that follows x in C_2 lies either in y or in some link that follows y in C_1. Similarly there exists a chain C_3 having a relation to C_2 analogous to the above indicated relation of C_2 to C_1 and such that every link of C_3 is a connected domain whose closure is a subset of some region of G_3. This process may be continued. Thus there exists an infinite sequence of chains $C_1, C_2, C_3, \cdots$ such that (1) if x is a link of C_{n+1}, $\bar{x}$ is a subset of some link of C_n, (2) if a link x of C_{n+1} lies in a link y of C_n then each link that follows x in C_{n+1}

lies either in y or in some link that follows y in C_n, (3) every link of C_n is a connected domain whose closure is a subset of some region of G_n.

Let C denote the common part of the point sets $C_1^*, C_2^*, C_3^*, \cdots$. It will be shown that C is an arc from A to B. That C is a compact continuum follows from Theorems 115 and 116 of Chapter I. It remains to show that every point of C except A and B is a cut point of C.

If X_1 and X_2 are two points of C there exists a number n such that no region of G_n contains both X_1 and X_2. But every link of C_n is a subset of some region of G_n. Thus, for every two points X_1 and X_2 belonging to C, there exists n such that X_1 and X_2 do not lie in the same link of C_n. Furthermore it is clear that if X_1 and X_2 do not lie in the same link of C_n and X_1 lies in a link of C_n that precedes one in which X_2 lies, then if $m \geqq n$ every link of C_m that contains X_1 precedes every link of C_m that contains X_2. The point X_1 will be said to *precede* the point X_2 if there exists an n such that every link of C_n that contains X_1 precedes every one that contains X_2. From the facts observed above it follows that if X_1 and X_2 are points of C then either X_1 precedes X_2 or X_2 precedes X_1, while if X_1 precedes X_2 then X_2 does not precede X_1. Furthermore if X_1 precedes X_2 and X_2 precedes X_3 then X_1 precedes X_3. For there exist numbers n_1 and n_2 such that X_1 and X_2 do not lie in the same link of C_{n_1} and X_2 and X_3 do not lie in the same link of C_{n_2}. Hence every link of $C_{n_1+n_2}$ that contains X_1 precedes every one that contains X_2 and every one that contains X_2 precedes every one that contains X_3. Hence every one that contains X_1 precedes every one that contains X_3. Therefore X_1 precedes X_3.

Suppose now that P is a point of C distinct from A and from B. The point set $C - P$ is the sum of the two sets S_A and S_B where S_A is the set of all points of C that precede P and S_B is the set of all those that follow P. It is clear that S_A contains A and S_B contains B. Suppose that P_A is a point of S_A. Then there exists a number n such that every link of C_n that contains P_A precedes every one that contains P and no link of C_n that contains P_A intersects any link of C_n that contains P. Suppose that some link y of the chain C_n contains P_A and also a point P_B of the set S_B. Since y precedes every link of C_n that contains P and has no point in common with any such link, it follows that P_B precedes P, which is contrary to hypothesis. Hence no link of C_n that contains P_A contains any point of S_B. But there is some link of C_n that contains P_A. Hence P_A is not a limit point of S_B. Similarly no point of S_B is a limit point of S_A. Therefore the point P is a cut point of C.

DEFINITION. If M is a point set, the point set K is said to be (everywhere) *dense in M* if and only if K is a subset of M and every point of M either belongs to K or is a limit point of K.

DEFINITION. The point P is said to be *accessible from* the point set M if and only if for every point X of M there exists an arc XP lying, except for P, wholly in M.

THEOREM 2. *Every component of a domain is a domain and if S is connected, K is a closed proper subset of S, and D is a component of the domain $S - K$, then K contains a boundary point of D. Furthermore, if β is the boundary of a connected domain D, the set of all points of β that are accessible from D is dense in β.*

PROOF. Suppose K is a component of the domain D. Suppose P is a point of K. There exists a connected domain R containing P and lying in D. Since R and K are connected subsets of D having the point P in common, therefore $K + R$ is a connected subset of D. But K is a *maximal* connected subset of D. Therefore R is a subset of K. Hence, by definition, K is a domain.

Suppose S is connected, K is a closed proper subset of S and D is a component of $S - K$. Since S is connected, one of the point sets D and $S - D$ contains a limit point of the other one. But since D is a domain it contains no limit point of $S - D$. Hence $S - D$ contains a point P which is a limit point of D. If P did not belong to K there would exist a connected domain R containing P but no point of K, and $D + R$ would be a connected subset of $S - K$ containing D and a point P not belonging to D, contrary to the hypothesis that D is a maximal connected subset of $S - K$.

Suppose β is the boundary of a connected domain D. Suppose R is a region containing a point P of β. The region R contains a connected domain R' containing P. The domain R' contains a point X of D. By Theorem 1, there exists an arc XP lying in R'. Let Y denote the first point of β on the arc XP. The interval XY of XP is an arc lying, except for Y, wholly in D. If A is any point of D distinct from X then, by Theorem 1, there exists an arc AX lying in D. Let B denote the first point, belonging to the closed point set $(XY)\cdot(AX)$, in the order from A to X on the arc AX. Let AB and BY denote intervals of AX and XY with endpoints as indicated. The point set $AB + BY$ is an arc from A to Y lying, except for Y, wholly in D. Thus every region that contains a point of β contains a point of β that is accessible from D.

That every component of a domain is itself a domain can not be proved on the basis of Axioms 0 and 1. This may be seen from the following example.

EXAMPLE 1. In a Cartesian plane, let O denote the origin, let A denote the point (2,0) and, for each n, let A_n denote the point $(2,1/n)$. Let S denote the point set obtained by adding together the straight line intervals OA, OA_1, OA_2, $OA_3, \cdots$. Interpret *region* to mean the common part of S and the interior of a circle which encloses at least one point of S. Then Axioms 0 and 1 are satisfied. But if D denotes the domain $S - O$, the component of D that contains A is $OA - O$ and this point set is not a domain.

By an example related to Example 2 of Chapter I as the above example is related to Example 1 of that chapter, it may be shown that it cannot be proved from Axioms 0 and 1 that if S is connected then every closed proper subset K of S contains a boundary point of every component of $S - K$. Let S denote the point set M_1 of Example 2 of Chapter I, interpret *region* to mean the common part of S and the interior of a circle which encloses at least one point of S and let K be defined as in the description of that example. The continuum K contains no limit point of the component H of $S - K$. Furthermore in this case $S - H$ is a connected domain and H is its boundary. But no point of H is accessible from $S - H$.

THEOREM 3. *If M is a closed point set and D is a component of $S - M$ then $S - M$ contains no boundary point of D.*

PROOF. Suppose P is a boundary point of D. By Theorem 2, D is a domain. Therefore P does not belong to D. Hence it is a limit point of D. By Theorem 29 of Chapter I, $D + P$ is connected. Thus if P were not a point of M, D would be a proper subset of a connected subset of $S - M$, contrary to the hypothesis that D is a component of $S - M$.

That Theorem 3 cannot be proved on the basis of Axioms 0 and 1, may be seen from Example 1. Let M denote the point O of that example and let D denote the point set $OA - O$. The boundary of D is the interval OA. Therefore it is not a subset of M. But D is a component of $S - M$.

THEOREM 4. *If G is a collection of point sets, M is the sum of all the point sets of G and N is the sum of their boundaries then every limit point of M either belongs to $M + N$ or is a limit point of N.*

PROOF. Suppose there exists a point P which is a limit point of M but which neither belongs to $M + N$ nor is a limit point of N. There exists a connected domain R containing P but no point of N. The domain R contains a point of some set g of the collection G. But P does not belong to g. Thus the point sets $g \cdot R$ and $R - g \cdot R$ exist. But their sum is the connected point set R. Therefore one of them contains a limit point of the other one. Every such point belongs to the boundary of g and therefore to N. But this involves a contradiction.

That Theorem 4 cannot be proved on the basis of Axioms 0 and 1 may be seen from Example 1. Let G denote the collection of point sets $(OA_1 - O)$, $(OA_2 - O)$, $(OA_3 - O), \cdots$. In this case N is the degenerate set whose only element is the point O. The point A is a limit point of M, but it does not belong to $M + N$ and it is not a limit point of N.

THEOREM 5. *If M is a closed point set and G is a collection of components of $S - M$ then there is no boundary point of G^* in $S - M$.*

PROOF. Suppose P is a boundary point of G^*. By Theorem 2, G^* is a domain. Hence P does not belong to G^*. Therefore, by Theorem 4, P either belongs to, or is a limit point of, N, the sum of the boundaries of the domains of the collection G. But, by Theorem 3, N is a subset of M. Hence P belongs to $\bar{M}$ and therefore to M since M is closed.

THEOREM 6. *If T is a closed point set, H and K are mutually exclusive subsets of $S - T$, and $S - T$ contains no arc from a point of H to a point of K, then $S - T$ is the sum of two mutually separated point sets one containing H and the other containing K.*

PROOF. Let G denote the set of all components of $S - T$ that intersect H. If one of them intersected K then, by Theorems 2 and 1, it would contain an arc from a point of H to a point of K, contrary to hypothesis. By hypothesis and Theorem 2, G^* is a domain. Hence it contains no limit point of $(S - T) - G^*$ and, by Theorem 5, T contains every limit point of G^* that does not belong to G^*. Therefore G^* and $(S - T) - G^*$ are mutually separated. But they contain H and K respectively.

That Theorem 6 cannot be proved on the basis of Axioms 0 and 1 may be seen with the aid of either of the two following examples. It is to be noted that, in Example 2(b), T consists of only one point.

EXAMPLE 2(a). In the space of Example 1, let B denote the point $(1,0)$, let C denote the point $(3/2,0)$ and let T denote the point set $O + C$. There is no connected subset of $S - T$ containing both A and B. But T does not separate A from B.

EXAMPLE 2(b). Let O denote the origin, let A denote the point $(1,0)$ and let B denote the point $(-1,0)$. For each n, let B_n denote the point $(-1,-1/n)$, let C_n denote the point whose abscissa is $(n + 1)/n$ and whose ordinate is $-1/n$ and let E_n denote the point whose abscissa is $(n + 1)/n$ and whose ordinate is $1/n$. Let BA denote the straight line interval whose end-points are A and B and, for each n, let α_n denote the arc from B_n to O which is the sum of the straight line intervals B_nC_n, C_nE_n and E_nO. Let S denote the point set $BA + \alpha_1 + \alpha_2 + \alpha_3 + \cdots$. Interpret *region* to mean the common part of S and the interior of a circle that encloses at least one point of S. Here (Figure 4) Axioms 0 and 1 are satisfied and there is no connected subset of $S - O$ containing both A and B. But $S - O$ is not the sum of two mutually separated point sets containing A and B respectively.

THEOREM 7. *If the points A and B are separated from each other by each point set of a monotonic collection G of closed and compact point sets then they are separated from each other by the common part of all the point sets of that collection.*

PROOF. Let K denote the common part of all the point sets of the collection G. Suppose T is a continuum containing A and B. Let H denote the collection of all point sets h such that h is the common part of T and some point set of the collection G. Then H is a monotonic collection of closed and compact point sets. Hence, by Theorem 37 of Chapter I, the point sets of the collection H have a point X in common. The point X belongs both to K and to T. Thus K intersects every continuum that contains both A and B. Therefore, by Theorem 6, K separates A from B.

THEOREM 8. *If the closed and compact point set M separates the point A from the point B, some subset of M is irreducible with respect to being a closed point set that separates A from B.*

Theorem 8 is a consequence of Theorem 39 of Chapter I and Theorem 7. Neither Theorem 7 nor Theorem 8 is a consequence of Axioms 0 and 1.

EXAMPLE 3. In the space of Example 1, let B denote the point $(1,0)$ and, for each n, let B_n denote the point $(1,1/2n)$. Let K_n denote the point set $B + B_n + B_{n+1} + \cdots$. For each n, K_n separates O from A but B, the common part of all the sets $K_1, K_2, K_3, \cdots$ does not do so.

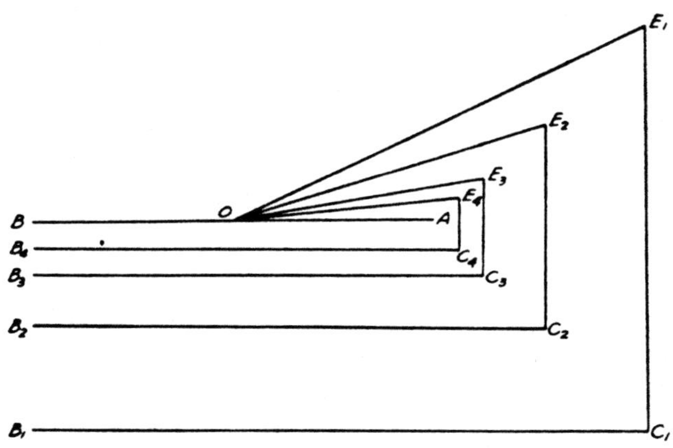

FIGURE 4

DEFINITION. The point set M is said to be *connected im kleinen* at the point O if and only if O belongs to M and, for every domain D with respect to M that contains O there exists a domain with respect to M which contains O and is a subset of a component of D.

DEFINITION. The point set M is said to be *locally connected* at the point O if and only if O belongs to M and every domain with respect to M that contains O contains a connected domain with respect to M containing O. If the point set M is locally connected at every one of its points it is said to be *locally connected* and if it is connected im kleinen at every one of them it is said to be *connected im kleinen*.

The point set M is connected im kleinen at O if and only if it is true that (1) O belongs to M and (2) if R is a region containing O there exists a region R' containing O such that every point of M that lies in R' can be joined to O by a connected subset of M lying in R.

DEFINITION. A connected im kleinen continuum is called a *continuous curve*.

THEOREM 9. *If D is a domain with respect to the point set M and M is connected im kleinen at every point of D then every component of D is a domain with respect to M.*

PROOF. Let T denote a component of the domain D with respect to the point set M. Suppose P is a point of T. By hypothesis, there exists a domain with respect to M which contains P and belongs to some component of D and therefore to T. Hence T is a domain with respect to M.

THEOREM 10. *The point set M is connected im kleinen at the point O if it is locally connected at that point; and if M is connected im kleinen at every point of some domain with respect to M that contains O then M is locally connected at O.*

PROOF. It is clear that if a point set is locally connected at a point it is connected im kleinen there. Suppose that the point set M is connected im kleinen at every point of N, some domain with respect to M that contains O. Suppose D is a domain with respect to M containing O. Let T denote the component of $D \cdot N$ that contains O. By Theorem 9, T is a domain with respect to M. But T is connected. Hence M is locally connected at O.

THEOREM 11. *If the continuum M is locally compact at the point O but is not connected im kleinen there then if R is a region containing O there exist a domain D containing O and lying in R, an infinite sequence of points $O_1, O_2, O_3, \cdots$ converging to O, and an infinite sequence of mutually exclusive continua $M_1, M_2, \cdots$ such that* (1) *$M \cdot \bar{D}$ is compact,* (2) *for each n, M_n is a component of $M \cdot \bar{D}$ containing O_n and a point of the boundary of D,* (3) *the sequence $M_1, M_2, \cdots$ converges to a subcontinuum of M which contains O but no point of the sequence $O_1, O_2, O_3, \cdots$.*

PROOF. There exists a domain D containing O such that (1) $M \cdot \bar{D}$ is a compact proper subset of M, and (2) if Q is a subdomain of D containing O then there exists a point of Q which can not be joined to O by a connected subset of $M \cdot \bar{D}$. Let $R_1, R_2, \cdots$ denote a sequence of regions closing down on O such that R_1 is a subset of D. There exists in R_1 a point O_1 which belongs to M but not to the component of $M \cdot \bar{D}$ that contains O. Let N_1 denote the component of $M \cdot \bar{D}$ that contains O_1. By Theorems 50 and 29 of Chapter I, N_1 contains a point of the boundary of D. There exists a number n_1 such that R_{n_1} contains no point of the continuum N_1. There exists, in R_{n_1}, a point O_2 belonging to M but not to the component of $M \cdot \bar{D}$ that contains O. Let N_2 denote the component of $M \cdot \bar{D}$ that contains O_2. The point set N_2 contains a point of the boundary of D. There exists a number n_2 such that R_{n_2} contains no point of N_2. The point set $M \cdot R_{n_2}$ contains a point O_3 not lying in the component of $M \cdot \bar{D}$ that contains O. Let N_3 denote the component of $M \cdot \bar{D}$ that contains O_3. This process may be continued. Thus there exist a sequence of points $O_1, O_2, O_3, \cdots$ converging to O, and a sequence of mutually exclusive continua $N_1, N_2, N_3, \cdots$ such that, for each n, N_n is a component of $M \cdot \bar{D}$ containing both O_n and a point of the boundary of D. By Theorems 58 and 59 of Chapter I, there is a subsequence of $N_1, N_2, N_3, \cdots$ which converges to some continuum. Clearly this sequence fulfills all the requirements of Theorem 11.

THEOREM 12. *Suppose M is a continuous curve or, indeed, any connected and connected im kleinen inner limiting set. If Σ is a space whose points are the points of M and, in the space Σ, the term "region" is interpreted to mean a connected domain with respect to M then, with respect to this interpretation of "point" and "region", Axioms* 0, 1 *and* 2 *are satisfied and "limit point" is invariant under this change.*

PROOF. By hypothesis M is the common part of the domains of a sequence $D_1, D_2, D_3, \cdots$ such that, for each n, D_n contains D_{n+1}. For each n, let G'_n denote the collection of all point sets h such that h is a component of the common part of M and some region g (in the old sense, in the original space) such that $\bar{g}$ is a subset of D_n and of some region of the collection G_n. Since M is connected im kleinen therefore, by Theorem 8, every point set of the collection G'_n is a connected domain with respect to M and therefore a "region" *in the new sense.* It is easy to see that the sequence $G'_1, G'_2, G'_3, \cdots$ satisfies all the conditions required of $G_1, G_2, G_3, \cdots$ *and that furthermore* the point P of M is a limit point (in the original sense) of the subset K of M if and only if every "region" *in the new sense* that contains P contains a point of K distinct from P. In other words, in the subspace of the original space whose points are the points of M, a point is a limit point in the new sense (that is in accordance with the new interpretation of region) of a point set if and only if it is a limit point, in the original sense, of that point set.

As an immediate consequence of Theorems 1 and 12, we have the following result.

THEOREM 13. *Every two points of a connected and connected im kleinen inner limiting set are the extremities of an arc lying in that set. In particular, every two points of a connected domain D with respect to a continuous curve are the extremities of an arc lying in D.*

THEOREM 14. *If D is a point set, M is a point set, α is a sequence of distinct components of $M \cdot D$ and O is a point belonging to the limiting set of α and lying in some domain with respect to M which is a subset of D then M is not connected im kleinen at O.*

PROOF. Let H denote a domain with respect to M lying in D and containing O and let N denote the component of H that contains O. Suppose M is connected im kleinen at O. Then there exists a region R containing O such that $R \cdot M$ is a subset of N and therefore of T, the component of $M \cdot D$ that contains O. But there exist two point sets x and y belonging to the sequence α and containing points of $R \cdot M$. Since x and y are different components of $M \cdot D$, $x \cdot R \cdot M$ and $y \cdot R \cdot M$ lie in different components of $M \cdot D$. This involves a contradiction.

THEOREM 15. *Suppose* (1) *the subset* M *of the connected point set* T *is closed with respect to* T, (2) A, B *and* O *are three points of* M *such that neither of the points* A *and* B *separates the other one from* O *in* T, (3) *if* H *is the set of all points* X *of* T *such that* $O + X$ *is separated from* $A + B$ *in* T *by some point of* M *and* P *is a point of* $T - H$ *which is a limit point of* H *then* $H + P$ *is connected im kleinen at* P. *Then there exists a point of* M *which separates* O *from* $A + B$ *in* T *but which is not itself separated from* $A + B$ *in* T *by any point of* M.

PROOF. Let K denote the set of all points X of M such that X separates O from $A + B$ in T. For every point P of K, let K_P denote some definite sect of T from P that contains O but neither A nor B and let K'_P denote the sect $T - (K_P + P)$.

If X and Y are two points of K and K_X contains K_Y then K_Y does not contain K_X. For since $X + K_X$ is closed with respect to T, X is distinct from Y, and Y is a limit point of K_Y, therefore Y is in K_X. But Y is not in K_Y. Hence K_Y does not contain K_X.

Suppose that X and Y are two points of K. If X belongs to K_Y then the connected point set $Y + K'_Y$ contains the point A of K'_X but not the point X and therefore $Y + K'_Y$ is a subset of K'_X and consequently K_Y contains K_X. If X belongs to K'_Y then the connected point set $Y + K_Y$ contains the point O of K_X but not the point X and therefore K_X contains K_Y. Thus either K_X contains K_Y or K_Y contains K_X.

If, for two points X and Y of the set K, the set K_X contains the sect K_Y then X will be said to *precede* Y in K. In view of the results established above it is clear that if X and Y are two points of K then (1) if X precedes Y, Y does not precede X, (2) if X precedes Y and Y precedes Z then X precedes Z, (3) either X precedes Y or Y precedes X.

Let H denote the set of all points Y such that Y belongs to K_X for some point X of K and suppose K is a subset of H.

If X and Y are two points of K and X precedes Y then by the *slice* XY, or the *slice* YX, of H will be meant the point set $(X + K_X) - K_Y$. The slice XY of H is connected and it is closed with respect to T. The points X and Y will be called its *endpoints*.

Suppose that $T - H$ contains two limit points of H. Then there exist an infinite sequence $t_1, t_2, t_3, \cdots$ of mutually exclusive slices of H and two points C and F belonging to $T - H$ such that (1) if R is a region containing either C or F then there exists a number δ such that, for every n greater than δ, t_n contains a point of R, (2) C is not a limit point of the sum of all the endpoints of the terms of the sequence $t_1, t_2, t_3, \cdots$. There exists a region R containing C but no endpoint of any slice belonging to the sequence t_1, t_2,

$t_3, \cdots$. There exists a domain L with respect to $H + C$ which contains C and which is a subset of W, the component of $(H + C) \cdot R$ which contains C. There exists a number m such that each of the point sets t_m and t_{m+1} intersects L. Since L is a connected subset of $H + C$ intersecting both t_m and t_{m+1} it follows that it contains an endpoint of one of them. Thus the supposition that $T - H$ contains two limit points of H leads to a contradiction.

Suppose X is a point of H. There exists a point E of K such that K_E contains X. But $T - H$ is a subset of $T - K_E$ and no point of K_E is a limit point of $T - K_E$. Hence X is not a limit point of $T - H$. But T is connected. It follows that $T - H$ contains only one limit point of H. Call this point Q. It separates O from $A + B$ in T. Suppose it does not belong to M. Then there exists a region of R_1 containing Q but no point of M. There exists a region R_2 containing Q such that $R_2 \cdot (Q + H)$ is a subset of Z, the component of $R_1 \cdot (Q + H)$ that contains Q. Suppose Y is a point of $H \cdot R_1$. Some point of K separates the point Y, in T, from $A + B$ and therefore from Q since the connected point set $Q + (T - H)$ contains $A + B$ and Q but not Y. But Z is a connected subset of T containing Y and Q and no point of K. This involves a contradiction. Hence Q belongs to M and therefore to K. No point of M separates Q from $A + B$ in T.

THEOREM 16. *Suppose* (1) *O and A are two points of the connected point set T, and* (2) *if L is a point set each point of which separates O from A in T and either H is the set of all points X such that some point of L separates $X + O$ from A in T or H is the set of all points X such that some point of L separates $X + A$ from O in T, then if P is a point of $T - H$ which is a limit point of H, $H + P$ is locally connected at P. Then the set of all points that separate O from A in T is compact.*

PROOF. Let K denote the set of all points X such that X separates O from A in T. For every point P of K let K_P denote some definite sect of T from P that contains O but not A and let K'_P denote the sect $T - (K_P + P)$. Let the statement that X precedes Y in K be defined, in terms of K_X and K_Y, in exactly the same way as in the above proof of Theorem 15. Suppose now that K is not compact. Then there exists a sequence α of distinct points $P_1, P_2, P_3, \cdots$, all belonging to K, such that (1) either, for every n, P_n precedes P_{n+1} or, for every n, P_{n+1} precedes P_n, (2) the point set $P_1 + P_2 + P_3 + \cdots$ has no limit point. Suppose that, for every n, P_{n+1} precedes P_n. Let H denote the set of all points Y such that Y belongs to K_X for some point X of the sequence α. For every two points X and Y of K such that Y precedes X let XY denote the connected point set $K_Y \cdot K'_X + Y + X$. It is easy to see that there exist a point B of $T - H$, a sequence of points $B_1, B_2, B_3, \cdots$ and an ascending sequence of positive integers $n_1, n_2, n_3, \cdots$ such that (1) B is the sequential limit point of the sequence $B_1, B_2, B_3, \cdots$, (2) for each m, B_m belongs to $P_{n_m} P_{n_{m+1}}$. There exists a region R_1 containing B but no point of the sequence $P_1, P_2, P_3, \cdots$. There exists a region R_2 containing B such that $R_2 \cdot (B + H)$ is a subset of the component of $R_1 \cdot (B + H)$ that contains B. But there exists a number j such that B_j and B_{j+1} belong to $R_2 \cdot (B + H)$ and every connected subset of T that contains B_j and B_{j+1} contains the point $P_{n_{j+1}}$ which does not belong to R_1. This involves a contradiction.

THEOREM 17. *If M is a locally compact continuum containing two points A and B such that A and B are separated in M by every point of $M - (A + B)$ then M is an arc from A to B.*

PROOF. By hypothesis and Theorems 11 and 10 of the present chapter and Theorems 106, 53 and 52 of Chapter I, M is a continuous curve. Hence, by Theorem 13, it contains an arc AB, with endpoints at A and B. Since every point of $M - (A + B)$ separates A from B in M, $M - (A + B)$ is a subset of AB. Hence M is AB.

THEOREM 18. *If A and B are two points of a locally compact continuum M and* (1) *$M - A$ and $M - B$ are connected but* (2) *if P is any point of M distinct from A and B, $M - P$ is the sum of two mutually separated connected point sets then M is an arc from A to B.*

PROOF. Suppose that M is not an arc from A to B. Then, by Theorem 17, there exists at least one point X of M which is separated from $A + B$ by some point of M. Let H denote the set of all such points X. By Theorems 106 and 53 of Chapter I and Theorems 11 and 15 of the present chapter, for each point X of H there exists a point O_X which separates X from $A + B$ in M but which is not itself separated from $A + B$ in M by any one point. Let N_1 denote the set of all points O_X for all points X of H. By Theorem 88 of Chapter I, from any given point of M there are only two sects of M. For each point X of N_1, let M_X denote the sect of M from X that contains neither A nor B. Let N_2 denote the point set $M - (H + N_1)$. Let N denote the set $N_1 + N_2$. One of the sets N_2 and $N_1 + H$ contains a limit point of the other one. Suppose first that N_2 contains a point O which is a limit point of $N_1 + H$. Then there exist a sequence of distinct points $X_1, X_2, X_3, \cdots$, belonging to N_1 and a sequence of points $P_1, P_2, P_3, \cdots$ such that O is the sequential limit point of the sequence $P_1, P_2, P_3, \cdots$ and such that either (1) for every n, P_n coincides with X_n or (2) for every n, P_n belongs to M_{X_n} and O is not a limit point of N_1. There exist two regions R_1 and R_2 containing O such that $\bar{R}_2$ is a subset of R_1, $M \cdot \bar{R}_1$ is compact, and $\bar{R}_1$ contains no point of N_1 unless O is a limit point of N_1. There exists a number n such that the points $P_n, P_{n+1}, P_{n+2}, \cdots$ are all in R_2. Now, for each point X of N_1, the continuum $M_X + X$ is noncompact. For if it were compact then, by Theorem 93 of Chapter I, $M_X + X$ would have a non-cut point distinct from X and every such point would be a non-cut point of M, contrary to hypothesis. It follows that, for each m, the continuum $M_{X_{n+m}} + X_{n+m}$ contains both a point of R_2 and a point of $S - \bar{R}_1$ and therefore, by

Theorems 50 and 29 of Chapter I, it contains a continuum k_m which contains at least one boundary point of R_1 and at least one boundary point of R_2 and lies wholly in the point set $\bar{R}_1 - R_2$. If, for each m, t_m denotes the component of $M \cdot (\bar{R}_1 - R_2)$ that contains k_m, the continua $t_1, t_2, t_3, \cdots$ are mutually exclusive. It follows, by Theorems 58, 59, 106 and 22 of Chapter I, that M has uncountably many non-cut points, contrary to hypothesis. Thus a contradiction is obtained in case N_2 contains a limit point of $H + N_1$.

Suppose secondly that $H + N_1$ contains a point O which is a limit point of N_2. Clearly O must belong to N_1. If X is a point of N_2 distinct from A and from B, and Y is a point belonging either to N_1 or to N_2, X will be said to precede Y or to follow Y according as Y belongs to the sect XB, or to the sect XA, of M. It may be easily proved that there exists a sequence of distinct points $X_1, X_2, X_3, \cdots$ all lying in N_2 and distinct from A and from B, such that O is the sequential limit point of this sequence and such that either (1) for every n, X_n precedes O and X_{n+1} or (2) for every n, X_n follows O and X_{n+1}. Suppose that, for every n, X_n precedes O and X_{n+1}. Each of the sections $X_1X_2, X_2X_3, X_3X_4, \cdots$ of M is a continuum, no two of them have in common any point other than a common endpoint and no one of them contains the point O. That O is the only limit point of the point set $X_1X_2 + X_2X_3 + \cdots$ that does not belong to that point set may be proved by an argument similar to that used, in the proof of Theorem 15, to show that $M - H$ does not contain two limit points of H. Now sect $OA = M - (M_O + O) = (\text{sect } X_1A + X_1X_2 + X_2X_3 + \cdots) + (B + V)$, where X_1X_2, $X_2X_3, \cdots$ denote sections of M with endpoints as indicated and V denotes the point set $M - (B + M_O + X_1A + X_1X_2 + X_2X_3 + \cdots)$. Neither of the two sets $(\text{sect } X_1A + X_1X_2 + X_2X_3 + \cdots)$ and $B + V$ contains a limit point of the other one. Hence the sect OA is not connected, contrary to hypothesis. Hence it is not true that, for every n, X_n precedes O and X_{n+1}. In an entirely similar way it may be proved that X_n does not follow O and X_{n+1} for every n. Thus a contradiction is obtained in case $H + N_1$ contains a limit point of N_2.

It follows that the set H does not exist. Therefore M is an arc from A to B.

Definition. An *open curve* is a locally compact continuum M such that if P is a point of M, $M - P$ is the sum of two mutually separated connected point sets.

Theorem 18′. *If A and B are two points of an open curve M, the section AB of M is an arc from A to B.*

Proof. Let K denote the section AB of M. It is easy to see that K is a continuum. Suppose O is any point of K distinct from A and from B.

Case I. Suppose that one of the sects of M from O contains both A and B. Let OX denote the other one. Let N denote the common part of the sect

OA and the section K. The point set N contains both A and B. Suppose it is not connected. Then it is the sum of two mutually separated point sets N_A and N_B where N_A contains A. Let AC denote the sect of M from A that does not contain B, and let BD denote the sect of M from B that does not contain C. If N_A contains B then the sect OA is the sum of two mutually separated point sets $AC + BD + N_A$ and N_B. If N_A does not contain B then the section OA is the sum of two mutually separated sets $AC + N_A$ and $BD + N_B$. In either case the sect OA is not connected. This involves a contradiction. Hence N is connected. But the sect OX is connected and $K - O$ is the sum of N and the sect OX. Thus $K - O$ is the sum of two mutually separated connected point sets.

Case II. Suppose the sects OA and OB have no point in common. Then the point sets $K \cdot (OA)$ and $K \cdot (OB)$ are mutually separated. It is easy to see that these sets are connected and that their sum is $K - O$.

Thus, both in Case 1 and in Case 2, $K - O$ is the sum of two mutually separated connected point sets. Furthermore $K - A$ and $K - B$ are both connected. Hence, by Theorem 18, K is an arc from A to B.

DEFINITION. If O is a point, a *ray* from O is a locally compact continuum M containing O such that (1) $M - O$ is connected and (2) if P is any point of M distinct from O then $M - P$ is the sum of two mutually separated connected point sets.

THEOREM 19. *If O is a point of an open curve M and A is a point of M distinct from O then the point set composed of the point O and the sect OA of M is a ray from O and there is no other ray from O lying in M and containing A.*

PROOF. Let N denote the point set composed of the point O and the sect OA of M. Clearly N is a continuum. The point set $N - O$ is a sect of M. Hence it is connected. Suppose P is a point of N distinct from O. Let K denote the section OP of M and let PX denote the sect of M from P that does not contain O. The point sets $K - P$ and PX are both connected. But their sum is $N - P$. Hence N is a ray from O.

Suppose there exists a ray L from O lying in M, containing A, and distinct from N. Let OB denote the sect of M from O distinct from OA. The point set $L - O$ is a connected subset of the sum of the two mutually separated point sets OA and OB and it contains the point A of OA. Hence $L - O$ is a subset of OA. But L is distinct from $O + OA$. Hence there exists a point Y belonging to OA but not to L. Of the two sects of M from Y, let YA denote the one that contains A and let YZ denote the other one. Since the connected point set L is a subset of the sum of the two mutually separated point sets YA and YZ, it is a subset of one of them. But it contains the point A of YA. Hence it is a subset of YA. Hence it is a subset of T, the section of M whose endpoints are A and Y. But, by Theorem 18′, T is an arc. Therefore L is an arc from A to some point C. The point set $L - C$ is connected, contrary to the definition of a ray.

Definition. If O and A are two points of the locally compact open curve M, the ray from O lying in M and containing A will be called the ray OA of M.

Theorem 20. *Every nondegenerate compact subcontinuum of an open curve is an arc.*

Proof. Suppose K is a nondegenerate compact subcontinuum of the open curve M. By Theorem 93 of Chapter I, K has two non-cut points A and B. Let T denote the section of M whose endpoints are A and B. By Theorem 18′, T is an arc from A to B. Of the two sects of M from A, let AB denote the one that contains B and let AC denote the other one. The sect AC contains no point of K. For if it did then $K - A$ would be the sum of the two mutually separated point sets $K \cdot (AC)$ and $K \cdot (AB)$. Hence K is a subset of the ray AB of M. Similarly it is a subset of the arc T. But it contains A and B, the endpoints of T. Hence K is identical with T.

Definition. If A and B are two points of the open curve M, the point X is said to precede the point Y, in the order from A to B, on M if X is distinct from Y and there exists a ray of M which is a subset both of the ray AB of M and of the ray XY of M. The point Y is said to follow the point X, in the order from A to B, on M if X precedes Y in that order on M.

It may be shown that (1) if X and Y are two points of the open curve M then either X precedes Y or Y precedes X in the order from A to B on M, (2) if X precedes Y in the order from A to B on M, then Y does not precede X in that order, (3) if X precedes Y and Y precedes Z in the order from A to B on M, then X precedes Z in that order.

Theorem 21. *Every arc, simple closed curve, open curve or ray is a continuous curve.*

Proof. Suppose M is either an arc, a simple closed curve, an open curve or a ray. Suppose P is a point of M and D is a domain containing P. There exists an arc AB lying in $M \cdot D$ and containing P and having, as one of its endpoints, a point A which is or is not identical with P according as P is or is not an endpoint or a starting point of M. There exists a region R containing P such that $M \cdot R$ is a subset of $AB - (A + B) + P$. But $AB - (A + B) + P$ is a connected domain with respect to $M \cdot D$. Hence M is connected im kleinen at P.

Theorem 22. *If no compact subcontinuum of the locally compact continuum M has more than two boundary points with respect to M then M is an arc, a simple closed curve, a simple open curve or a ray.*

Proof. If H is a continuum of condensation of the continuum N then every point of H is a boundary point of H with respect to N. It follows that M has no continuum of condensation. Hence, by Theorem 11, M is connected im kleinen. Therefore, by Theorem 13, every two points of M are the endpoints of an arc lying in M. Let G denote the set of all arcs that lie in M. If AB is an arc of the set G, no point of AB distinct from A and B is a limit point of $M - AB$. For if there should exist such a point P there would exist two points A' and B' in the order $AA'PB'B$ on AB, and the interval $A'B'$ of AB would be a subcontinuum of M with three boundary points with respect to M, these points being A', P and B'.

If g_1 and g_2 are two arcs of the collection G with a common point, the point set $g_1 + g_2$ is either an arc or a simple closed curve. For, otherwise, one of the arcs g_1 and g_2 (call it g) would contain a point which is not an endpoint of g but which is a limit point of $M - g$.

Let A and B denote two definite points of M and let AB denote a definite arc of the set G having A and B as endpoints. Let K_C ($C = A,B$) denote the set of all points X of M such that C and X are the extremities of an arc belonging to the set G and containing no point except C in common with AB. An arc from C to X satisfying these conditions will be called the arc CX. If neither K_A nor K_B exists then M is the arc AB. Suppose that K_A exists. If X_1 and X_2 are points of K_A, X_1 is said to *precede* X_2 (and X_2 is said to *follow* X_1) if and only if AX_1 is a subset of AX_2. Of any two points of K_A one of them precedes the other one. The set $K_A + A + B$ is closed. For suppose there exist two points O and O' distinct from A and belonging to $\bar{K}_A - K_A$. Then there exist two sequences of points $A_1, A_2, A_3, \cdots$ and $A_1', A_2', A_3', \cdots$ belonging to K_A such that, for every n, A_n precedes A_n' and A_n' precedes A_{n+1} and such that O is the sequential limit point of $A_1, A_2, A_3, \cdots$ while O' is the sequential limit point of $A_1', A_2', A_3', \cdots$. It follows, by an argument similar to that used in a similar connection in the proof of Theorem 15, that there exists a continuum K' containing O such that if Y is a point of K' then every region that contains Y intersects infinitely many of the arcs $A_1A_2, A_2A_3, A_3A_4, \cdots$ where, for every n, A_nA_{n+1} denotes the interval of the arc AA_{n+1} whose endpoints are A_n and A_{n+1}. For no value of n does the arc A_nA_{n+1} contain a limit point of the sum of the arcs of the sequence $A_{n+2}A_{n+3}, A_{n+3}A_{n+4}, \cdots$. Hence K' does not intersect any of the arcs $A_1A_2, A_2A_3, \cdots$. It follows that K' is a continuum of condensation of M. Thus the supposition that there exist two limit points of $K_A + A$ that do not belong to it leads to a contradiction. It follows that either $K_A + A + B$ is closed or there exists only one point P, distinct from B, belonging to $\bar{K}_A - K$. In the latter case, P is the sequential limit point of a sequence of points $P_1, P_2, P_3, \cdots$ belonging to K_A such that if, for each n, P_nP_{n+1} denotes the interval of AP_{n+1} whose endpoints are P_n and P_{n+1} then, for every n, P_n precedes P_{n+1} and the point set $P + AP_1 + P_1P_2 + P_2P_3 + \cdots$ is an arc from A to P. It follows that P either coincides with B or belongs to K_A. But this is contrary to supposition. Hence the set $K_A + A + B$ is closed.

There are several cases to be considered.

Case I. Suppose that K_A is not compact and that K_B does not exist. The set $K_A + A$ contains a countably infinite set of distinct points $A_1, A_2,$

$A_3, \cdots$ with no limit point. Let A'_1 denote A_1. Let A'_2 denote the first point of the sequence $A_1, A_2, A_3, \cdots$ that follows A_1, A'_3 the first one that follows A'_2, etc., in the set K_A. The sequence $A'_1, A'_2, A'_3, \cdots$ is an infinite subsequence of $A_1, A_2, A_3, \cdots$ and, for every n, A'_n precedes A'_{n+1} in K_A. For each n, let $A'_nA'_{n+1}$ and AA'_n denote arcs of the set G with endpoints as indicated. Suppose O is a point of M not belonging to the point set $AB + AA'_1 + A'_1A'_2 + A'_2A'_3 + \cdots$. There exists an arc AO lying in M. Since K_B does not exist, the arc AO does not contain B. It follows that O belongs to K_A. Hence, for each n, one of the arcs AA'_n and AO contains the other one. But O does not belong to AA'_n. Hence, for every n, AA'_n is a subset of AO. Thus the noncompact point set $A'_1 + A'_2 + A'_3 + \cdots$ is a subset of the arc AO. This involves a contradiction. Hence $M = AB + AA'_1 + A'_1A'_2 + \cdots$. But this point set is a ray.

Case II. Suppose that K_A is noncompact and that K_B exists and is compact. Then K_A and K_B are mutually exclusive and there does not exist more than one point of K_B having the property that it is not followed by any other point of K_B. If P is any point of K_B not having this property then $(K_B + B) - P$ is not connected. Hence the point set $K_B + B$ is an arc. That $(K_B + B) + K_A$ is a ray follows by an argument similar to that employed in Case I.

Case III. Suppose K_A and K_B are both noncompact. In this case the closed sets K_A and K_B are mutually exclusive and the set $K_B + AB + K_A$ is an open curve.

Case IV. Suppose that K_A is compact, that K_B either is compact or does not exist and that K_A and K_B are mutually exclusive. In this case M is clearly an arc.

Case V. Suppose that K_A is compact and that K_B and K_A have a point in common. In this case $K_A + A + B$ is an arc from A to B and M is a simple closed curve.

If the term simple continuous curve is applied only to point sets which are either arcs, *simple closed curves*, open curves or rays then it is clear, in view of the above results, that these could have been defined as follows:

DEFINITION. A *simple continuous curve* is a nondegenerate locally compact continuum M such that no subcontinuum of M has more than two boundary points with respect to M.

DEFINITION. A *simple closed curve* is a nondegenerate locally compact continuum M such that every nondegenerate proper subcontinuum of M has only two boundary points with respect to M.

DEFINITION. An *open curve* is a nondegenerate locally compact continuum M such that every nondegenerate compact subcontinuum of M has only two boundary points with respect to M.

DEFINITION. An *arc* is a nondegenerate compact continuous curve M containing a point A such that every proper subcontinuum of M that contains A has only one boundary point with respect to M.

DEFINITION. A *ray* is a nondegenerate locally compact continuous curve M containing a point A such that every compact subcontinuum of M that contains A has only one boundary point with respect to M.

THEOREM 23. *If β is the boundary of the closed point set M and D is a component of $M - \beta$ then D is a domain and, if D has a boundary, its boundary is a subset of β.*

PROOF. It is clear that $M - \beta$ is a domain. Hence, by Theorem 2, D is a domain. Suppose P is a boundary point of D. The point P belongs to $M - D$ and $D + P$ is connected. Therefore if P does not belong to β then D is a proper subset of the connected subset $D + P$ of $M - \beta$, contrary to the supposition that D is a maximal connected subset of $M - \beta$.

THEOREM 24. *If β is the boundary of the continuum M and, for every component D of $M - \beta$, T_D is a connected point set lying in M and containing the boundary of D and T is the sum of all the point sets T_D for all such components D, then $\beta + T$ is connected.*

PROOF. Suppose $\beta + T$ is the sum of two mutually separated point sets H_1 and H_2. Suppose D is a component of $M - \beta$. Since the connected point set T_D is a subset of $H_1 + H_2$ therefore, by Theorem 28 of Chapter I, T_D is a subset either of H_1 or of H_2. If there exists a component D of $M - \beta$ such that T_D is a subset of H_i, let H_i' denote H_i plus the sum of all such point sets D. Otherwise let H_i' denote H_i. By Theorem 23, the components of $M - \beta$ are all domains and, by Theorem 4, every limit point of their sum belongs either to one of them or to the closure of the sum of their boundaries. But, since they are domains, no one of them contains a limit point of its complement and, by Theorem 23, the boundary of every one of them is a subset of β. It follows that if the point P of H_2' is a limit point of H_1' then P belongs to H_2 and is a limit point of H_1. This is impossible. Similarly H_1' contains no limit point of H_2'. But $H_1' + H_2'$ is the connected point set M. This involves a contradiction.

THEOREM 25. *If β is the boundary of the continuum M and every component of $M - \beta$ has a connected boundary, then β is connected.*

Theorem 25 is a consequence of Theorems 23 and 24.

THEOREM 26. *If β, the boundary of the continuum M, is the sum of two mutually exclusive closed point sets β_1 and β_2 then there exists a component D of $M - \beta$ such that $\overline{D}$ intersects both β_1 and β_2.*

PROOF. The closure of every component of $M - \beta$ intersects β and therefore either β_1 or β_2. Let M_i denote the set of all points that belong either to β_i or to some component of $M - \beta$ whose closure intersects β_i. Since their sum is the connected point set M, M_1 and M_2 are not mutually separated. Suppose they are mutually exclusive. Then one of them contains a limit point of the other one. But every component of $M - \beta$ is a domain and no domain contains a limit point of its complement. Hence either β_1 contains a limit point of M_2 or β_2 contains a limit point of M_1. Suppose β_1 contains a point P which is a limit point of M_2. Let Q denote the collection of all components of $M - \beta$ whose closures intersect β_2. Since it is not a limit point of β_2, P must be a limit point of Q^*. Hence, by Theorem 4, P belongs to the closure of the sum of the boundaries of the domains of the collection Q. Since the boundary of every domain of Q lies in the closed point set β_2, it follows that P belongs to β_2 and therefore that β_1 intersects β_2. The same result is obtained in case β_2 contains a limit point of M_1. Thus the supposition that M_1 and M_2 are mutually exclusive leads to a contradiction. It follows that there is a component of $M - \beta$ whose closure intersects β_1 and β_2.

THEOREM 27. *If M is a closed point set such that every component of $S - M$ has a connected boundary and K is a continuum containing the boundary of M then $M \cdot K$ is a continuum.*

PROOF. Suppose $M \cdot K$ is the sum of two mutually separated point sets H_1 and H_2. For each point P of K not belonging to M, let β_P denote the boundary of the component of $S - M$ that contains P. For each such P, β_P is connected and therefore is a subset either of H_1 or of H_2. Let H_i' denote a point set such that the point P belongs to it if and only if either P belongs to H_i or P belongs to $K \cdot (S - M)$ and β_P is a subset of H_i. The point sets H_1' and H_2' are mutually exclusive. Suppose the point P of H_1' is a limit point of H_2'. Since no complementary domain of M whose boundary belongs to H_1 contains a point of H_2' therefore P belongs to H_1. Hence it is not a limit point of H_2. Therefore it is a limit point of the sum of all complementary domains of M whose boundaries are subsets of H_2. But P neither belongs to one of these domains nor is a limit point of the sum of their boundaries. Thus the supposition that H_1' contains a limit point of H_2' leads to a contradiction. Similarly H_2' contains no limit point of H_1'. So H_1' and H_2' are mutually separated. But their sum is the continuum K. Thus the supposition that $M \cdot K$ is not connected has led to a contradiction.

THEOREM 28. *If D is a domain and every component of $S - \bar{D}$ has a connected boundary and M is a continuum containing β, the boundary of D, then $M \cdot (D + \beta)$ is a continuum.*

PROOF. Suppose $M \cdot (D + \beta)$ is the sum of two mutually separated point sets K_1 and K_2. By Theorem 28 of Chapter I, each boundary of a complementary domain of $\bar{D}$ is a subset either of K_1 or of K_2. For each i ($i = 1,2$) let K_i' denote a point set such that the point P belongs to it if and only if P belongs either to K_i or to the common part of M and some complementary domain of $\bar{D}$ whose boundary is a subset of K_i. It may be seen, with the help of Theorem 3, that K_1' and K_2' are mutually separated. Since their sum is the continuum M, this involves a contradiction.

THEOREM 29 (HARLAN CROSS MILLER). *Every closed and compact subset of a continuum is a subset of a compact continuum.*

PROOF. Suppose the closed and compact point set K is a subset of the continuum M. For each n, let G'_n denote a collection such that g belongs to it if and only if g is a connected domain which is a subset of some region of G_n. There exist a finite subcollection H'_1 of G'_1 covering K and a coherent subcollection T_1 of G'_1 covering M. By Theorem 165 of Chapter I, for each pair of domains of H'_1 there exists a simple chain of domains of T_1 intersecting both domains of that pair. It follows that there exists a finite coherent subcollection H_1 of $H'_1 + T_1$ covering K. There exists an infinite sequence $H_1, H_2, H_3, \cdots$ such that H_1 is as indicated and, for each n, H_{n+1} is a finite coherent subcollection of G'_{n+1} covering K such that the closure of each domain of H_{n+1} is a subset of some domain of H_n. By Theorem 115 and 116 of Chapter I, the common part of the point sets $H^*_1, H^*_2, H^*_3, \cdots$ is a compact continuum containing K.

THEOREM 30. *If the boundary of the connected domain D is a compact point set β and every component of the complement of $\bar{D}$ has a connected boundary then there exists a compact continuum containing β and lying in $D + \beta$.*

PROOF. The compact closed point set β is a subset of the continuum $D + \beta$. Hence, by Theorem 29, β is a subset of some compact continuum M. By Theorem 28, $M \cdot (D + \beta)$ is a continuum. This continuum is a compact subset of $D + \beta$ containing β.

THEOREM 31. *No compact point set intersects infinitely many components of S.*

PROOF. Suppose there exist a compact point set M and an infinite collection G of components of S such that each element of G intersects M. For each element g of G let P_g denote a definite point of $M \cdot g$. Let K denote a point set such that X belongs to it if and only if X is P_g for some element g of G. The point set K has a limit point P. There exists a region R containing P and therefore containing two points A and B of M. But R is connected. Hence the component of S that contains A also contains B. This involves a contradiction.

THEOREM 32. *If M is a continuum and K is a closed point set lying in M and containing β, the boundary of M, and, for every component D of $M - \beta$, $K \cdot \bar{D}$ is a continuum containing the boundary of D then K is a continuum.*

PROOF. Suppose, on the contrary, that K is the sum of two mutually exclusive closed point sets K_1 and K_2. If D is a component of $M - \beta$ then, since it is connected, $K \cdot \bar{D}$ is a subset either of K_1 or of K_2. For each i $(i = 1,2)$ let K'_i denote a point set such that a point belongs to it if and only if it belongs either to K_i or to some component D of $M - \beta$ such that $K \cdot \bar{D}$ is a subset of K_i. With the help of Theorems 2 and 4 it may be seen that the point sets K_1' and K_2' are mutually exclusive and closed. But their sum is M. This involves a contradiction.

THEOREM 33. *If M is a continuum with a boundary β such that every component of $M - \beta$ has a connected boundary and K is a continuum intersecting β then $M \cdot K + \beta$ is a continuum.*

PROOF. By Theorem 25, β is connected. Hence so is $K + \beta$. If $M \cdot K$ is K or a subset of β then $M \cdot K + \beta$ is $K + \beta$ or β. If $M \cdot K$ is neither K nor a subset of β then $(K + \beta) - \beta$ is the sum of two mutually separated point sets $M \cdot K - M \cdot K \cdot \beta$ and $K - M \cdot K$. Hence $(M \cdot K - M \cdot K \cdot \beta) + \beta$ is connected. But this point set is identical with $M \cdot K + \beta$.

THEOREM 34. *If the continuum K contains β, the boundary of the continuum M and, for every component D of $M - \beta$ and component E of $S - \bar{D}$, the boundary of E is connected then $M \cdot K$ is connected.*

PROOF. Suppose $M \cdot K$ is the sum of two mutually exclusive closed point sets H and L. The point sets $H \cdot \beta$ and $L \cdot \beta$ exist and their sum is β. By Theorem 26, there exists a component D of $M - \beta$ whose closure intersects both $H \cdot \beta$ and $L \cdot \beta$. Let Q_H denote a collection such that q belongs to it if and only if q is a component of $S - \bar{D}$ whose boundary intersects H and let Q_L denote a collection satisfying the same conditions except for the substitution of L in place of H. Suppose q is an element of Q_H and β_q is its boundary. Since the connected subset β_q of β intersects H and is a subset of $H + L$ it contains no point of L and is a subset of H. Similarly L contains the boundary of every element of Q_L. Every point of K not belonging to $\bar{D}$ belongs either to a domain of the collection Q_H or to one of the collection Q_L.

Let K_H denote the point set $H \cdot \bar{D} + K \cdot Q_H^*$ and let K_L denote $L \cdot \bar{D} + K \cdot Q_L^*$. Since the domain Q_H^* contains no point of Q_L^*, it contains no limit point of that point set. Suppose the point P of $H \cdot \bar{D}$ is a limit point of K_L. Since H and L are mutually separated, P is not a limit point of $L \cdot \bar{D}$. Suppose P is a limit point of $K \cdot Q_L^*$. Then it is a limit point of Q_L^*. But it belongs to no domain of the collection Q_L. Therefore, by Theorem 4, it is a limit point of the sum of the boundaries of the domains of this collection. Hence it is a limit point of L. This involves a contradiction. So K_H contains no limit point of K_L. Similarly, K_L contains no limit point of K_H and K_H and K_L are mutually separated. Thus the supposition that $M \cdot K$ is not connected has led to a contradiction.

THEOREM 35. *If M is a closed point set and β, the boundary of M, is compact then there exists a closed and compact point set M' lying in M, containing β, and such that if D is any component of $M - \beta$ such that every component of $S - \bar{D}$ has a connected boundary then $M' \cdot \bar{D}$ is a continuum containing the boundary of D.*

PROOF. By Theorem 31, there exists a finite collection H such that h belongs to H if and only if it is a component of S intersecting β. If h is a domain of the collection H, $h \cdot \beta$ is identical with the boundary of $h \cdot M$. Since it is a closed and compact subset of the continuum h, $h \cdot \beta$ is a subset of some compact continuum C_h. Let M' denote a point set such that P belongs to it if and only if P belongs to M and to C_h for some domain h of the collection H. Suppose D is a component of $M - \beta$ such that every component of the complement of $\bar{D}$ has a connected boundary. If D intersects some component x of S not belonging to H then D is x and $\bar{D}$ is x and every component of the complement of $\bar{D}$ is a component of S and therefore has no boundary. So D intersects some domain h of the collection H. Hence, since it is connected, it is a subset of h. The point set D is a connected domain whose boundary γ is a subset of $\beta \cdot h$ and is therefore compact. The compact continuum C_h contains $\beta \cdot h$ and therefore γ. Hence, by Theorem 28, $\bar{D} \cdot C_h$ is a continuum. Hence so is $\bar{D} \cdot M'$. But M' is a closed and compact subset of M and it contains β.

THEOREM 36. *If every point of a region lies in a domain lying in that region and having a compact boundary then if M and N are two mutually exclusive closed point sets and one of them is compact there exist two domains D_M and D_N containing M and N respectively such that each of them either is closed or is bounded by a compact point set and $\bar{D}_M$ and $\bar{D}_N$ are mutually exclusive.*

Theorem 36 may be established by an argument closely related to that employed to prove Theorem 16 of Chapter I and differing from it largely in the use of domains with compact boundaries instead of certain regions.

THEOREM 37. *If every point of a region lies in a domain lying in that region and having a compact boundary then if H is a component of a closed and compact point set M and K is a closed point set containing no point of H there exists a connected domain D which contains H but no point of K and which either is closed or is bounded by a compact point set containing no point of $M + K$.*

PROOF. With the aid of Theorem 36, in the place of Theorem 16 of Chapter I, by an argument similar to that used to prove Theorem 52 of that chapter, it may be shown that there exists a domain Z containing H but no point of K and bounded by a compact point set containing no point of $M + K$. Let D denote the component of Z that contains H. By Theorem 2, the connected point set D is a domain. With the aid of Theorem 3, it may be seen that the boundary of D is compact. It clearly fulfills the remaining requirements of the conclusion of Theorem 37.

THEOREM 38. *If every point of a region lies in a domain lying in that region and having a compact boundary then if the boundary of the connected domain D is a compact point set β there exists a compact continuum containing β and lying in $D + \beta$.*

PROOF. Suppose D is a connected domain whose boundary is the connected point set M. There exists a sequence $W_1, W_2, W_3, \cdots$ satisfying, with respect to M, all the conditions of Theorem 117 of Chapter I. For each n, let D_n denote W_n^*. For each n, D_n is the sum of a finite number of regions of the collection G_n of Axiom 1. Furthermore the sequence of domains $D_1, D_2, D_3, \cdots$ closes down on M.

Let A denote some definite point of D. Suppose x is a component of M. With the help of Theorem 37, it may be seen that there exists a connected domain I_x containing x and bounded by a compact point set J_x such that $\bar{I}_x$ is a subset of D_1 not containing A, and J_x contains no point of M. It follows that there exists a finite collection H_1 of mutually exclusive connected domains which irreducibly fulfills the condition that if x is a component of M there exists a domain belonging to H_1 and containing x and such that its closure lies in D_1 and its boundary is compact and contains no point of M. There exists a finite collection H_2 of connected domains irreducibly fulfilling the condition that if x is a component of M there exists a domain h belonging to H_2 and containing x and such that (1) $\bar{h}$ lies in D_2 and in the domain of H_1 that contains x and (2) the boundary of h is compact and contains no point of M. This process may be continued. It follows that there exists a sequence $H_1, H_2, H_3, \cdots$ such that, for each n, H_{n+1} is a finite collection of mutually exclusive connected domains which irreducibly fulfills the condition that if x is a component of M there exists a domain h belonging to H_{n+1} and containing x such that (1) $\bar{h}$ lies in D_{n+1} and in the domain of H_n that contains x and (2) the boundary of h is a compact point set containing no point of M. If, for some n, h is a domain of the collection H_n there exists a finite collection V_h of mutually exclusive connected domains such that (1) V_h covers the common part of D and the boundary of h and (2) every domain of V_h is a subset of D and (if $n > 2$) of the domain of H_{n-1} that contains h.

For each n, let Z_n denote the set of all domains z such that, for some domain h of the collection H_n, z belongs to V_n. There exists a finite set T_1 of arcs such that (1) every arc of T_1 lies in D and joins A to a point of some domain of the collection Z_1, (2) each domain of the collection Z_1 contains a point of some arc of the collection T_1.

There exists a sequence $T_1, T_2, T_3, \cdots$ such that T_1 is as indicated and, for each n, T_{n+1} is a finite collection of arcs such that (1) if x and y are domains belonging to H_n and H_{n+1} respectively and $\bar{y}$ is a subset of x, and x' and y' are domains belonging to V_x and V_y respectively, and there exists an arc t containing a point of y' and lying wholly in x except for one of its endpoints which belongs to x' and to the boundary of x then, for these particular domains x' and y', there is only one such arc t belonging to T_{n+1}, (2) if t is an arc of T_{n+1}, there exist two domains x and y belonging to H_n and H_{n+1} respectively, such that $\bar{y}$ is a subset of x and such that t intersects some domain of V_y and lies wholly in x except for one of its endpoints, which belongs to the boundary of x and to some domain of V_x.

For each n, there exists a finite set L_n of arcs such that (1) if z is a domain of Z_n intersecting both an arc x of T_n and an arc y of T_{n+1} then there is an arc of L_n lying wholly in z and intersecting both x and y, (2) if q is an arc of L_n there exists a domain z of Z_n such that q lies wholly in z and intersects both some arc of T_n and some arc of T_{n+1}.

The point set $M + (T_1^* + T_2^* + \cdots) + (L_1^* + L_2^* + \cdots)$ is a compact subcontinuum of $D + M$ containing M.

Theorem 38 does not remain a consequence of these axioms if, in the statement of its hypothesis, the words "If every point of a region lies in a domain lying in that region and having a compact boundary" are replaced by "If P is a point of a region R and X is a point distinct from P there exists a finite point set lying in R and separating P from X". Consider the following example.

EXAMPLE 4. Let E denote a Cartesian space of three dimensions. In this space, for each positive integer n, let C_n denote a circle tangent to the X-axis at the origin O and with its center at the point $(0,1/n,0)$. Let K_1 and K_2 denote two spheres tangent to the Y-axis at O and with centers at $(-1,0,0)$ and $(1,0,0)$ respectively. For each n, let A_n and B_n denote the points $(0,2/n,0)$ and $(0,2/n,1)$ respectively and let A_nB_n, B_nB_{n+1} and A_nA_{n+1} denote straight line intervals with endpoints as indicated. Let S denote the point set obtained by adding together all the circles of the sequence C_1, C_2, $C_3, \cdots$ and all the straight line intervals A_1B_1, A_2B_2, $A_3B_3, \cdots$, B_1B_2, B_3B_4, $B_5B_6, \cdots$ and A_2A_3, A_4A_5, $A_6A_7, \cdots$. Let D denote the set of all points of S which lie outside both of the spheres K_1 and K_2. Let Σ denote the subspace of E whose points are the points of S. In the space Σ, D is a connected domain whose boundary is a compact and countable point set containing the point O and having that point as its only limit point. Not only is the point O not accessible from D but it does not belong to any nondegenerate compact subcontinuum of $\bar{D}$.

Theorem 38 does not remain true in every space satisfying Axioms 0, 1 and 2 if, in the statement of its hypothesis, the words "If every point of a region lies in a domain lying in that region and having a compact boundary" are replaced by "If S has no cut point and for every two points A and B every region that contains A contains a simple closed curve that separates A from B". Consider the following example.

EXAMPLE 4(a). In a Cartesian space E of three dimensions, let K_1 and K_2 denote two congruent right circular cones, with a common vertex O but no other point in common, such that their axes are at an acute angle with each other. Let T_1, T_2, $T_3, \cdots$ denote a sequence of mutually exclusive right circular cylinders converging to O such that no one of them contains O and, for each n, K_1 encloses one base of T_n and K_2 encloses the other one. Let C_1, C_2, $C_3, \cdots$ denote a sequence of mutually exclusive cylinders with circular bases such that (1) for each n, one base of C_n is enclosed by T_n and its other base is enclosed by T_{n+1}, (2) C_n does not intersect K_1 or K_2 or any cylinder of the sequence $T_1, T_2, T_3, \cdots$ except T_n and T_{n+1}, (3) if n is even C_n is a right circular cylinder, (4) if n is odd C_n is a bent cylinder such that the set of all points of C_n that are neither within T_n nor within T_{n+1} is connected, (5) the sequence $C_2, C_4, C_6, \cdots$ converges to O and the sequence $C_1, C_2, C_3, \cdots$ converges to a straight line interval OB having no point except O in common with K_1 or K_2 or any point set of the sequence $T_1, T_2, T_3, \cdots$ or the sequence $C_1, C_2, C_3, \cdots$. Let S denote a point set such that P belongs to it if and only if either (1) P belongs to the lateral surface of one of the cones K_1 and K_2 but is not enclosed by any cylinder of the sequence $T_1, T_2, T_3, \cdots$ or (2) P belongs to the lateral surface of a cylinder of one of the sequences $T_1, T_2, T_3, \cdots$ and $C_1, C_2, C_3, \cdots$ but is not enclosed by a cylinder of one of these sequences or by K_1 or K_2. It is to be noted that no point of the interval OB except O belongs to S. Let D denote the set of all points of S that do not belong to K_1 or K_2 and let Σ denote the subspace of E whose points are the points of S. The space Σ is connected and satisfies Axioms 0, 1 and 2 and S has no cut point and if A and B are two points of S every region in Σ that contains A contains a simple closed curve that separates A from B. But D is a connected domain whose boundary is a compact point set which is not a subset of any compact continuum lying in $\bar{D}$.

Definition. The collection G of point sets is said to be *semi-contracting* if and only if it is true that if M is a point set such that each point of M belongs to some point set of the collection G but no two points of M belong to the same point set of that collection then M is compact.

Theorem 39. *If every point of a region lies in a domain lying in that region and having a compact boundary then if the domain D is bounded by a compact point set β and no component of D is closed, the components of D form a semi-contracting collection.*

Proof. Suppose the components of D do not form a semi-contracting collection. Then there exist a sequence of points $P_1, P_2, P_3, \cdots$ and a sequence $D_1, D_2, D_3, \cdots$ of distinct components of D such that (1) for each n, P_n belongs to D_n and (2) the point set $P_1 + P_2 + P_3 + \cdots$ has no limit point. Since no component of D is closed therefore, for each n, the continuum $\bar{D}_n$ contains a point of the compact point set β. Hence the limiting set of the sequence $D_1, D_2, D_3, \cdots$ exists. But it is a subset of β. Hence it is a compact point set N. The point set $P_1 + P_2 + P_3 + \cdots$ is closed and it contains no point of N. Hence, by Theorem 36, there exists a domain W containing N and bounded by a compact point set T such that $W + T$ contains no point of the sequence $P_1, P_2, P_3, \cdots$. There exists an ascending sequence of positive integers $n_1, n_2, n_3, \cdots$ such that, for each j, D_{n_j} contains a point of W. But, for each j, D_{n_j} contains also the point P_{n_j} which does not belong to W. Hence, since it is connected, D_{n_j} contains a point X_j of T, the boundary of W. Since T is compact, the point set $X_1 + X_2 + X_3 + \cdots$

has a limit point O. Since T is closed, O belongs to T. But it belongs to N, and N and T are mutually exclusive. Thus the supposition that Theorem 39 is false has led to a contradiction.

Theorem 39 does not remain a consequence of these axioms if, in the statement of its hypothesis, the supposition "If every point of a region lies in a domain lying in that region and having a compact boundary" is replaced by the supposition "If A and B are two points every region that contains A contains a simple closed curve that separates A from B". Consider the following example.

EXAMPLE 5. In a Cartesian plane E, let O denote the origin and, for each n, let P_n denote the point $(1/n,0)$. Let $C_1, C_2, C_3, \cdots$ denote a sequence of circles, each lying without all of the others, such that, for each n, the center of C_n is at P_n. For each n, let r_n denote the radius of C_n and let $C_{n1}, C_{n2}, C_{n3}, \cdots, C_{nn}$ denote n circles with center at P_n such that, for every positive integer j less than $n + 1$, the radius of C_{nj} is r_n/j. For each n, let J_n denote a circle with center at O and enclosing the circle C_n but such that if $i < n$ then C_i lies wholly without J_n. Let D denote the sum of the interiors of the circles $C_1, C_2, C_3, \cdots$. Let Σ denote a space whose points are the points, in the ordinary sense, of the plane E but in which a point set R is a "region" if and only if either (1) R is the interior of a circle which does not enclose O or (2) for some positive integer n, R is the set of all points which lie within J_n but without all of the circles $C_{n1}, C_{n+1,2}, C_{n+2,3}, \cdots, C_{n+k,k+1}, \cdots$. In the space Σ, the boundary of the domain D is compact but the point set $P_1 + P_2 + P_3 + \cdots$ has no limit point and therefore the set of all components of D is not semi-contracting.

THEOREM 40. *If every point of a region lies in a domain lying in that region and having a compact boundary and M is a closed point set whose boundary is a compact point set β then there exists a closed and compact point set M' lying in M, containing β, and such that if D is any component of $M - \beta$ with a boundary then $\overline{D}\cdot M'$ is a continuum containing the boundary of D.*

PROOF. By Theorem 38, if D is a component of $M - \beta$, there exists a compact continuum T_D lying in $\overline{D}$ and containing β_D, the boundary of D. Let T denote the set of all continua T_D for all components D of $M - \beta$. If H is the collection of all components of $M - \beta$, the boundary of the domain H^* is a subset of the compact point set β. Hence, by Theorem 39, H is semi-contracting. Hence T is semi-contracting. It follows that $T^* + \beta$ satisfies all the conditions required of M' in the statement of Theorem 40.

THEOREM 41. *If every point of a region lies in a domain lying in that region and having a compact boundary and M is a continuum whose boundary is compact then M is compactly connected.*

PROOF. By Theorems 39 and 40 there exists a compact continuum K lying in M and containing β. Suppose P_1 and P_2 are two points of M. If the point P_i ($i = 1$ or 2) does not belong to K then it belongs to a component D_i of $M - \beta$ and there exists an arc t_i lying in $\bar{D}_i$ and joining P_i to some point which belongs to the boundary of D_i and therefore to K. Let T_i denote P_i or t_i according as P_i does or does not belong to K. The point set $K + T_1 + T_2$ is a compact subcontinuum of M joining P_1 to P_2.

THEOREM 41′. *Suppose that if A and B are two points every region that contains A contains a closed and compact point set that separates A from B. Then if R is a region containing the point A but no point of the closed and compact point set M, there exists a connected domain which contains A and which is bounded by a compact subset of R.*

PROOF. For each point X of M there exists, in R, a closed and compact point set T_X such that $S - T_X$ is the sum of two mutually separated point sets D_X and E_X containing A and X respectively. There exists a finite point set $X_1, X_2, X_3, \cdots, X_n$ such that M is a subset of $E_{X_1} + E_{X_2} + \cdots + E_{X_n}$. The component of $D_{X_1} \cdot D_{X_2} \cdot D_{X_3} \cdots D_{X_n}$ that contains A is bounded by a subset β of $T_{X_1} + T_{X_2} + \cdots + T_{X_n}$. The point set β is a closed and compact subset of R.

THEOREM 42. *The sum of a finite number of closed and connected im kleinen point sets is itself connected im kleinen.*

PROOF. Suppose G is a finite set of closed and connected im kleinen point sets. If P is a point of G^* then, since G is finite and all its elements are closed, there exists a region U containing P but containing no point of any point set of the collection G that does not contain P. Let H denote the set of all point sets of the collection G that contain P. Suppose R is a region containing P. For each point set h of the set H there exists a region R_h, containing P, such that $h \cdot R_h$ is a subset of T_h, the component of $h \cdot R$ that contains P. Let L denote the set of all regions R_h for all point sets h of the set H and let N denote the set of all point sets T_h for all elements h of H. Since the elements of N are all connected and they all contain the point P therefore their sum N^* is connected. Since the set L is finite there exists a region V containing P and lying in U and in every region of the set L. The point set $G^* \cdot V$ is identical with $H^* \cdot V$. But $H^* \cdot V$ is a subset of every R_h and therefore of N^*, the component of $H^* \cdot R$ that contains P. Hence, by definition, G^* is connected im kleinen at P.

THEOREM 43. *If a continuum is the sum of a finite number of continuous curves it is a continuous curve.*

Theorem 43 is a consequence of Theorem 42.

It is not, however, true that if a compact continuum in the plane is the sum of a finite number of connected and connected im kleinen point sets then it is a continuous curve.

EXAMPLE 6. In a Cartesian plane let H denote the interval of the Y-axis whose extremities are $(1,0)$ and $(-1,0)$ and let K denote the graph of $y = \sin 1/x$ $(0 < x \leqq 1)$. Here H and K are both connected and connected im kleinen but their sum, though a continuum, is not a continuous curve. It is not connected im kleinen at any point of H.

It does not follow from these axioms, indeed it is not even true in the plane, that if the common part of two continuous curves is a continuum then it is a continuous curve.

EXAMPLE 7. In a Cartesian plane, let O and A denote the points $(0,0)$ and $(1,0)$ respectively and, for each n, let O_n and A_n denote the points $(0,1/n)$ and $(1,1/n)$ respectively. Let OA and OO_1 denote straight line intervals with endpoints as indicated and, for each n, let O_nA_n denote the straight line interval whose endpoints are O_n and A_n. For each positive integer n and *odd* positive integer i less than 2^n, let B_{in} denote the point $(i/2^n,1/n)$. Let $B_1, B_2, B_3, B_4, B_5, B_6, B_7, B_8, B_9, B_{10},\cdots$ denote $B_{11}, B_{12}, B_{32}, B_{13}, B_{33}, B_{53}, B_{73}, B_{14}, B_{34}, B_{54},\cdots$ respectively. For each n, let C_n denote the projection of B_n upon OA. Let T denote the continuum $OA + OO_1 + O_1A_1 + O_2A_2 + O_3A_3 + \cdots$, let H denote $T + B_1C_1 + B_3C_3 + B_5C_5 + B_7C_7 + \cdots$ and let K denote $T + B_2C_2 + B_4C_4 + B_6C_6 + \cdots$. The point sets $H + T$ and $K + T$ are continuous curves and T, their common part, is a continuum. But T (Figure 5) is not connected im kleinen at any point of OA except O.

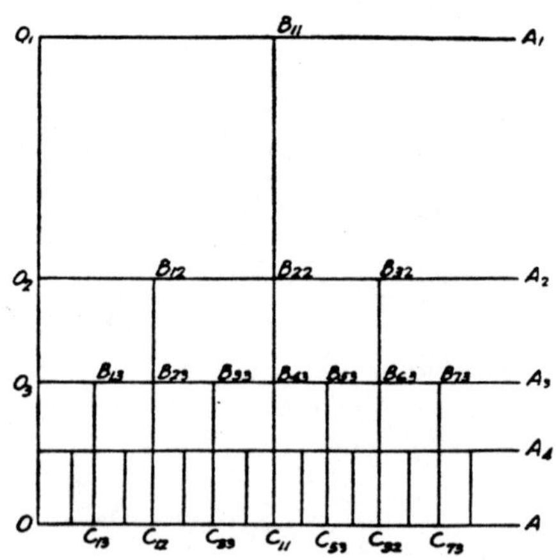

FIGURE 5

It is not true in the plane that if the compact continuous curve L is a subset of the compact continuous curve M and $\overline{(M - L)}$ is a continuum then $\overline{(M - L)}$ is a continuous curve.

EXAMPLE 8. Using the figure and notation of Example 7, let L denote $OA + B_1C_1 + B_2C_2 + B_3C_3 + B_4C_4 + \cdots$ and let M denote $L + T$. Here not only is L a continuous curve but so is every one of its subcontinua. It is to be noted however that M has a continuum of condensation.

THEOREM 44. *If O is a point of a continuum M whose boundary β is compact and every component of the complement of M has a connected boundary, but M is not connected im kleinen at O then there exist a domain D containing O and a sequence $M_1, M_2, M_3, \cdots$ of mutually exclusive subcontinua of M such that (1) for each n, M_n is a component of $M \cdot \bar{D}$ containing a point of D and a point of the boundary of D, (2) the sequence $M_1, M_2, M_3, \cdots$ converges to a point set that contains a subcontinuum of M containing O and a point of the boundary of D.*

PROOF. There exist a domain D containing O and a sequence of distinct points $P_1, P_2, P_3 \cdots$ belonging to $M \cdot D$ such that (1) this sequence converges to O and (2) no component of $M \cdot \bar{D}$ contains more than one of the points $O, P_1, P_2, P_3, \cdots$. There exists a compact continuum T containing $\beta + O + P_1 + P_2 + P_3 + \cdots$. Let L denote $T \cdot M$. By hypothesis and Theorem 27, L is connected. For each n, L contains P_n and O but no connected subset of D containing P_n and O. Hence it contains a point of $S - D$. But L is compact. Hence L_n, the component of $L \cdot \bar{D}$ that contains P_n, contains a point of the boundary of D. The point sets $L_1, L_2, L_3, \cdots$ are mutually exclusive subsets of the closed and compact point set $L \cdot \bar{D}$. It follows that there exists an increasing sequence of positive integers $n_1, n_2, n_3, \cdots$ such that $L_{n_1}, L_{n_2}, L_{n_3}, \cdots$ converges to a compact continuum N containing O and lying in $L \cdot \bar{D}$ and intersecting the boundary of D. For each j, let T_j denote the component of $M \cdot \bar{D}$ that contains L_{n_j}. Some subsequence $M_1, M_2, M_3, \cdots$ of $T_1, T_2, T_3, \cdots$ converges to a point set containing N.

Theorem 44 does not hold true in every space satisfying Axioms 0 and 1. Consider the following example.

EXAMPLE 9. In a Cartesian plane Z, let O, A, B, C, D, and E denote the points $(0,0)$, $(0,-1)$, $(1,-1)$, $(2,-1)$, $(2,1)$ and $(1,-2)$ respectively and let AC, CD and BE denote straight line intervals with endpoints as indicated. Let K denote the set of all points of the graph of $y = \sin(1/x)$ whose abscissae lie between 0 and 2, let M denote the point set $K + O + AC + CD$ and let S denote $M + BE$. Let Σ denote the subspace of Z whose points are the points of S. In this space the continuum M is not locally connected at O. Its boundary consists of the single point B, and the only component of $S - M$ has B as its boundary. But M has no continuum of condensation.

That Theorem 44 does not remain true in every space satisfying Axioms 0, 1 and 2 if the requirement that every component of the complement of M have a connected boundary is omitted may be seen with the help of Example 4. Let M denote the closure of the domain D of that example.

THEOREM 45. *If M is a continuous curve and N is a subcontinuum of M whose boundary is compact and whose boundary with respect to M is a compact point set L with no continuum of condensation and every component of $S - N$ has a connected boundary, then N is a continuous curve. In particular, if M is a continuous curve and L is a compact continuum intersecting M and having no continuum of condensation, and T is a component of $M - L \cdot M$ such that the boundary of $\bar{T}$ is compact and every component of $S - \bar{T}$ has a connected boundary then $\bar{T}$ is a continuous curve and if $M - L \cdot M$ is the sum of two mutually separated point sets H and K and T is a component of $\bar{H}$ or of $H + L \cdot M$ having these same two properties then T is a continuous curve.*

PROOF. Suppose N is not a continuous curve. Then, with the help of Theorem 44, it may be shown that there exist two domains D_1 and D_2 and a sequence of mutually exclusive continua M', M_1, M_2, $M_3, \cdots$ such that $\bar{D}_2$ is a subset of D_1 and such that if D denotes $D_1 - \bar{D}_2$ then (1) for every n, M_n is a component of $N \cdot \bar{D}$ containing at least one point of the boundary of D_1 and at least one point of the boundary of D_2, (2) the sequence M_1, M_2, $M_3, \cdots$ converges to M'. If P is a point of $M' \cdot D$ and R is a region containng P then there exists a region R_1 containing P such that every point of $M \cdot R_1$ can be joined to P by an arc lying in $M \cdot D \cdot R$. But there exists a number n such that R_1 contains a point X of M_n. There exists an arc XP lying wholly in $M \cdot D \cdot R$. Let Y denote the first point, in the order from X to P, that XP has in common with M'. Let XY denote the interval of XP whose endpoints are X and Y. The point set $M_n + (XY - Y)$ is a connected subset of $M \cdot \bar{D}$ containing M_n and points not belonging to M_n. But M_n is a component of $N \cdot \bar{D}$. Therefore $XY - Y$ is not a subset of N. Hence $XY - Y$ contains a point Z belonging to L but not to M'. Thus if P is any point of $M' \cdot D$, every region R containing P contains a point of L not belonging to $M' \cdot D$. But it may be shown that $M' \cdot D$ contains a nondegenerate continuum. Every such continuum is a continuum of condensation of L. But this is contrary to hypothesis. The remainder of Theorem 45 may be established with the help of Theorem 3.

THEOREM 46. *If the locally compact continuum M is not connected im kleinen at every one of its points then M contains a nondegenerate continuum K such that it is not connected im kleinen at any point of K.*

Theorem 46 is a consequence of Theorems 11 and 14.

THEOREM 47. *If T is a totally disconnected closed subset of a locally compact continuous curve M and $M - T$ is the sum of the two mutually separated point sets H and K then $H + T$ is a continuous curve provided it is connected.*

PROOF. It is clear that $H + T$ is connected im kleinen at every point of H and thus, if it is not connected im kleinen at each of its points, the set of all points of $H + T$ at which $H + T$ is not connected im kleinen is a subset of T and therefore contains no nondegenerate continuum, contrary to Theorem 46.

DEFINITION. The point P of a continuous curve M is said to be an *endpoint* of M if and only if P is an endpoint of every arc lying in M and containing P.

THEOREM 48. *No continuous curve is separated by the omission of one of its endpoints.*

PROOF. Suppose the point P is a cut point of the continuous curve M. Then $M - P$ is the sum of two mutually separated point sets H and K. Let A and B denote points belonging to H and K respectively. By Theorem 47 of the present chapter and Theorem 60 of Chapter I, the point sets $H + P$ and $K + P$ are continuous curves. Hence there exist arcs AP and BP lying in $H + P$ and $K + P$ respectively. The point set $AP + BP$ is an arc containing P but with its endpoints at A and B which are distinct from P. Hence P is not an endpoint of M.

DEFINITION. A continuous curve is said to be *acyclic* if and only if it contains no simple closed curve. An acyclic continuous curve is also called a *dendron*.

THEOREM 48′. *If A and B are points of an acyclic continuous curve M there exists only one arc lying in M and having A and B as its endpoints.*

PROOF. By Theorem 1, there exists an arc AXB lying wholly in M. Suppose that there exists another arc AYB lying in M and with the same endpoints. One of these arcs contains a point not belonging to the other one. Suppose AYB contains a point P not belonging to AXB. Let C denote the first point that the interval AYB has in common with AXB and let D denote the first point that the interval PB of AYB has in common with AXB. Let CD denote the interval of AXB with endpoints at C and D and let PC and PD denote intervals of AYB with endpoints as indicated. The point set $PC + PD + CD$ is a simple closed curve lying in M, contrary to the hypothesis that M is a dendron.

THEOREM 49. *If M is a nondegenerate continuous curve every point of M which is neither a cut point nor an endpoint of M lies on some simple closed curve which is a subset of M.*

PROOF. Suppose the point P of M is neither a cut point nor an endpoint of M. There exists an arc APB lying in M. Since $M - P$ is connected, therefore, by Theorem 13, it contains an arc AXB. The point set $AXB + APB$ contains a simple closed curve containing P.

THEOREM 50. *Every point of a nondegenerate acyclic continuous curve M is either a cut point or an endpoint of M and if AB is an arc lying in M and P is a point of AB distinct from A and from B then P separates A from B in M.*

PROOF. That every point of M is either a cut point or an endpoint of M is an immediate consequence of Theorem 49. Suppose P does not separate A from B in M. Then, by Theorems 12 and 6, A and B belong to the same component of $M - P$ and therefore, by Theorems 9 and 13, $M - P$ contains an arc from A to B, contrary to Theorem 48.

THEOREM 51. *If M is an acyclic continuous curve, every nondegenerate connected subset of M is arcwise connected.*

PROOF. Suppose A and B are points of the connected subset K of the acyclic continuous curve M. There exists an arc AB lying in M. Let O denote a point of AB not belonging to K. By Theorem 50, O separates A from B in M. But K contains A and B and it is a connected subset of $M - O$. Thus the supposition that Theorem 51 is false leads to a contradiction.

THEOREM 52 (JOHN S. RINEY). *Every connected subset of an acyclic continuous curve is connected im kleinen.*

PROOF. Suppose R is a region containing the point O of the connected subset K of the dendron M. There exists a region R' containing O such that $M \cdot R'$ is a subset of D_O, the component of $R \cdot M$ that contains O. Suppose A is a point of $K \cdot R' - O$. By Theorem 51, K contains an arc α from A to O and D_O contains an arc α' from A to O. By Theorem 48, α' is identical with α. So α contains A and O and is a subset of $K \cdot D_O$ and therefore of the component of $R \cdot K$ that contains O. So K is connected im kleinen.

THEOREM 53 (MAZURKIEWICZ). *If AB is an arc lying in the compact acyclic continuous curve M there exists an arc which lies in M and contains AB but which is not a proper subset of any other arc lying in M.*

PROOF. If neither A nor B is a cut point of M then, by Theorem 50, AB is not a subset of any other arc lying in M. Suppose B is a cut point of M. Then $M - B$ is the sum of two mutually separated point sets H and K where H contains A. Since $AB - B$ is connected it is a subset of H. By Theorem 93 of Chapter I, the continuum $K + B$ has a non-cut point Y distinct from B. But, by Theorem 47 of the present chapter and Theorem 60 of Chapter I, $K + B$ is a continuous curve. Hence it contains an arc BY. The arcs AB and BY have only the point B in common. Hence $AB + BY$ is an arc AY. It is easy to see that Y is an endpoint of M as well as of $K + B$. It follows that if A is a non-cut point of M then AY is not a subset of any other arc lying in M. If A is a cut point of M then it follows, by the same argument, that there exists an arc AX joining A to some endpoint X of M and having only the point A in common with AY. The point set $AX + AY$ is an arc from X to Y containing AB and not lying in any other arc which is a subset of M.

THEOREM 54. *Suppose O is a point and $J_1, J_2, J_3, \cdots$ is a sequence of simple closed curves such that* (1) *for each n, J_n and J_{n+1} have at least two points in common,* (2) *either O belongs to some simple closed curve of this sequence or, for every region R containing O, there exists a number δ such that if $n > \delta$ then J_n is a subset of R. Then if A is a point of J_1 the point set $O + J_1 + J_2 + J_3 + \cdots$ contains a simple closed curve containing A and O.*

PROOF. If there exists no n such that J_n contains O then let α denote the sequence $J_1, J_2, J_3, \cdots$. If there exists a number n such that J_n contains O, then let m denote the smallest such number and let α denote the sequence $J_1, J_2, J_3, \cdots, J_m$. In either case, if no element of α contains both A and O let J'_1 denote the last element of α that contains A. If no element of α that contains two points of J'_1 contains O, let J'_2 denote the last element of α that has at least two points in common with J'_1. Let t_1 denote the interval of J'_1 which contains A and which has only its endpoints in common with J'_2. If no element of α that contains two points of $J'_2 + t_1$ contains O, let J'_3 denote the last element of α that has at least two points in common with the point set $J'_2 + t_1$. There exists an arc t_2 containing A, lying in $J'_2 + t_1$ and having only its endpoints in common with J'_3. If no element of α that contains two points of $J'_3 + t_2$ contains O, let J'_4 denote the last element of α that has two points in common with the point set $J'_3 + t_2$. There exists an arc t_3 containing A, lying in $J'_3 + t_2$ and having only its endpoints in common with J'_4. In case α is finite this process terminates with an element $J'_{m'+1}$ of α and an interval $t_{m'}$ containing A and lying in $O + J_1 + J_2 + \cdots + J_m$ and having only its endpoints in common with $J'_{m'+1}$ such that J_m has two points in common with the point set $J'_{m'+1} + t_{m'}$. It is easy to see that, in this case, $J'_{m'+1} + t_{m'} + J_m$ contains a simple closed curve containing A and O.

In case α is infinite there exist an infinite sequence of simple closed curves $J'_1, J'_2, J'_3, \cdots$ and an infinite sequence of arcs $t_1, t_2, t_3, \cdots$ such that (1) for each n, J'_{n+1} follows J'_n in the sequence α, (2) J'_n contains A if and only if $n = 1$, (3) the arc t_n is a subset of $J'_1 + J'_2 + \cdots + J'_n$ containing A and having only its endpoints in common with J'_{n+1}, (4) the common part of t_n and t_{n+1} is an arc containing A, (5) J'_{n+2} is the last element of α that has more than one point in common with the continuum $J'_{n+1} + t_n$. For each n, the common part of all the arcs of the sequence $t_n, t_{n+1}, t_{n+2}, \cdots$ is an arc γ_n containing A. For each n, γ_n is a subset of γ_{n+1}. The point set $O + \gamma_1 + \gamma_2 + \gamma_3 + \cdots$ is a simple closed curve containing A and O.

THEOREM 55. *If S is connected, M is a continuous curve and N is a point set obtained by adding to M one or more of its complementary domains then N is itself a continuous curve.*

PROOF. It is easy to see that N is a continuum and that it is connected im kleinen at every point of $N - M$ and at every point of M which is not a limit point of the domain $N - M$. Suppose R is a region containing a point P of M which is a limit point of $N - M$. There exists a connected domain R'

containing P such that $M \cdot R'$ is a subset of K, the component of $M \cdot R$ that contains P. Suppose X is a point of $N - M$ belonging to R'. There exists an arc XP lying entirely in R'. Let O denote the first point that the arc XP has in common with M and let XO denote the interval of XP from X to O. The point set $K + XO$ is a connected subset of $N \cdot R$ containing X and P. Hence N is connected im kleinen at P. It follows that N is a continuous curve.

THEOREM 56. *Suppose* (1) *P and O are two points,* (2) *AO is an arc,* (3) *$P_1, P_2, P_3, \cdots$ is a sequence of points converging to O,* (4) *$R_1, R_2, R_3, \cdots$ is a sequence of regions closing down on O such that R_1 does not contain P,* (5) *Q is an infinite sequence of arcs $PP_1, PP_2, PP_3, \cdots$ all lying in $S - AO$,* (6) *for each n, $m_{n1}, m_{n2}, m_{n3}, \cdots$ is the sequence, arranged in order of magnitude, of all numbers m such that P_m belongs to R_n,* (7) *for each n and i, q_{ni} is the interval of $PP_{m_{ni}}$ whose endpoints are $P_{m_{ni}}$ and the last point, in the order from P to $P_{m_{ni}}$, that $PP_{m_{ni}}$ has in common with the boundary of R_n,* (8) *for each n, Q_n is the sequence $q_{n1}, q_{n2}, q_{n3}, \cdots$,* (9) *there exists a sequence $Q'_1, Q'_2, Q'_3, \cdots$ such that, for each n, Q'_n is a subsequence of Q_n and each arc of Q'_{n+1} is an interval of some arc of Q'_n and such that if x and y are two arcs of Q'_n then there exists an arc lying in R_n and containing both a point of x and a point of y but no point of AO. Then there exists an arc from P to O having only the point O in common with AO.*

PROOF. For each n, let t_n denote an arc belonging to the sequence Q'_n and containing an arc h_n of the sequence Q'_{n+1}. For each n, there exists an arc k_n lying in $R_n \cdot (S - AO)$ and containing both a point of h_n and a point of t_{n+1}. Let α denote an arc of the sequence $PP_1, PP_2, PP_3, \cdots$ that contains t_1. The point set $O + \alpha + (t_1 + k_1) + (t_2 + k_2) + \cdots$ is a compact continuous curve containing P and O. Hence it contains an arc PXO. The arcs PXO and OA have only the point O in common.

THEOREM 57. *Suppose conditions* 1 – 8 *of the hypothesis of Theorem 56 are satisfied and PO is an arc having AO as a proper subset. Suppose furthermore that there exists an infinite sequence α of mutually exclusive closed and compact point sets $T_1, T_2, T_3, \cdots$ such that* (1) *for each n, T_n is a subset of R_n that intersects all but a finite number of the arcs of the sequence Q_n but does not contain O. Then there exists a simple closed curve containing both O and some segment of PO that contains A.*

PROOF. There exists a point P_1 of T_1 such that every region that contains P_1 intersects infinitely many arcs of the sequence Q_1. Let D_1 denote a region containing P_1, lying in R_1, and containing no point of PO unless P_1 belongs to PO. Let Q'_1 denote the set of all arcs of Q_1 that intersect D_1. There exists a point P_2 of T_2 such that every region that contains P_2 intersects infinitely many arcs of Q_2 that are subsets of arcs of Q'_1. There exists a region D_2 containing P_2, lying in R_2, and containing no point of PO

unless P_2 belongs to PO. This process may be continued indefinitely. Thus there exist infinite sequences $P_1, P_2, P_3, \cdots, D_1, D_2, D_3, \cdots$ and $Q'_1, Q'_2, Q'_3, \cdots$ such that P_1, P_2, D_1, D_2 and Q'_1 are as indicated and, for each n, (1) P_{n+1} is a point of T_{n+1} such that every region that contains it intersects infinitely many arcs of Q_{n+1} that are subsets of arcs of Q'_n, (2) D_{n+1} is a region containing P_{n+1}, lying in R_{n+1}, and containing no point of PO unless P_{n+1} belongs to PO, (3) Q'_{n+1} is the set of all arcs of Q_{n+1} that intersect D_{n+1}.

Case I. Suppose there exists a positive integer j such that if $n \geqq j$, P_n belongs to $S - PO$. Then if $n > j$ and x and y are arcs of the sequence Q'_n, x and y both intersect D_n and therefore there is an arc intersecting x and y and lying in D_n and therefore in R_n. It follows, by Theorem 56, that there exists an arc PXO from P to O having only O in common with AO. The sum of the arcs PO and PXO contains a simple closed curve containing both O and some segment of PO that contains A.

Case II. Suppose that for infinitely many numbers n, P_n belongs to PO. Then there exists an infinite sequence $n_1, n_2, n_3, \cdots$ such that, for each m, $P_{n_{m+1}}$ lies between P_{n_m} and O on the arc AO. For each n, let A_m denote P_{n_m}, let W_m denote Q_{n_m} and let R'_m denote R_{n_m}. There exists a sequence of mutually exclusive regions $K_1, K_2, K_3, \cdots$ such that, for each m, K_m is a subset of D_{n_m} containing A_m, and every point of $K_m \cdot (OA)$ precedes every point of $K_{m+1} \cdot (OA)$ in the order from A to O on AO. There exists a sequence of arcs $q_1, q_3, q_5, q_7, \cdots$ such that, for each odd number m, q_m belongs to W_m and intersects both K_m and K_{m+3}. For each odd number m there exist two arcs h_m and t_m lying in K_m and K_{m+3} respectively and each intersecting both q_m and AO. The point set $h_m + q_m + t_m$ contains an arc $B_mE_mC_m$ which has, in common with AO, only its endpoints B_m and C_m which belong to K_m and K_{m+3} respectively. For every odd number m, the points B_m, B_{m+2}, B_{m+4}, B_{m+6} and C_m, C_{m+2}, C_{m+4}, C_{m+6} lie on AO in the order $AB_mB_{m+2}C_mB_{m+4}C_{m+2}B_{m+6}C_{m+4}C_{m+6}$. For each odd m, let $B_mB_{m+2}C_m$ denote the interval of AO whose endpoints are B_m and C_m. The arc $B_1E_1C_1$ intersects an arc PP_i of the sequence Q. Let X denote the first point that the arc PP_i has in common with the point set $B_1E_1C_1 + B_3E_3C_3 + B_5E_5C_5 + \cdots$. There exists an odd number k such that X belongs to $B_kE_kC_k$. Let Y denote the first point, in the order from X to P, that the interval XP of PP_i has in common with PO. Let XY denote the interval of XP whose endpoints are X and Y, and let XC_k denote the interval of $B_kE_kC_k$ whose endpoints are X and C_k and let YAC_k denote the interval of PO whose endpoints are Y and C_k. Let J_1 denote the simple closed curve $YX + XC_k + YAC_k$. For each positive integer m greater than 1, let J_m denote the simple closed curve $B_{k+2(m-1)}E_{k+2(m-1)}C_{k+2(m-1)} + B_{k+2(m-1)}B_{k+2m}C_{k+2(m-1)}$. By Theorem 54, $O + J_1 + J_2 + J_3 + \cdots$ contains a simple closed curve containing A and O. Every such curve contains a segment of PO containing A.

THEOREM 58. *Suppose O and P are two points, PO is an arc from P to O. A is a point of PO distinct from P and AO is an interval of PO with endpoints at A and O. Suppose that every region R that contains O contains a closed and compact point set that separates O from $S - \bar{R}$ and that every region containing O contains a point X of OA which is joined to P by an arc containing no point of AO except X. Then there exists a simple closed curve containing O and some segment of PO that contains A.*

PROOF. There exists a sequence of regions $R_1, R_2, R_3, \cdots$ closing down on O such that R_1 does not contain P. For each n, the point set $S - AO$ contains an arc PP_n from P to some point P_n of R_n. There exists a sequence $Q_1, Q_2, Q_3, \cdots$ satisfying the conditions described in (6), (7) and (8) of the hypothesis of Theorem 56. It follows, by Theorem 57, that there exists a simple closed curve containing O and some segment of PO that contains A.

THEOREM 59. *If O and B are two points such that no point separates them from each other but if R is a region containing one of them R contains a closed and compact point set that separates that one from $S - \bar{R}$ then some simple closed curve contains both O and B.*

PROOF. There exists an arc BO with endpoints at B and O. Let A denote a point of this arc distinct from B and O. Let H denote the set of all arcs that have only their endpoints in common with BO. If h is an arc of the set H, the segment of BO whose endpoints are the endpoints of h will be said to be subtended by h. Let Q denote the set of all segments q such that q is subtended by some arc of H.

Suppose O is not an endpoint of any arc of the set H.

Case I. Suppose there exists a point F such that every segment of BO with one endpoint at O contains an endpoint of some arc of the set H whose other endpoint follows F in the order from O to B on OB. Then it follows, with the help of Theorem 58, that there exists a simple closed curve β containing both O and some segment of BO that contains F. If A does not belong to β, the interval AF of BO is covered by a finite subset Q' of Q. For each segment q of the set Q' let h_q denote some definite arc of the set H subtending q. Let W denote the set of all simple closed curves $q + h_q$ for all segments q of the set Q'. There exists a finite sequence of simple closed curves $J_1, J_2, \cdots, J_m$ such that (1) J_m is β, J_1 contains A and, for each n less than m, J_n is a curve of the set W, (2) for each n less than m, J_n and J_{n+1} have more than one point in common. By Theorem 54, the point set $J_1 + J_2 + \cdots + J_m$ contains a simple closed curve containing O and A.

Case II. Suppose that for every point P of $BO - O$ there exists a point X_P between P and O on BO such that no segment of the set Q contains both P and X_P. Then there exists an infinite sequence α of points $A_1, A_2, A_3, \cdots$ converging to O such that (1) A_1 is A, (2) for each n, A_n precedes A_{n+1} in the order from A to O on the interval AO of BO, and no segment of the collection

Q contains both A_n and A_{n+1}. No arc of the set H that subtends a segment of the set Q containing one point of α intersects an arc of H subtending a segment of Q containing *another* point of α.

Case II(a). Suppose there exists a region D containing O such that every subinterval of AO that contains O contains a segment of the collection Q which is subtended by some arc of the set H that is not a subset of $\bar{D}$. Then there exist an infinite sequence of mutually exclusive segments $q_1, q_2, q_3, \cdots$ of the collection Q converging to O, and a sequence of mutually exclusive arcs $h_1, h_2, h_3, \cdots$ of the collection H such that, for each n, h_n subtends q_n and intersects $S - \bar{D}$ and furthermore there exist two points of the sequence $A_1, A_2, A_3, \cdots$ each separating q_n from q_{n+1} in AO. There exists a point C such that the interval CO of AO is a subset of D. There exists a region D_1 lying in D and containing O but no point of the interval AC of AO. There exists a closed and compact point set K lying in D_1 and separating O from $S - \bar{D}_1$. There exists a point F belonging to K such that every region that contains F intersects infinitely many arcs of the sequence $h_1, h_2, h_3, \cdots$.

Suppose F belongs to AO. There exists a connected domain D_F containing F such that $(OB)\cdot\bar{D}_F$ is a subset of $CO - O$. There exist two positive integers i and j such that each of the points A_i and A_{i+1} separates the closure of the segment q_j from $(AO)\cdot\bar{D}_F$ in AO, and the arc h_j contains a point P_i of D_F. The domain D_F contains an arc FP_i. The point set $h_j + FP_i$ contains an arc t having in common with AO only its endpoints, which belong to q_j and D_F respectively. The arc t belongs to Q and subtends a segment of the set H containing both A_i and A_{i+1}. This involves a contradiction.

Suppose F does not belong to AO. Then there exists a connected domain D_F containing F such that $\bar{D}_F$ contains no point of AO. There exist three positive integers i, j and k such that each of the points A_k and A_{k+1} separates q_i from q_j on CO and h_i and h_j contain points P_i and P_j respectively of D_F. There exists an arc P_iP_j lying wholly in D_F. The point set $h_i + h_j + P_iP_j$ contains an arc t having in common with AO only its endpoints which belong to q_i and q_j respectively. The arc t subtends a segment belonging to Q and containing both A_k and A_{k+1}. This involves a contradiction.

Case II(b). Suppose every region D containing O contains a subinterval I of AO containing O such that D contains every arc of the set H which subtends a segment of the collection Q which is a subset of I. Then there exists a sequence $q_1, q_2, q_3, \cdots$ of segments of the set Q such that (1) q_n contains A if, and only if, $n = 1$, (2) if $m > n$, q_m intersects q_n if, and only if, $m = n + 1$, (3) every point of AO except O belongs to some segment of this sequence, (4) for each n, q_n is subtended by some arc h_n of the collection H. For each n, let J_n denote the simple closed curve $q_n + h_n$. By Theorem 54, the point set $O + J_1 + J_2 + J_3 + \cdots$ contains a simple closed curve containing A and O.

Thus, in every case, there exists a simple closed curve J_O containing A

and O. Similarly there exists a simple closed curve J_B containing A and B. If J_O and J_B have more than one point in common, their sum contains a simple closed curve containing O and B. Suppose they have only A in common. Since A does not separate O from B, there exists an arc OZB not containing A. The point set $J_O + J_B + OZB$ contains a simple closed curve containing O and B.

THEOREM 59′. *If no connected domain has a cut point and the points O and B belong to the connected domain D, then there exists a simple closed curve lying in D and containing O and B.*

PROOF. There exists an arc BO in D. Let A denote a point of $BO - (B + O)$. Every region that contains a point X of the interval AO of BO, distinct from O, contains a connected domain D_X lying in D and containing X but neither B nor O, and the connected domain $D_X - X$ contains an arc A_XO_X such that $(A_XO_X)\cdot(BO) = A_X + O_X$ and X lies between A_X and O_X on BO. It follows that, using the same notation as in Case II(b) of the proof of Theorem 59, there exist infinite sequences $q_1, q_2, q_3, \cdots$ and $h_1, h_2, h_3, \cdots$ satisfying conditions (1) to (4) described there and such that every region that contains O contains all but a finite number of the simple closed curves $q_1 + h_1, q_2 + h_2, \cdots$ and these curves all lie in D. It follows, as in the proof of Theorem 59, that there exists a simple closed curve lying in D and containing both B and O.

THEOREM 60. *In order that the point A of the locally compact continuous curve M be an endpoint of M it is necessary and sufficient that A be a non-cut point of M such that, for every point B of M distinct from A, there exists a point of M that separates A from B in M.*

Theorem 60 may be easily proved with the aid of Theorems 48, 50 and 59.

THEOREM 61. *If M is a locally compact continuous curve, the set of all endpoints of M, if it has any, is totally disconnected.*

PROOF. Suppose, on the contrary, that there exists a nondegenerate connected point set T such that every point of T is an endpoint of the locally compact continuous curve M. Let A and B denote two points of T. By Theorems 12 and 59, there exists a point X that separates A from B in M. Since T is a connected subset of M containing A and B therefore X belongs to T. Hence X is an endpoint of M, contrary to Theorem 48.

THEOREM 62. *In order that the locally compact continuum M be an acyclic continuous curve it is necessary and sufficient that every nondegenerate subcontinuum of M contain uncountably many cut points of M.*

PROOF. This condition is sufficient. For suppose M is a locally compact continuum such that every nondegenerate subcontinuum of M contains uncountably many cut points of M. Suppose M is not a continuous curve.

Then there exists a point O belonging to M but at which M is not connected im kleinen. There exists a region Q containing O such that $M \cdot \bar{Q}$ is compact. There exists a region R containing O such that $\bar{R}$ is a subset of Q. By Theorem 11, there exist a domain D containing O and lying in R, an infinite sequence of points $O_1, O_2, O_3, \cdots$ converging to O, and an infinite sequence of mutually exclusive continua $M_1, M_2, M_3, \cdots$ such that (1) $M \cdot \bar{D}$ is compact, (2) for each n, M_n is a component of $M \cdot \bar{D}$ containing O_n and a point of the boundary of D, (3) the sequence $M_1, M_2, M_3, \cdots$ converges to a nondegenerate subcontinuum K of M which contains O and which has no point in common with any of the continua $M_1, M_2, M_3, \cdots$. By hypothesis, K contains an uncountable set H of cut points of M. For each point X of the set H, $M - X$ is the sum of two mutually separated point sets T_X and K_X where K_X contains infinitely many of the continua $M_1, M_2, M_3, \cdots$ and therefore contains $K - X$. As in the proof of Theorem 106 of Chapter I, let G denote the collection of all point sets T_X for all points X of H. By Theorem 22 of Chapter I, $M \cdot \bar{Q}$ contains a countable point set N such that every point of $M \cdot \bar{Q}$ either belongs to N or is a limit point of N. Suppose X is a point of H. The region Q contains X, and X is a limit point of T_X. Therefore Q contains a point of T_X. Thus each point set g of the collection G contains a point P_g belonging to $M \cdot Q$. There exists a region W containing P_g but no point of $M - g$. The region W contains a point P'_g belonging to N. The point P'_g belongs to g. Thus each point set of the collection G contains a point of N. But G is an uncountable collection of mutually exclusive point sets and N is countable. Thus the supposition that M is not a continuous curve involves a contradiction. If M were not acyclic it would contain a simple closed curve J. But, by Theorem 109 of Chapter I, no simple closed curve lying in M contains uncountably many cut points of M. It follows that M is a dendron.

This condition is also sufficient. For suppose M is a locally compact acyclic continuous curve and that K is a nondegenerate subcontinuum of M. By Theorem 52, K is itself an acyclic continuous curve. Hence it contains an arc t. By Theorem 50, every point of t except its endpoints is a cut point of M. Hence K contains uncountably many cut points of M.

DEFINITIONS. The point set M is said to be connected *in the strong sense*, or *strongly connected*, if and only if it is true that for every two points X and Y of M there exists a continuum which contains both X and Y and which is a subset of M. The point set M is said to be *arcwise connected* if every two points of M are the extremities of an arc lying in M.

THEOREM 63. *If K is the set of all non-cut points of the compact continuous curve M then in order that M be acyclic it is necessary and sufficient that, for every subset T of K, $M - T$ be strongly connected.*

PROOF. This condition is sufficient. For suppose, on the contrary, that M contains a simple closed curve J but that if T is any subset of K then

$M - T$ is strongly connected. Let K' denote the point set $M - K$. Clearly every subset of M that contains K' is connected. It follows that every point of M is a limit point of K'. Let A and B denote two points of J. There exists, on J, a point C distinct from A and B and not belonging to K'. Let R_A and R_B denote two domains, containing A and B respectively, such that $\bar{R}_A$, $\bar{R}_B$ and C are mutually exclusive. There exist two domains D_A and D_B, containing A and B respectively, such that every point of M belonging to D_A or to D_B is joined to A or to B by a subcontinuum of M lying in R_A or R_B. But, since every point of M is a limit point of K', there exist two points A' and B' belonging to K' and lying in D_A and D_B respectively. Hence there exist two continua M_A and M_B, both subsets of M, such that M_A contains A and A' and lies wholly in R_A while M_B contains B and B' and lies wholly in R_B. Since K' is connected in the strong sense, it contains a continuum H which contains both A' and B'. Let E and F, respectively, denote the first points which the intervals CA and CB of the arc ACB of the curve J have in common with the closed point set $H + M_A + M_B$. The continuum $H + M_A + M_B$ contains a continuum L which is irreducible from E to F. Let ECF denote the interval of the arc ACB which contains C and has E and F as its endpoints. Consider the point set $L + ECF$. If P is any point of L distinct from E then either $L - P$ is connected or it is the sum of two connected point sets L_E and L_F containing E and F respectively. In either case, $(L + ECF) - P$ is connected. Also, if P is any point of ECF distinct from F then $(L + ECF) - P$ is connected. Hence $L + ECF$ does not contain uncountably many cut points of M. But L contains uncountably many points of H and H is a subset of K'. Thus the supposition that the condition of Theorem 63 is not sufficient has led to a contradiction.

This condition is also necessary. For suppose that K is the set of all non-cut points of the compact dendron M. Let T denote a subset of K and let A and B denote any two points of $M - T$. There exists an arc AB lying in M. By Theorem 50, no point of AB belongs to T. Thus $M - T$ is arcwise connected.

THEOREM 64. *If A is a point of the compact continuum M and the set of all points X such that M is irreducible from A to X is a continuum K and D and R are two domains, each containing points of K, such that $\bar{D}$ and $\bar{R}$ have no point in common then there exists an infinite sequence of mutually exclusive continua $M_1, M_2, M_3, \cdots$ such that* (1) *for each n, M_n contains a point of $\bar{D}$ and a point of $\bar{R}$ and is a component of $(M - K)\cdot E$, where E denotes the point set $S - (D + R)$,* (2) *the sequence $M_1, M_2, M_3, \cdots$ converges to a subset of K.*

PROOF. Let B denote a point belonging to $K \cdot R$. Clearly, if N is a component of $M \cdot (S - D)$ containing a point of $M - K$, then N is a subset of $M - K$. Thus B is not a limit point of any component of $(M - K)\cdot(S - D)$. But B is a limit point of $M - K$. It follows that B is the sequential limit point of an infinite sequence of distinct points $P_1, P_2, P_3, \cdots$ lying in R such that if i is distinct from j, P_i and P_j belong to different components of $(M - K)\cdot(S - D)$. For each n, let T_n denote the component of $(M - K)\cdot(S - D)$ that contains P_n. For each n, there exists a continuum M'_n which intersects both $\bar{D}$ and $\bar{R}$ and is a component of the common part of T_n and the point set $S - (D + R)$. Let $M_1, M_2, M_3, \cdots$ denote some convergent subsequence of $M'_1, M'_2, M'_3, \cdots$.

Theorem 65. *If B and C are two mutually exclusive subcontinua of a compact continuum M and A is a point of $M - (B + C)$ and every subcontinuum of M that contains A and intersects one of the continua B and C also intersects the other one then, for every two domains D_B and D_C containing B and C respectively and such that $\bar{D}_B$ and $\bar{D}_C$ have no point in common, there exists an infinite sequence of mutually exclusive continua $M_1, M_2, M_3, \cdots$ such that* (1) *for each n, M_n intersects both $\bar{D}_B$ and $\bar{D}_C$ and is a component of the set of all points common to M and $S - (D_B + D_C)$,* (2) *the point set $A + M_1 + M_2 + M_3 + \cdots$ is a subset of a strongly connected subset of $M - (B + C)$,* (3) *the sequence $M_1, M_2, M_3, \cdots$ converges to a continuum containing no point that is joined to A by a strongly connected subset of $M - (B + C)$.*

Proof. Let K denote the set of all points X such that X is joined to A by a subcontinuum of M containing no point of $B + C$. If D is a domain containing $B + C$ but not A then the boundary of D contains a point of the component of $M - M\cdot D$ that contains A. It follows that either B or C contains a limit point of K. But if only one of them does so then $\bar{K}$ is a subcontinuum of M intersecting only one of these continua, contrary to hypothesis. Hence each of them contains a limit point of K. Suppose now that D_B and D_C are domains, containing B and C respectively, such that $\bar{D}_B$ and $\bar{D}_C$ are mutually exclusive. If T is a component of $K\cdot(S - D_B)$ then it is also a component of $M\cdot(S - D_B)$. For suppose X is a point of T. Let N denote the component of $M\cdot(S - D_B)$ that contains X. There exists a subcontinuum L of M containing X and A but no point of $B + C$. The continuum $N + L$ contains A but no point of B. Hence, by hypothesis, it does not intersect C. Therefore N is a subset of K. Hence it is identical with T. But C contains a point F which is a limit point of K. The point F is the sequential limit point of a sequence of distinct points $P_1, P_2, P_3, \cdots$, all belonging to $K\cdot D_C$ such that, if i is distinct from j, P_i and P_j belong to different components of $M\cdot(S - D_B)$. For each n, let t_n denote the component of $M\cdot(S - D_B)$ that contains P_n. For each n, there exists a continuum r_n which intersects both $\bar{D}_B$ and $\bar{D}_C$ and is a component of the common part of t_n and the point set $S - (D_B + D_C)$. There exists an infinite ascending sequence of numbers $n_1, n_2, n_3, \cdots$ such that if, for each i, M_i denotes r_{n_i} then the sequence $M_1, M_2, M_3, \cdots$ converges to a continuum M'. There exists an infinite subsequence of the sequence $t_{n_1}, t_{n_2}, t_{n_3}, \cdots$ converging to a continuum Q which contains F and is a subset of $S - D_B$. The continuum Q contains no point of K. For if it contained a point belonging to K, and therefore to a continuum Z containing A and lying in $M - B$, then F would be joined to A by a continuum $Q + Z$ lying in $M - B$, contrary to hypothesis. But M' is a subset of Q. Therefore M' does not intersect K. Thus $M_1, M_2, M_3, \cdots$ satisfies all of the requirements of Theorem 65.

THEOREM 66. *If the compact continuum M is not connected im kleinen there exist two domains D and R, a continuum M', and a sequence of continua $M_1, M_2, M_3, \cdots$ such that* (1) *$\bar{D}$ and $\bar{R}$ are mutually exclusive,* (2) *for each n, M_n intersects both $\bar{D}$ and $\bar{R}$ and is a component of the common part of M and the point set $S - (D + R)$,* (3) *the sequence $M_1, M_2, M_3, \cdots$ converges to M',* (4) *there exists a strongly connected subset of $M - M'$ containing all of the continua of the sequence $M_1, M_2, M_3, \cdots$.*

PROOF. By Theorem 11, there exist a point O of M, a domain D_1 containing O, and a sequence of distinct continua $N_1, N_2, N_3, \cdots$ such that (1) for each n, N_n is a component of $\bar{D}_1 \cdot M$ containing a point of the boundary of D_1, (2) the sequence $N_1, N_2, N_3, \cdots$ converges to a continuum N containing O but intersecting no point set of this sequence. There exist three domains D_2, D_3 and D_4, all containing O, such that $\bar{D}_2$ is a subset of D_1, $\bar{D}_3$ is a subset of D_2 and $\bar{D}_4$ is a subset of D_3. There exist two sequences of distinct continua $U_1, U_2, U_3, \cdots$ and $V_1, V_2, V_3, \cdots$ such that (1) for each n, U_n is a component of $N_n \cdot (\bar{D}_1 - D_2)$ intersecting both the boundary of D_1 and the boundary of D_2, while V_n is a component of $N_n \cdot (\bar{D}_3 - D_4)$ intersecting both the boundary of D_3 and the boundary of D_4, (2) the sequence $U_1, U_2, U_3, \cdots$ converges to a continuum U and $V_1, V_2, V_3, \cdots$ converges to a continuum V, U and V being subcontinua of N. Suppose first that, for each n, there is a subcontinuum of M containing N_n and intersecting only one of the point sets U and V. There exist an infinite sequence of distinct numbers $n_1, n_2, n_3, \cdots$, and an infinite sequence $t_1, t_2, t_3, \cdots$ of subcontinua of M such that (1) for each j, t_j intersects N_{n_j} and (2) either, for each j, t_j intersects U but not V or, for each j, t_j intersects V but not U. If, for each j, t_j intersects U then, for each j, let M_j denote V_{n_j} and let M', D and R denote V, D_4 and $S - \bar{D}_3$ respectively. If, for each j, t_j intersects V then let M_j denote U_{n_j} and let M', D and R denote U, D_2 and $S - \bar{D}_1$ respectively. In the first case $U + t_1 + t_2 + \cdots + N_1 + N_2 + \cdots$, and in the second case $V + t_1 + t_2 + \cdots + N_1 + N_2 + \cdots$, is a strongly connected subset of $M - M'$ containing all the continua of the sequence $M_1, M_2, M_3, \cdots$.

Suppose, secondly, that there exists a number m such that every subcontinuum of M that contains N_m and a point of one of the continua U and V also contains a point of the other one. In this case the truth of Theorem 66 follows readily with the aid of Theorem 65.

THEOREM 67. *If the locally compact continuum M is not connected im kleinen at the point O of M then there exists a closed subset K of $M - O$ such* *that if G is a finite set of continua filling up M then at least one of them contains both O and a point of K.*

PROOF. Since M is not connected im kleinen at O, there exist a domain D, an infinite sequence of points $O_1, O_2, O_3, \cdots$ and an infinite sequence of continua $M_1, M_2, M_3, \cdots$ satisfying, with respect to M and O, conditions (1), (2) and (3) of Theorem 11. Let K denote the closed point set $M - M \cdot D$. Suppose H is a finite set of continua filling up M. There exists a continuum h belonging to the set H and containing infinitely many points of the sequence $O_1, O_2, O_3, \cdots$ and therefore containing O, the sequential limit point of this sequence. Let O_k denote a point of $O_1, O_2, O_3, \cdots$ belonging to h. The point set h is a subcontinuum of M containing both O_k and O. But M_k, the component of $M \cdot \bar{D}$ that contains O_k, does not contain O. Hence h contains a point of $M - M \cdot D$, that is to say, a point of K.

THEOREM 68. *If H is a finite collection of mutually exclusive closed subsets of the compact continuous curve M, there exists a finite collection G of connected domains with respect to M such that* (1) *each point of M belongs to some point set of the collection G,* (2) *if g is a point set of the collection G, the continuum $\bar{g}$ does not intersect more than one point set of the collection H.*

PROOF. For each point P of M there is a connected domain D_P with respect to M containing P such that $\bar{D}_P$ does not intersect more than one point set of the collection H. Let Q denote a collection such that q belongs to it if and only if q is D_P for some point P of M. The point set M is covered by some finite subcollection G of Q.

THEOREM 69. *In order that the compact continuum M be a continuous curve it is necessary and sufficient that for every finite collection H of mutually exclusive closed subsets of M there exist a finite collection G of subcontinua of M such that $G^* = M$ and such that no continuum of G intersects two point sets of the collection H.*

Theorem 69 is an immediate consequence of Theorems 67 and 68.

THEOREM 70. *If the continuum M is locally compact at the point O then in order that M be connected im kleinen at O it is sufficient that for every subcontinuum K of M not containing O there exist a finite collection of continua filling up M such that no one of them contains both O and a point of K.*

PROOF. Suppose D is a domain with respect to M containing O such that $\bar{D}$ is compact. If D is M then it is connected. Suppose it is not M. Then it has a boundary L with respect to M. For each point X of L there exists a finite collection G_X of continua filling up M such that no one of them contains both O and X. Let T_X denote the sum of all the continua of G_X that contain X. Let G denote the collection of all the continua T_X for all points X of L. It is clear that, for each X, T_X contains a domain with respect to M containing X. Hence, since it is closed and compact, L is covered by a finite subset H of G. Let N denote the sum of all the continua of the collection H. The closed point set N does not contain O.

Let K denote the component of $(\overline{M - N})\cdot\bar{D}$ that contains O. Suppose O is a limit point of $M - K$. Then it is a limit point of $\bar{D}\cdot(\overline{M - N}) - K$. Since K is a component of $\bar{D}\cdot(\overline{M - N})$, O is not a limit point of a connected subset of $\bar{D}\cdot(\overline{M - N}) - K$. Hence there exist a sequence of points $P_1, P_2, P_3, \cdots$ and a sequence of distinct continua $K_1, K_2, K_3, \cdots$ such that (1) for each n, K_n is a component of $\bar{D}\cdot(\overline{M - N})$ distinct from K, (2) for each n, P_n belongs to K_n, (3) the point O is the sequential limit point of the sequence $P_1, P_2, P_3, \cdots$. The collection H is finite. It follows that there exist a continuum g of the collection H, and an infinite subsequence α of the sequence $K_1, K_2, K_3, \cdots$, such that every continuum of the sequence α has a point in common with g. Let Q_1 denote the collection of all continua x of H such that there exists a subcontinuum of M intersecting g and x but not containing O. There exists a finite collection G' of continua filling up M such that no continuum of G' contains both O and a point of Q_1^*. Let T denote the sum of all the continua of G' that contain O. The continuum T contains a domain with respect to M containing O. Hence it intersects two continua h and k of the sequence α. But h and k are components of $\bar{D}\cdot(\overline{M - N})$ not containing O. Therefore T contains a point P of $h\cdot N$. There exists a continuum x of H containing P. The continuum h intersects both x and g and does not contain O. Hence x belongs to Q_1. So T intersects Q_1^*. Thus the supposition that O is a limit point of $M - K$ has led to a contradiction. Hence K contains a domain with respect to M containing O. But K is a connected subset of $\bar{D}$. Therefore M is connected im kleinen at O.

That the condition of Theorem 70 is not necessary may be seen from the following example.

EXAMPLE 10. In a Cartesian plane, let O denote the origin, let A denote the point $(1,0)$ and, for each n, let A_n denote the point $(1,1/n)$. Let OA, $OA_1, OA_2, OA_3, \cdots$ denote straight line intervals with endpoints as indicated. Let M denote the sum of all of these intervals. Let K denote the set whose only element is the point A. Let α denote the sequence $A_1, A_2, A_3, \cdots$. The continuum M is connected im kleinen at the point O. But suppose G is a finite collection of continua filling up M. Some continuum g of the collection G contains infinitely many points of α and therefore contains A, the sequential limit point of α. Since O separates every two points of α from each other in M and g is a connected subset of M containing two points of α therefore g contains O. So some element of G contains O and intersects K.

However the following theorem holds true.

THEOREM 71 (W. A. WILSON). *In order that the compact continuum M be a continuous curve it is necessary and sufficient that if H and K are two mutually* *exclusive subcontinua of M then M be the sum of two continua such that neither of them intersects both H and K.*

Proof. That this condition is sufficient is an immediate consequence of Theorem 70. It is also necessary. For suppose M is a compact continuous curve and that H and K are two mutually exclusive subcontinua of M. By Theorem 69, there exists a finite collection G' of continua filling up M such that no one of them intersects both H and K. Let M_H denote the sum of all the continua g of G' such that there exists a connected subset of M which contains g and H and which is the sum of a number of continua of G' that have no point in common with K. Let M_K denote the sum of all continua g of G' such that there exists a connected subset of M which contains g and K and which is the sum of a number of continua of G' that do not intersect H. Suppose x_1 is a continuum of G'. If x_1 intersects H or K, let t_1 denote x_1. If x_1 does not intersect H or K, since the collection G' is finite and M is connected therefore there exists a continuum x_2 of the collection G' distinct from x_1 and intersecting x_1. If x_2 intersects H or K, let t_2 denote $x_1 + x_2$. If x_2 does not intersect $H + K$, there exists a continuum x_3 of G' distinct from x_1 and from x_2 but intersecting $x_1 + x_2$. If x_3 intersects H or K, let t_3 denote $x_1 + x_2 + x_3$. This process may be continued. The collection G' being finite, it follows that, if x_1 does not intersect H or K, there exists a set of continua $x_2, x_3, \cdots, x_n$, all belonging to G', such that if t_n denotes their sum then (1) t_n is a continuum, (2) the point set $x_1 + x_2 + x_3 + \cdots + x_{n-1}$ intersects neither H nor K, (3) x_n intersects $H + K$. If t_n intersects H then x_1 is a subset of M_H. If t_n intersects K then x_1 is a subset of M_K. Of course, for some x_1 there may conceivably exist various selections of $x_2, x_3, \cdots, x_n$ and thus some x_1 may be a subset both of M_H and of M_K. But M_H contains no point of K, M_K contains no point of H and $M_H + M_K = M$.

Theorem 72 (W. L. Ayres). *In order that the compact continuum M be a continuous curve it is necessary and sufficient that if H and K are two mutually exclusive closed subsets of M then H be separated from K in M by a subset of M which is the sum of a finite number of continua.*

Proof. This condition is sufficient. For suppose M is a compact continuum having this property. Suppose O is a point of M and K is a subcontinuum of M not containing O. By hypothesis, there exists a point set L such that (1) $M - L$ is the sum of two mutually separated point sets M_O and M_K containing O and K respectively, (2) L is the sum of a finite number of subcontinua of M. Let G denote the set of all point sets that are components either of $M_O + L$ or of $M_K + L$. Since L is the sum of a finite number of continua therefore the collection G is finite. But it fills up M and clearly no one of its elements contains both O and a point of K. Therefore, by Theorem 70, M is a continuous curve.

This condition is also necessary. For suppose that H and K are two mutually exclusive closed subsets of the compact continuous curve M.

There exist two mutually exclusive domains with respect to M, D_H and D_K, containing H and K respectively. For each point P of $M - (H + K)$, let D_P denote a connected domain with respect to M containing P such that $\bar{D}_P$ has no point in common with $H + K$. Let Q denote the set of all point sets D_P for all points P of $M - (H + K)$. Let W denote the collection of point sets whose elements are D_H, D_K and the point sets of the collection Q. The continuum M is covered by some finite subcollection G of the collection W. To each point set of the collection G that belongs to Q add its boundary and let T denote the sum of the finite number of continua thus obtained. Let H_1 and K_1 denote the point sets $D_H \cdot (M - T)$ and $D_K \cdot (M - T)$ respectively. These point sets are mutually separated subsets of $M - T$ containing H and K respectively. Furthermore $M - T = H_1 + K_1$. In other words T separates H from K in M. But T is the sum of a finite number of continua.

Another proof of the necessity of this condition is as follows. Suppose H and K are two mutually exclusive closed subsets of the compact continuous curve M. There exists a domain D_1 with respect to M containing H such that $\bar{D}$ contains no point of K. There exists a domain D_2 with respect to M containing H such that $\bar{D}_2$ is a subset of D_1. Let D denote the point set $\bar{D}_1 - D_2$. With the aid of Theorems 41 and 42 of Chapter I, it may be shown that at least one component of D intersects both β_1, the boundary of D_1, and β_2, the boundary of D_2. Let G denote the set of all such components of D. Suppose G is infinite. Then some infinite sequence α of distinct elements of G converges to a continuum which intersects both β_1 and β_2 and therefore contains a point O of the domain $D_1 - \bar{D}_2$ with respect to M. Since M is connected im kleinen at O this is in contradiction with Theorem 14. Therefore G is finite. Let L denote G^*. Suppose L does not separate H from K in M. Then, by Theorems 12 and 6, there exists an arc t from A to B lying wholly in $M - L$. The arc t contains an arc t' lying in H and intersecting both β_1 and β_2. The arc t' is a subset of some continuum of the set G. But this involves a contradiction.

THEOREM 73. *If the locally compact continuum M is not connected im kleinen at the point O of M then there exists a nondegenerate subcontinuum T of M containing O such that no two points of T are separated from each other in M by a finite subset of M.*

PROOF. By Theorem 11, there exist a connected domain D containing O, an infinite sequence of points $O_1, O_2, O_3, \cdots$ converging to O and an infinite sequence of mutually exclusive continua $M_1, M_2, M_3, \cdots$ such that (1) for each n, M_n is a component of $M \cdot \bar{D} - O$ containing O_n and a point of the boundary of D, (2) $M_1, M_2, M_3, \cdots$ converges to a nondegenerate subcontinuum T of M which contains O. Suppose X and Y are points of T and N is a finite subset of M. There exists a number i such that no one of the continua $M_i, M_{i+1}, M_{i+2}, \cdots$ intersects N. Suppose $M - N$ is the sum of two mutually separated point sets H and K containing X and Y respectively. By Theorem 28 of Chapter I, each continuum of the sequence M_i, M_{i+1}, $M_{i+2}, \cdots$ is a subset either of H or of K. Hence there exists an infinite subsequence α of this sequence such that either H or K contains every continuum of α. Let W denote the sum of all the continua of the sequence α and let Q denote H or K according as W is a subset of H or a subset of K. Since X and Y are limit points of W belonging to $M - N$ therefore they both belong to Q contrary to the supposition that they belong to H and K respectively.

THEOREM 74. *The nondegenerate compact continuum M is a dendron if and only if every two points of M are separated from each other in M by some point.*

PROOF. That every two points of a nondegenerate dendron M are separated from each other in M by some point is a consequence of Theorems 48 and 50.

Suppose every two points of the compact continuum M are separated from each other in M by some point. By Theorem 73, M is a continuous curve. Suppose it contains a simple closed curve J. By hypothesis some point separates two points of J from each other in M and therefore in J contrary to Theorem 108 of Chapter I.

THEOREM 75. *Every two points of a compact continuous curve M are separated from each other in M by the sum of a finite number of subcontinua of M.*

Theorem 75 is a corollary of Theorem 72.

DEFINITION. If P is a point of a continuum M and every point of M distinct from P is separated from P in M by a finite point set then M is said to be *regular at P*.

THEOREM 76. *A compact continuum is connected im kleinen at the point P if it is regular at P.*

Theorem 76 is a consequence of Theorem 73.

THEOREM 77. *If O is a point of a compact continuum M and every domain D with respect to M containing O contains a domain with respect to M containing O and bounded, with respect to M, by a finite subset of D then M is regular at O.*

PROOF. If B is a point of M distinct from O the point set $M - B$ is a domain with respect to M containing O and therefore $M - B$ contains a domain with respect to M containing O and bounded, with respect to M, by a finite subset β of $M - B$. The point set β separates O from B in M. Hence M is regular at O.

THEOREM 78. *If L is a closed proper subset of the compact continuum M and M is regular at every point of L then every domain D with respect to M containing L contains a domain with respect to M containing L and bounded, with respect to M, by a finite subset of D.*

PROOF. Suppose D is a domain with respect to M containing L. There exists a domain D_1 with respect to M which contains L and is a proper subset of D. The point sets L and $M - D_1$ are mutually exclusive and closed. Therefore, by Theorem 67 of Chapter I, there exists a finite subset T of M such that $M - T$ is the sum of two mutually separated point sets H and K containing L and $M - D_1$ respectively. The point set H is a domain with respect to M and its boundary with respect to M is a subset of the finite point set T. The point set T is necessarily a subset of D_1 and therefore of D.

It might, at first sight, be thought that, by an argument analogous to that used in proving Theorem 78 with the use of Theorem 67 of Chapter I, one could show, with the use of Theorem 72, that if O is a point of a compact continuous curve M then every domain with respect to M containing O contains a domain with respect to M containing O and bounded by the sum of a finite number of continua. That this is not true may be seen from Example 8, M and O being as indicated in the description of that example.

A continuum is said to be a (Menger) *regular curve* if and only if it is regular at every one of its points.

THEOREM 79. *No compact regular curve contains two mutually exclusive closed point sets H and K and infinitely many mutually exclusive continua each intersecting both H and K.*

PROOF. Suppose H and K are two mutually exclusive closed subsets of a compact regular curve M and G is an infinite collection of mutually exclusive subcontinua of M each intersecting H and K. Let $g_1, g_2, g_3, \cdots$ denote an infinite sequence of distinct continua of the collection G. For each n, let A_n and B_n denote points of g_n belonging to H and K respectively. There exists an infinite ascending sequence of positive integers $n_1, n_2, n_3, \cdots$ such that $A_{n_1}, A_{n_2}, A_{n_3}, \cdots$ converges to a point A of H, while $B_{n_1}, B_{n_2}, B_{n_3}, \cdots$ converges to a point B of K. By hypothesis there exists a finite subset T of M such that $M - T$ is the sum of two mutually separated point sets L and N containing A and B respectively. Since the continua of the sequence $g_{n_1}, g_{n_2}, g_{n_3}, \cdots$ are mutually exclusive, not more than a finite number of them intersect T. Hence there exists a positive integer j such that, for every i greater than j, g_{n_i} is a subset of $L + N$. But every connected subset of $L + N$ is a subset either of L or of N. Therefore there exists a subsequence $m_1, m_2, m_3, \cdots$ of the sequence $n_{j+1}, n_{j+2}, n_{j+3}, \cdots$ such that either L or N (say L) contains every continuum of the sequence $g_{m_1}, g_{m_2}, g_{m_3}, \cdots$. Thus L contains every point of the sequence $A_{m_1}, A_{m_2}, A_{m_3}, \cdots$ and also every point of the sequence $B_{m_1}, B_{m_2}, B_{m_3}, \cdots$. But N contains the sequential limit point of the second of these sequences. Therefore N contains a limit point of L. But this involves a contradiction.

THEOREM 80. *No compact regular curve contains uncountably many mutually exclusive nondegenerate continua.*

Theorem 80 is a consequence of Theorem 166 of Chapter I and Theorem 79 of the present chapter.

There exists, in the plane, a compact continuous curve which does not contain uncountably many mutually exclusive nondegenerate continua but which is not a regular curve. Consider the compact continuous curve M of Example 8.

THEOREM 81. *If a compact continuum M contains two mutually exclusive closed point sets H and K and the elements of an infinite collection W of mutually exclusive continua each intersecting H and K then M has a continuum of condensation.*

PROOF. By Theorems 58 and 59 of Chapter I, there exists a sequence α of distinct continua of the collection W converging to a continuum T. Since T intersects both H and K it is nondegenerate. Suppose first that some continuum g of α contains a nondegenerate subcontinuum N of T. Then, since the continua of α are mutually exclusive, no one of them, except g, contains a point of N. But if β denotes the sequence formed by the omission of g from α then every point of N is a sequential limit point of a sequence of distinct points each belonging to some continuum of the sequence β. Hence, in this case, N is a continuum of condensation of M.

Suppose, secondly, that no continuum of the sequence α contains a nondegenerate subcontinuum of T. Suppose there exists a point P of T which is not a limit point of $M - T$. Then there exists a region R containing P such that $M \cdot R$ is a subset of T. The region R intersects some continuum g of the sequence α. There exists a nondegenerate subcontinuum N of g lying in R and therefore in T. This involves a contradiction. Therefore if no continuum of α contains a nondegenerate subcontinuum of T then T is a continuum of condensation of M.

Thus, in any case, M has a continuum of condensation.

That the continuum T referred to in the first paragraph of the above proof of Theorem 81 is not necessarily a continuum of condensation of M may be seen from the following example.

EXAMPLE 11. In the plane, let M denote the point set consisting of a circle J plus its interior I. There exist two mutually exclusive arcs H and K and a sequence $t_1, t_2, t_3, \cdots$ of mutually exclusive arcs, all lying in I, such that, for each n, t_n intersects both H and K and each point of M is at a distance less than $1/n$ from some point of t_n. Let W denote the collection of all arcs of this sequence α. The continuum M is the sequential limiting set of α but it is not a continuum of condensation of itself.

DEFINITION. If the arc AB is a subset of the point set M and no point of AB except its endpoints is a limit point of $M - AB$ then the point set $AB - (A + B)$ will be called a *free segment* with respect to M.

THEOREM 82. *If M is a compact continuum with no continuum of condensation, H is a countable collection of totally disconnected closed subsets of M*

and A and B are two points of $M - H^*$ then A and B are separated from each other in M by a finite subset of $M - H^*$.

PROOF. By Theorem 11, every subcontinuum of M is a continuous curve. Every arc lying in M contains a free segment with respect to M. Otherwise it would be a continuum of condensation of M, contrary to hypothesis. Let K denote the set of all points P of M, if there are any, such that P belongs to no free segment with respect to M. The point set K is closed. Hence if it contained a nondegenerate connected point set it would contain a nondegenerate continuum and that continuum would be a continuous curve and would contain an arc and therefore a free segment with respect to M contrary to the supposition that it is a subset of K. Therefore K is a totally disconnected closed subset of M.

Suppose A and B are two points of $M - H^*$. If either A or B belongs to a free segment with respect to M then, with the aid of Theorem 57 of Chapter I, it may be seen that A and B are separated from each other in M by a point set lying in that segment and consisting of two points of $M - H^*$. Suppose both A and B belong to K. Then, by Theorem 42 of Chapter I, K is the sum of two mutually exclusive closed point sets K_A and K_B containing A and B respectively. Let G denote the set of all free segments with respect to M with one endpoint in K_A and the other one in K_B. There exists an arc AB in M. Every such arc contains at least one segment belonging to the collection G. Therefore G exists. If there were infinitely many segments in G then, by Theorem 81, M would have a continuum of condensation, contrary to hypothesis. Hence the set G is finite. In each segment of this collection select just one point not belonging to H^* and let T denote the finite set of points so selected. Every arc lying in M and joining a point of K_A to a point of K_B contains a segment of the collection G and therefore a point of T. But M is a continuous curve. Therefore, by Theorems 12, 6, 2 and 1, T separates A from B in M. Hence M is a regular curve.

THEOREM 83. *If M is a locally compact and compactly connected subcontinuum of the continuous curve T and O is a point of M at which M is not connected im kleinen then there exists a subcontinuum K of M containing O such that K is not connected im kleinen at O and no point of K separates any two points of K from each other in T.*

PROOF. By hypothesis and Theorem 11, there exist a domain D containing O, a sequence α of distinct points $P_1, P_2, P_3, \cdots$ converging to O, a continuum L and a sequence of mutually exclusive continua $M_1, M_2, M_3, \cdots$ such that (1) $M \cdot \bar{D}$ is compact, (2) the continuum L contains O, (3) for each n, M_n is a component of $M \cdot \bar{D}$ containing P_n and a point of the boundary of D, (4) the sequence $M_1, M_2, M_3, \cdots$ converges to L and no continuum of this sequence contains a point of L.

Let A denote some definite point of L distinct from O. The point O is not separated from A in M by any point of M. Hence it is not separated from

A in T by any point of M. It follows that if a point of $M - A$ separates a point of M from O in T then it separates it from $A + O$ in T. Therefore, by Theorem 15, if the point P of M is separated from O in T by some point of M and O does not separate P from A in T, and A does not separate P from O in T, then there exists, in M, a point X_P which separates O from P in T but which is not itself separated from O in T by any point of M. Let N denote the set of all points P of M such that no point separates P from O in T. That N is a continuum may be established by a very slight modification of Case I of the argument to prove Theorem 123 of Chapter I. Let Z denote the component of $T - O$ that contains $L - O$ and let K denote the common part of N and $Z + O$. The point set K is a continuum.

Suppose K is connected im kleinen at O. Then there exists a number m such that K contains no point of the sequence $P_{m+1}, P_{m+2}, P_{m+3}, \cdots$ but every point of this sequence belongs to Z and no point of it is separated from O in T by A. For each n, let B_n denote P_{m+n} and let O_n denote the point X_{B_n}. For each n, O_n belongs to K and $T - O_n$ is the sum of two mutually separated point sets U_n and V_n containing B_n and O respectively. Let H_n denote the point set $M \cdot U_n + O_n$. For each n, H_n is a continuum containing O_n and B_n. There exists an ascending sequence of positive integers $n_1, n_2, n_3, \cdots$ such that $O_{n_1}, O_{n_2}, O_{n_3}, \cdots$ converges to some point F and $H_{n_1}, H_{n_2}, H_{n_3}, \cdots$ are mutually exclusive. Let R denote a region containing F. Suppose infinitely many of the continua $H_{n_1}, H_{n_2}, H_{n_3}, \cdots$ intersect $S - R$. Let R_1 denote a region containing F such that $\bar{R}_1$ is a subset of R. There exists an ascending subsequence $m_1, m_2, m_3, \cdots$ of the sequence $n_1, n_2, n_3, \cdots$ such that, for every i, O_{m_i} belongs to R_1 and H_{m_i} contains a point of $S - R$. For each i, H_{m_i} contains a continuum H_i' lying wholly in $\bar{R} - R_1$ and containing both a point of $\bar{R}_1$ and a point of the boundary of R. The limiting set of the sequence $H_1', H_2', H_3', \cdots$ contains a point F' lying in $R - \bar{R}_1$. For each i, let K_i denote the component of $T \cdot (\bar{R} - R_1)$ that contains H_i'. The continua of the sequence $K_1, K_2, K_3, \cdots$ are mutually exclusive and its limiting set contains the point F'. Therefore, by Theorem 14, T is not connected im kleinen at F'. But this is contrary to hypothesis. It follows that each region that contains F contains all but possibly a finite number of the continua $H_{n_1}, H_{n_2}, H_{n_3}, \cdots$. Therefore F is the sequential limit point of the sequence $B_{n_1}, B_{n_2}, B_{n_3}, \cdots$. Hence F is identical with O. Therefore there exists a region Q containing F such that $Q \cdot K$ is a subset of the component of $K \cdot D$, and therefore of W the component of $M \cdot \bar{D}$, that contains F. There exists a positive integer δ such that, for every i greater than δ, H_{n_i} is a subset of Q. But H_{n_i} is a continuum containing B_{n_i} and the point O_{n_i} of K. Hence $B_{n_{\delta+1}}$ and $B_{n_{\delta+2}}$ both belong to W. But they belong respectively to $M_{m+n_{\delta+1}}$ and $M_{m+n_{\delta+2}}$, two distinct components of $M \cdot \bar{D}$. Thus the supposition that K is connected im kleinen at O has led to a contradiction.

THEOREM 84. *If S is connected and the closed point set K separates the point A from the point B then some closed subset of K is irreducible with respect to the property of being a point set that separates A from B.*

PROOF. Let D_A denote the component of $S - K$ that contains A, let D_B denote the component of $S - \bar{D}_A$ that contains B and let β denote the boundary of D_B. The closed point set β separates A from B and it is a subset of K. Suppose T is a proper subset of β. Let P denote some point of $\beta - T$. The point sets $D_A + P$ and $D_B + P$ are connected and they both contain P. Hence their sum is a connected point set containing both A and B but no point of T. Therefore T does not separate A from B in M.

THEOREM 85. *If S is connected, K is a closed and compact point set and the closed subset H of $S - K$ does not separate any two points of K from each other then there exists a continuum containing K but no point of H.*

PROOF. For each point P of K there exists a connected domain R_P containing P and such that $\bar{R}_P$ contains no point of H. Let U denote the collection of all such domains R_P for all points P of K. The closed and compact point set K is covered by a finite subcollection W of the collection U. To each region of the collection W add its boundary and let V denote the finite collection of continua so obtained. For each continuum v of the collection V let P_v denote a definite point of v. For each two continua x and y of V, let P_xP_y denote a definite arc from P_x to P_y containing no point of H. Let T denote the sum of all such arcs. The point set $V^* + T$ is a continuum containing K but no point of H.

THEOREM 86. *Suppose that S has no cut point and that if A and B are two points every region that contains A contains a closed and compact point set separating A from B. Then if the point O does not belong to the closed and compact point set M and D is a domain containing O there exists a connected domain which contains O but no point of M and which is bounded by a compact point set that lies in D and contains no point of M.*

PROOF. For each point X of M there exists in D a closed and compact point set T_X such that $S - T_X$ is the sum of two mutually separated point sets D_X and E_X containing O and X respectively. Since M is closed and compact and, for each X, D_X is a domain therefore there exists a finite set $X_1, X_2, \cdots, X_n$ of points of M such that M is a subset of $E_{X_1} + E_{X_2} + \cdots + E_{X_n}$. The point set $T_{X_1} + T_{X_2} + \cdots + T_{X_n}$ is a closed and compact subset T of D and $S - T$ is the sum of the two mutually separated domains $D_{X_1} \cdot D_{X_2} \cdot D_{X_3} \cdots D_{X_n}$ and $E_{X_1} + E_{X_2} + \cdots + E_{X_n}$ which contain O and M respectively. The component of $D_{X_1} \cdot D_{X_2} \cdot D_{X_3} \cdots D_{X_n}$ that contains O is bounded by a subset β of T. The point set β is a closed and compact subset of D containing no point of M.

THEOREM 87. *Suppose that S has no cut point and that if A and B are two points every region that contains A contains a closed and compact point set separating A from B. Then if P is a point of a closed and compact point set M and D is a domain containing P there exists a connected domain I containing P such that β, the boundary of I, is compact and $I \cdot M + \beta$ is a subset of D.*

PROOF. The point set $M \cdot (S - D)$ is a closed and compact subset of $S - P$. Hence, by Theorem 86, there exists a connected domain I containing P such that (1) the boundary of I is a subset of D and (2) I contains no point of $M \cdot (S - D)$ and therefore $I \cdot M$ is a subset of D.

THEOREM 88. *Suppose that S has no cut point and that if A and B are two points every region that contains A contains a closed and compact point set separating A from B. Then if H and K are two mutually exclusive closed and compact point sets and L is a closed point set containing no point of H and G is a collection of regions covering H there exists a domain D containing H such that (1) $\bar{D}$ contains no point of K and the boundary of D is a compact point set containing no point of L, (2) D is the sum of a finite number of mutually exclusive connected domains Q such that Q is the sum of a finite number of connected domains each bounded by a closed and compact subset of some region of G.*

PROOF. If P is a point of H then, by Theorem 87, there exists a connected domain D_P containing P such that β_P, the boundary of D_P, is compact and $\beta_P + (H + K) \cdot D_P$ is a subset of the common part of $S - (K + L)$ and some region of the collection G. Let W_1 denote the collection of all domains D_P for all points P of H. The point set H is covered by a finite subcollection W_2 of the collection W_1. Let D denote the sum of all the domains of the collection W_2.

THEOREM 89. *If S has no cut point, ω is a point, M is a closed and compact point set not containing ω, and $W_1, W_2, W_3, \cdots$ is a sequence satisfying, with respect to M, all the requirements of Theorem 117 of Chapter I then, for each closed and compact point set T which does not intersect M, there exists a number δ such that if a closed point set K separates a point P of M from ω and lies in some region of W_δ which contains P then K separates P from T.*

PROOF. Suppose there exists a closed and compact point set T for which there is no such number δ. Then, for each positive integer n, there exists a region g_n belonging to W_n and containing both a point P_n of M and a closed point set which separates P_n from ω but not from T. There exists an ascending sequence of positive integers $n_1, n_2, n_3, \cdots$ such that if, for each i, P'_i denotes the term of the sequence $P_1, P_2, P_3, \cdots$ whose subscript is n_i, then $P'_1, P'_2, P'_3, \cdots$ converges to a point P of M. For each i, let W'_i denote the term of the sequence $W_1, W_2, W_3, \cdots$ whose subscript is n_i and let g'_i denote the term of $g_1, g_2, g_3, \cdots$ with the same subscript. By hypothesis, the domain $S - P$ is connected. Hence, by Theorem 29, there exists a continuum L containing $T + \omega$ and lying in $S - P$. For each n, there exists a closed point set K_n, lying in g'_n, such that $S - K_n$ is the sum of two mutually separated point sets one containing ω and the other one containing both P'_n and some point of T, and therefore of L. Since L is connected it contains a point X_n belonging to K_n and therefore to g'_n. Since, for each n, X_n belongs to a region of W'_n, and therefore to a region of W_n, some subsequence α of the sequence $X_1, X_2, X_3, \cdots$ converges to a point X of M. Since each point of α belongs to L therefore X belongs also to L. Hence it is distinct from P. Therefore there exists a number m such that if x, y and z are regions each belonging to one of the collections $W_m, W_{m+1}, W_{m+2}, \cdots$, and x intersects y and y intersects z and x contains P then z does not contain X. There exist regions x and z belonging to W_m and containing P and X respectively. There exists a number j, greater than m, such that P'_j belongs to x and X_j belongs to z. The region g'_j contains P'_j and X_j and lies in some region y of W_m. Since x intersects y and y intersects z, the region z does not contain X. This involves a contradiction.

THEOREM 90. *Suppose that S has no cut point and that if A and B are two points every region that contains A contains a closed and compact point set separating A from B. Then if M is a closed and compact point set there exists a sequence of domains $D_1, D_2, D_3, \cdots$ such that* (1) *for each n, $\bar{D}_{n+1}$ is a subset of D_n,* (2) *for each n, the boundary of D_n is compact,* (3) *if D is a domain containing M there exists a number δ such that, for every n greater than δ, the boundary of D_n is a subset of D,* (4) *if D is a domain containing M and bounded by a compact point set then, for some n, D_n is a subset of D,* (5) *M is the common part of the domains of this sequence.*

Theorem 90 may be established with the aid of Theorem 117 of Chapter I and Theorems 88 and 89 of the present chapter.

THEOREM 91. *Under the hypothesis of Theorem* 90, *no domain whose boundary is compact has uncountably many components.*

PROOF. Suppose there exists a domain D bounded by a compact point set M and having uncountably many components. There exists a sequence of domains $D_1, D_2, D_3, \cdots$ satisfying, with respect to M, all the conditions of Theorem 90. If x is a component of D there exists a positive integer n_x such that $\bar{D}_{n_x}$ does not contain x. There exists a positive integer k such that, for uncountably many components x of D, n_x is k. For each such component x of D, the common part of x and β_k, the boundary of D_k, exists and is a domain with respect to β_k. But since β_k is compact there do not exist uncountably many mutually exclusive domains with respect to it. This involves a contradiction.

It is not true that if, for every two points A and B, every region that contains A contains a closed and compact point set that separates A from B, then either no domain with a compact boundary has uncountably many components or, for each point P, there exists a sequence of domains D_1, D_2, $D_3, \cdots$, each having a compact boundary, such that the set whose only element is P is their common part. Consider the following example.

EXAMPLE 12. Suppose E is a Cartesian space of three dimensions. Interpret the word point to mean a point of E and call a point set M a region if and only if either M is the interior of a sphere with center at O or M is a segment, not containing O, of some straight line through O. There exists no countable collection of domains with compact boundaries such that O is their common part. Furthermore the set of all points distinct from O is a domain whose boundary is the set whose only element is O and this domain has uncountably many components.

THEOREM 92. *Every simple link of a locally compact continuous curve is itself a continuous curve.*

PROOF. Suppose, on the contrary, that there exists a continuum K which is not a continuous curve but which is a simple link of a locally compact continuous curve M. By Theorem 11, there exist a domain D and an infinite sequence of mutually exclusive continua M_1, M_2, $M_3, \cdots$ such that (1) $M \cdot \bar{D}$ is compact, (2) for each n, M_n is a component of $K \cdot \bar{D}$ containing points of D and at least one point of the boundary of D, (3) the sequence M_1, M_2, $M_3, \cdots$ converges to a nondegenerate continuum T that has no point in common with any continuum of this sequence. Let A and B denote two distinct points of $T \cdot D$. There exist two mutually exclusive regions R_A and R_B containing A and B respectively and lying wholly in D. There exist regions D_A and D_B containing A and B respectively and such that $M \cdot D_A$ and $M \cdot D_B$ are subsets of components of $M \cdot R_A$ and $M \cdot R_B$ respectively. There exists a number k such that the continuum M_k contains points A' and B' belonging to D_A and D_B respectively. There exist arcs $A'A$ and $B'B$ lying in $M \cdot R_A$ and $M \cdot R_B$ respectively. The arc $A'A$ contains an arc A_1A_2 which has in common with $T + M_k$ only its endpoints A_1 and A_2 which belong to T and M_k respectively. Similarly the arc $B'B$ contains an arc B_1B_2 which has in common with $T + M_k$ only its endpoints B_1 and B_2 which belong to T and M_k respectively. Since M_k is a component of $K \cdot \bar{D}$ not containing A_1 therefore the arc A_1A_2 contains a point X not belonging to K. Let XA_1 and XA_2 denote intervals of A_1A_2 with endpoints as indicated. By Theorems 27, 53 and 121 of Chapter I, the continuum T contains a point P which is a proper point of M with respect to G, the collection of all degenerate subsets of M. Suppose Y is a point of $M - (P + X)$. If Y does not belong to $T + XA_1$ then $T + XA_1$ is a connected subset of $M - Y$. If Y belongs to XA_1 but not to T then $B'B + XA_2 + M_k + T$ is a connected subset of $M - Y$. If Y belongs to T and $M - Y$ is the sum of two mutually separated point sets H and L containing P and X respectively then the connected point set $\bar{M}_k + s(A_1A_2) + s(B_1B_2)$ is a subset of H and all but a finite number of the continua

M_1, M_2, M_3, $\cdots$ lie in L and therefore, since Y is not both A_1 and B_1, either A_1 or B_1 is in L which is impossible since each of them is a limit point of H. Thus, in any case, Y does not separate X from P in M. Therefore, by Theorem 124 of Chapter I, X belongs to K. Thus the supposition that Theorem 92 is false leads to a contradiction.

THEOREM 93. *If A and B are two points of a locally compact continuous curve M then in order that there exist no point that separates A from B in M it is necessary and sufficient that there exist a simple closed curve lying in M and containing both A and B.*

PROOF. The necessity of this condition is a consequence of Theorems 12 and 59. Its sufficiency is a consequence of Theorems 108 and 28 of Chapter I.

DEFINITIONS (G. T. WHYBURN). The continuous curve M is said to be *cyclically connected* if and only if every two points of M lie together in some simple closed curve which is a subset of M. A subset E of a continuous curve M is said to be a *cyclic element* of M if and only if either (1) E is a cut point or an endpoint of M or (2) E is a nondegenerate cyclically connected continuous curve which is not a subset of any other cyclically connected continuous curve which is a subset of M.

THEOREM 94. *Every simple link of a compact continuous curve M is a cyclic element of M.*

PROOF. Suppose K is a simple link of M. By Theorem 92, K is a continuous curve. Let G denote the collection of all degenerate subsets of M. There exists a proper point P of M with respect to G such that M_{GP} is identical with K. If K is degenerate, then, by Theorem 134 of Chapter I, the only element of K is either a cut point of M or a non-cut point O of M such that each point of M distinct from O is separated from O, in M, by some point. In the latter case, by Theorem 60, O is an endpoint of M. Hence, in either case, K is, by definition, a cyclic element of M.

Suppose K is nondegenerate. Let Y and Z denote any two distinct points of K. By Theorem 128 of Chapter I, no point separates Y from Z in M and therefore, by Theorem 93, there is a simple closed curve J lying in M and containing both Y and Z. By Theorems 27, 53 and 104 of Chapter I, the curve J contains a point O which is a proper point of M with respect to G. By Theorem 93, J is a subset of M_{GO}. Hence M_{GP} and M_{GO} have both Y and Z in common. Therefore, by Theorem 130 of Chapter I, M_{GO} is identical with M_{GP}. Therefore J is a subset of M_{GP}. Hence M_{GP}, that is to say K, is cyclically connected. Suppose N is a cyclically connected continuous curve containing K and lying in M. Let X denote any point of N distinct from P. Since it is cyclically connected, N contains a simple closed curve containing both X and P. Therefore, by Theorem 93, no point separates X from P in M. Hence X belongs to M_{GP}. Therefore N is identical with K. Therefore K is a cyclic element of M.

THEOREM 95. *Every nondegenerate cyclic element of a compact continuous curve M is a simple link of M.*

PROOF. Suppose K is a nondegenerate cyclic element of M. The continuum K contains a point P which is a proper point of M with respect to G, the collection of all degenerate subsets of M. Suppose X is any point of K distinct from P. There exists a simple closed curve lying in K and containing both P and X. Hence no point separates X from P in M. Therefore X belongs to M_{GP}. Thus K is a subset of M_{GP}. But, by Theorem 94, M_{GP} is a cyclic element of M. Hence K is identical with M_{GP}.

THEOREM 96. *If K is a degenerate cyclic element of a compact continuous curve M then in order that K be a simple link of M it is necessary and sufficient that it not be a subset of any nondegenerate cyclic element of M.*

PROOF. This condition is sufficient. For suppose P, the only element of K, does not belong to any nondegenerate cyclic element of M. By Theorem 127 of Chapter I, there exists a simple link K' of M containing P. By Theorem 95, K' is a cyclic element of M. Hence K' is degenerate and therefore it is identical with K, and K is a simple link of M.

This condition is also necessary. For suppose, on the contrary, that K is a simple link of M and that it is a subset of a nondegenerate cyclic element K' of M. By Theorem 95, K' is a simple link of M. Thus the simple link K of M is a proper subset of the simple link K' of M, contrary to Theorem 131 of Chapter I.

THEOREM 97. *Every endpoint of a locally compact continuous curve M is a simple link of M.*

Theorem 97 is a consequence of Theorem 60.

THEOREM 98. *No compact continuous curve has uncountably many nondegenerate simple links.*

PROOF. Suppose there exists an uncountable set G of nondegenerate simple links of the compact continuous curve M. By Theorem 118 of Chapter I, there exists an element g_1 of G containing two points A and B such that if a finite collection of regions properly covers $A + B$ then there exist uncountably many elements of G each intersecting every region of that collection. Let R_A and R_B denote two mutually exclusive connected domains with respect to M containing A and B respectively. There exists a continuum g_2 belonging to G, distinct from g_1, and containing a point A_2 of R_A and a point B_2 of R_B. There exist two arcs AA_2 and BB_2 lying wholly in R_A and R_B respectively. There exists a point P of g_1 such that g_1 is the set of all points X of M such that X is not separated from P in M by any one point. There exists a point C belonging to g_2 but not to g_1. There exists a point Y separating C from P in M. But, by Theorem 94, g_1 and g_2

are cyclic elements of M. Hence Y is not a cut point of g_1 or of g_2. Hence either $(g_1 + g_2 + AA_2) - Y$ or $(g_1 + g_2 + BB_2) - Y$ is a connected subset of $M - Y$. This involves a contradiction.

DEFINITION. A *simple closed surface* is a nondegenerate compact continuous curve M having no cut point and such that (1) if the simple closed curve J is a subset of M, $M - J$ is the sum of two mutually separated connected point sets each having J as its boundary with respect to M and (2) if D is a connected domain with respect to M and AB is an arc lying wholly in D except possibly for the point B then $(D + B) - AB$ is connected.

DEFINITION. A *cactoid* is a compact continuous curve M such that every nondegenerate simple link of M is a simple closed surface.

DEFINITION. A cactoid M is said to be *aspiculate* if and only if it does not contain any arc t such that no point of t except its endpoints is a limit point of $M - t$. An aspiculate cactoid M is said to be *simple* if and only if no two cut points of M are separated in M by infinitely many different points.

CHAPTER III

CONSEQUENCES OF AXIOMS 0, 1, 2, 3 AND 4

AXIOM 3. *If O is a point, $S - O$ is nondegenerate and connected.*

AXIOM 4. *If J is a simple closed curve, $S - J$ is the sum of two mutually separated connected point sets such that J is the boundary of each of them.*

NOTATION. Unless otherwise indicated, AC denotes an arc whose endpoints are A and C and ABC denotes an arc having A and C as its endpoints and containing the point B which is distinct from A and C. If AC or ABC or α is an arc then $s(AC)$ or $s(ABC)$ or $s(\alpha)$ denotes the segment of that arc whose endpoints are its endpoints. Thus $s(AC)$ denotes $AC - (A + B)$ and $s(ABC)$ denotes $ABC - (A + B)$. If ABC is an arc, $s(ABC)$ will sometimes be referred to as the *segment* ABC. If A, B and C are three points of a simple closed curve J, then by the *arc* ABC *of* J is meant the arc ABC which is a subset of J and by the *segment* ABC *on* J is meant the segment AC of the arc ABC of J.

If M is a closed point set not identical with S then, by Theorem 2 of Chapter I, every component of $S - M$ is a domain.

DEFINITION. If M is a closed point set distinct from S, every component of $S - M$ is called a *complementary domain* of M.

It follows from Axiom 4, and Theorem 28 of Chapter I, that *each simple closed curve has only two complementary domains.*

DEFINITION. If J is a simple closed curve and M is a point set lying wholly in one of its complementary domains, then the complementary domain of J which contains M will be called the *exterior* of J with respect to M and the complementary domain of J which contains no point of M will be called the *interior* of J with respect to M.

CONVENTION. Select some definite point and call it *the point at infinity* and denote it by the symbol ω.

DEFINITION. The point set M is said to be *bounded* if and only if the point at infinity is neither a point of M nor a limit point of M. If J is a bounded simple closed curve, then the interior of J with respect to ω will be called the *interior* of J and the exterior of J with respect to ω will be called the *exterior* of J.

THEOREM 1. *Space is connected.*

PROOF. By Axioms 0, 1 and 3, there exist three points A, B and C and $S - A$ and $S - B$ are connected. But these two point sets have the point C in common and their sum is S. Hence S is connected.

THEOREM 2. *If the boundaries of the connected domains D_1 and D_2 are mutually exclusive continua, but D_1 and D_2 have a point in common, and $S - \bar{D}_1$ and $S - \bar{D}_2$ have a point in common, then either D_1 is a subset of D_2 or D_2 is a subset of D_1.*

PROOF. Suppose neither of the domains D_1 and D_2 is a subset of the other one. Since the connected point set D_1 intersects both D_2 and $S - D_2$ therefore it intersects β_2, the boundary of D_2. But β_2 is connected and it contains no point of β_1, the boundary of D_1. It follows that β_2 is a subset of D_1. Similarly β_1 is a subset of D_2. Let E_1 and E_2 denote the components of $S - \bar{D}_1$ and $S - \bar{D}_2$ respectively that contain the point P. If E_1 is a subset of E_2 then $E_2 + \beta_2$ contains the boundary of E_1 and therefore a point of β_1 contrary to the fact that β_1 is a subset of D_2. Suppose E_1 is not a subset of E_2. Then it is a connected point set intersecting both E_2 and $S - E_2$ and therefore it contains a point belonging to the boundary of E_2 and therefore to β_2, contrary to the fact that β_2 is a subset of D_1. Thus the supposition that neither of the domains D_1 and D_2 is a subset of the other one leads to a contradiction.

THEOREM 3. *If I_1 and I_2 are the interiors of the bounded simple closed curves J_1 and J_2 respectively and J_2 is a subset of $\bar{I}_1$, then I_2 is a subset of I_1.*

PROOF. The point set $S - \bar{I}_1$ contains at least one point, namely ω, that does not belong to I_2. If it contains also a point of I_2 then $S - \bar{I}_1 = S_1 + S_2$ where S_1 is a subset of I_2 but no point of S_2 belongs to I_2. But $S - \bar{I}_1$ is connected and no point of the domain I_2 can be a limit point of a point set S_2 containing no point of I_2. Hence S_2 contains a limit point of S_1. Thus S_2 contains a point of the boundary of I_2, that is to say, a point of J_2. But this is contrary to hypothesis. It follows that I_2 is a subset of $\bar{I}_1$. If I_2 contained a point of the boundary of I_1 then it would contain a point of $S - \bar{I}_1$. It follows that I_2 is a subset of I_1.

THEOREM 4. *No region is a subset of an arc.*

PROOF. Suppose, on the contrary, that the region R is a subset of the arc AB. Then R contains a point O belonging to AB but distinct from A and B. By Axiom 3, $S - O$ is connected. It is also a domain. Hence, by Theorem 1 of Chapter II, there exists an arc AXB lying wholly in $S - O$. There exist on AB two points A_1 and B_1 such that A_1 is the last point that AXB has in common with the interval AO of the arc AB, and B_1 is the first point, following A_1 on AXB, that AXB has in common with the interval OB of AB. The interval A_1OB_1 of AB and the interval B_1A_1 of AXB together constitute a simple closed curve J. There exists a region R_1 containing O but no point of the closed point set $AA_1 + B_1B$, where AA_1 and B_1B are intervals of AB or degenerate point sets. The regions R and R_1 have in common a region R_2 containing O. Since R_2 is a subset of R that contains no point of $AA_1 + B_1B$ it must be a subset of A_1OB_1 and therefore of J. But O is on the boundary of D, one of the complementary domains of J. Therefore, since R_2 contains O, R_2 must contain a point of D. Thus the supposition that R is a subset of AB leads to a contradiction.

THEOREM 5. *If the simple closed curve J does not contain the point at infinity and the pair of points A and B separates the points C and D on J and AXB is an arc such that $s(AXB)$ is a subset of I, the interior of J, then* (1) *I, the interior of the closed curve $AXBCA$, is a subset of I,* (2) *$s(ADB)$ is entirely without the boundary of I_1 and* (3) *I_1 has no point in common with I_2, the interior of $AXBDA$.*

PROOF. That I_1 is a subset of I is a consequence of Theorem 2. Hence I_1 contains no point of ADB.

Suppose that I_1 and I_2 have a point in common. Since the boundary of I_2 contains points that are neither in I_1 nor on the boundary of I_1, I_2 is not a subset of I_1. Hence $I_2 = S_1 + S_2$, where S_1 is a subset of I_1 but no point of S_2 belongs to I_1. The point set S_1 can not contain a limit point of S_2. Hence, since I_2 is connected, S_2 must contain a point P which is a limit point of S_1. Clearly P must be on the boundary of S_1. Thus I_2 contains a point of AXB or of ACB. But this is contrary to hypothesis and the fact that ACB is without the boundary of I_2. It follows that I_1 and I_2 have no point in common.

THEOREM 6. *Under the same hypothesis as in Theorem 5, $I = s(AXB) + I_1 + I_2$.*

PROOF. Suppose it is not true that $I = s(AXB) + I_1 + I_2$. Then, by hypothesis and Theorem 5, $I = s(AXB) + I_1 + I_2 + Y$, where the point sets $s(AXB)$, I_1, I_2 and Y are mutually exclusive. Let E denote a point of Y. By Theorem 1 of Chapter II, there exists an arc EX lying entirely in I. There exists a point O (Figure 6) which is the first point that EX has in common with AXB. Let OE denote the interval of EX whose endpoints are O and E. Now (1) $OE - O$ is connected, (2) every point of $OE - O$ belongs to I_1 or I_2 or Y, (3) no one of these three sets contains a point or limit point of either of the other two. It follows that since E belongs to Y therefore $OE - O$ is a subset of Y. Let F denote a point of I_1. Let AZO and BWO denote intervals of the arc AXB with endpoints as indicated. There exists a connected domain T containing Z but no point of the closed point set $F + OWB + J$. The domain T contains a point G of I_1. There is an arc FG in I_1, and an arc GZ in T. The point set $FG + GZ$ contains an arc FH such that H is in T and on the boundary of I_1 (and therefore on AZO) while $FH - H$ is a subset of I_1. Similarly there exists an arc FK such that K is a point of OWB and $FK - K$ is a subset of I_1. The point set $FH + FK$ contains an arc HP_1K. The segment $s(HP_1K)$ is a subset of I_1. Likewise there exist a point H_1 between A and H, a point K_1 between B and K, and an arc $H_1P_2K_1$ such that $s(H_1P_2K_1)$ is a subset of I_2. Let U denote the simple closed curve $H_1P_2K_1 + K_1K + KP_1H + HH_1$. By Theorem 3, N, the interior of U, is a subset of I. But P_1 is on U and is therefore a limit point of N. Moreover P_1 belongs to I_1. It follows that N must contain a point of I_1. Similarly P_2 is a limit point of N, and N contains a point of I_2. Since D and E are without $AXBCA$ therefore there exists an arc ED lying without that simple closed curve. There exists on ADB a point L which is the first point that ED has in common with ADB. The point set $EL - L$ lies in I and contains no point of AXB. It follows that $EL - L$ is a subset of Y. But $EO - O$ is a subset of Y. Hence there exists an arc OL such

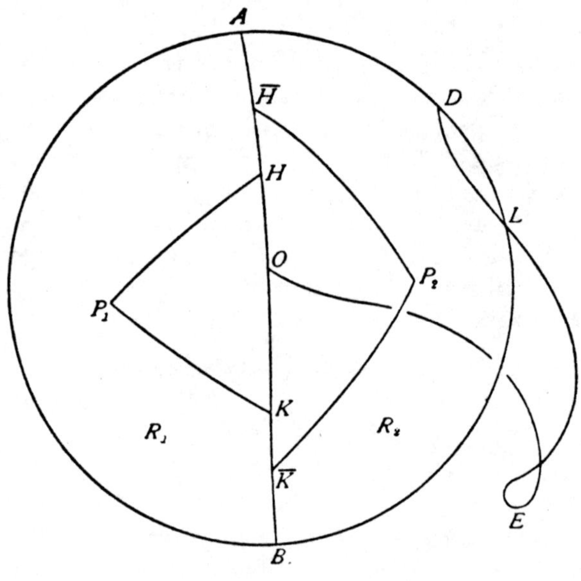

FIGURE 6

that $s(OL)$ is a subset of Y. Hence OL contains no point of U. But L is in the exterior of U. Therefore O, and consequently $s(HOK)$ is in the exterior of U. Since J is in the exterior of U and the exterior of U and the connected point sets $AH_1 - H_1$ and $BK_1 - K_1$ intersect J and contain no point of U therefore these two point sets lie in the exterior of U. It follows that N contains no point of AXB. Hence every point of N belongs to I_1, I_2 or Y. But these three point sets are mutually separated and it has been shown that N intersects both I_1 and I_2. Hence N is not connected. Thus the supposition that Y exists leads to a contradiction. It follows that $I = I_1 + I_2 + s(AXB)$.

THEOREM 7. *If the pair of points A and B separates the points C and D from each other on the simple closed curve J and AXB is an arc lying, except* *for its endpoints, in the complementary domain I of J then AXB separates C from D in $I + J$.*

Theorem 7 is a consequence of Theorems 5 and 6.

THEOREM 8. *If A and B are points and ACB, ADB and AFB are three bounded arcs no two of which have any point in common except A and B, then the interiors of the simple closed curves $ACBFA$, $ACBDA$ and $ADBFA$ are not mutually exclusive.*

PROOF. Denote the curves $ACBFA$, $ACBDA$ and $ADBFA$ by J, J_1, J_2, their interiors by I, I_1, I_2 and their exteriors by E, E_1, E_2 respectively. Suppose that I, I_1 and I_2 are mutually exclusive. Then E contains $s(ADB)$, I_1 and I_2. The domain E contains the point at infinity and no one of the sets I_1, I_2 and ADB does so. Hence E contains points not belonging to any one of these sets. Let Y denote the set of all such points. It is clear that Y is composed of all points that are common to E, E_1 and E_2. Let P (Figure 7) denote a point of Y. There exists an arc PD in E. Let O

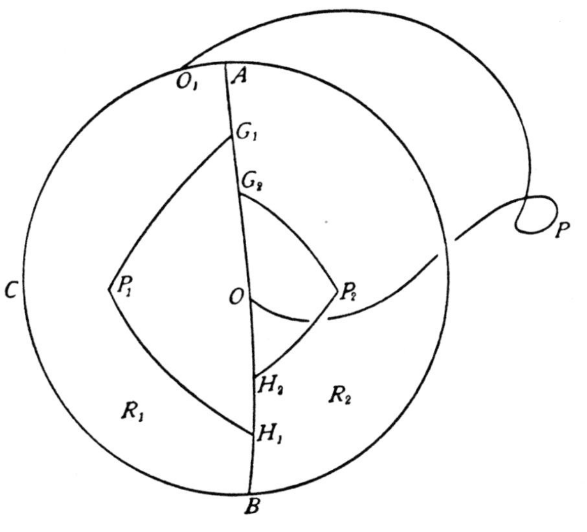

FIGURE 7

denote the first point that PD has in common with ADB. The point set $PO - O$ is a subset of Y. Similarly there exists an arc PO_1 such that O_1 belongs to ACB and $PO_1 - O_1$ is a subset of Y. It can be shown (compare proof of Theorem 6) that there exist points G_1, G_2, H_1, H_2 lying on ADB in the order $AG_1G_2OH_2H_1B$ and arcs $G_1P_1H_1$, $G_2P_2H_2$ such that $s(G_1P_1H_1)$ and $s(G_2P_2H_2)$ lie in I_1 and I_2 respectively. Let J', J_1', J_2' denote the simple closed curves $G_1P_1H_1H_2P_2G_2G_1$, $G_1P_1H_1OG_1$, $G_2P_2H_2OG_2$ respectively. Let I', I'_1, I'_2 denote their respective interiors. It follows from Theorem 3 that I'_1 and I'_2 are subsets of I_1 and I_2 respectively.

There are three cases to be considered.

Case I. Suppose that O lies in I'. Then, by Theorem 6, $I' = I'_1 + I'_2 + s(G_2OH_2)$. It follows that if O is a limit point of a point set M then M contains a point of I'_1 or of I'_2 or of $s(G_2OH_2)$. But O is a limit point of $PO - O$, and $PO - O$ contains no point of $I'_1 + I'_2 + s(G_2OH_2)$. Thus the supposition that O is in I' leads to a contradiction.

Case II. Suppose that A is in I'. Then J lies in I'. But O_1 belongs to J and PO_1 contains no point of J'. Hence P is in I'. Similarly every other point of Y is in I'. But Y contains ω. Hence Y is not a subset of I'. Thus the supposition that A is in I' leads to a contradiction.

Case III. Suppose that neither O nor A is in I'. Then no point of J or of ADB is in I'. But P_1 is a limit point of I' and P_1 belongs to I_1. Hence I' contains a point of I_1 but no point of the boundary of I_1. Hence it is a subset of I_1. Similarly, it is a subset of I_2, contrary to the fact that I_1 and I_2 are mutually exclusive. Thus the supposition that I' contains neither O nor A leads to a contradiction.

It follows that I, I_1 and I_2 are not mutually exclusive.

THEOREM 9. *If the pair of points A and B separates the points C and D from each other on the bounded simple closed curve J and AXB is a bounded arc such that $s(AXB)$ is without J then* (1) *either D is without $AXBCA$ or C is without $AXBDA$,* (2) *if D is without $AXBCA$ then C is within $AXBDA$ and the interior of $AXBDA = s(ACB) +$ the interior of $ACBDA +$ the interior of $AXBCA$.*

PROOF. First suppose that D is not without $AXBCA$. Then the segment $s(ADB)$ is within $AXBCA$. Hence, by Theorem 5, C is without $AXBDA$.

Suppose, secondly, that D is without $AXBCA$. Then the segment $s(ADB)$ is without $AXBCA$. Let I and I_1 denote the interiors of $ACBDA$ and $AXBCA$ respectively. Suppose that I and I_1 have a point in common. Then $I_1 = M + M_1$ where M is a subset of I but no point of M_1 belongs to $\bar{I}$. The point sets M and M_1 are mutually separated. But I_1 is connected. Thus a contradiction is obtained. It follows that if D is without $AXBCA$ then the interiors of $AXBCA$ and $ACBDA$ are mutually exclusive. If, at the same time, C were without $AXBDA$ it would follow that the interiors of $ACBDA$, $AXBDA$ and $AXBCA$ are mutually exclusive. But this would be contrary to Theorem 8. It is thus established that if D is without $AXBCA$ then C is within $AXBDA$. It follows, by Theorem 6, that in this case the interior of $AXBDA = s(ACB) +$ the interior of $AXBCA +$ the interior of $ACBDA$.

THEOREM 10. *If AXB, ACB and ADB are three bounded arcs no two of which have in common any point except their endpoints A and B, and I_1, the* *interior of the simple closed curve $AXBCA$, is a subset of E, the exterior of the simple closed curve $ACBDA$, then $I_1 + s(ACB)$ is a subset of the interior of $AXBDA$.*

PROOF. Since I_1 is a subset of E, $s(AXB)$ contains no point of the interior of $ACBDA$. But it contains no point of $ACBDA$. Hence it is a subset of E. Therefore, by Theorem 9, either D is without $AXBCA$ or C is without $AXBDA$ and if C is without $AXBDA$ then D is within $AXBCA$ and the interior of $AXBCA = s(ADB) +$ the interior of $AXBDA +$ the interior of $ACBDA$ and thus the interior of $AXBCA$ has that of $ACBDA$ as a subset, contrary to the supposition that it is a subset of the exterior of $ACBDA$.

THEOREM 11. *If, on the bounded simple closed curve J, the pair of points A and B separates the points C and D, and AXB and CYD are bounded arcs, and $s(AXB)$ and $s(CYD)$ are either both within or both without J, then $s(AXB)$ and $s(CYD)$ have at least one point in common.*

PROOF. *Case* I. Suppose that $s(AXB)$ and $s(CYD)$ are both within J. By Theorem 6, $I = I_1 + I_2 + s(AXB)$, where I, I_1 and I_2 are the interiors of J, $AXBCA$ and $AXBDA$ respectively. There exists a region K containing C but no point of $\bar{I}_2$. Since C is a limit point of $s(CYD)$, K contains a point P of $s(CYD)$. The point P is in I_1 while D is without $AXBCA$. Hence the interval PD of the arc CYD contains a point of the boundary of I_1. Every such point must be on $s(AXB)$.

Case II. Suppose that $s(AXB)$ and $s(CYD)$ are both without J. Then, by Theorem 9, either C is without $AXBDA$ or D is without $AXBCA$. Suppose that C is without $AXBDA$. Then, by Theorem 9, $I_1 = I + s(ADB) + I_2$. The point D is a limit point of $s(CYD)$. Hence I_1 contains a point P belonging to $s(CYD)$. But $s(CYD)$ contains no point of $\bar{I}$. Hence P belongs to I_2. Therefore the interval CP of the arc CYD has a point in common with the boundary of I_2. Every such point must be on $s(AXB)$.

THEOREM 12. *Under the hypothesis of Theorem 11, the interiors of $AXBDA$ and $CYDBC$ have a point in common.*

THEOREM 13. *If J and C are bounded simple closed curves, O is a point of J but not of C, A_1 and A_2 are points common to C and J and A_1XA_2 is an arc of C such that $s(A_1XA_2)$ lies within J, then there exist two points O_1 and O_2 distinct from O such that if A_1O, A_2O, O_1O and O_2O denote intervals of the arc A_1OA_2 of J, with endpoints as indicated, then (1) O_1 and O_2 lie on the arcs A_1O and A_2O respectively, (2) there exists an arc O_1YO_2 of the curve C such that $s(O_1YO_2)$ is within J, (3) if B_1 and B_2 are points on the arcs O_1O and O_2O respectively such that there exists on C, from B_1 to B_2, an arc which, except for its endpoints, lies entirely within J, then B_1 is O_2 and B_2 is O_2.*

PROOF. Let M_1 denote the set of all points P on A_1O such that P can be joined to some point of A_2O by an arc of C which, except for its endpoints, lies within J and let M_2 denote the set of all points P on A_2O such that P can be joined to some point of A_1O by such an arc of C. No two segments of a simple closed curve have a point in common unless one of them contains an endpoint of the other one. It follows that if x and y are two segments on C each lying wholly within J but having one endpoint in M_1 and the other one in M_2 then x and y have no point in common. There exist, on A_1O and A_2O respectively, points F_1 and F_2 such that the arc F_1OF_2 of J contains no point of C. The interval A_1F_1 of A_1O and the interval A_2F_2 of A_2O are mutually exclusive closed point sets and M_1 is a subset of A_1F_1 and M_2 is a subset of A_2F_2. Hence, by Theorem 110 of Chapter I, M_1 and M_2 are finite point sets. Therefore there exist two points O_1 and O_2 such that O_1 is the last point of M_1 on A_1O and O_2 is the last point of M_2 on A_2O. Suppose there exists no segment on C lying within J and with O_1 and O_2 as endpoints. Then there exist two points X_1 and X_2 and two segments O_1X_2 and O_2X_1 on C such that (a) X_1 and X_2 belong to A_1O and A_2O respectively and (b) the segments O_1X_2 and O_2X_1 lie wholly within J. But, in this case, the pair of points X_1 and O_2 separates X_2 from O_1 on J and therefore, by Theorem 11, O_1X_2 and O_2X_1 have a point in common. But this involves a contradiction. Hence there exists a segment on C lying wholly within J and having O_1 and O_2 as its endpoints. It follows that the points O_1 and O_2 satisfy all the requirements of Theorem 13.

THEOREM 14. *If J and C are bounded simple closed curves, O is a point on J but not on C, A_1 and A_2 are points common to C and J, and A_1XA_2 is an arc of C such that the interior of the simple closed curve $A_1XA_2OA_1$ is a subset of the exterior of J then there exist two points O_1 and O_2, distinct from O, such that* (1) *on the arc O_1OO_2 of J, O is between O_1 and A_2 and between O_2 and A_1,* (2) *there is, on the curve C, an arc O_1YO_2 such that $s(O_1YO_2)$ and the interior of the simple closed curve $O_1YO_2OO_1$ are subsets of the exterior of J,* (3) *if the exterior of J contains $s(B_1ZB_2)$ and the interior of the simple closed curve $B_1ZB_2OB_1$ where B_1 and B_2 are on that arc of J from O_1 to O_2 which does not contain O, and B_1ZB_2 is an arc of C, then one of the points B_1 and B_2 is O_1 and the other one is O_2.*

PROOF. Let N denote the simple closed curve $A_1OA_2XA_1$. Since the interior of N is a subset of the exterior of J therefore J is not N and it contains no point of the interior of N. Hence J contains a point F_1 without N. There exists an arc $F_1\omega$ lying in the exterior of N. This arc contains an arc $F\omega$ having only the point F in common with J. Let A_1F and A_2F denote intervals of the arc A_1FA_2 of J with endpoints as indicated. Let W denote the set of all intervals t of A_1XA_2 such that t lies wholly without J except for its endpoints, which belong to J. Suppose each arc of the set W either has both of its endpoints on A_1F or has both of them on A_2F. For each i ($i = 1$ or 2), let M_i denote either A_iF or A_iF plus every arc of W whose endpoints belong to A_iF according as there is not or is at least one such arc of W. With the help of Theorem 110 of Chapter I, it may be shown that M_1 and M_2 are mutually separated. But their sum contains the arc A_1XA_2 and this arc intersects both of them, contrary to Theorem 28 of Chapter I. It follows that there exists an interval $A_1'X'A_2'$ of A_1XA_2 lying wholly in the exterior of J except for its endpoints, which belong to A_1F and A_2F respectively. Let β denote the simple closed curve which is the sum of the arc $A'X'A_2'$ and the arc $A_1'OA_2'$ of J. Since $F\omega$ contains no point of N, F is without β. It follows, with the help of Theorem 9, that the interior of β is a subset of the exterior of J. There exists a connected domain R containing O but no point of the arc $A_1'FA_2'$ of J. The domain R contains a point Z without J. There exists an arc $Z\omega$ lying in the exterior of J. There exists an arc OZ in R. Let Q denote the continuum $OZ + Z\omega$. Let H denote the set of all arcs XY such that (a) the points X and Y belong to the arc $A_1'FA_2'$ (b) XY lies, except for its endpoints, wholly in the exterior of J and (c) the interior of the simple closed curve formed by XY and the arc XOY of J is a subset of the exterior of J. Suppose there are infinitely many arcs in the set H. Then there exists, on A_1FA_2, a point P such that every region that contains P contains an endpoint, distinct from P, of some arc of the set H. There exists a region R_1 containing P but no point of Q and there exists a region R_2 containing P such that $\bar{R}_2$ is a subset of R_1. If XY is an arc of the set H then, by Theorem 9, O is within the simple closed curve formed by XY and the interval of $A_1'FA_2'$ whose endpoints are X and Y. But ω is without this curve and the continuum Q contains both O and ω. Hence, since it contains no point of $A_1'FA_2'$, Q must contain a point of XY. Thus every arc of the set H contains a point without R_1. But infinitely many arcs of the set H intersect R_2. Each of these arcs intersects each of the two mutually exclusive closed point sets $\bar{R}_2$ and $S - R_1$ and no two of them intersect each other. But this is in contradiction with Theorem 110 of Chapter I. Hence the collection H is finite. The truth of Theorem 14 may now be established by an argument analogous to that employed in a similar connection in the proof of Theorem 13.

THEOREM 15. *If the bounded simple closed curve C encloses the point O of the bounded simple closed curve J but does not enclose the interior of J then, if D denotes either the interior or the exterior of J, there exists a simple closed curve Q containing the point O and such that* (1) *every point of Q belongs either to C or to J,* (2) *the interior of Q is a subset both of D and of I, the interior of C,* (3) *if P is a point of J lying without C then P is separated from O, on J, by a pair of points which are the extremities of a segment on C which lies wholly in D and is also a segment on Q.*

PROOF. Suppose that D is the interior of J. The domain I contains points of E, the exterior of J. Hence, since I is connected and E is not a subset of I, E contains a point of C. Since the connected point set D contains both a point of I and a point not belonging to I, therefore D contains a point of C, the boundary of I. Thus C contains a point within J as well as a point without J and therefore it has at least two points in common with J. It follows that there exist, on J, two points A and B belonging to C such that no point of C, except A and B, is on the arc AOB of J. If X is a point of C that lies in D then C and J have, in common, two points A'_X and B'_X such that the arc $A'_X X B'_X$ of C lies, except for its endpoints, in D. By Theorem 13, there exists, on C, an arc $A_X B_X$ such that (1) A_X is A'_X or is on J in the order $B'_X A'_X A_X O$, (2) B_X is B'_X or is on J in the order $A'_X B'_X B_X O$, (3) $s(A_X B_X)$ is in D, (4) if Y and Z are points belonging respectively to the interval $A_X O$ of the arc $B'_X A'_X O$ of J and the interval $B_X O$ of the arc $A'_X B'_X O$ of J and such that if an arc of C from Y to Z lies, except for its endpoints, in D, then $Y = A_X$ and $Z = B_X$. It is clear that the arc $A_X B_X$ is completely determined by X. For every X on C that lies in D construct the corresponding $A_X B_X$. Let Q denote the point set which is composed of all the arcs $A_X B_X$ so constructed together with every point F on J which has the property that for no X is F separated from O on J by the pair of points A_X and B_X. It may be easily proved that Q is a simple closed curve. By Theorem 3, I', the interior of Q, is a subset of D. Since it contains the point O of Q, the domain I contains a point of I'. But I' is connected and contains no point of C, the boundary of I. Hence I' is a subset of I. Suppose now that P is a point of J lying without C. Then P does not belong to Q. Hence there exists on C a point X such that P is separated from O, on J, by $A_X + B_X$. But A_X and B_X are the endpoints of a segment on C which lies wholly in D and is also a segment on Q.

A similar argument may be given in case D is the exterior of J.

THEOREM 16. *If O and P are points of J, the boundary of the simple domain D, and C is a simple closed curve separating O from P then C contains an arc t separating O from P in $\bar{D}$ and lying wholly in D except for its endpoints, whose sum separates O from P on J.*

PROOF. Since P does not belong to C there exists a region R containing P but no point of C. The region R contains a point ω not belonging to $\bar{D}$. With respect to ω regarded as the point at infinity, D is the interior of C and P lies without C. Hence, by Theorem 15, P is separated from O on J by a pair of points which are the extremities of a segment on C which lies in D. By Theorem 7, the closure of this segment separates O from P in $\bar{D}$.

THEOREM 17. *If J_1 and J_2 are two bounded simple closed curves whose interiors I_1 and I_2 have a point O in common then there exists a simple closed curve J such that* (1) *every point of J belongs either to J_1 or to J_2, and* (2) *$I_1 + I_2$ is a subset of I, the interior of J.*

PROOF. *Case* I. Suppose J_1 and J_2 do not have more than one point in common. Since I_1 and I_2 are distinct there is a point of one of them that does not belong to the other one. Suppose I_2 contains a point not belonging to I_1. Then since I_2 is connected and intersects both I_1 and $S - I_1$, it contains a point A of J_1. If J_1 contains a point B of E_2 then each arc of J_1 from A to B contains a point of J_2, contrary to the hypothesis that J_1 and J_2 do not have two points in common. Hence J_1 is a subset of $I_2 + J_2$ and therefore, by Theorem 3, I_1 is a subset of I_2. In this case J is J_2. If I_1 contains a point not belonging to I_2, J is J_1.

Case II. Suppose J_1 and J_2 (Figure 8) have at least two points in common. If every point of J_1 is within or on J_2 then J is J_2. Suppose the point P of J_1 is without J_2 and some point of J_2 is without J_1. Then there exist two points A' and B', common to J_1 and J_2, such that the arc $A'PB'$ of J_1 lies, except for its endpoints, entirely without J_2. The points A' and B'

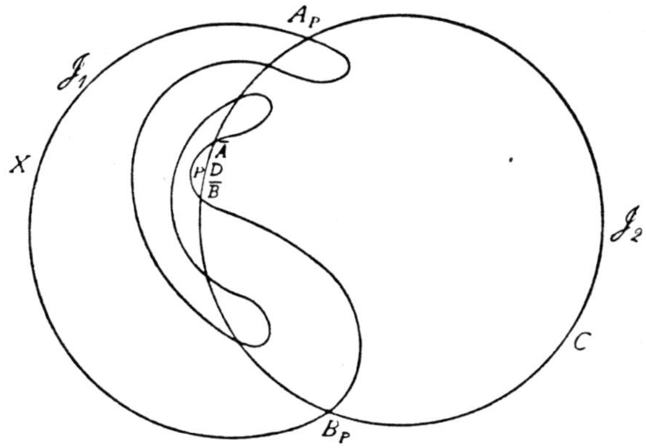

FIGURE 8

are the extremities of two arcs of J_2. By Theorem 9, one of these arcs, $A'CB'$ is such that if $A'DB'$ denotes the other one then C is without $A'PB'DA'$ and D is within $A'PB'CA'$ and the interior of $A'PB'CA'$ = $A'DB' - (A' + B')$ + the interior of $A'PB'DA' + I_2$ so that the interior of $A'PB'DA'$ is a subset of the exterior of J_2. By Theorems 9 and 14, there exist, on the arc $B'CA'$, points A_P and B_P such that the pair of points A' and A_P does not separate B' from B_P on this arc and furthermore (1) there exists an arc A_PXB_P of J_1, lying entirely without J_2 except for its endpoints, such that if C' is a point on J_2 that is separated from D on J_2 by the pair of points A_P and B_P then $A_PXB_PC'A_P$ encloses I_2, (2) if Z and W are points on A_PC' and B_PC' respectively and there is, on J_1, an arc ZVW, lying entirely without J_2, except for its endpoints, such that $ZVWCZ$ encloses I_2, then $Z = A_P$ and $W = B_P$. The curve $A_PXB_PC'A_P$ contains or encloses every point of the curve $PA'A_PC'B_PB'P$ and every point of the curve J_2. For every point P which is on J_1 but without J_2, construct the corresponding arc

A_PXB_P. Call this arc t_P and let k_P denote the interval of J_2 from A_P to B_P which has the property that the interior of the simple closed curve $t_P + k_P$ is entirely without J_2. Let J denote the point set composed of all such arcs t_P together with the set of all points Y on J_2 such that for no P is Y on the arc k_P. It may be proved that J is a simple closed curve satisfying all the requirements of Theorem 17.

THEOREM 18. *If the sum of the interiors of the elements of the finite set G of bounded simple closed curves is a connected point set then there exists a simple closed curve J such that* (1) *every point of J belongs to some curve of the set G.* (2) *the interior of J contains the interiors of all the curves of the set G.*

THEOREM 19. *If the point O of the bounded simple closed curve J is within every curve of the finite set G of bounded simple closed curves, and the interior of J is not a subset of the interior of any curve of the set G then there exist simple closed curves J_1 and J_2, containing O, such that* (1) *every point on J_1 or on J_2 belongs to J or to some curve of the set G,* (2) *every point within J_1 is without J while every point within J_2 is within J, and* (3) *every point within J_1 or within J_2 is within every curve of the set G.*

THEOREM 20. *If the point O of the bounded simple closed curve J is without every member of the finite set G of bounded simple closed curves and every curve of G has at least one point in the exterior of J then there exists a simple closed curve J' containing O such that* (1) *every point of J' belongs either to J or to some curve of the set G, and* (2) *every point within J' is within J and without every curve of the set G.*

DEFINITION. A complementary domain of a simple closed curve will be called a *simple domain* and the closure of a simple domain will be called a *simple disk*.

THEOREM 21. *No simple domain has a cut point.*

PROOF. Suppose I, the interior of the simple closed curve J, has a cut point O. There exists an arc OA lying in I except for the point A which belongs to J. There exist (1) a component D_1 of $I - O$ containing $s(OA)$ and (2) a component D_2 of $I - O$ distinct from D_1. Let B denote a point of D_2. There exists, in I, an arc BO. The segment $s(BO)$ is a connected subset of $I - O$ containing the point B of D_2. Hence $s(BO)$ is a subset of D_2 and therefore it contains no point of OA. Hence $OA + BO$ is an arc from A to B. Since $S - O$ is a connected domain it contains an arc α from B to A. Let C denote the first point of J in the order from B to A on α and let BC denote the interval of α whose endpoints are B and C. Since $BC - C$ is a connected subset of $I - O$ containing the point B of D_2 therefore $BC - C$ is a subset of D_2 and therefore contains no point of $OA - A$. Let F denote the last point in the order from B to C that BC has in common with OB and let OF and FC denote intervals of OB and BC respectively with endpoints as indicated. If AC denotes A or an arc of J with endpoints at A and C according as C is, or is not, identical with A, then $OA + OF + FC + AC$ is a simple closed curve whose interior D is a subset of $I - O$. Every point of $s(OA)$ is a limit point of D and so is every point of $s(FC)$. Hence $D + s(OA) + s(BC)$ is a connected subset of $I - O$. But it intersects both D_1 and D_2. This involves a contradiction.

THEOREM 22 (J. M. SLYE). *If J_1 and J_2 are two mutually exclusive simple closed curves there exists a connected domain whose boundary is $J_1 + J_2$.*

PROOF. Since it is a connected subset of their sum, J_2 is a subset of one of the complementary domains of J_1. Call this domain I_1, let ω denote some point in E_1, the other complementary domain of J_1, regard ω as the point at infinity and let I_2 and E_2 denote the interior and exterior respectively of J_2 from this point of view. Let A_1 and A_2 denote two points belonging to J_1 and J_2 respectively. There exists an arc A_1A_2. This arc contains an arc B_1B_2 having only B_1 in common with J_1 and only B_2 in common with J_2. Let D denote $I_1 - (I_2 + J_2)$. The segment $s(B_1B_2)$ is a subset of D. Suppose D is not connected. Then there exists a component K of D not containing $s(B_1B_2)$. Since it is a subset of D, K contains no limit point of $J_1 + E_1$. By Theorem 2, K is a domain. Hence it contains no limit point of $D - K$. It follows that K contains no limit point of $(D - K) + J_1 + E_1$. Since K is a component of D, $D - K$ contains no limit point of K. But the sum of K and the point set $(D - K) + J_1 + E_1$ is the connected set E_2. Hence some point L_1 of $J_1 + E_1$ is a limit point of K. The exterior of J_1 contains no limit point of K. So L_1 belongs to J_1. Similarly some point L_2 of J_2 is a limit point of K. There exist two mutually exclusive connected domains D_1 and D_2 containing L_1 and L_2 respectively such that D_1 contains no point of J_2 and D_2 contains no point of J_1. There exist two points X_1 and X_2 of K belonging to D_1 and D_2 respectively. Since K is a connected domain, there exists an arc X_1X_2 lying in K. There exist arcs L_1X_1 and L_2X_2 lying in D_1 and D_2 respectively. The point set $L_1X_1 + X_1X_2 + L_2X_2$ contains an arc C_1C_2 lying in D except for its endpoints C_1 and C_2 which belong to J_1 and J_2 respectively. The segment $s(C_1C_2)$ is a connected subset of D that intersects K. Hence it is a subset of K.

If C_2 is distinct from B_2, let F_2 denote C_2. Suppose C_2 is identical with B_2. Let X and Y denote two points belonging to $s(C_1C_2)$ and J_2 respectively. By Theorem 21 of this chapter and Theorem 1 of Chapter II, there exists, in $I - C_2$, an arc XY from X to Y. This arc contains an arc F_1F_2 which has only F_1 in common with C_1C_2 and only F_2 in common with J_2. Since $F_1F_2 - F_2$ is a connected subset of I intersecting the component K of I, therefore it is a subset of K. Hence it has no point in common with B_1B_2. The sum of the arc F_1F_2 and the interval of C_1C_2 whose endpoints are C_1 and F_1 is an arc C_1F_1 having only C_1 on J_1 and only F_2 on J_2. Whether or not C_2 is identical with B_2, the arcs B_1B_2 and C_1F_2

have no point in common other than B_2 and there exist an arc F_2B_2 of J_2 and a subcontinuum B_1C_1 of J_1 such that (1) B_1C_1 is an arc from B_1 to C_1 if B_1 is distinct from C_1 and is the set whose only element is B_1 if C_1 is identical with B_1, (2) the point set $B_1C_1 + B_1B_2 + B_2F_2 + C_1F_2$ is a simple closed curve whose interior I is a subset of D. Every point of $s(B_1B_2)$ is a limit point of I and so is every point of $s(C_1F_2)$. Hence $I + s(B_1B_2) + s(C_1F_2)$ is a connected subset of D. But $s(B_1B_2)$ and $s(C_1F_2)$ are subsets of different components of D. Thus the supposition that D is not connected has led to a contradiction.

THEOREM 23. *If I is a simple domain, the subspace, of the original space, whose points are the points of I satisfies Axioms* 0 *and* 1 – 4.

PROOF. By Theorem 12 of Chapter II, this space Σ' satisfies Axioms 0, 1 and 2. By Theorem 21, it satisfies Axiom 3. Let J denote the boundary of I and let ω denote some point of $S - (J + I)$. With respect to this point ω, I is the interior of J. Suppose J' is a simple closed curve lying in I. Let I' denote the interior of J' and let E' denote $I - (J' + I')$. By Theorem 22, E' is connected. Suppose P is a point of J'. The point P is a limit point of $S - J'$ but not of $J + E$. Hence it is a limit point of $(S - J') - (J + E)$, that is to say of E'. Thus $I - J'$ is the sum of two mutually separated connected point sets I' and E' and, in Σ', it is the boundary of each of them. Hence Axiom 4 holds true in the space Σ'.

THEOREM 24. *If G is a finite collection of mutually exclusive simple disks, $S - G^*$ is connected.*

PROOF. Suppose this proposition is not a consequence of Axioms 0 and 1–4. Then it follows from Theorem 22 that there exists a positive integer k greater than 1 such that if, for each space Σ, S_Σ denotes the set of all points of Σ then (1) if, in a space Σ satisfying these axioms, G is a collection of k mutually exclusive simple disks, $S_\Sigma - G^*$ is connected, but (2) statement (1) does not hold true if k is replaced by $k + 1$. There exists a space Σ satisfying Axioms 0 and 1–4 in which there is a collection G of $k + 1$ mutually exclusive simple disks, $g_1, g_2, g_3, \cdots, g_{k+1}$ such that $S - G^*$ is not connected. Let Σ' denote the subspace of Σ whose points are the points of $S_\Sigma - g_{k+1}$. By Theorem 23, Σ' satisfies Axioms 0 and 1–4. Therefore $(S - g_{k+1}) - (g_1 + g_2 + \cdots + g_k)$ is connected. But this point set is identical with $S - (g_1 + g_2 + \cdots + g_{k+1})$. This involves a contradiction.

THEOREM 25. *If M is a finite point set, $S - M$ is connected.*

PROOF. Suppose this proposition is not a consequence of Axioms 0 and 1–4. Then it follows from Axiom 3 that there exists a positive integer k such that if, for each space Σ, S_Σ denotes the set of all points of Σ then (1) if, in a space Σ satisfying these axioms, M is a set of not more than k points, $S_\Sigma - M$ is connected but (2) statement (1) does not remain true if k is replaced by $k + 1$. There exists a space Σ, satisfying Axioms 0 and 1–4, in which there is a set H of $k + 1$ points such that $S_\Sigma - H$ is not connected. Let A denote a point of H. By Axiom 3, $H - A$ exists. The point set $S_\Sigma - (H - A)$ is connected. Let X denote a point of $H - A$. There exists an arc AX. Let B denote the first point of $H - A$ in the order from A to X on AX and let AB denote the interval of AX whose endpoints are A and B. Let D denote the component of $S_\Sigma - H$ that contains $s(AB)$ and let C denote a point of $S_\Sigma - H$ not belonging to D. Since $S_\Sigma - (H - B)$ and $S_\Sigma - (H - A)$ are connected there exist arcs CA and CB lying in $S_\Sigma - (H - A)$ and $S_\Sigma - (H - B)$ respectively. If A' is the first point of AB in the order from C to A on CA the point set $CA' - A'$ is a connected subset of $S_\Sigma - H$ containing the point C of D. But A' is a limit point of $CA' - A'$ and it is not B. Suppose A' is not A. Then it belongs to the component of $S - H$ that contains C. But it belongs to $s(AB)$ and therefore to D. Hence A' is A. So CA has only A in common with AB. Similarly CB has only B in common with AB. But $CA + CB$ contains an arc AYB. The point set $AYB + AB$ is a simple closed curve. Let I denote its interior. The point set I is a simple domain in which there are not more than $k - 1$ points of H. Hence, by Theorem 23, the set of all points of I that are not in H is a connected point set W. The point set W has $AYB + AB$ as its boundary. So $W + AYB + AB$ is a connected subset of $S - H$. But $s(AYB)$ and $s(AB)$ lie in different components of $S - H$. This involves a contradiction.

THEOREM 26. *If D is a connected domain bounded by two mutually exclusive simple closed curves J_1 and J_2, A_1 and B_1 are points of J_1, A_2 and B_2 are points of J_2 and A_1A_2 and B_1B_2 are two mutually exclusive arcs lying, except for their endpoints, wholly in D, then there exist two connected domains D_1 and D_2, two arcs $A_1X_1B_1$ and $A_1Y_1B_1$ of J_1 and two arcs $A_2X_2B_2$ and $A_2Y_2B_2$ of J_2 such that* (1) *D_1 is bounded by $A_1X_1B_1 + B_1B_2 + A_2X_2B_2 + A_1A_2$, D_2 is bounded by $A_1Y_1B_1 + B_1B_2 + A_2Y_2B_2 + A_1A_2$ and* (2) *$D = D_1 + D_2 + s(A_1A_2) + s(B_1B_2)$.*

PROOF. The curve J_1 is the sum of two arcs $A_1X_1B_1$ and $A_1Y_1B_1$, and J_2 is the sum of two arcs A_2UB_2 and A_2VB_2. Let A_2WB_2 denote the arc $A_1A_2 + A_1XB_1 + B_1B_2$. No two of the arcs A_2UB_2, A_2VB_2 and A_2WB_2 have any point in common except A_2 and B_2. Let Q denote the component of $S - J_1$ that contains J_2. By Theorem 2, Q contains a complementary domain Q_1 of the curve $A_2UB_2 + A_2WB_2$, a complementary domain Q_2 of the curve $A_2VB_2 + A_2WB_2$ and a complementary domain Q_3 of the curve J_2. Since Q_3 contains no point of A_2WB_2 therefore either U is in Q_2 or V is in Q_1. In the first case let X_2, Y_2, D_1 and D_3 denote U, V, Q_1 and Q_2 respectively and in the second case let them denote V, U, Q_2 and Q_1 respectively. In either case, $D_3 = s(A_2X_2B_2) + D_1 + Q_3$ and $Q = D_3 + s(A_1A_2Y_2B_2B_1) + D_2$ where $A_1A_2Y_2B_2B_1$ denotes the arc $A_1A_2 + A_2Y_2B_2 + B_2B_1$ and D_2 is one of the complementary domains of the curve formed by the arcs $A_1Y_1B_1$ and $A_1A_2Y_2B_2B_1$. Hence $Q = D_1 + D_2 + s(A_1A_2) + s(B_1B_2) + J_2 + Q_3$ and therefore $D = D_1 + D_2 + s(A_1A_2) + s(B_1B_2)$.

THEOREM 27. *If H and K are two mutually exclusive compact continua and for every two points X and Y belonging to H and K respectively there exists a simple closed curve separating X from Y then there exists a simple closed curve separating H from K.*

PROOF. Suppose X is a point of H. If Y is a point of K there exists a simple closed curve J_{XY} separating X from Y. Let G_X denote a collection such that g belongs to it if and only if, for some point Y of K, g is the complementary domain of J_{XY} that contains Y. By Theorem 4 of Chapter I, K is covered by some finite subcollection G'_X of G_X. With respect to X regarded as the point at infinity, each domain of the collection G'_X is the interior of the simple closed curve which constitutes its boundary. Hence, by Theorem 18, there exists a simple closed curve J_X such that every point of J_X belongs to the boundary of some domain of G'_X and the interior of J_X contains every domain of the collection G'_X. The curve J_X separates X from K.

Let ω denote some definite point of K. Let G denote a collection such that g belongs to it if and only if, for some point X of H, g is the complementary domain of J_X that contains X and is therefore the interior of J_X with respect to ω regarded as the point at infinity. The point set H is covered by some finite subcollection G' of G. Since H is connected and each domain of the collection G' contains a point of H therefore the sum of all the domains of G' is connected. Hence, by Theorem 18, there exists a simple closed curve J such that every point of J belongs to the boundary of some domain of G' and the interior of J contains every domain of G' and therefore contains H. The simple closed curve J does not intersect K, and E, the exterior of J, contains ω, a point of K. Since K is connected, it follows that E contains K. Hence J separates H from K.

THEOREM 28. *If D is a simple domain and each of the two mutually exclusive continua H and K is a subset of $\bar{D}$ and each of them intersects J, the boundary of D, then no pair of points of H separates two points of K from each other on J.* (Proved by Harlan Cross Miller without the First Edition assumption that H and K are compact.)

PROOF. Suppose, on the contrary, that two points C and F of K are separated from each other on J by the pair of points A and B of H. Let ACB and AFB denote the arcs of J from A to B that contain C and F respectively. With respect to some point ω of $S - (D + J)$ regarded as the point at infinity, D is the interior of J. Let E denote its exterior with respect to ω. There exist two mutually exclusive domains D_C and D_F such that

(1) D_C contains, and each component of it intersects, $K\cdot(ACB)$, (2) D_F contains, and each component of it intersects, $K\cdot(AFB)$, (3) D_C does not intersect $H + AFB$ and D_F does not intersect $H + ACB$. For each point P of K not belonging to J let D_P denote a definite connected domain containing P but no point of $J + H$. Let G denote a collection such that g belongs to it if and only if g is a component of D_C or of D_F or is D_P for some point P of $D\cdot K$. The connected domain G^* contains an arc CF from C to F. Let C_1 denote the last point in the order from C to F that this arc has in common with ACB and let F_1 denote the first point following C_1 that it has in common with AFB. The segment C_1F_1 (Figure 9) of the arc CF is connected and it contains no point of J. Hence it is a subset either of D or of

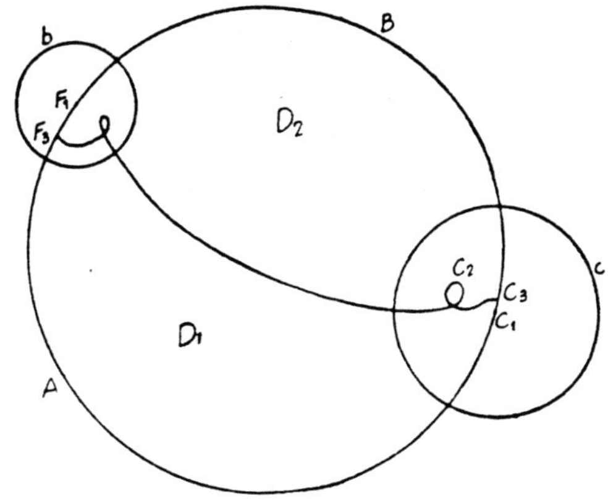

FIGURE 9

E. If it is a subset of E it is a subset of the sum of the two mutually separated point sets $E\cdot D_C$ and $E\cdot D_F$. But it intersects both of these point sets. This involves a contradiction. Hence $s(C_1F_1)$ is a subset of D. The pair of points C_1 and F_1 separates A from B on J. Let J_1 denote the simple closed curve formed by C_1F_1 and the arc C_1AF_1 of J and let J_2 denote the one formed by C_1F_1 and the arc C_1BF_1 of J. By Theorem 6, D is the sum of D_1, D_2 and $s(C_1F_1)$ where D_1 and D_2 are the interiors of J_1 and J_2 respectively. The continuum H is a subset of $\bar{D}_1 + \bar{D}_2$ intersecting both $\bar{D}_1$ and $\bar{D}_2$ but containing no point of their common part. This involves a contradiction.

That Theorem 28 does not remain true even in the plane if the supposition that H and K are compact continua is replaced by the supposition that they are connected point sets may be shown by an example indicated in Figure 10. Here, of the two points of the arc t indicated by dots, one belongs to H and the other belongs to K.

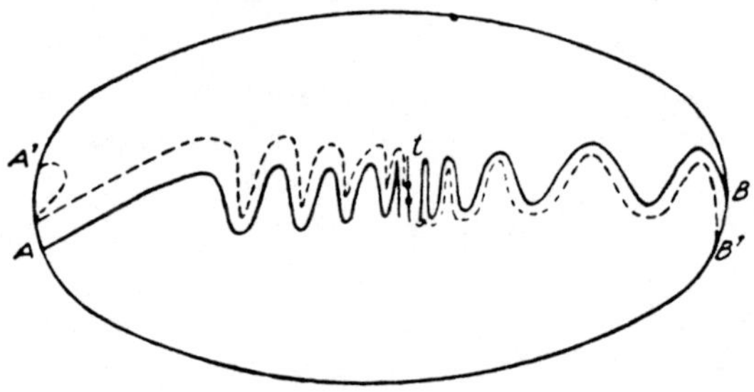

FIGURE 10

THEOREM 29. *If I is a complementary domain of the simple closed curve J then $I + J$ is a continuous curve.*

PROOF. Clearly the continuum $I + J$ is connected im kleinen at every point of I. Suppose R is a connected domain containing a point P of J. There exists a connected domain R' containing P and lying in R such that $J \cdot R'$ is a subset of APB. Suppose X is a point of $R' \cdot (I + J)$ distinct from P. If X belongs to J, the interval XP of APB is a connected subset of $I + J$ lying wholly in R. If X belongs to I, there exists an arc XP lying in R'. Let Y denote the first point the arc XP has in common with J. The interval XY of XP is a subset of $R' \cdot (I + J)$ and therefore of R. Since Y belongs to $J \cdot R'$, it is a point of APB. Let t denote P or the interval YP of APB according as Y is or is not identical with P. The point set $XY + t$ is an arc from X to P lying in R. Thus if X is any point of $R' \cdot (I + J)$ there is an arc XP lying in $R \cdot (I + J)$. Therefore $I + J$ is connected im kleinen at the point P.

THEOREM 30. *If the pair of points A and B separates the points C and D from each other on the simple closed curve J, I is one of the complementary domains of J, and M is a continuum lying in $I + J$ and containing both C and D but not containing either A or B, then M separates A from B in $I + J$.*

PROOF. Suppose M does not separate A from B in $\bar{I}$. By Theorem 29, $I + J$ is a continuous curve. Hence, by Theorem 2 of Chapter II, A and B lie in the same component of $(I + J) - M$. Therefore, by Theorem 12 of Chapter II, there exists an arc AXB in $(I + J) - M$. But this is contrary to Theorem 28.

THEOREM 31. *If D is a connected domain bounded by two mutually exclusive simple closed curves J_1 and J_2 and M is a subcontinuum of $\bar{D}$ containing a point of J_2 and at least two points of J_1, but not containing all of J_1, then there exists an arc AB of J_1 containing $M \cdot J_1$ and such that A and B belong to M and M intersects every subcontinuum of $\bar{D}$ that intersects both AB and J_2.*

Proof. Suppose $A'XB'$ and $A'YB'$ are two intervals of J_1 having only their endpoints in common and such that A' and B' belong to M. Suppose that $\bar{D}$ contains two continua H and K such that H intersects $A'XB'$ and J_2, K intersects $A'YB'$ and J_2 and neither H nor K intersects M. Let E and F denote points of $H \cdot J_2$ and $K \cdot J_2$ respectively. There exist connected domains e and f, containing E and F respectively, such that neither of them intersects $M + J_1$. Let U denote the complementary domain of J_2 that does not contain J_1. The domain U contains points L and T belonging to e and f respectively and there exist an arc LT lying in U and two arcs EL and TF lying in e and f respectively. Let I denote the complementary domain of J_1 that contains J_2. The point set $H + EL + LT + TF + K$ is a continuum lying wholly in $\bar{I}$ and containing two points that are separated from each other on J_1 by the pair of points A' and B' of M. Hence, by Theorem 28, M and N are not mutually exclusive. Therefore M intersects either H or K. Thus a contradiction has been obtained on the supposition that neither of the arcs $A'XB'$ and $A'YB'$ has the property Q that M intersects every subcontinuum of $\bar{D}$ that intersects both that arc and J_2. Therefore, of any two arcs of J_1 whose endpoints belong to M and which have only their endpoints in common, at least one has property Q.

Case I. Suppose every arc of J_1 has property Q. Since M does not contain all of J_1 there exists an arc of J_1 with endpoints in M and containing $M \cdot J_1$. In this case, let AB denote any such arc.

Case II. Suppose not every arc of J_1 has property Q. Then there exists a subcontinuum t of $\bar{D}$ containing a point O of J_1 and a point of J_2 but containing no point of M. Let A and B denote the endpoints of the component of $J_1 - M \cdot J_1$ that contains O. Let AB denote the arc of J_1 from A to B which does not contain O. Since the arc AOB of J_1 does not have property Q, the arc AB does have this property. But AB contains $M \cdot J_1$.

Theorem 32. *If O is a point and K is a closed and compact point set not containing O then there exists a region R containing O such that no two points of K are separated from each other by any point set that lies wholly in R.*

Proof. For each point X of K there exists a domain D_X containing X such that $\bar{D}_X$ does not contain O. Let G denote the collection of all such domains for all points X of K. The point set K is covered by some finite collection of domains $D_1, D_2, \cdots, D_k$ all belonging to G. By Axiom 3 and Theorem 13 of Chapter II, for each i less than n, there exists an arc t_i lying in $S - O$ and joining a point of D_i to a point of D_{i+1}. Let T denote the point set obtained by adding together the point sets $\bar{D}_1, \bar{D}_2, \bar{D}_3, \cdots, \bar{D}_n$ and the arcs $t_1, t_2, t_3, \cdots, t_{n-1}$. The point set T is both closed and connected and it does not contain O. There exists a region R containing O but no point of T. If R contained a point set N separating two points of K from each other then, since T is a connected point set containing K, N and therefore R would contain a point of T. But this would involve a contradiction.

If the supposition that K is compact is removed from the hypothesis of Theorem 32, the resulting proposition is not a consequence of Axioms 0 and 1–4. To see this let O denote the origin and let K denote the point set $P_1 + P_2 + P_3 + \cdots$ in the space Σ of Example 5 of Chapter II.

THEOREM 33 (C. W. VICKERY). *Suppose D is a connected domain such that if B is a point of the boundary of D and AB is an arc lying, except for B, wholly in D then $D - (AB - B)$ is connected. Then if P is a point of D, $D - P$ is connected.*

PROOF. There exists an arc PB from P to some point B of the boundary of D and lying, except for B, wholly in D. By hypothesis, $D - (PB - B)$ is connected. Suppose X is a point of $PB - (P + B)$. Suppose H is a region containing X. There exists a region K containing X and lying in $H \cdot D$. By Theorem 4, K is not a subset of PB. Hence, since it is a subset of D, it contains a point of $D - (PB - B)$. But the latter point set is connected. Hence the sum of the point sets $D - (PB - B)$ and $PB - (B + P)$ is connected, that is to say $D - P$ is connected.

THEOREM 34. *Suppose D is a connected domain whose boundary is a connected point set β such that if B is a point of β and AB is an arc lying, except for B, wholly in D then $D - (AB - B)$ is connected. Then if O is a point of D and ω is a point not belonging to D there exists a simple closed curve lying in D and separating O from ω.*

PROOF. The domain D contains a point X distinct from O. There exists an arc OX lying wholly in D. Let C denote a point of OX distinct from O and from X. By Theorem 33 of the present chapter and Theorem 13 of Chapter II, there exists an arc OC_1X lying wholly in $D - C$. The point set $OX + OC_1X$ contains a simple closed curve J_1. Suppose J_1 does not separate O from ω. Let U denote the component of $S - \beta$ that contains ω. Let B_1 denote some point of J_1 distinct from O. By Axiom 3 and Theorem 13 of Chapter II, there exists, in the domain $S - O$, an arc $B_1\omega$. This arc contains an arc BC (Figure 11) which has only the point B in common with J_1 and only the point C in common with β. Let A_1 denote a point of J_1 distinct from B and from O. By hypothesis, the domain $D - (BC - C)$ is connected. Hence it contains an arc OA_1. Let A denote the first point that the arc OA_1 has in common with J_1 and let N denote either O or the interval of OA_1 whose endpoints are O and A according as O is or is not identical with A. The curve J_1 is the sum of two arcs AF_1B and AH_1B. There exists a connected domain R_1 containing F_1 but no point of the closed point set $N + AH_1B + BC + (S - D)$. The domain R_1 contains a point L not in $J_1 + I$ where I denotes the complementary domain of J_1 that does not contain ω. There exists, in R_1, an arc LF_1. Let F denote the first point that LF_1 has in common with J_1. There exists, in D, an arc LH_1 which has no point in common with $N + AF_1B + BC$. Let H denote the first point that LH_1 has in common with J_1. The point set obtained by adding together the intervals LF and LH of LF_1 and LH_1 respectively contains an arc FZH which, except for its endpoints, is a subset of $D - (J_1 + I)$. Let HAF and HBF denote the arcs of J_1 from H to F that contain A and B respectively. The point set $\alpha + U + BC$ is a connected set containing ω and B and having no point in common with $FZH + HAF$. Therefore B is not separated from ω by $FZH + HAF$. Furthermore Z is not separated from ω by $HAF + HBF$. It follows, by Theorem 9, that A is separated from ω by the simple closed curve J formed by adding together the arcs HZF and HBF. But N is a connected point set containing O and A but no point of J. Therefore J separates A from ω.

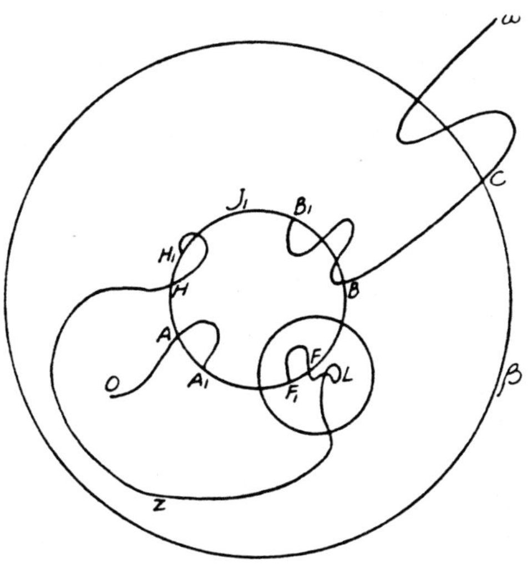

FIGURE 11

THEOREM 35. *If D_1 and D_2 are simple domains containing the point O and each of the mutually exclusive arcs A_1B_1 and A_2B_2 has one endpoint on J_1, the boundary of D_1, and its other endpoint on J_2, the boundary of D_2, but no point, except its endpoints, in common with $O + J_1 + J_2$, then there exists a simple domain D containing O and bounded by a subset of $J_1 + J_2 + A_1B_1 + A_2B_2$ that contains either A_1B_1 or A_2B_2.*

THEOREM 36. *If two bounded simple closed curves have more than one point in common, their sum contains a simple closed curve that encloses both of their interiors.*

Chapter IV

CONSEQUENCES OF AXIOMS 0 AND 1–5

Axiom 5 (F. B. Jones). *If A and B are two points every region that contains A contains a compact continuum that separates A from B.*

Theorem 1. *No point separates any connected domain.*

Proof. Suppose O is a point of a connected domain D such that $D - O$ is the sum of two mutually separated point sets D_A and D_B containing the points A and B respectively. By Theorem 13 of Chapter II and Axiom 3, there exists an arc AXB lying in $S - O$. There exists a connected domain R lying in D and containing O but no point of the arc AXB. There exists an arc from A to B lying in D and therefore containing O. Call one such arc AOB. The domain R contains a compact continuum K that separates O from A. Since AXB is a connected subset of $S - K$ containing A, K separates O from AXB and therefore from B. Hence the continuum K intersects the intervals AO and BO of the arc AOB. But $AO - O$ and $BO - O$ are subsets of D_A and D_B respectively. Thus K intersects both D_A and D_B. But it is a subset of $D_A + D_B$. Since K is connected and D_A and D_B are mutually separated this involves a contradiction.

Theorem 2. *Every connected domain is cyclically connected.*

Theorem 2 is a consequence of Theorems 12 and 59′ of Chapter II and Theorem 1.

Theorem 3. *If A and B are two points there exists a simple closed curve separating A from B.*

Proof. By Theorem 2 there exist two arcs AXB and AYB having only A and B in common. There exist two arcs X_1Y_1 and X_2Y_2 lying in different complementary domains of $AXB + AYB$ except that X_1 and X_2 belong to $s(AXB)$ and Y_1 and Y_2 belong to $s(AYB)$. The sum of the arcs X_1Y_1 and X_2Y_2, the interval X_1X_2 of AXB and the interval Y_1Y_2 of the arc AYB is a simple closed curve. It separates A from B.

Theorem 4. *Every two mutually exclusive compact continua are separated from each other by some simple closed curve.*

Theorem 4 is a consequence of Theorem 3 of this chapter and Theorem 27 of Chapter III.

Theorem 5. *If A and B are two points every region that contains A contains a simple closed curve that separates A from B.*

Proof. By Theorem 2 there exist two arcs AX_1B and AX_2B having only A and B in common. Let J denote their sum. Suppose R is a region containing A. There exists a region R' containing A such that R not only contains $\bar{R}'$ but also contains every point that precedes a point of $\bar{R}'$ in the order from A to B either on AX_1B or on AX_2B. By Axiom 5 there exists, in R', a compact continuum M that separates A from B. The continuum M intersects both AX_1B and AX_2B. Let I and E denote the complementary domains of the simple closed curve J. There exist points A_1, B_1, A_2 and B_2, not belonging to M, such that A_1 precedes B_1 in the order from A to B on AX_1B, A_2 precedes B_2 in the order from A to B on AX_2B, the interval A_1B_1 of AX_1B contains $M \cdot (AX_1B)$ and the interval A_2B_2 of AX_2B contains $M \cdot (AX_2B)$ and, finally, A_1B_1 and A_2B_2 are subsets of R. Suppose the closed and compact point set $M \cdot \bar{I}$ contains no continuum intersecting both AX_1B and AX_2B. Then, by Theorem 44 of Chapter I, it is the sum of two mutually exclusive closed point sets H_1 and H_2 containing $M \cdot (AX_1B)$ and $M \cdot (AX_2B)$ respectively. The point sets $H_1 + A_1B_1$ and $H_2 + A_2B_2$ are mutually exclusive compact continua. Hence, by Theorem 4, there exists a simple closed curve J' separating them from each other. By Theorem 16 of Chapter III and Theorem 28 of Chapter I, the curve J' contains an arc α separating $H_1 + A_1B_1$ from $H_2 + A_2B_2$ and lying, except for its endpoints, wholly in I. One endpoint of α lies between A_1 and A_2 on the arc A_1AA_2 of J and the other one lies between B_1 and B_2 on the arc B_1BB_2 of J. Hence $\alpha + s(A_1AA_2) + s(B_1BB_2)$ is a connected point set containing both A and B. But it contains no point of M, contrary to the supposition that M separates A from B.

So $M \cdot \bar{I}$ contains a continuum K that intersects both AX_1B and AX_2B. By Theorems 43 and 47 of Chapter I there exists a connected subset K' of $K \cdot I$ such that $\bar{K}'$ contains a point C_1 of $M \cdot (AX_1B)$ and a point C_2 of $M \cdot (AX_2B)$. Let D denote the component of $I \cdot R$ that contains K'. There exist two mutually exclusive connected domains D_1 and D_2 lying in R and containing C_1 and C_2 respectively such that $D_1 \cdot J$ is a subset of $s(A_1B_1)$ and $D_2 \cdot J$ is a subset of $s(A_2B_2)$. There exist points X_1 and X_2 in $D_1 \cdot D$ and $D_2 \cdot D$ respectively and arcs C_1X_1, X_1X_2 and C_2X_2 lying in D_1, D and D_2 respectively. The sum of these three arcs contains an arc Z_1Z_2 lying in R and lying wholly in I except for its endpoints Z_1 and Z_2 which belong to $s(AX_1B)$ and $s(AX_2B)$ respectively. The interval AZ_1 of AX_1B and the interval AZ_2 of AX_2B are both subsets of R. Similarly there exists, in R, an arc W_1W_2 such that W_1 belongs to $s(AX_1B)$, W_2 belongs to $s(AX_2B)$ and the interval AW_1 of AX_1B and the interval AW_2 of AX_2B are both subsets of R and $s(W_1W_2)$ is a subset of E. If, for each n less than 3, W_nZ_n denotes W_n or the interval W_nZ_n of AX_nB according as W_n is or is not identical with Z_n then the simple closed curve $Z_1Z_2 + W_1Z_1 + W_1W_2 + W_2Z_2$ lies wholly in R and separates A from B.

THEOREM 6. *If D is a complementary domain of a simple closed curve J, O is a point of J and R is a connected domain containing O, and C and F are points of J distinct from O then there exists an arc AXB such that* (1) *the points*

A and B lie on J in the order $CAOBF$, (2) *the simple closed curve formed by the arc AXB and the arc AOB of J is a subset of R,* (3) *AXB lies, except for its endpoints, wholly in D.*

PROOF. There exist two points Y and Z, lying on J in the order $CYOZF$, such that the arc YOZ of J is a subset of R. Let t denote the other arc of J whose endpoints are Y and Z. Let H denote the common part of the domain R and $S - t$. By Theorem 5, there exists a simple closed curve Q lying in H and separating O from Y. Since t is a connected set containing Y but no point of Q therefore Q separates O from t. By Theorem 16 of Chapter III, there exists an arc EFL of Q separating O from Y in $\bar{D}$ and lying wholly in D except for its endpoints E and L whose sum separates O from Y on J. One of the points E and L lies between O and Y on the arc ZOY. Call this one A and let B denote the other one. The point B lies on the arc OZY of J. But since B belongs to Q it does not lie on t. Hence it lies between O and Z on OZY. Thus the points A and B lie on J in the order $CAOBF$.

THEOREM 7. *If CF is an arc, O is a point of CF distinct from C and F, R is a connected domain containing O, and K is a closed and compact point set not containing O then there exist, in R, two mutually exclusive arcs $C_1X_1F_1$ and $C_2X_2F_2$ such that* (1) *the points C_1, C_2, F_1 and F_2 lie on the arc CF in the order $CC_1C_2OF_2F_1F$,* (2) *neither $C_1X_1F_1$ nor $C_2X_2F_2$ has any point, except its endpoints, in common with CF,* (3) *the interval C_1OF_1 of CF is a subset of R, and the simple closed curve J formed by the arcs $C_1X_1F_1$ and $C_2X_2F_2$ and the intervals C_1C_2 and F_1F_2 of CF separates O from $K + C + F$,* (4) *D, the complementary domain of J that contains O, is the sum of D_1, D_2 and the segment C_2OF_2 of CF, D_1 and D_2 being mutually exclusive connected domains bounded by $C_1X_1F_1 + C_1OF_1$ and $C_2X_2F_2 + C_2OF_2$ respectively.*

PROOF. By Axiom 3 and Theorem 1 of Chapter II, there exists an arc CYF lying in $S - O$. Let E denote the last point that CYF has in common with the interval CO of CF and let L (Figure 12) denote the first point, following E, that CYF has in common with the interval OF of CF. The interval EY_1L of CYF has no point except E and L in common with CF.

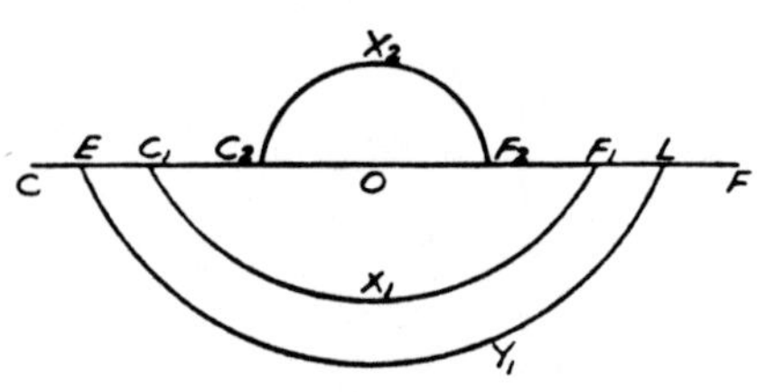

FIGURE 12

Let α denote the interval CE of CF or the point C according as E is or is not distinct from C. Let β denote the interval FL or the point F according as L is or is not distinct from F. By Theorem 32 of Chapter III, there exists a region Q containing O but containing no point that separates two points of $K + \alpha + \beta$ from each other. Let W denote one of the complementary domains of the simple closed curve formed by the arc EY_1L and the interval EOL of CF. The regions Q and R have, as a common subset, a connected domain R_1 containing O. By Theorem 6, there exists an arc $C_1X_1F_1$ such that (1) the points C_1 and F_1 lie on CF in the order EC_1OF_1L, (2) the closed curve J_1, formed by the arc $C_1X_1F_1$ and the interval C_1OF_1 ot CF, is a subset of R_1, (3) $C_1X_1F_1$ lies, except for its endpoints, wholly in W. There is a region H containing O but containing no point of the closed point set $CC_1 + C_1X_1F_1 + F_1F$ where CC_1 and F_1F denote intervals of CF. The domains H and R_1 have, as a common subset, a connected domain R_2 containing O. Let D_1 denote the complementary domain of J_1 that does not contain C. By Theorem 6, there exists an arc $C_2X_2F_2$ such that (1) the points C_2 and F_2 lie on CF in the order $EC_1C_2OF_2F_1L$, (2) the simple closed curve J_2, formed by the arc $C_2X_2F_2$ and the interval C_2OF_2 of CF, is a subset of R_2, (3) $C_2X_2F_2$ lies, except for its endpoints, wholly in the domain $S - \bar{D}_1$. Let D_2 denote the complementary domain of J_2 that does not contain C. Since J_1 and J_2 have only the arc C_2OF_2 in common therefore the connected point set $J_1 - C_2OF_2$ lies wholly in one of the complementary domains of J_2. It follows, with the help of Theorem 2 of Chapter III, that either D_1 and D_2 are mutually exclusive or one of them is a subset of the other one. Since $C_2X_2F_2$ is not a subset of $\bar{D}_1$, therefore D_2 is not a subset of D_1. Suppose D_1 is a subset of D_2. Then C_1 belongs to $\bar{D}_2$. But it does not belong to J_2. Hence C_1 belongs to D_2. But the interval CC_1 of CF has no point in common with J_2. Hence C belongs to D_2. But this involves a contradiction. Therefore D_1 is not a subset of D_2. It follows that D_1 and D_2 are mutually exclusive. Let J denote the simple closed curve formed by the arcs $C_1X_1F_1$ and $C_2X_2F_2$ and the intervals C_1C_2 and F_1F_2 of CF. Let D denote the complementary domain of J that contains O. By Theorem 6 of Chapter III, $D = D_1 + D_2 + s(C_2OF_2)$. The point C belongs neither to D_1 nor to D_2 nor to C_2OF_2. Hence it does not belong to D. Hence, since J does not separate C from any point of $K + F$, D contains no point of $K + C + F$.

THEOREM 8. *If CF is an arc, O is a point of CF distinct from C and from F, R is a connected domain containing O, K is a closed and compact point set not containing O, and OX is an arc having no point except O in common with CF then there exists in R an arc AYB such that* (1) *the points A and B lie on CF in the order $CAOBF$,* (2) *the interval AOB of CF lies in R, and AYB has no point except A and B in common with CF,* (3) *if J denotes the simple closed curve formed by the arcs AOB and AYB and I denotes the complementary* *domain of J that does not contain C, then I contains no point of $K + F$, but there exists, on the arc OX, a point H distinct from O such that the segment OH of OX lies wholly in I.*

PROOF. Let $C_1X_1F_1$, $C_2X_2F_2$, D_1, D_2 and D denote arcs and domains satisfying, with respect to CF, K and O, all the conditions of Theorem 7. The domain D contains the point O. Hence it contains an interval OH of the arc OX. But $D = D_1 + D_2 + s(C_2OF_2)$ and $OH - O$ contains no point of C_2OF_2. Hence $OH - O$ is a subset of $D_1 + D_2$. But D_1 and D_2 are mutually separated and $OH - O$ is connected. Hence $OH - O$ is a subset either of D_1 or of D_2. Let AYB (Figure 13) denote $C_1X_1F_1$ or $C_2X_2F_2$ according as $OH - O$ is a subset of D_1 or a subset of D_2.

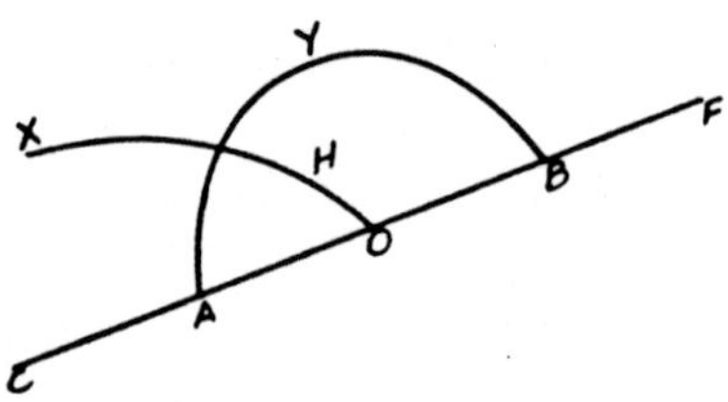

FIGURE 13

THEOREM 9. *If β is the boundary of the domain Q, the arc CF is a subset of β, and no point of $s(CF)$ is a limit point of $\beta - CF$ then CF is a subset of the boundary of some component of Q and if Z is any component of Q whose boundary contains a point of $s(CF)$ then the boundary of Z contains CF.*

PROOF. Let X denote a point of Z, let O denote a point of $s(CF)$ belonging to γ, the boundary of Z, and let K denote the set whose only element is X. Suppose R is a connected domain containing O but no point of $\beta - CF$. There exist arcs and domains $C_1X_1F_1$, $C_2X_2F_2$, D_1, D_2 and D_3 and a simple closed curve J satisfying all the conditions of Theorem 7. The point set $Z + O$ is connected and it contains both X and O. But J separates O from X. Hence one of the segments $s(C_1X_1F_1)$ and $s(C_2X_2F_2)$ contains a point of Z. But these segments are both connected subsets of $S - \beta$. Hence one of them is a subset of Z. Therefore γ contains either $C_1 + F_1$ or $C_2 + F_2$. Thus every region that contains a point O of $\gamma \cdot s(CF)$ contains a point of $\gamma \cdot s(CF)$ preceding O in the order from C to F on CF and also a point of $\gamma \cdot s(CF)$ following O in that order. But $\gamma \cdot s(CF)$ is closed. Hence it is CF.

THEOREM 10. *Suppose that either* (1) *the arc CF lies, except for its endpoints, wholly in the connected domain D or* (2) *CF is a subset of β, the boundary of D, and no point of CF, except its endpoints, is a limit point of $\beta - CF$. Then* (1) *if Y is a point of D, not belonging to CF, there exists an arc YF lying,* *except for* F, *wholly in* D *and having only the point* F *in common with* CF, (2) *there exists an arc with endpoints at* C *and* F *and lying, except for* C *and* F, *wholly in* D *and having only* C *and* F *in common with* CF.

PROOF. There exists an arc YO joining Y to some point O of $s(CF)$ and such that $YO - O$ lies in D and contains no point of CF. There exists an infinite sequence of regions $R_1, R_2, R_3, \cdots$ closing down on the point F. There exists a connected domain R containing O but no point of the bounddary of D that does not belong to the arc CF. By Theorem 8, there exists, in R, an arc $A_1Y_1B_1$ such that (1) the points A_1 and B_1 lie on CF in the order CA_1OB_1F, (2) the interval A_1OB_1 of CF lies in R, and $A_1Y_1B_1$ has no point except A_1 and B_1 in common with CF, (3) if J_1 denotes the simple closed curve formed by the arcs A_1OB_1 and $A_1Y_1B_1$ and I denotes the complementary domain of J_1 that does not contain C, then I_1 does not contain F or Y and there exists, on the arc YO, a point H_1 distinct from O such that the segment OH_1 of YO lies in I_1. If CF lies, except for its endpoints, wholly in D then R contains a point of D but no boundary point of D and therefore, since it is connected, R is a subset of D. Hence, in this case, $s(A_1Y_1B_1)$ is a subset of D. Suppose that CF is a subset of β. It will be shown that, in this case also, $s(A_1Y_1B_1)$ is a subset of D. The point C does not belong to $\bar{I}_1$ and it belongs to the boundary of D. Hence there exist points of D not belonging to $\bar{I}_1$. But the segment OH_1 is a subset of D and thus D contains points of I_1. Thus the connected point set D intersects both I_1 and $S - I_1$. Hence D intersects J_1, the boundary of I_1. But D contains no point of CF. Hence it intersects $s(A_1Y_1B_1)$. But $s(A_1Y_1B_1)$ is a subset of R and therefore it contains no point of the boundary of D not belonging to CF. Hence, since it is connected, $s(A_1Y_1B_1)$ is a subset of D. Thus, in either case, $A_1Y_1B_1$ lies, except possibly for its endpoints, wholly in D.

If L and N are two points of the arc CF distinct from C and F and L precedes N in the order from C to F on CF and W is a domain containing the interval LN of CF then LN is said to be *subtended* in $W \cdot D$ by the set of arcs $C_1X_1F_1, C_2X_2F_2, \cdots, C_nX_nF_n$ provided (1) L lies between C_1 and F_1, and N lies between C_n and F_n, on CF, (2) the points $C_1, F_1, C_2, F_2, \cdots, C_n, F_n$ are in the order $CC_1C_2F_1C_3F_2C_4F_3\cdots C_nF_{n-1}F_n$ on CF, (3) except for its endpoints, $C_iX_iF_i$ lies in $W \cdot D$ and has no point in common with CF $(1 \leqq i \leqq n)$, (4) if J_i denotes the simple closed curve formed by the arc $C_iX_iF_i$ and the interval $C_iF_{i-1}F_i$ of CF $(1 < i \leqq n)$ and D_i denotes the complementary domain of J_i that does not contain C, then D_i does not contain F and there exists, on $C_iX_iF_i$, a point H_i such that the interval F_iH_i of $C_iX_iF_i$ lies, except for the point F_i, wholly in D_{i+1} $(1 \leqq i < n)$.

Suppose that L is a point of CF distinct from C and F, W is a domain containing L, and $C_1X_1F_1$ is an arc lying, except for its endpoints, in $W \cdot D$ and having, in common with CF, only the points C_1 and F_1, which lie on CF in the order CC_1LF_1F. Suppose furthermore that if J_1 denotes the simple closed curve formed by the arcs $C_1X_1F_1$ and the interval C_1F_1 of CF, and

D_1 denotes the complementary domain of J_2 that does not contain C then D_1 does not contain F. Then, if N is a point, distinct from F, following L on CF, such that the interval LN of CF is a subset of W, there exist arcs $C_2X_2F_2, \cdots, C_nX_nF_n$ such that the set of arcs $C_1X_1F_1, C_2X_2F_2, \cdots, C_nX_nF_n$ subtends LN in $W \cdot D$. For suppose this is not the case. Then there exists, on the segment $s(LF)$ of CF, a point T which is the first point X of this segment such that the interval LX of CF is not subtended in $W \cdot D$ by a set of arcs of which the first one is $C_1X_1F_1$. There exists a connected domain E containing T but no point of the boundary of D that does not belong to CF. By Theorem 7, there exist, in $W \cdot E$, two mutually exclusive arcs $A_1Z_1B_1$ and $A_2Z_2B_2$ such that (1) the points A_1, A_2, B_1 and B_2 lie on CF in the order $CA_1A_2TB_2B_1F$, (2) neither $s(A_1Z_1B_1)$ nor $s(A_2Z_2B_2)$ has any point in common with CF, (3) the interval A_1TB_1 of CF is a subset of W, and if J denotes the simple closed curve formed by the arcs $A_1Z_1B_1$, $A_2Z_2B_2$ and the intervals A_1A_2 and B_1B_2 of CF then I, the complementary domain of J that contains T, is the sum of I_1, I_2 and the segment A_2TB_2 of CF, I_1 and I_2 being mutually exclusive domains bounded by $A_1Z_1B_1 + A_1TB_1$ and $A_2Z_2B_2 + A_2TB_2$ respectively, (4) $\bar{I}$ contains neither C nor F. The interval LA_2 of CF is subtended in $W \cdot D$ by a finite set G of arcs of which the first one is $C_1X_1F_1$. It follows, with the help of an argument largely similar to one employed in the proof of Theorem 8, that LT is subtended in $W \cdot D$ by a finite set Q of arcs composed of a subset of G (beginning with $C_1X_1F_1$) together with either $A_1Z_1B_1$ or $A_2Z_2B_2$ and possibly one additional arc, no additional arc being necessary unless A_1 or A_2 coincides with an end-point of some arc of the set G. But this is contrary to the definition of T.

There exists a sequence of points N_1, N_2, $N_3, \cdots$ lying between O and F on the arc CF such that, for every i, N_i precedes N_{i+1} and the interval N_iF lies wholly in R_i. There exists a finite set of arcs U_1 subtending the interval NN_1 in D, $A_1Y_1B_1$ being the first arc of U_1. There exists a finite set U_2 of arcs subtending N_1N_2 in $D \cdot R_1$, the first arc of U_2 being the last arc of U_1, and the arcs of U_1 and those of U_2 constituting a set V_2 of arcs subtending NN_2, the ith arc of U_1 being the ith arc of V_2 and the ith arc of U_2 being the $(n_1 + i)$th arc of V_2, where n_1 is one less than the number of arcs in U_1. This process may be continued. Thus there exists a sequence U_1, U_2, $U_3, \cdots$ such that, for each n, (1) U_{n+1} is a finite set of arcs subtending, in $D \cdot R_n$, the interval N_nN_{n+1} of CF, (2) the arcs of the sets $U_1, U_2, \cdots, U_n$ form a single set V_n, of arcs, subtending NN_n, the ith arc of U_{j+1} $(1 \leqq j \leqq n)$ being the $(n_1 + n_2 + \cdots + n_j + i)$th arc of V_n, where n_j is one less than the number of arcs in U_j.

Let V denote the infinite sequence constituted by the arcs of all the sets $U_1, U_2, U_3, \cdots$, the ith arc of U_{j+1} being the $(n_1 + n_2 + \cdots + n_j + i)$th arc of V. The first arc of V is $A_1Y_1B_1$. For each n, let $A_nY_nB_n$ denote the nth arc of V. Let M denote the continuum obtained by adding together the point F and all the arcs of the sequence V. Neither A_n nor B_n is a limit point of $(YO + M) - A_nY_nB_n$. It follows that, for each n, there exist, on $A_nY_nB_n$, two points A'_n and B'_n distinct from A_n and B_n such that the interval $A'_nB'_n$ contains every point of $A_nY_nB_n$ which is a limit point of $M - A_nY_nB_n$. Let M' denote the point set $F + A'_1B'_1 + A'_2B'_2 + \cdots$. This point set is a continuum having only the point F in common with CF. The arc $A'_1B'_1$ contains a point Y' of the arc YO. Let YY' denote the interval of YO with endpoints as indicated. The continuum $YY' + M'$ is locally connected. It follows that it contains an arc YF. This arc lies wholly in D and has only the point F in common with CF. Similarly there exists an arc YC having only C in common with CF and lying, except for C, wholly in D. The point set $YC + YF$ contains an arc t from C to F. The arc t lies, except for its endpoints, wholly in D and has only C and F in common with CF.

THEOREM 11. *Suppose that either* (1) *the arc CF lies, except for its endpoints, wholly in the domain D or* (2) *CF is a subset of β, the boundary of D, and no point of CF, except its endpoints, is a limit point of $\beta - CF$. Suppose, furthermore, that Y is a point of D that can be joined to some point O of $s(CF)$ by an arc lying, except for O, wholly in D and containing no point except O in common with CF. Then* (1) *there exists an arc YF lying, except for F, wholly in D and having only the point F in common with CF and* (2) *there exists an arc CLF lying, except for C and F, wholly in the component of D that contains Y and having only C and F in common with CF.*

PROOF. The point O is a boundary point of Z, the component of D that contains Y. Hence, by Theorem 9, CF is a subset either of Z or of the boundary of Z. Therefore, by Theorem 10, there exists an arc YF having only F in common with CF and lying, except for F wholly in Z and therefore wholly in D. That there exists an arc CLF satisfying condition (2) follows by an argument given in the proof of Theorem 10.

THEOREM 12. *If the point O does not belong to the closed and compact point set M and D is a domain containing O then there exists a simple domain containing O but no point of M and bounded by a simple closed curve lying in D and containing no point of M.*

PROOF. Let ω denote some definite point distinct from O. By Axiom 3, the domain $S - O$ is connected. Hence, if X is a point of M, there exists an arc $X\omega$ lying wholly in $S - O$. By Theorem 5, there exists a simple closed curve J_X lying in the domain $D\cdot(S - X\omega)$ and separating O from ω. Since $X\omega$ is a connected subset of $S - J_X$, J_X separates O from X. For each point X of M, let E_X denote the complementary domain of J_X that contains X. Let G denote the collection of all domains E_X for all points X of M. The closed and compact point set M is covered by a finite subcollection G' of G. Let Q denote the collection of all simple domains q such that $q = S - \bar{g}$ for some domain g of the collection G'. If q is a domain of the finite collection Q then q contains O and $S - \bar{q}$ contains ω. Hence, by Theorem 18 of Chapter III, there exists a simple closed curve W containing O such that (1) W is a subset of every domain of the collection Q', (2) every point of the boundary of W belongs to the boundary of some domain of the collection Q'. The domain W contains O, its boundary is a subset of D and $\overline{W}$ contains no point of M.

THEOREM 13. *If H and K are two mutually exclusive closed point sets and β, the boundary of H, is compact and A and B are points belonging to H and K respectively then there exists a simple closed curve separating A from B and containing no point of H or of K.* (Proved by John M. Slye without the First Edition supposition that the components of $H - \beta$ form a semi-contracting set.)

PROOF. Regard B as the point at infinity. For each point P of $\beta + A$ there exists a simple closed curve separating A from B and containing no point of K. Since $\beta + A$ is closed and compact there exists a finite collection G of such curves such that $\beta + A$ is covered by W, the set of all their interiors (with respect to B as the point at infinity). Let Z denote the set of all components of W^*. For each element z of Z let G_z denote the set of all elements g of G such that g is the boundary of some element of W which is a subset of z. Since the sum of the interiors of the curves of the set G_z is the connected point set z therefore, by Theorem 18 of Chapter III, there exists a simple closed curve J'_z which is a subset of G^*_z that encloses the interiors of the curves of G_z. Let U' denote a set such that u belongs to it if and only if u is J'_z for some element z of Z. There exists a subcollection U of the collection U' such that (1) if I is the interior of a curve of the collection U', I is a subset of the interior of some curve of the collection U but (2) if I is the interior of a curve of the collection U, I is not a *proper* subset of the interior of any curve of the collection U'.

The interiors of the curves of the collection U are mutually exclusive. Suppose the curves of the collection U are not all mutually exclusive. Let T denote a point set such that P belongs to it if and only if P is common to at least two curves of this collection. There exists a finite collection V of simple closed curves such that (1) every point of T is in the interior of some curve of V, (2) no curve of V intersects K, (3) if x is a curve of V, there exists a point P of T such that x encloses P but neither encloses nor contains any point of any element of U that does not contain P, (4) if the elements x and y of V intersect each other some element of U intersects both the interior of x and the interior of y. Let N denote a point set such that P belongs to it if and only if P is in the interior either of some curve of the collection U or of some curve of the collection V. For each component x of N there exists a simple closed curve J_x which is a subset of $U^* + V^*$ that lies in $\bar{x}$ and encloses x. Let Q' denote a collection such that q belongs to it if and only if q is J_x for some component x of N. There exists a subcollection Q of Q' such that

(1) if I is the interior of a curve of the collection Q', I is a subset of the interior of some curve of the collection Q, (2) if I is the interior of a curve of the collection Q, I is not a *proper* subset of the interior of any curve of the collection Q'.

Every simple closed curve of the collection Q lies wholly in the exterior of every other one and every point of $\beta + A$ lies in the interior of some one of them. Therefore, by Theorem 24 of Chapter III, the common part of the exteriors of the curves of this collection is a connected point set D. Since the domain D is connected and every point of Q^* is a limit point of D therefore $D + Q^*$ is connected. But $D + Q^*$ contains the point B of $S - H$ and contains no point of β, the boundary of H. Therefore $D + Q^*$ is a subset of $S - H$. Therefore if J denotes the curve of the collection Q whose interior contains A then J contains no point of H. But it contains no point of K and it separates A from B.

THEOREM 14. *If the points A and B belong to different components of the closed and compact point set M and L is a closed point set containing no point of M, there exists a simple closed curve separating A from B and containing no point of M or of L.*

PROOF. By Theorem 44 of Chapter I, M is the sum of two mutually exclusive closed point sets H and K containing A and B respectively. Since L is a closed point set not intersecting K, $M + L$ is the sum of two mutually exclusive closed point sets $H + L$ and K containing A and B respectively. Hence, by Theorem 13, there exists a simple closed curve fulfilling the required conditions.

THEOREM 15. *If P is a point of the closed and compact point set M and D is a domain containing P there exists a simple domain I containing P such that if J denotes the boundary of I then $I \cdot M + J$ is a subset of D.*

PROOF. The point set $M \cdot (S - D)$ is a closed and compact subset of $S - P$. Hence, by Theorem 12, there exists a simple domain I containing P such that (1) the boundary of I is a subset of D, (2) I contains no point of $M \cdot (S - D)$ and therefore $I \cdot M$ is a subset of D.

THEOREM 16. *If H and K are two mutually exclusive closed and compact point sets and L is a closed point set containing no point of H there exists a domain D containing H such that (1) $\bar{D}$ contains no point of K, (2) D is the sum of a finite number of mutually exclusive connected domains Q such that every complementary domain of $\bar{Q}$ is bounded by a simple closed curve containing no point of L.*

PROOF. If P is a point of H then, by Theorem 15, there exists a simple domain D_P containing P such that, if J_P denotes the boundary of D_P, then $D_P \cdot (H + K) + J_P$ is a subset of the domain $S - (K + L)$ and therefore $D_P + J_P$ contains no point of K and J_P contains no point of L. Let G_1'

denote the collection of all domains D_P for all points P of H. The point set H is covered by some finite subcollection G'_2 of the collection G'_1. Let D denote the sum of all the domains of the collection G'_2. With the help of Theorem 18 of Chapter III, it may be seen that D satisfies all the requirements of Theorem 16.

THEOREM 16(a). *If H and K are two mutually exclusive closed and compact point sets and L is a closed point set containing no point of H and G is a collection of regions covering H there exists a domain D containing H such that* (1) *$\bar{D}$ contains no point of K and the boundary of D is a compact point set containing no point of L,* (2) *D is the sum of a finite number of mutually exclusive connected domains Q such that* (a) *every complementary domain of $\bar{Q}$ is bounded by a simple closed curve,* (b) *Q is the sum of a finite number of domains each bounded by a simple closed curve lying in some region of G.*

If, in the statement of Theorems 15 *and* 16(a), *"simple closed curve" is replaced by "closed and compact point set" and "simple domain" is replaced by "domain with a compact boundary", the resulting propositions are consequences of Axioms* 0, 1 *and* 2 *and the axiom obtained by replacing "compact continuum" by "closed and compact point set" in the statement of Axiom* 5. This may be proved with the help of Theorem 41′ of Chapter II.

THEOREM 16(b). *If M is a closed and compact point set there exists a sequence of domains $D_1, D_2, D_3, \cdots$ such that* (1) *for each n, D_n contains M, and $\bar{D}_{n+1}$ is a subset of D_n and the boundary of D_n is compact,* (2) *for each n, D_n is the sum of a finite number of mutually exclusive connected domains Q such that every complementary domain of $\bar{Q}$ is bounded by a simple closed curve,* (3) *if D is a domain containing M there exists a number δ such that, for every n greater than δ, the boundary of D_n is a subset of D,* (4) *if D is a domain containing M and bounded by a compact point set then, for some n, D_n is a subset of D,* (5) *M is the common part of the domains of this sequence.*

Theorem 16(b) may be established with the aid of Theorem 117 of Chapter I, Theorem 89 of Chapter II and Theorem 16(a).

THEOREM 17. *If H is a component of the closed and compact point set M, K is a closed point set having no point in common with H and T is a continuum having no point in common with H, then there exists a domain D containing H but no point of T and bounded by a simple closed curve containing no point of $M + K + T$.*

PROOF. Let A and B denote points belonging to H and T respectively. By Theorem 52 of Chapter I, there exists a domain I containing H but no point of $K + T$ and such that there is no point of $M + K + T$ on the boundary of I. Let L denote $M \cdot I$ and let N denote $K + T$ or $K + T + M \cdot (S - I)$ according as M does or does not lie wholly in I. The point set $M + K + T$ is the sum of the two point sets L and N, which contain H and $K + T$ respectively. Since L and N are closed and mutually exclusive and L is compact therefore, by Theorem 13, there exists a simple domain D containing A but not B and bounded by a simple closed curve containing no point of $L + N$, that is to say of $M + K + T$. Since D contains A and H is a connected point set containing A but no point of the boundary of D, H is a subset of D. Since the boundary of D contains no point of the continuum T and D does not contain the point B of T therefore D contains no point of T.

THEOREM 18. *If to the hypothesis of Theorem* 17 *there be added a stipulation requiring that K be compact and that H do not separate any two points of $T + K$ from each other then, under the hypothesis so obtained, there exists a simple domain containing H but no point of $K + T$ and bounded by a simple closed curve containing no point of $M + K + T$.*

PROOF. Let B denote a point of T. By Theorem 85 of Chapter II, there exists a continuum L lying in $S - H$ and containing $B + K$. The point set $L + T$ is a continuum containing no point of H. Hence, by Theorem 17, there exists a domain D containing H but no point of $L + T$ and bounded by a simple closed curve containing no point of $M + L + T$.

That Theorem 18 does not remain true on the omission of the supposition that H does not separate any two points of $B + K$ from each other may be seen from the example where, in a Euclidean plane, M is a circle, T is the set whose only element is the center of M, and K is the point set consisting of T plus some point in the exterior of M.

THEOREM 19. *If J is the boundary of a simple domain D and A, X, B, Y, F, Z, C and W are points lying on J in the order indicated, and H and K are mutually exclusive closed point sets such that H is compact and $J \cdot H$ and $J \cdot K$ are subsets of the arcs AXB and CZF respectively of the curve J then there exists an arc, with endpoints on the segments $s(AWC)$ and $s(BYF)$ respectively, lying, except for its endpoints, wholly in D and containing no point of $H + K$.*

PROOF. The point sets $H + AXB$ and $K + CFD$ are closed, compact and mutually exclusive. Hence, by Theorem 13, there exists a simple closed curve Q separating A from C and containing no point of either of these point sets. By Theorem 16 of Chapter III, there exists an arc of Q which lies, except for its endpoints, wholly in D and which separates A from C on J. Clearly, the endpoints of this arc lie respectively on the segments $s(AWC)$ and $s(BYF)$.

THEOREM 20 (JANISZEWSKI). *If the closed point sets H and K are mutually exclusive, or their common part is a continuum, and neither H nor K separates the point A from the point B and H is compact then $H + K$ does not separate A from B.*

PROOF. Suppose, on the contrary, that $H + K$ separates A from B. Let S_1 and S_2 denote the components of $S - (H + K)$ that contain A and B respectively. There exist arcs AXB and AYB such that AXB contains no point of H and AYB contains no point of K. Let X_1 and X_2 denote the first points that AXB and AYB respectively have in common with the boundary of S_2. There exist simple domains D_1 and D_2, containing X_1 and X_2 respectively, such that $\bar{D}_1$ and $\bar{D}_2$ have no point in common and neither of them contains A or B or a point common to H and K. Let J_1 and J_2 denote the boundaries of D_1 and D_2 respectively. The domain S_2 contains two points P_1' and P_2' lying in D_1 and D_2 respectively. There exists an arc $P_1'P_2'$ lying wholly in S_2. The arc $P_1'P_2'$ contains an arc P_1P_2 containing no point of $\bar{D}_1 + \bar{D}_2$ except P_1 and P_2 which belong to J_1 and J_2 respectively. Let Z_1 denote the first point of J_1 on the arc AXB and let Z_2 denote the first point of J_2 on AYB, in each case in the order from A to B. Let AZ_1 and AZ_2 denote intervals of AXB and AYB respectively with endpoints as indicated. Let A_2 denote the last point that AZ_1 has in common with AZ_2. Let A_2Z_1 and A_2Z_2 denote intervals of AZ_1 and AZ_2 respectively with endpoints as indicated and let Z_1Z_2 denote the arc $A_2Z_1 + A_2Z_2$. There exists, on A_2Z_1, a point A_1 distinct from A_2 and Z_2 such that the interval A_2A_1 of A_2Z_1 contains no point of K. For each n less than 3 let P_nZ_n denote P_n or an arc of J with endpoints at P_n and Z_n according as P_n is or is not identical with Z_n. The point set $Z_1Z_2 + P_1Z_1 + P_1P_2 + P_2Z_2$ is a simple closed curve J. Since $H \cdot K$ either does not exist or is connected and contains no point of J, therefore there exists a complementary domain D of J that contains no point common to H and K. Let H_1 and K_1 denote the point sets $H \cdot \bar{D}$ and $K \cdot \bar{D}$ respectively, if they exist. The point sets $H_1 \cdot J$ and $K_1 \cdot J$ are subsets of the arcs $A_2Z_2P_2$ and $A_1Z_1P_1$ respectively of the simple closed curve J. Furthermore, H_1 and K_1 are closed and mutually exclusive. Hence, by Theorem 19, there exists an arc with one endpoint on the segment $s(A_2A_1)$ of the arc A_2Z_1 and its other endpoint on the segment $s(P_1P_2)$ (Figure 14), lying, except for its endpoints, wholly in D and containing no point of $H_1 + K_1$ and therefore containing no point of $H + K$. But the endpoints of this arc belong to different components of $S - (H + K)$. Thus the supposition that $H + K$ separates A from B has led to a contradiction.

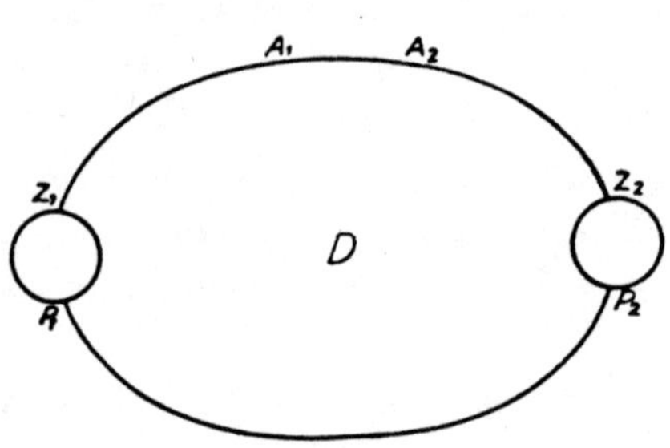

FIGURE 14

THEOREM 21. *No arc separates S.*

PROOF. Suppose there exists an arc AB that separates two points C and F from each other. Let T_1 denote the point set consisting of the point A, together with all points X of AB distinct from A, if there are any, such that the interval AX of AB does not separate C from F. Let T_2 denote the point set $AB - T_1$. It is easy to see that every point of T_1 precedes every point of T_2 on AB in the order from A to B. It follows that there exists a point O which is either the last point of T_1 or the first point of T_2 in that order. Suppose O is the first point of T_2. Then there exists an infinite sequence of points $O_1, O_2, O_3, \cdots$ converging to O and lying on AB in the order $AO_1O_2O_3\cdots$. For each n, the intervals AO_n and O_nO of AB have only the point O_n in common and their sum separates C from F. But, since O_n belongs to T_1, AO_n does not separate C from F. Therefore, by Theorem 20, O_nO separates C from F. But O is the common part of all the intervals $O_1O, O_2O, O_3O, \cdots$. Hence, by Theorem 7 of Chapter II, O separates C from F.

Suppose, on the other hand, that O is the last point of T_1. There exists a sequence of points $O_1, O_2, O_3, \cdots$ converging to O and lying between O and B in the order $BO_1O_2O_3\cdots$. In this case, the intervals AO and OO_n have only the point O in common and their sum separates C from F but AO does not. Therefore each of the intervals $OO_1, OO_2, OO_3, \cdots$ separates C from F and therefore O, their only common point, separates C from F.

Thus, in either case, O separates C from F. But this is contrary to Axiom 3.

THEOREM 22 (JANISZEWSKI). *If H and K are two continua and β, the boundary of $H \cdot K$ is compact and $H \cdot K$ is not connected, then $H + K$ separates S.*

PROOF (KURATOWSKI). Let M and T denote $H + K$ and $H \cdot K$ respectively. Suppose that M does not separate S. By hypothesis T is the sum of two mutually exclusive closed point sets L and N. Let A and B denote points belonging to L and N respectively. By Theorem 13, there exists a simple closed curve J separating A from B but containing no point of T. Let H_1 and K_1 denote $H \cdot J$ and $K \cdot J$ respectively. Both H_1 and K_1 exist since A and B are separated from each other by J, and H and K are connected point sets each containing both A and B. There exist two mutually exclusive closed point sets H_2 and K_2 such that (1) each of them is either an arc of J or the sum of a finite number of arcs of J, (2) H_2 contains H_1, and K_2 contains K_1, (3) $H_1 + K_1$ contains no limit point of $J - (H_2 + K_2)$. Let W denote the point set $J - (H_2 + K_2)$. The point set W has only a finite number of components and each of them is a segment of an arc of J. Suppose no two points of W are separated from each other by M. In each component of W select just one point, join every two of the points so selected by an arc containing no point of M and let V denote the compact continuum obtained by adding to $\overline{W}$ all the arcs so constructed. The connected point set K contains both A and B and has no point in common with $H_2 + V$. Hence $H_2 + V$ does not separate A from B. Similarly $K_2 + V$ does not separate A from B. Furthermore $H_2 + V$ and $K_2 + V$ are closed and compact and their common part is the continuum V. Hence, by Theorem 20, $(H_2 + V) + (K_2 + V)$ does not separate A from B. But this point set is identical with $J + V$ which separates A from B. Thus the supposition that $H + K$ does not separate any two points from each other has led to a contradiction.

THEOREM 23. *If the boundary β of a connected domain D is a compact subset of a continuum which has no point in common with D then β is itself a continuum.*

PROOF. Suppose, on the contrary, that β is the sum of two mutually exclusive closed point sets H and K. Then, by Theorem 13, there exists a simple closed curve J separating a point of H from a point of K but containing no point either of H or of K. Let I and E denote the complementary domains of J. Since each of these domains contains a point of β each of them contains a point of D. But D is connected. Hence J has a point in common with D. Hence, since it contains no boundary point of D, J is a subset of D. But, by hypothesis, there exists a continuum N containing both H and K but no point of D. Since N contains both a point of I and a point of E, it too contains a point of J. Thus J is not a subset of D. This involves a contradiction.

THEOREM 24. *If the boundary of a complementary domain of a continuum is compact it is itself a continuum.*

PROOF. If D is a complementary domain of a continuum M then M is a continuum containing the boundary of D but having no point in common with D. Hence, by Theorem 23, if the boundary of D is compact it is a continuum.

DEFINITION. If D is a connected domain and the point set M is a subset of the component E of $S - \overline{D}$ then the boundary of E will be called the *outer boundary of D with respect to M.*

THEOREM 25. *If D and E are two mutually exclusive connected domains and the outer boundary of D with respect to E is compact then it is a continuum and no proper subset of it separates D from E.*

PROOF. Let β denote the outer boundary of D with respect to E. The compact point set β is the boundary of a complementary domain E' of the continuum $\overline{D}$. Hence, by Theorem 24, β is a continuum. Suppose K is a proper subset of β. Let P denote a point of $\beta - K$. Since P is a limit point of each of the connected point sets D and E' therefore $D + E' + P$ is connected. But this point set contains no point of K. Hence K does not separate A from B.

THEOREM 25(a). *If the boundary of the connected domain D is a compact point set β there exists a compact continuum containing β and lying in $D + \beta$.*

Theorem 25(a) is a consequence of Theorem 30 of Chapter II and Theorem 25 of the present chapter.

THEOREM 25(b). *If the boundary of the continuum M is compact then M is compactly connected.*

PROOF. By Theorem 29 of Chapter II, the boundary of M is a subset of a compact continuum K. If β is the boundary of a component D of $S - M$ then, by Theorem 3 of Chapter II, β is compact and therefore, by Theorem 24 of the present chapter, it is connected. Hence, by Theorem 27 of Chapter II, $M \cdot K$ is a continuum. That M is compactly connected follows as in the proof of Theorem 41 of Chapter II.

THEOREM 25(c). *If M is a continuous curve and N is a subcontinuum of M whose boundary is compact and whose boundary with respect to M is a compact point set with no continuum of condensation then N is a continuous curve.*

Theorem 25(c) may be easily established with the aid of Theorem 45 of Chapter II and Theorem 24 of the present chapter.

THEOREM 26. *If the points A and B are separated from each other by the closed and compact point set M then they are separated from each other by a continuum which is a subset of M and which has no proper subset that separates A from B.*

PROOF. Let D denote the complementary domain of M that contains A and let β denote the outer boundary of D with respect to B. That β is a subset of M is a consequence of Theorem 3 of Chapter II. But β separates A from B and, by Theorem 25, it is a continuum with no proper subset that separates A from B.

THEOREM 27. *If D is a domain bounded by a simple closed curve β, A and B are points of $D + \beta$ and K is a closed and compact point set that separates A from B in $D + \beta$, then A and B are separated from each other in $D + \beta$ by a continuum which is a subset of K.*

PROOF. If A and B are separated in S by K then, by Theorem 26, they are separated in S by a continuum T which is a subset of K and it is clear that they are separated in $D + \beta$ by T. Suppose A and B are not separated in S by K. Then there exists an arc AB lying wholly in $S - K$. Let A' and B' respectively denote the first and the last points of β on the arc AB in the order from A to B. The points A' and B' are separated from each other in $D + \beta$ by K. Hence K contains two points X and Y of β such that $X + Y$ separates A' from B' on β. Let $A'XB'$ and $A'YB'$ denote the arcs of β from A' to B' that contain X and Y respectively. Suppose K contains no continuum containing both a point of $A'XB'$ and a point of $A'YB'$. Then, by Theorem 44 of Chapter I, K is the sum of two mutually exclusive closed point sets H and N containing $K\cdot(A'XB')$ and $K\cdot(A'YB')$ respectively. It follows, with the help of Theorem 19, that there exists an arc AZB lying wholly in $D + \beta$ and containing no point of $H + N$. This involves a contradiction. Hence K contains a continuum T containing both a point of $A'XB'$ and a point of $A'YB'$. By Theorem 30 of Chapter III, T separates A' from B' in $D + \beta$. It follows that it separates A from B in $D + \beta$.

THEOREM 28. *If the outer boundary, with respect to some point, of a connected domain D is compact then it is not disconnected by the omission of any one of its subcontinua.*

PROOF. Suppose that D is a domain, E is a complementary domain of $\bar{D}$, and the compact point set β is the boundary of E. Suppose that β contains a continuum T such that $\beta - T$ is the sum of two mutually separated point sets H and K. Then $H + T$ and $K + T$ are compact continua whose common part is the continuum T. But, by Theorem 25, neither $H + T$ nor $K + T$ separates D from E. Hence, by Theorem 20, $(H + T) + (K + T)$ does not separate D from E. Thus the supposition that Theorem 28 is false leads to a contradiction.

THEOREM 29. *If D is a connected domain bounded by two mutually exclusive simple closed curves J_1 and J_2, M is a closed and compact point set lying in $\bar{D}$, H is a component of M intersecting J_1 and J_2, and K is a closed subset of $\bar{D}$ containing no point of H and such that every component of K intersects J_1 and J_2 then there exist two arcs A_1A_2 and B_1B_2 lying on J_1 and J_2 respectively and two mutually exclusive arcs A_1B_1 and A_2B_2 lying, except for their endpoints, wholly in D, such that if J denotes the simple closed curve formed by the arcs A_1A_2, A_2B_2, B_1B_2 and A_1B_1 and I denotes the component of $S - J$ that lies in D then I contains $H\cdot D$ but $A_1B_1 + A_2B_2$ contains no point of M and $\bar{I}$ contains no point of K and thus $A_1B_1 + A_2B_2$ separates H from K in $\bar{D}$.*

PROOF. With the help of Theorem 31 of Chapter III it is clear that there exist two arcs C_1C_2 and F_1F_2 lying on J_1 and J_2 respectively and containing $H\cdot J_1$ and $H\cdot J_2$ respectively, such that neither C_1C_2 nor F_1F_2 contains any point of K and such that none of the points C_1, C_2, F_1 and F_2 belongs to H. It follows that $K + (J_1 - s(C_1C_2)) + (J_2 - s(F_1F_2))$ is a compact continuum containing no point of H. Hence, by Theorem 17, there exists a simple domain U_1 containing H such that $\bar{U}_1$ contains no point of K and the boundary of U_1 contains no point of M. By Theorem 18 of Chapter III, there exists a simple domain U_2 containing a point of H and such that (1) U_2 is a subset of $D\cdot U_1$, (2) every boundary point of U_2 belongs either to J_1, to J_2, or to the boundary of U_1. Let β denote the boundary of U_2. The simple closed curve β intersects $H\cdot J_1$ and $H\cdot J_2$. Hence it contains two arcs A_1B_1 and A_2B_2 each having one endpoint on J_1 and the other endpoint on J_2. It is easy to see that β does not have a third arc having these properties, that A_1B_1 and A_2B_2 are subsets of the boundary of U_1 and that they have in common no point except possibly one or both of their endpoints. Let A_1A_2 and B_1B_2 (Figure 15) denote arcs of J_1 and J_2 respectively with endpoints as indicated and each containing a point of H. Let J denote the simple closed curve formed by the arcs A_1A_2, A_2B_2, B_1B_2 and A_1B_1 and let I denote the component of $S - J$ that lies in D. The point set $H \cdot D$ is a subset of I. Suppose that I contains a point of a continuum T which is a component of K containing both a point of J_1 and a point of J_2. There exists an arc XY lying wholly in U_2 except for the points X and Y which belong to $D \cdot (A_1B_1)$ and $D \cdot (A_2B_2)$ respectively. The continuum T contains no point of A_1B_1 or of A_2B_2. It follows that it is a subset of $\bar{I}$. Hence,

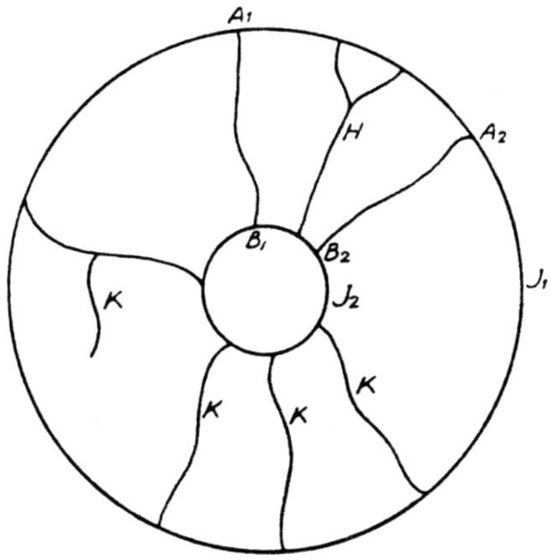

FIGURE 15

as may be seen with the help of Theorem 6 of Chapter III, T contains a point of the arc XY and therefore of U_2. But T is a subset of K, and $\bar{U}_2$ contains no point of K. Thus the supposition that I contains a point of T leads to a contradiction.

To see that Theorem 29 does not remain true, in every space satisfying these axioms, if the supposition that every component of K intersects both J_1 and J_2 is omitted, consider the example indicated by Figure 16.

THEOREM 30. *If O is a point of the simple closed curve J, R is a connected domain containing O, D is a complementary domain of J and K is a closed and compact subset of $\bar{D} - O$, then there exists, in R, an arc AXB such that* (1) *A and B are on J,* (2) *except for its endpoints, AXB lies wholly in D,* (3) *of the two arcs of J from A to B, the one which contains O lies in R,* (4) *AXB separates K from O in $\bar{D}$.*

PROOF. Let P denote some point of J distinct from O. By Theorem 15, there exists, in R, a simple closed curve Q separating O from $K + P$. There exist, on J, two points Y and Z such that the interval YOZ of J is a subset of R containing no point of Q and therefore no point of $K + P$. Let T denote the continuum obtained by adding Q to the arc YPZ of J. By Theorem 6, there exists, in the domain $R \cdot (S - T)$, an arc AXB such that (1) the points A and B lie on J in the order $ZAOBY$, (2) AXB lies, except for its endpoints, wholly in D. Let U denote the component of $S - Q$ that contains $K + P$. The arc AXB separates O from P in $\bar{D}$. But $\bar{U}$ is a connected point set containing P but no point of AXB or of AOB. It follows, with the aid of Theorem 6 of Chapter III, that AXB separates O from $\bar{U} \cdot D$ in $\bar{D}$. But K is a subset of $\bar{U} \cdot D$.

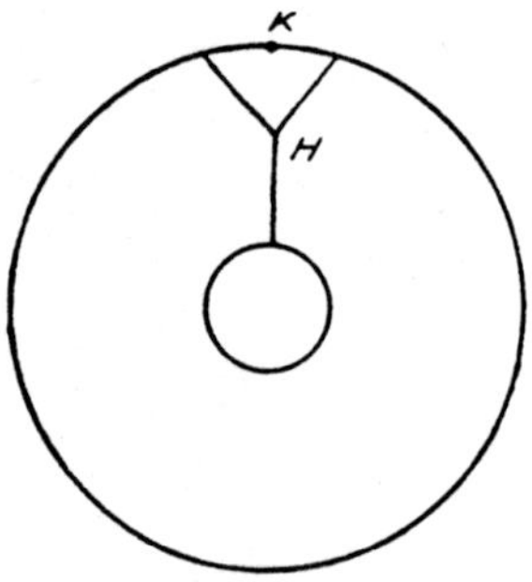

FIGURE 16

DEFINITION. If the arcs A_1B_1 and A_2B_2 have only the points B_1 and B_2, respectively, in common with the arc CF and neither B_1 nor B_2 is an endpoint of CF and J is a simple closed curve containing a segment of CF of which the interval B_1B_2 of CF is a subset, then A_1B_1 and A_2B_2 are said to *abut on CF from the same side with respect to J*, or to *abut on CF from opposite sides with respect to J*, according as it is, or is not, true that one of the complementary domains of J contains both a segment of A_1B_1 with one endpoint at B_1 and a segment of A_2B_2 with one endpoint at B_2.

The statement that the arcs A_1B_1 and A_2B_2 *abut on the arc CF from the same side* means that there exists a simple closed curve J such that they abut on CF from the same side with respect to J; and the statement that they *abut on CF from opposite sides* means that they abut on CF from opposite sides with respect to some simple closed curve.

THEOREM 31. *If J_1 and J_2 are two simple closed curves with a common arc AB then there exist two mutually exclusive domains D and E such that* (1) *each of these domains is a component of $S - (J_1 + J_2)$ and each of them is bounded by a simple closed curve containing AB,* (2) *D and E belong to different components of $S - J_1$ and also to different components of $S - J_2$,* (3) *if M is* *any connected subset of* $S - (J_1 + J_2)$ *such that some point of* AB, *distinct from* A *and* B, *is a limit point of* M *then* M *is a subset either of* D *or of* E.

PROOF. Let O denote some point of AB distinct from A and B. There exists a connected domain R containing O but no point of $J_1 + J_2$ that does not belong to $s(AB)$. By Theorem 4 of Chapter III there exists in R a point P not belonging to AB. Let D_1 and D_2 denote the complementary domains of J_1 and J_2 respectively that contain the point P and let E_1 and E_2 denote $S - \bar{D}_1$ and $S - \bar{D}_2$ respectively. By Theorem 17 of Chapter III, there exists a domain D containing P such that D is a subset of D_1 and of D_2 and the boundary of D is a simple closed curve which is a subset of $J_1 + J_2$. There exists an arc PX joining P to some point X of AB and lying wholly in R and having only the point X in common with AB. The point set $PX - X$ is connected and contains no point either of J_1 or of J_2. Furthermore, it has the point P in common with D_1 and D_2. Hence it is a connected subset of D_1 and of D_2 and therefore of D. But X is a limit point of $PX - X$. Hence X is a boundary point of D. But X belongs to AB and the boundary of D is a simple closed curve J such that, if Y is a point of AB, there is a region containing Y but containing no point of J that does not belong to AB. It follows that AB is a subset of J. By Theorem 10, there exists an arc ACB lying, except for its endpoints, wholly in D. Of the two domains into which D is divided by $s(ACB)$, let I denote the one which is bounded by the simple closed curve formed by the arcs AB and ACB. Let I_i $(i = 1,2)$ denote the domain $D_i - (ACB + I)$. There exists a region Q containing O but no point of $\bar{I}_1 + \bar{I}_2$. The region Q contains a point T not belonging to $\bar{I}$. The point T belongs both to E_1 and to E_2. Hence, by Theorem 17 of Chapter III, there exists a domain E containing T such that E is a subset both of E_1 and of E_2 and the boundary of E is a simple closed curve which is a subset of $J_1 + J_2$. The boundary of E contains AB.

Suppose now that M is a connected subset of $S - (J_1 + J_2)$ such that the point L of $s(AB)$ is a limit point of M. There exists a region W containing L but no point of $\bar{I}_1 + \bar{I}_2$. The region W contains a point Z belonging to M. The point Z belongs either to I or to E. Hence, since M is connected and contains no point of the boundary of E, M is a subset either of I or of E. Hence it is a subset either of D or of E.

THEOREM 32. *If the arcs* A_1B_1 *and* A_2B_2 *abut on the arc* CF *from the same side with respect to the simple closed curve* J_1, *and* J_2 *is a simple closed curve containing a segment of* CF *of which the interval* B_1B_2 *is a subset then* A_1B_1 *and* A_2B_2 *abut on* CF *from the same side with respect to* J_2.

PROOF. Let C_1F_1 denote an interval of CF belonging both to J_1 and to J_2 and such that $s(CF)$ contains B_1B_2. By hypothesis, there exists a component D_1 of $S - J_1$ which contains segments of A_1B_1 and of A_2B_2 with endpoints at B_1 and B_2 respectively. By Theorem 31, there exist two domains D and E such that (1) each of these domains is a component of

$S - (J_1 + J_2)$ and each of them is bounded by a simple closed curve containing C_1F_1, (2) D and E belong to different components of $S - J_2$, (3) if M is any connected subset of $S - (J_1 + J_2)$ such that some point of C_1F_1 is a limit point of M then M is a subset of D or of E, (4) D is a subset of D_1. There exist, on $s(A_1B_1)$ and $s(A_2B_2)$, points T_1 and T_2 respectively such that neither $s(T_1B_1)$ nor $s(T_2B_2)$ contains a point of $J_1 + J_2$. Each of the connected point sets $s(T_1B_1)$ and $s(T_2B_2)$ lies wholly either in D or in E. But neither of them lies wholly in E. For E is a subset of $S - \bar{D}_1$ and $s(T_1B_1)$ and $s(T_2B_2)$ are both subsets of D_1. Hence $s(T_1B_1)$ and $s(T_2B_2)$ are subsets of D. But D is a subset of one of the complementary domains of J_2. Thus A_1B_1 and A_2B_2 abut on CF from the same side with respect to J_2.

THEOREM 33. *If A_1B_1, A_2B_2, A_3B_3 and CD are arcs and A_1B_1 and A_2B_2 abut on CD from the same side and A_2B_2 and A_3B_3 abut on CD from the same side, then A_1B_1 and A_3B_3 abut on CD from the same side.*

PROOF. By Axiom 3 and Theorem 10, there exists a simple closed curve J that contains CD. Let T_1, T_2 and T_3 denote points of A_1B_1, A_2B_2 and A_3B_3 such that no one of the point sets $s(T_1B_1)$, $s(T_2B_2)$ and $s(T_3B_3)$ contains a point of J. Let D denote the complementary domain of J that contains $s(T_2B_2)$. By Theorem 32, A_1B_1 and A_2B_2 abut on CD from the same side with respect to J and so do A_2B_2 and A_3B_3. Hence both $s(T_1B_1)$ and $s(T_3B_3)$ lie in D and therefore A_1B_1 and A_3B_3 abut on CD from the same side.

THEOREM 34. *If B_1 and B_2 are non-endpoints of the arc CF and A_1B_1 and A_2B_2 are arcs having only B_1 and B_2 respectively in common with CF then* (1) *either A_1B_1 and A_2B_2 abut on CF from the same side or they abut on CF from opposite sides,* (2) *if they abut on CF from the same side they do not abut on CF from opposite sides,* (3) *if they abut on CF from opposite sides, every simple closed curve that contains a segment of CF which contains the interval B_1B_2 of CF separates some segment of A_1B_1 with one endpoint at B_1 from some segment of A_2B_2 with one endpoint at B_2,* (4) *if they abut on CF from the same side, then if J is any simple closed curve containing a segment of CF which contains the interval B_1B_2 of CF, one of the complementary domains of J contains both a segment B_1X_1 of A_1B_1 and a segment B_2X_2 of A_2B_2,* (5) *if A_1B_1 and A_2B_2 abut on CF from the same side and XY is an arc containing some segment of CF that contains B_1B_2, and XY contains no point of A_1B_1 except B_1 and no point of A_2B_2 except B_2 then A_1B_1 and A_2B_2 abut on XY from the same side,* (6) *if A_1B_1 and A_2B_2 abut on CF from opposite sides and A_1B_1 and A_3B_3 abut on CF from opposite sides then A_2B_2 and A_3B_3 abut on CF from the same side.*

DEFINITION. The arc TL is said to *cross the arc AF at the point B* if B is common to AF and TL and the intervals TB and LB of TL abut on AF

from different sides. The arc TL is said to *cross* the arc AF *at the interval* BC if the interval BC is a subset of the common part of AF and TL and the intervals TB and LC (or the intervals TC and LB) of TL abut on AF from different sides.

THEOREM 35. *If the arc TL crosses the arc AF at a point or interval then AF crosses TL at the same point or interval.*

PROOF. Suppose that B is a point common to $s(AF)$ and $s(TL)$, C is a point common to $s(AF)$ and $s(TL)$, either B is identical with C or B precedes C in the order from A to F on AF, and TB and LC abut on AF from opposite sides. By Theorem 10 and Axiom 3, there exists an arc AXF having only the points A and F in common with AF. Let J denote the simple closed curve formed by the arcs AF and AXF. There exist, on TB and LC respectively, points T_1 and L_1 (Figure 17) such that neither the

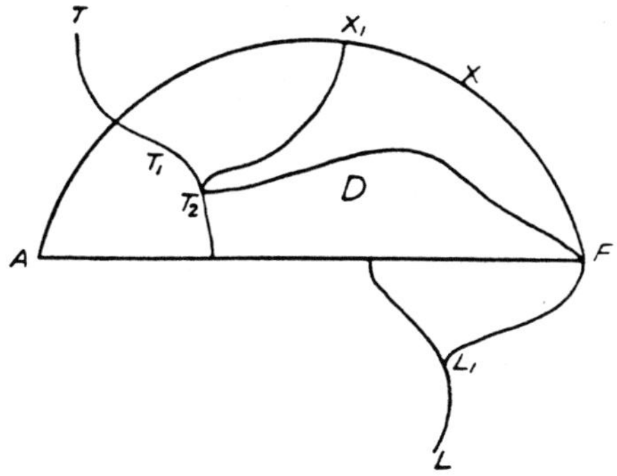

FIGURE 17

interval BT_1 of TB nor the interval CL_1 of LC has any point in common with AXF. By Theorem 34, CL_1 and BT_1 lie, except for the points B and C, in different complementary domains of J. Let D and E denote the complementary domains of J that contain BT_1 and CL_1 respectively. There exists an arc T_1X_1 lying wholly in D except for X_1 which is a point of AXF. Let T_2 denote the first point of T_1X_1 in the order from B to T_1 on the arc BT_1. Let BT_2 denote the interval of BT_1 whose endpoints are B and T_2 and let T_2X_1 denote the interval of T_1X_1 whose endpoints are T_2 and X_1. Let BX_1 denote the arc $BT_2 + T_2X_1$. Let X_1F and BF denote intervals of AXF and of AF respectively with endpoints as indicated. Let J_1 denote the simple closed curve $BX_1 + X_1F + BF$. Let I_1 denote the complementary domain of J_1 which is a subset of D. Let T_2BF denote the arc $BF + BT_2$. The arc T_2BF is a subset of the boundary of the domain $I_1 - I_1 \cdot (TB)$ and no point of T_2BF, except T_2 and F, is a limit point of the remainder of the boundary of this domain. Hence, by Theorem 11, there exists an arc T_2F lying wholly in I_1 except for the points T_2 and F. The arc T_2F lies, except for F, wholly in D and has only the point T_2 in common with TB. Similarly there exists an arc L_2F lying, except for L_2, wholly in E and having only the point L_2 in common with LC and such that the interval CL_2 of LC is a subinterval of CL_1. By Theorem 6 of Chapter III, $D = s(BT_2F) + D_1 + D_2$ and $E = s(CL_2F) + E_1 + E_2$ where $BT_2F = BT_2 + T_2F, CL_2F = CL_2 + L_2F$, D_1 is bounded by BT_2F plus the interval BF of AF, D_2 by $BT_2F + AXF$ plus the interval AB of AF, E_1 by CL_2F plus the interval FC of AF, and E_2 by $CL_2F + AXF$ plus the interval AC of AF. Let Q denote the simple closed curve formed by the arcs BT_2F and CL_2F together with either the point B or the interval BC of AF according as B is or is not identical with C. Let I denote the component of $S - Q$ that contains $s(FC)$. With the help of Theorem 6 of Chapter III, it may be seen that $I = D_1 + E_1 + s(FC)$. But AB has no point in common with D_1, E_1 or FC. Hence it contains no point of $\bar{I}$. Thus AB and FC abut on TL from opposite sides with respect to the simple closed curve Q.

THEOREM 36. *If the connected domain D is bounded by a continuum M and AB is an arc lying wholly in D except for its endpoints which belong to M then (1) $D - s(AB)$ is the sum of two mutually exclusive connected domains D_1 and D_2, and AB is a subset both of the boundary of D_1 and of the boundary of D_2, (2) if Y_1 and Y_2 are points of $s(AB)$ and X_1Y_1 and X_2Y_2 are arcs such that the point sets $X_1Y_1 - Y_1$ and $X_2Y_2 - Y_2$ either both lie in D_1 or both lie in D_2, then X_1Y_1 and X_2Y_2 abut on AB from the same side.*

PROOF OF (1). Let O denote some point of $s(AB)$. By Theorem 8, there exist in D two arcs $A_1X_1B_1$ and $A_2X_2B_2$ having in common with AB only the points A_1 and B_1 and the points A_2 and B_2 respectively (Figure 18) and such that (1) the points A_1, B_1, A_2 and B_2 lie on AB in the order $AA_1A_2OB_2B_1B$, (2) if J denotes the simple closed curve formed by the arcs $A_1X_1B_1$ and $A_2X_2B_2$ and the intervals A_1A_2 and B_1B_2 of AB, I denotes the

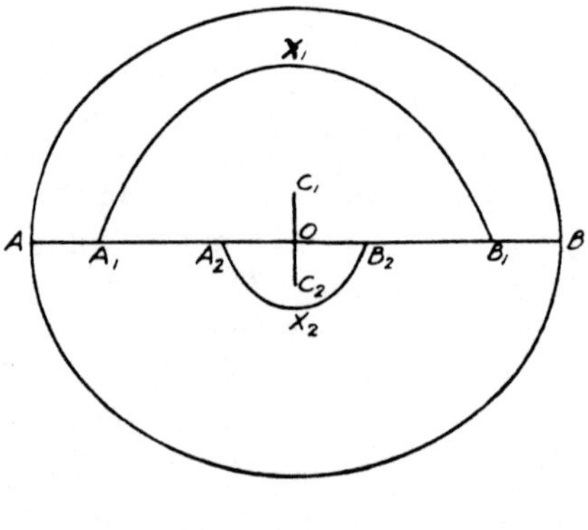

FIGURE 18

complementary domain of J that contains O, and A_1OB_1 and A_2OB_2 denote intervals of AB then J separates O from A and $I = I_1 + I_2 + s(A_2OB_2)$ where I_1 and I_2 are simple domains bounded by $A_1X_1B_1 + A_1OB_1$ and $A_2X_2B_2 + A_2OB_2$ respectively. Let E denote $S - \bar{I}$. Since A belongs to E but not to D therefore E is not a subset of D. But since the boundary of D is connected and J lies in D therefore either I or E is a subset of D. Hence I is a subset of D. By Theorem 10, there exist arcs C_1O and C_2O lying, except for O, in I_1 and I_2 respectively. Let D_1 and D_2 denote the components of $D - s(AB)$ that contain C_1 and C_2 respectively. Suppose D_1 has a point in common with D_2. Then it is identical with D_2 and there exists an arc C_1YC_2 lying wholly in D_1 and therefore lying wholly in $D - s(AB)$. The point set $C_1OC_2 + C_1YC_2$ contains a simple closed curve Q containing an interval $C_1'OC_2'$ of C_1OC_2. The arc C_1OC_2 crosses AB at O. Hence, by Theorem 35, AB crosses C_1OC_2 at O. Hence, by Theorem 32, the intervals AO and BO abut on C_1OC_2 from opposite sides with respect to Q. But neither AO nor BO contains any point of Q except O. Hence A and B are in different complementary domains of Q. But the continuum M contains both A and B. Therefore M contains a point of Q. But Q is a subset of D and D is a component of $S - M$. Thus the supposition that D_1 has a point in common with D_2 involves a contradiction.

Suppose now that P is a point of $D - AB$. Since the point sets $I_1 + s(A_1X_1B_1)$ and $I_2 + s(A_2X_2B_2)$ are connected subsets of $D - s(AB)$ containing C_1 and C_2 respectively they are subsets of D_1 and D_2 respectively. Hence if P belongs to one of these point sets it belongs to D_1 or to D_2. Suppose it belongs to neither of them. Then it does not belong to I. By Theorem 10, there exists an arc PO lying in D and having only the point O in common with AB. Since O belongs to I and P belongs to $S - \bar{I}$ therefore PO contains a point of J. Let T denote the first point of J in the order from P to O on PO and let PT denote the interval of PO whose endpoints are P and T. The point T belongs either to $s(A_1X_1B_1)$ or to $s(A_2X_2B_2)$ and therefore either to D_1 or to D_2. But PT is a connected subset of $D - AB$ and therefore P and T belong to the same component of $D - AB$. Hence P belongs either to D_1 or to D_2.

The point O belongs both to the boundary of D_1 and to the boundary of D_2. Let O' denote some point of AB other than O. There exist two mutually exclusive connected domains D_1' and D_2' such that $D - AB = D_1' + D_2'$ and such that O' belongs to the boundary of each of them. Since D_1 and D_2 are mutually separated connected point sets, and D_1' and D_2' are mutually separated connected point sets and, furthermore, $D_1 + D_2 = D_1' + D_2'$, therefore one of the domains D_1 and D_2 is identical with D_1' and the other one is identical with D_2'. Therefore O' is a boundary point both of D_1 and of D_2.

THEOREM 37. *If the arc AB is a subset of the continuum M and no point of $s(AB)$ is a limit point of $M - AB$ and K is a closed and compact subset of $S - AB$, then* (1) *AB is a subset of the boundary of some component of $S -$*

$(M + K)$ and if there exist two components D_1 and D_2 of $S - (M + K)$ such that $s(AB)$ contains both a boundary point of D_1 and a boundary point of D_2 then AB is a subset both of the boundary of D_1 and of the boundary of D_2 but $s(AB)$ contains no boundary point of any other component of $S - (M + K)$, (2) if O is a point of $s(AB)$ there exists a simple domain I which is the sum of a segment of AB containing O and two mutually exclusive simple domains I_1 and I_2 which are subsets of $S - (M + K)$, the domains I_1 and I_2 being subsets of different components of $S - (M + K)$ in case O is a limit point of two different components of $S - (M + K)$.

PROOF. Let N denote $M + K$. Suppose O and O' are two points of $s(AB)$. By Theorem 7, there exist, in $S - (N - s(AB))$, two mutually exclusive arcs $A_1X_1B_1$ and $A_2X_2B_2$ such that (1) A_1, A_2, B_1 and B_2 lie on AB in the order $AA_1A_2OB_2B_1B$, (2) neither of the arcs $A_1X_1B_1$ and $A_2X_2B_2$ has any point, except its endpoints, in common with AB, (3) if J denotes the simple closed curve formed by the arcs $A_1X_1B_1$ and $A_2X_2B_2$ and the intervals A_1A_2 and B_1B_2 of AB then J separates O from $A + B + K$, (4) I, the complementary domain of J that contains O is the sum of I_1, I_2 and the segment A_2OB_2 of AB, I_1 and I_2 being mutually exclusive domains bounded by $A_1X_1B_1 + A_1OB_1$ and $A_2X_2B_2 + A_2OB_2$ respectively (Figure 19). Suppose the domain I_1 contains a point belonging to N. Then, since

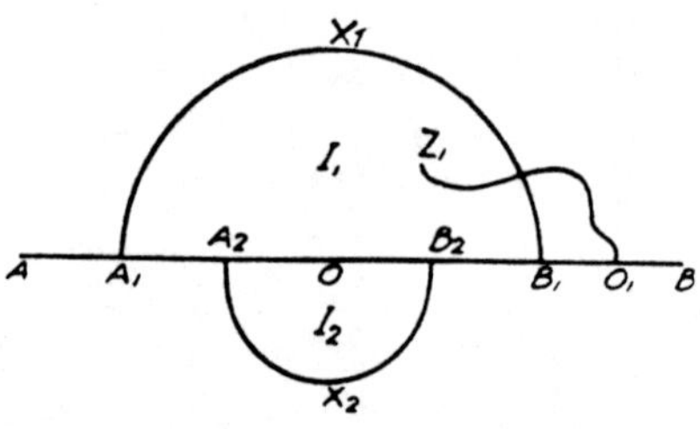

FIGURE 19

it contains no point of K, it contains a point of M. Since I_1 does not contain A, and J_1, the boundary of I_1, contains no point of the connected point set $AA_1 - A_1$ therefore I_1 contains no point of AA_1. Similarly it contains no point of B_1B; and it contains no point of A_1OB_1 since this arc is a subset of its boundary. Hence $I_1 \cdot M$ is a subset of $M - AB$. But J_1 contains no limit point of $M - AB$ and $S - (I_1 + J_1)$ contains no limit point of I_1. Hence $M - I_1 \cdot M$ contains no limit point of $I_1 \cdot M$. Furthermore, I_1 contains no limit point of $S - I_1$ and therefore $I_1 \cdot M$ contains no limit point of $M - I_1 \cdot M$. But, since A belongs to M but not to I_1, $M - I_1 \cdot M$ exists. Therefore M is not connected. Thus the supposition that I_1 contains a point of N leads to a contradiction. In a similar way it may be shown that I_2 contains no point of N. Let D_1 denote the component of $S - N$ that contains I_1 and let D_2 denote the one containing I_2. Let Z_1 denote a point of I_1. By Theorem 10, there exists an arc Z_1O lying, except for O, wholly in I_1 and therefore having only O in common with N. It follows, by Theorem 11, that there exists an arc Z_1O' having only O' in common with N. The point set $Z_1O' - O'$ is a connected subset of $S - N$ containing Z_1. Hence it is a subset of D_1. Hence O' is a boundary point of D_1. In a similar way it may be shown to be a boundary point of D_2. Since $I_1 + I_2 + s(A_2OB_2)$ contains a point of every point set of which O is a limit point, therefore O is not a boundary point of any component of $S - N$ except D_1 and D_2. Thus O' is a boundary point of every component of $S - N$ of which O is a boundary point. Similarly O is a boundary point of every one of which O' is a boundary point. Hence AB is a subset of the boundary of D_1 and of the boundary of D_2 and it contains no boundary point of any other complementary domain of N. But either D_2 is identical with D_1 or D_1 and D_2 are mutually exclusive. The truth of Theorem 37 is therefore established.

THEOREM 38. *If H and K are two mutually exclusive compact continua, AB is an arc having only A in common with H and only B in common with K and CF is an arc having only C in common with H and only F in common with K, and, furthermore, $s(AB)$ and $s(CF)$ have no point in common and D_H is the component of $S - K$ that contains H, D_K is the component of $S - H$ that contains K, H' is $S - D_K$ and K' is $S - D_H$ then, if M denotes the point set $H' + K' + AB + CF$, $S - M$ is the sum of two mutually exclusive connected domains D_1 and D_2 and $AB + CF$ is a subset both of the boundary of D_1 and of the boundary of D_2.*

PROOF. The common part of the continua $AB + H'$ and $K' + CF$ is compact and not connected. Hence, by Theorem 22, M separates S. Let D denote a complementary domain of M and let β denote its boundary. Since M separates S, β separates D from some point O of $S - M$. Suppose there exists a point of AB that does not belong to β. Then there exist two points A_1 and B_1 lying on AB in the order AA_1B_1B and such that the interval A_1B_1 of AB contains no point of β. It follows that if AA_1 and BB_1 denote intervals of AB with endpoints as indicated then β is a subset of the continuum $H' + K' + CF + AA_1 + BB_1$. But, with the aid of Theorem 21 and repeated applications of Theorem 20, it may be proved that this continuum does not separate D from O. Hence β does not separate D from O. Thus the supposition that AB is not a subset of β leads to a contradiction. Hence AB, and likewise CF, is a subset of every point set which is the boundary of a complementary domain of M. Hence, by Theorem 37, M does not have more than two complementary domains. But it has at least two. It follows that it has only two complementary domains D_1 and D_2 and that $AB + CF$ is a subset both of the boundary of D_1 and of the boundary of D_2.

THEOREM 39. *If AB and CF are two mutually exclusive arcs and H, K and L are three mutually exclusive compact continua and $H\cdot(AB) = A$, $H\cdot(CF) = C$, $K\cdot(CF) = F$ and $K\cdot(AB) = B$, $L\cdot(AB)$ and $L\cdot(CF)$ are connected and $H' = S - D_H$, $K' = S - D_K$ and $L' = S - D_L$, where D_H, D_K and D_L are complementary domains of H, K and L respectively containing $K + L$, $H + L$ and H respectively, D is a complementary domain of $H + K + AB + CF$ and L is a subset of $\bar{D}$, then $D = D\cdot L' + D_1 + D_2$, where D_1 and D_2 are mutually exclusive complementary domains of $H' + L' + AA_1 + CC_1$ and of $K' + L' + BB_1 + FF_1$ respectively, AA_1 and BB_1 denoting intervals of AB, and CC_1 and FF_1 denoting intervals of CF such that $(AA_1)\cdot L = A_1$, $(BB_1)\cdot L = B_1$, $(CC_1)\cdot L = C_1$ and $(FF_1)\cdot L = F_1$.*

PROOF. Let M, M_1 and M_2 denote the continua $H' + K' + AB + CF$, $H' + L' + AA_1 + CC_1$ and $K' + L' + BB' + FF'$ respectively. Let E denote $S - (M + D)$. By Theorem 38, M_1 has only two complementary domains. One of these domains contains E. It follows that the other one, D_1, is a subset of D. Similarly M_2 has only two complementary domains and one of them, D_2, is a subset of D. Suppose P is a point of $D - D\cdot L'$. Let X denote a point of $M - (H' + K' + L')$. Since no one of the continua H', K' and L' separates S and neither X nor Y belongs to any one of them, therefore, by Theorem 20 of the present chapter and Theorems 2 and 1 of Chapter II, there exists an arc PX containing no point of any one of the compact point sets H, K and L. It is easy to see that PX contains no point of H', K' or L'. Let O denote the first point that PX has in common with M. The point O lies in one of the segments $s(AA_1)$, $s(BB_1)$, $s(CC_1)$ and $s(FF_1)$. Suppose first that it belongs to CC_1. By Theorem 36, O is a limit point of D_1 and there exists a domain I containing O such that $I = I_1 + I_2 + s(YZ)$ where $s(YZ)$ is a segment of CC_1 and I_1 and I_2 are simple domains which are subsets of $D\cdot(S - L')$ and $E\cdot(S - L')$ respectively. Since D_1 has no point in common with E or with $s(YZ)$, therefore it contains no point of $I_2 + s(YZ)$. It follows that I_1 intersects D_1. But I_1 does not intersect the boundary of D_1. Hence D_1 contains I_1. Also O is a limit point of $PO - O$, and $PO - O$ contains no point of $I_2 + YZ$. Hence I_1, and therefore D_1, intersects $PO - O$. But $PO - O$ is a connected subset of $S - M_1$. It follows that it is a subset of D_1. Thus, in case O belongs to $s(CC_1)$, P belongs to D_1. A similar argument would show that, if O belongs to AA_1, P belongs to D_1 and if O belongs to $s(FF_1)$ or $s(BB_1)$ then P belongs to D_2. Thus every point of $D - D\cdot L'$ belongs either to D_1 or to D_2. It follows that $D = D_1 + D_2 + D\cdot L'$.

THEOREM 40. *If H and K are two mutually exclusive compact continua and AB and CF are two mutually exclusive arcs such that AB has only A in common with H and only B in common with K and CF has only C in common with H and only F in common with K and D_H is the component of $S - K$ that contains H and D_K is the component of $S - H$ that contains K, H' is $S - D_K$ and K'*

is $S - D_H$, M is the point set $H' + K' + AB + CF$ and XY is an arc having, in common with M, only the points X and Y, which belong to H and K respectively and D is the complementary domain of M that contains $s(XY)$, then $D - s(XY)$ is the sum of two mutually exclusive connected domains D_1 and D_2 (Figure 20) such that AB and CF are subsets of the boundaries of D_1 and D_2 respectively, but the boundary of D_1 contains no point of $s(CF)$ and the boundary of D_2 contains no point of $s(AB)$.

PROOF. By Theorem 38, $S - (M + D)$ is a connected domain E. By the same theorem, $S - (AB + XY + H' + K')$ is the sum of two mutually exclusive domains. Since E contains no point of the boundary of either of these domains it is a subset of one of them. Call this one E_1 and call the other one D_1. By Theorem 38, every point of CF is a limit point of E. Hence no point of CF belongs to D_1. Therefore D_1 contains no point of M. Hence it is a subset of D. Since D_1 is a complementary domain of a continuum that contains no point of $s(CF)$ therefore $s(CF)$ contains no boundary point of D_1. Similarly if D_2 denotes the complementary domain of $S - (CD + XY + H' + K')$ that does not contain E then D_2 is a subset of D and no point of $s(AB)$ is a boundary point of D_2. By Theorem 38, $AB + XY$ and $CF + XY$ are subsets of the boundaries of D_1 and D_2 respectively.

Suppose now that P is any point whatsoever of $D - s(XY)$. There exists an arc PO lying wholly in D and having, in common with $s(XY)$, only the point O. The connected point set $PO - O$ contains no point of $M + XY$. Hence it lies wholly in some one component of $S - (M + XY)$. Call this component D_3. Since O is a limit point of $PO - O$ it is a boundary point of D_3. But it is a boundary point both of D_1 and of D_2. Hence, by Theorem 37, D_3 is identical either with D_1 or with D_2. Thus every point of $D - s(XY)$ belongs either to D_1 or to D_2. Hence $D - s(XY) = D_1 + D_2$.

NOTE. As indicated in Figure 20, it is possible, under the hypothesis of Theorem 40, that B should be a boundary point of D_2.

THEOREM 41. *If the connected domain D is bounded by the two mutually exclusive simple closed curves J_1 and J_2 and H and K are compact and mutually*

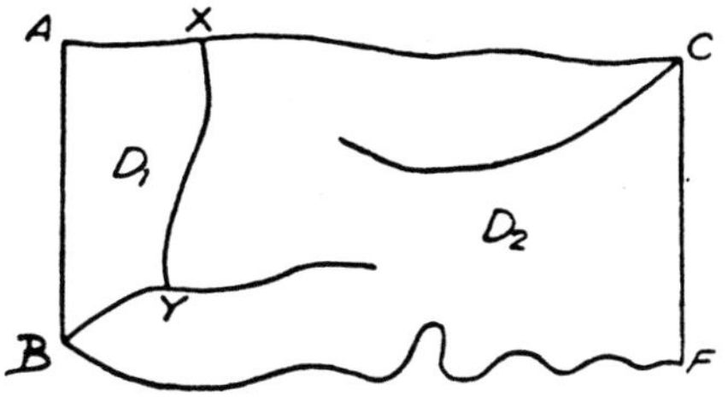

FIGURE 20

exclusive subcontinua of $\bar{D}$ such that $H \cdot J_1$, $H \cdot J_2$, $K \cdot J_1$ and $K \cdot J_2$ are connected and H' is the point set obtained by adding to H all of its complementary domains that are subsets of D and K' is the point set having the same relationship to K, then $D = D \cdot H' + D \cdot K' + D_1 + D_2$ where D_1 is a complementary domain of a continuum formed by H', K' and two arcs t_1 and t_2, and D_2 is a complementary domain of a continuum formed by H', K' and two other arcs f_1 and f_2 such that $t_1 + f_1 + H \cdot J_1 + K \cdot J_1 = J_1$, $t_2 + f_2 + H \cdot J_2 + K \cdot J_2 = J_2$ and $t_i \cdot H$, $t_i \cdot K$, $f_i \cdot H$, $f_i \cdot K$ $(i = 1,2)$ are all degenerate.

PROOF. It is easy to see that H' and K' are mutually exclusive continua. By Theorem 29, there exist arcs AB and CF, lying, except for their endpoints, wholly in D, such that (1) A and C separate $J_1 \cdot H$ from $J_1 \cdot K$ on J_1 and B and F separate $J_2 \cdot H$ from $J_2 \cdot K$ on J_2, (2) of the two simple domains into which $s(AB) + s(CF)$ separates D, one, I_1, contains $D \cdot H$ and therefore $D \cdot H'$; and the other, I_2, contains $D \cdot K$ and therefore $D \cdot K'$. The domain I_1 is bounded by $AB + CF + AX_1C + BX_2F$ and I_2 is bounded by $AB + CF + AY_1C + BY_2F$, AX_1C and AY_1C being arcs of J_1, and BX_2F and BY_2F being arcs of J_2. By Theorem 39, $I_1 = H' \cdot D + U_1 + U_2$ and $I_2 = K' \cdot D + V_1 + V_2$ where U_1, U_2, V_1 and V_2 (Figure 21) are complementary

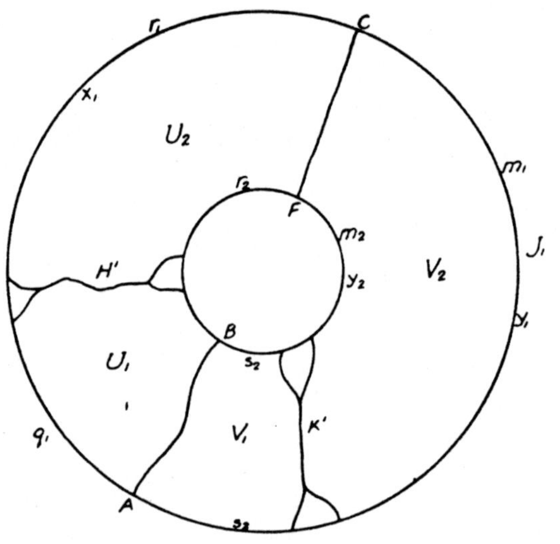

FIGURE 21

domains of $AB + H' + q_1 + q_2$, $CF + H' + r_1 + r_2$, $AB + K' + s_1 + s_2$ and $CF + K' + m_1 + m_2$ respectively, where q_1 is an interval of AX_1C with one endpoint at A and the other at a point of H', r_1 is an interval of AX_1C, with one endpoint at C, and the other at a point of H', s_1 is an interval of AY_1C with one endpoint at A and the other at a point of K', m_1 is an interval of AY_1C with one endpoint at C and the other one at a point of K', q_2 is an interval of BX_2F with one endpoint at B and the other at a point of H', r_2 is an interval of BX_2F with one endpoint at F and the other one at a point of H', s_2 is an interval of BY_2F with one endpoint at B and the other at a point of K' and m_2 is an interval of BY_2F with one endpoint at F and the other at a point of K', each of the intervals q_1, q_2, r_1, r_2, s_1, s_2, m_1 and m_2 having only one endpoint in common with $H' + K'$. Let D_1 denote $U_1 + s(AB) + V_1$, let D_2 denote $U_2 + s(CF) + V_2$, and let t_1, t_2, f_1 and f_2 denote $q_1 + s_1$, $q_2 + s_2$, $m_1 + r_1$ and $m_2 + r_2$ respectively. The point sets D_1, D_2, t_1, t_2, f_1 and f_2 satisfy all the conditions required in the statement of the conclusion of Theorem 41.

THEOREM 42 (MARIE TORHORST). *If the boundary of a complementary domain of a continuous curve is compact it is a continuous curve.*

PROOF. Suppose, on the contrary, that there exists a complementary domain W of a locally connected continuum N such that α, the boundary of N, is a compact point set which is not a continuous curve. By Theorem 24, N is a continuum. With the aid of Theorem 15, it may be shown that Theorem 66 of Chapter II remains true if it is stipulated that the domains D and R be simple domains. It follows that there exist a connected domain D (bounded by two simple closed curves J_1 and J_2), a continuum M, and an infinite sequence of continua $M_1, M_2, M_3, \cdots$ such that (a) the continua M, $M_1, M_2, M_3, \cdots$ are mutually exclusive, (b) for each n, M_n is a component of $N \cdot \bar{D}$ and M_n intersects both J_1 and J_2, (c) M is the sequential limiting set of the sequence $M_1, M_2, M_3, \cdots$.

By Theorem 31 of Chapter III, for each n there exist continua t_n and f_n lying on J_1 and J_2 respectively and containing $M_n \cdot J_1$ and $M_n \cdot J_2$ respectively and continua t and f lying on J_1 and J_2 respectively and containing $M \cdot J_1$ and $M \cdot J_2$ respectively such that $M + t + f$, $M_1 + t_1 + f_1$, $M_2 + t_2 + f_2, \cdots$ are mutually exclusive continua. Let M' denote the point set obtained by adding to $M + t + f$ all of its complementary domains that are subsets of D and, for each n, let M_n' denote the point set obtained by adding to $M_n + t_n + f_n$ all of its complementary domains that are subsets of D. It may be seen that the continua M', M_1', M_2', $M_3', \cdots$ are all mutually exclusive. Let T denote $D \cdot M'$ and, for each n, let T_n denote $D \cdot M_n'$. Of the two domains into which $T_1 + T$ separates D there is one, D_1, that contains infinitely many of the point sets $T_1, T_2, T_3, \cdots$. For each n such that T_n is a subset of D_1, T_n separates D_1 into two domains D_n and I_n such that D_n is bounded by a subset of $M_n' + M' + A_nA + B_nB$, where A_nA and B_nB are arcs of J_1 and J_2 respectively such that A_nA has only A in common with M' and only A_n in common with M_n' and B_nB has only B in common with M' and only B_n in common with M_n'. With the aid of the fact that M is the sequential limiting set of $M_1, M_2, M_3, \cdots$ it is easy to see that no I_n contains infinitely many point sets of the sequence $T_1, T_2, T_3, \cdots$.

It follows that there exists a number n_2 such that I_{n_2} contains no point set of this sequence. The domain D_{n_2} contains every T_n, except T_{n_2}, that lies in D_1. There exists a number n_3 such that D_{n_3} contains every T_n except T_{n_3} that lies in D_{n_2}. This process may be continued indefinitely. Thus there exists an infinite sequence of positive integers $n_1, n_2, n_3, \cdots$ such that $n_1 = 1$ and, for each i, D_{n_i} contains $T_{n_{i+1}} + D_{n_{i+1}}$. Let X and Y denote two points of $D \cdot M$ and let R_X and R_Y (Figure 22) denote two mutually exclusive

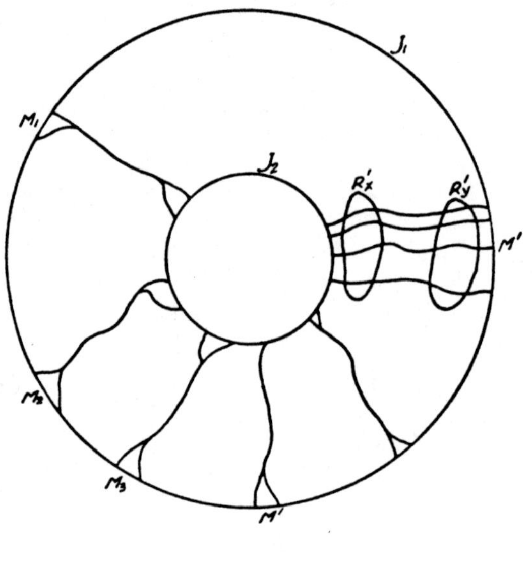

FIGURE 22

regions which are subsets of D and which contain X and Y respectively. There exist regions R'_X and R'_Y, containing X and Y respectively and lying in R_X and R_Y respectively, such that every point of M that belongs to R'_X can be joined to X by an arc lying in $M \cdot R_X$ and every point of M that belongs to R'_Y can be joined to Y by an arc lying in $M \cdot R_Y$. There exists a number j such that M_{n_j} contains a point X_j of R'_X and a point Y_j of R'_Y. There exist arcs XX_j and YY_j lying wholly in $M \cdot R_X$ and $M \cdot R_Y$ respectively. Let Q_j denote the domain $D_{n_j} - (T_{n_{j+2}} + D_{n_{j+2}})$. The arcs XX_j and YY_j contain intervals Z_jZ_{j+2} and W_jW_{j+2} respectively lying wholly in Q_j except that Z_j and W_j belong to M_{n_j}, and Z_{j+2} and W_{j+2} belong to $M_{n_{j+2}}$. For each i greater than 1, let A'_{n_i} denote the first point of M_{n_i} in the order from A_1 to A on A_1A and let B'_{n_i} denote the first point of M'_{n_i} in the order from B_1 to B on B_1B. By Theorem 39, $Q_j - s(Z_jZ_{j+2})$ is the sum of two mutually exclusive connected domains U_j and V_j such that $A_{n_j}A'_{n_{j+2}} + Z_jZ_{j+2}$ and $B_{n_j}B_{n_{j+2}} + Z_jZ_{j+2}$ are subsets of the boundary of U_j and the boundary of

V_j respectively. The segment $s(W_jW_{j+2})$ (Figure 23) is a subset either of U_j or of V_j. Suppose it is a subset of V_j. Then, by Theorem 39, $V_j - s(W_jW_{j+2})$ is the sum of two mutually exclusive connected domains I_j and L_j such that $Z_jZ_{j+2} + W_jW_{j+2}$ and $B_{n_j}B'_{n_{j+2}} + W_jW_{j+2}$ are subsets of the boundary of I_j and the boundary of L_j respectively. The continuum $M_{n_{j+1}}$ is a subset of $Q_j + s(A_{n_j}A'_{n_{j+2}}) + s(B_{n_j}B'_{n_{j+2}})$, and it contains both a point of $s(A_{n_j}A'_{n_{j+2}})$ and a point of $s(B_{n_j}B'_{n_{j+2}})$. But $Q_j + s(A_{n_j}A'_{n_{j+2}}) + s(B_{n_j}B'_{n_{j+2}}) = (s(A_{n_j}A'_{n_{j+2}}) + U_j) + (s(Z_jZ_{j+1}) + V_j + s(B_{n_j}B'_{n_{j+2}}))$, and $s(A_{n_j}A'_{n_{j+2}})$ and $s(Z_jZ_{j+2}) + V_j + s(B_{n_j}B'_{n_{j+2}})$ are mutually separated. Hence U_j contains a point of $M_{n_{j+1}}$. Also, $Q_j + s(A_{n_j}A'_{n_{j+2}}) + s(B_{n_j}B'_{n_{j+2}}) = (s(B_{n_j}B'_{n_{j+2}}) + L_j) + (s(A_{n_j}A'_{n_{j+2}}) + s(Z_jZ_{j+2}) + U_j + I_j)$ and the point sets $s(B_{n_j}B'_{n_{j+2}})$ and $s(A_{n_j}A'_{n_{j+2}}) + s(Z_jZ_{j+2}) + U_j + I_j$ are mutually separated. Hence L_j intersects $M_{n_{j+1}}$. But $Q_j - I_j$ is the sum of the mutually separated point sets $U_j + Z_jZ_{j+2}$ and $L_j + s(W_jW_{j+2})$. The

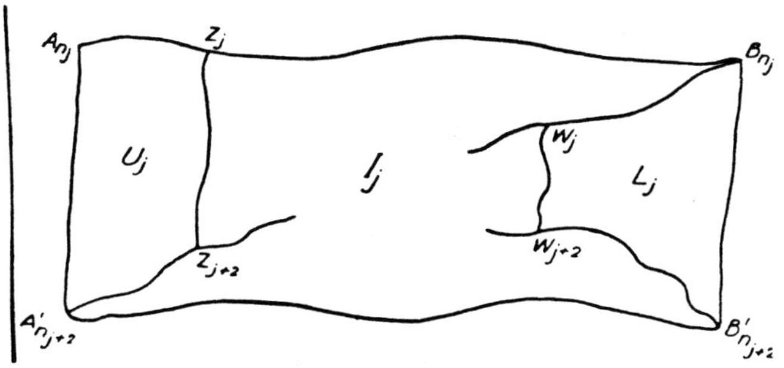

FIGURE 23

continuum $M_{n_{j+1}}$ intersects each of these point sets and it is a subset of their sum plus I_j. So both U_j and I_j intersect $M_{n_{j+1}}$. But every point of $M_{n_{j+1}}$ is a limit point of W. Hence W intersects both U_j and I_j. But U_j and I_j are mutually exclusive and W is connected. Therefore W contains a point of the boundary of I_j. But the boundary of I_j is a subset of N. Hence W intersects N. But W is a component of $S - N$. Thus the supposition that the boundary of W is not a continuous curve has led to a contradiction.

THEOREM 43. *If D and E are two complementary domains of the compact continuous curve M, the outer boundary of D with respect to E is a simple closed curve.*

PROOF. Let K denote the boundary of D and let T denote the outer boundary of D with respect to E. Since no point separates S, T contains two points A and B. By Theorem 42, T is a continuous curve. Hence, by Theorem 13 of Chapter II, it contains an arc AB. Since, by Theorem 21,

AB does not separate S, $T - AB$ exists. By Theorem 28, it is connected. If AB contained only one limit point of $T - AB$ that point would be a cut point of T, contrary to Theorem 28. So AB contains two points X and Y that are limit points of $T - AB$. Since it is a continuous curve, T contains two mutually exclusive connected point sets T_X and T_Y containing X and Y respectively, these point sets being domains with respect to T. The point sets T_X and T_Y contain points X_1 and Y_1 respectively belonging to $T - AB$. By Theorems 13 and 2 of Chapter II, $T - AB$ contains an arc X_1Y_1, T_X contains an arc X_1X and T_Y contains an arc YY_1. Let XY denote the interval of AB with endpoints as indicated. The arcs XY and X_1Y_1 are mutually exclusive and so are the arcs XX_1 and YY_1. It follows that the point set $XY + X_1Y_1 + XX_1 + YY_1$ contains a simple closed curve J. The curve J is a subset of T. Suppose that $T - J$ exists. Then it is connected and J contains two points X_2 and Y_2 that are limit points of $T - J$. By an argument analogous to the one used above to prove the existence of J it follows that there exists an arc CWF lying wholly in $T - J$ except for the points C and F which belong to J. Let CW_1F and CW_2F denote the arcs of J whose endpoints are C and F. Let I denote the component of $S - J$ that contains $s(CWF)$ and let Q denote the other component of $S - J$. Let I_1 and I_2 denote the components of $I - s(CWF)$ that are bounded by $CW_1F + CWF$ and $CWF + CW_2F$ respectively. Since I contains the point W of the boundary of D and $I = I_1 + I_2 + s(CWF)$, and CWF contains no point of D, therefore either I_1 or I_2 intersects D. Similarly, since $I_1 + s(CW_1F) + Q$ is a domain containing W_1, and W_1 (Figure 24) is a boundary point of D and no point of $s(CW_1F)$ belongs to D therefore

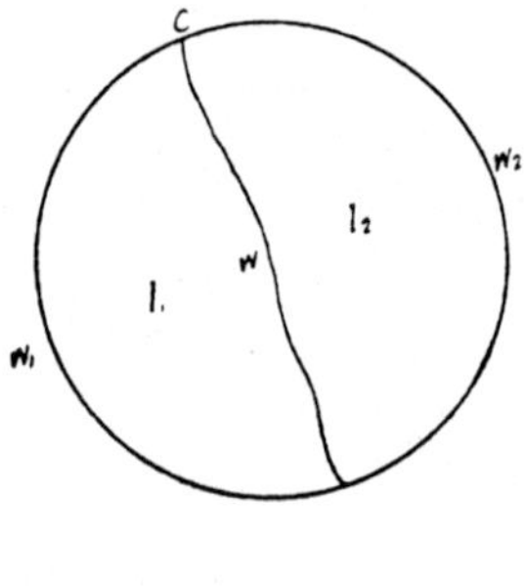

FIGURE 24

either I_1 or Q contains a point of D. In a similar way it may be shown that either I_2 or Q intersects D. It follows that, of the three domains I_1, I_2 and Q, there are two that intersect D. But these three domains are mutually separated and D is a subset of their sum. Hence D is not connected. Thus the supposition that T is not J has led to a contradiction.

THEOREM 44. *Suppose that H and K are two compact continua such that* (a) *K does not separate any two points of H from each other,* (b) *T, the common part of H and K, is totally disconnected,* (c) *$K - T$ is connected. Then there exists a simple closed curve which separates $H - T$ from $K - T$ and which contains T but no point of $(H + K) - T$.*

PROOF. By Theorem 117 of Chapter I, there exists a sequence of domains $D_1, D_2, D_3, \cdots$ such that (a) for each n, $\bar{D}_{n+1}$ is a subset of D_n, (b) T is the common part of the point sets $D_1, D_2, D_3, \cdots$, (c) if $P_1, P_2, P_3, \cdots$ is a sequence of points such that, for each n, P_n belongs to D_n, some subsequence of this sequence has a sequential limit point and every such sequential limit point belongs to T, (d) not all points of K belong to D_1. For each n, let K_n denote the point set $K \cdot (S - \bar{D}_n)$. By Theorem 113 of Chapter I, there exists a sequence $G'_1, G'_2, G'_3, \cdots$ such that (1) for each n, G'_n covers K, G'_n is a subcollection of G_n, and G'_{n+1} is a subcollection of G'_n, (2) if L and N are two mutually exclusive closed subsets of K there exists a positive integer δ such that no region of G'_δ intersects both L and N. There exists a finite set G''_1 of simple domains such that (a) if g is a domain of G''_1 then the point set consisting of $g \cdot K$ and the boundary of g is a subset of some region of the set G'_1, (b) every point of $\bar{K}_1$ belongs to some domain of the set G''_1 and every domain of G''_1 intersects $\bar{K}_1$, (c) no point of $H + \bar{D}_2 \cdot K$ is in, or on the boundary of, any domain of G''_1. There exists a finite set G''_2 of simple domains, covering $\bar{K}_2 - K_1$, such that (a) each domain of G''_2 intersects $\bar{K}_2 - K_1$, (b) if g is a domain of G''_2 then the point set consisting of $g \cdot K$ and the boundary of g is a subset of some region of G'_2, (c) no point of $H + \bar{D}_3 \cdot K$ is in, or on the boundary of, any region of G''_2. There exists a finite set G''_3 of simple domains covering $\bar{K}_3 - K_2$ such that (a) each domain of G''_3 intersects $\bar{K}_3 - K_2$, (b) if g is a domain of G''_3 then the point set consisting of $g \cdot K$ and the boundary of g is a subset of the common part of D_1 and some region of G'_3, (c) no point of $H + \bar{D}_4 \cdot K$ or of $(S - D_1) \cdot K$ is in any domain of G''_3. This process may be continued. Thus there exists an infinite sequence $G''_1, G''_2, G''_3, \cdots$ such that, for every n, (a) G''_{n+2} is a finite set of simple domains such that, if g is any one of them, the point set consisting of $g \cdot K$ and the boundary of g is a subset of the common part of D_n and some region of the set G'_{n+2}, (b) G''_{n+2} covers $\bar{K}_{n+2} - K_{n+1}$ and each domain of G''_{n+2} intersects $\bar{K}_{n+2} - K_{n+1}$, but no point of $H + K \cdot \bar{D}_{n+3}$ or of $(S - D_n) \cdot K$ is in, or on the boundary of, any domain of G''_{n+2}.

Let G'' denote the collection of all domains g such that, for some n, g belongs to G''_n. Let D denote the point set obtained by adding together all the simple domains of the collection G''. With the aid of the fact that $K - T$ is connected, it may be seen that the domain D is connected. No point of $D \cdot K$ belongs to infinitely many domains of the collection G''. It follows that if W denotes the collection of all domains g of G'' such that g is not a subset of any other domain of G'', then W covers D. Each domain w of the collection W is bounded by a simple closed curve J_w. Let N denote the point set obtained by adding together T and all the curves J_w for all domains w of the collection W. The point set N is closed and compact and it contains the boundary of D. Suppose N is not connected. Then it is the sum of two mutually exclusive closed and connected point sets N_1 and N_2 and, by Theorem 13, there exists a simple closed curve J which contains no point of N but which separates some point of N_1 from some point of N_2. Since every point of N is a limit point of D it follows that J separates some two points of D from each other. But D is connected. Therefore J contains a point of D. But J is a connected point set containing no point of N and therefore no point of the boundary of D. Hence J is a subset of D. It follows that it is a subset of some domain Q of the collection W. Since J is a subset of Q therefore, by Theorem 2 of Chapter III, Q contains a component Z of $S - J$. Since J separates some two points of N from each other therefore Z contains a point A of N. The point A belongs to the boundary t of some domain V of the collection W. Since the simple closed curve t contains a point of Z but no point of J, the boundary of Z, therefore t is a subset of Z and therefore of Q. But both V and Q belong to W and therefore their exteriors have H in common. Therefore V is a subset of Q. But no domain of the collection W is a subset of another domain of that collection. Thus the supposition that N is not connected leads to a contradiction. Hence N is a continuum.

Suppose P is a point of $N - T$. There exists a region R containing P such that $\bar{R}$ contains no point of T. There exists a positive integer m such that if $n > m$ then no domain of the set G''_n intersects $N \cdot R$. Let N_m denote the sum of the boundaries of all the domains of the set W that belong to a collection of the set $G''_1, G''_2, G''_3, \cdots, G''_m$. The point set N_m is the sum of a finite number of simple closed curves. But every simple closed curve is locally connected and the sum of a finite number of locally connected continua is also locally connected. Hence N_m is locally connected at the point P. Since N_m is a subset of N containing every point of N that belongs to R, it follows that N is locally connected at P.

Thus N is locally connected at every point of $N - T$. But T is a totally disconnected subset of N. It follows, by Theorems 11 and 14 of Chapter II, that N is locally connected at every one of its points. Hence it is a compact continuous curve.

Select some point O of $H - T$ and let E denote the complementary domain of N that contains O. The domain E contains no point of D. For if it did then, since it contains a point O not belonging to D, it would contain a point of the boundary of D and therefore a point of N, since the boundary of D is a subset of N. Thus E is a complementary domain of $\bar{D}$. Let β denote the outer boundary of E with respect to D. By Theorems 42 and 43, β is a simple closed curve separating D from E. Let S_D and S_E denote the components of $S - \beta$ that contain D and E respectively.

The point set $H - T$ is a subset of S_E. For suppose it is not. Then, since it contains no point of β, it contains a point U of S_D. By hypothesis, K does not separate O from U. Hence there exists an arc OU having no point in common with K. This arc contains an arc ACB which has no point in common with K and which has in common with H only its endpoints A and B which lie in S_D and S_E respectively. Let I denote the component of $S - H$ that contains $s(ACB)$. By Theorems 24 and 36, $I - s(ACB)$ is the sum of two mutually separated connected domains I_1 and I_2. Let M_1 denote the set of all points X such that X lies on some segment of ACB whose extremities are the extremities of a segment on AB which lies wholly in I_1. With the help of Theorem 110 of Chapter I and Theorem 36 of the present chapter and Theorem 13 of Chapter III, it may be shown that if X is a point of M_1 then β contains two points A_X and B_X lying on ACB in the order AA_XXB_XB and such that (a) there is a segment t_X on β lying wholly in I_1 and having A_X and B_X as its endpoints, (b) no other segment on β lies wholly in I_1 and has its endpoints on the intervals AA_X and BB_X respectively of the arc ACB. Thus with each point X of M_1 we have associated a definite segment t_X. Let M_2 denote the sum of all segments t_X for all points X of M_1. Let t denote the point set $M_2 + (ACB - M_1)$. This point set is an arc with endpoints at A and B. The point set $I - (t - A - B)$ is the sum of two mutually separated domains Z_1 and Z_2 where Z_1 is a subset of I_1. No segment on β lies in Z_1 and has both of its endpoints on t. By Theorem 36, A and B belong to S_D and S_E respectively. Hence Z_1 contains both points of S_D and points of S_E. Hence, since it is connected, Z_1 intersects β. Similarly Z_2 intersects β. Let L and L' denote points belonging to $\beta \cdot Z_1$ and $\beta \cdot Z_2$ respectively. The point L belongs to $\beta \cdot Z_1$ and the point L' belongs to $\beta \cdot (S - \bar{Z}_1)$. Therefore L belongs to some arc q of β which lies wholly in Z_1 except for its endpoints, which belong to the boundary of Z_1. Since the endpoints of q do not both belong to t, one of them belongs to $H - H \cdot t$. Hence q contains an arc PL lying wholly in Z_1 except for the point P which belongs to H but not to t. Since every point of PL except P belongs to the boundary of some domain of the collection W, but the point P does not do so, therefore P is the sequential limit point of some sequence of distinct points $P_1, P_2, P_3, \cdots$ lying on PL and such that no two of them belong to the boundary of the same domain of the set W. There exists a region R containing P but no point of ACB. There exists a region R' containing P such that $\bar{R}'$ is a subset of R. There exists a number δ such that if $n > \delta$ then no region of G'_n intersects both $K \cdot \bar{R}'$ and $K \cdot (S - R)$. There exist two numbers k and m such that $m > \delta$ and P_k is a point of R' lying on the boundary of a simple domain g belonging to the collection W and to G'_m. Let α denote the boundary of g. The point set $\alpha + g \cdot K$ is a continuum containing P_k and lying wholly in R and therefore having no point in common with ACB. Let X denote a point of $g \cdot K$. The continuum $\alpha + g \cdot K$ contains both X and P_k. But it contains no point of H or of ACB and therefore no point of the boundary of I_1. But P_k belongs to Z_1 and therefore to I_1. Therefore so does X. But X belongs to $K - T$. Thus $K - T$ intersects I_1. Similarly it intersects I_2. But, by hypothesis, $K - T$ is connected. Hence it intersects the boundary of I_1. But the boundary of I_1 is a subset of $ACB + H$, and $K - T$ does not intersect either ACB or H. Thus the supposition that $H - T$ is not a subset of S_E leads to a contradiction. But $K - T$ is a subset of S_D. Therefore J separates $H - T$ from $K - T$. Since T is a totally disconnected subset of H therefore every point of T is a limit point of $H - T$ and therefore of S_E. With the help of the fact that S_D contains $K - T$ it follows, in the same way, that every point of T is a limit point of S_D. Hence T is a subset of J. The truth of Theorem 44 is therefore established.

THEOREM 45. *If K is the boundary of a complementary domain D of a compact continuous curve M, J is the outer boundary of D with respect to some point of $S - \bar{D}$ and H is a component of $K - J$ then J contains only one limit point of H.*

PROOF. By Theorem 51 of Chapter I, J contains at least one limit point of H. Suppose it contains two such points, A and B. There exist two mutually exclusive regions R_A and R_B containing A and B respectively. Let O denote a point of H. By Theorems 2, 3 and 12 of Chapter II, there exist arcs OA' and OB' lying wholly in H except for A' and B' which belong to $J \cdot R_A$ and $J \cdot R_B$ respectively and are therefore distinct. By Theorems 21, 13 and 43 of Chapter II, the point set $OA' + OB'$ contains an arc $A'O'B'$. This arc lies, except for its endpoints, wholly in $I \cdot K$ where I is the component of $S - J$ that contains D. The curve J is the sum of two arcs $A'XB'$ and $A'TB'$. By Theorem 6 of Chapter III, $I - s(A'O'B')$ is the sum of two mutually separated connected domains I_1 and I_2, bounded by the simple closed curves $A'O'B' + A'XB'$ and $A'O'B' + A'YB'$ respectively. The point X is a boundary point of D but not of I_2. It follows that I_1 contains a point of D. Similarly I_2 contains a point of D. But $A'O'B'$ contains no point of D and therefore D is the sum of the two mutually separated point sets $D \cdot I_1$ and $D \cdot I_2$. Thus the supposition that J contains two limit points of H leads to a contradiction.

Theorem 45 does not remain true even in the plane if the stipulation requiring that M be a compact continuous curve is replaced by the stipulation that it be a compact continuum even though the requirement that $\bar{H} \cdot J$ be degenerate is replaced by the weaker requirement that it be a continuum. The truth of this statement may be seen with the help of an example indicated in Figure 25. In this example, N is a compact continuum, $S - N$ is the sum of two mutually exclusive domains I and E both bounded by N; ACB is an arc lying wholly in I except for its endpoints which belong to N; H is $s(ACB)$ and K is $H + N$; M is identical with K and $I - H$ is the sum of two mutually exclusive domains D and D' such that D is bounded by $N + ACB$ and E is the sum of two connected domains E_1 and E_2. Here K is the boundary of D, J is the outer boundary of D with respect to E_1 and H is a component of $K - J$. But $\bar{H} \cdot J$ consists of the two points A and B.

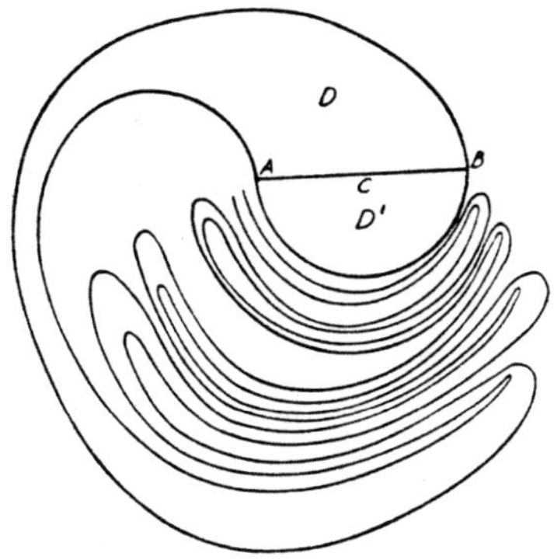

FIGURE 25

THEOREM 46. *If the nondegenerate compact continuous curve M has no cut point and K is the boundary of a complementary domain of M then K is a simple closed curve.*

PROOF. Suppose D is a complementary domain of M. Since M has no cut point it contains a simple closed curve T. If $\bar{D}$ were S, D would intersect both complementary domains of T, and therefore T. So $S - \bar{D}$ exists. Let J denote the outer boundary of D with respect to some point of $S - \bar{D}$. Suppose K, the boundary of D, is distinct from J. Let H denote a component of $K - J$. By Theorem 45, J contains only one limit point of H. Let P denote that point. The point sets H and $M - (H + P)$ are mutually separated and their sum is $M - P$, contrary to hypothesis.

THEOREM 47. *If every complementary domain of the compact continuum M is bounded by a simple closed curve and every two points of M are separated from each other in M by a continuum then no two boundaries of complementary domains of M have two points in common.*

PROOF. Suppose D_1 and D_2 are two complementary domains of M whose respective boundaries, J_1 and J_2, have two points A and B in common. The curve J_1 is the sum of two arcs AXB and AYB. Some continuum K separates A from B in M and therefore intersects both AXB and AYB. The continuum K is a subset of $S - D_1$. Hence, by Theorem 30 of Chapter III, K separates A from B in $S - D_1$. But $D_2 + A + B$ is a connected subset of $S - D_1$. This involves a contradiction.

THEOREM 48. *Every continuum which is irreducible with respect to the property of being a closed point set which separates S and which is a subset of* *the boundary of a complementary domain D of a continuum, is the outer boundary of D with respect to some point of $S - \bar{D}$.*

PROOF. Suppose that the continuum H is a subset of K, the boundary of a complementary domain D of a continuum and that H separates S but no closed proper subset of H does so. Let I denote the component of $S - H$ that contains D and let E denote another component of $S - H$. If E contained a point X of K then, since X would be a boundary point of D, E would contain a point of D. Hence E contains no point of K. But the boundary of E is a subset of H and therefore, since it separates S, it is identical with H. Hence, by definition, H is the outer boundary of D with respect to E.

THEOREM 49. *Every simple closed curve which is a subset of the boundary of a complementary domain D of a continuum is the outer boundary of D with respect to some point of $S - \bar{D}$.*

Theorem 49 is a consequence of Theorem 48, Axiom 4 and Theorem 21.

THEOREM 50. *If the boundary of the connected domain I is a compact continuum H, T is a totally disconnected closed subset of H, K is a compact continuum containing T such that $K - T$ is a subset of I and A and B are points of $H - T$ and $K - T$ respectively, there exists a simple closed curve β separating A from B and such that $\beta - \beta \cdot T$ is a subset of $I - (K - T)$.*

PROOF. Let $D_1, D_2, D_3, \cdots, G'_1, G'_2, G'_3, \cdots, G''_1, G''_2, G''_3, \cdots, K_1, K_2, K_3, \cdots$ and G'' be as described in the first paragraph, and the first sentence of the second paragraph, of the proof of Theorem 44. Let D' denote the sum of all the domains of the collection G'' and let D denote the component of D' that contains A. Let W denote the collection of all domains g of G'' such that g is not a subset of any other domain of G''. For each domain w of the collection W let J_w denote the boundary of w. Let N' denote the point set obtained by adding together all the curves J_w for all domains w of the collection W and let N denote the point set $\bar{N}'$. That N is a compact continuous curve follows almost as in the proof of Theorem 42. Let E denote the complementary domain of N that contains B. The outer boundary of E is a simple closed curve β separating A from B. The point set $\beta - \beta \cdot T$ is a subset of $I - (K - T)$.

It is not true even in the plane that, under the hypothesis of Theorem 50, there exists a simple closed curve β separating $K - T$ from A and such that $\beta - \beta \cdot T$ is a subset of $I - (K - T)$. Consider the following example.

EXAMPLE 1. In a Cartesian plane let T, A, C and E denote the points (0,0), (2,0), (2,2) and (0,2) respectively. For each positive integer n, let F_n denote the point $(1,1/n)$ and let T_n denote a point lying half way between the point of the interval TF_n whose abscissa is $1/n$ and the point of the interval TF_{n+1} with the same abscissa. Let H (Figure 26) denote the continuum composed of the straight line intervals TA, AC, CE and ET together with all the straight line intervals of the sequence TF_1, TF_2, $TF_3, \cdots$. Let K denote the sum of all the intervals of the sequence TT_1, TT_2, $TT_3, \cdots$. Let I denote the compact complementary domain of the continuum H. There exists no simple closed curve separating A from $K - T$ and lying, except for the point T, wholly in I.

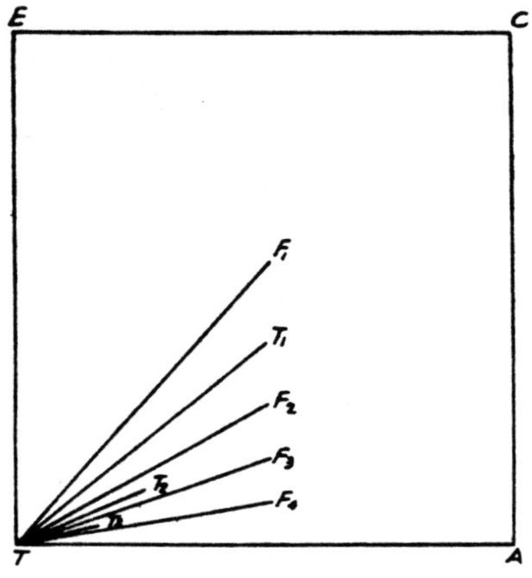

FIGURE 26

THEOREM 51. *If K is a closed subset of the continuous curve M, N is a component of $M - K$ whose closure is not M, and $T = \bar{N} \cdot K$ then T separates M and, indeed, $M - T$ is the sum of two mutually separated point sets of which one is N.*

THEOREM 52. *If K is the boundary of a complementary domain D of a compact continuous curve M and the arc AB is a subset of K then no component of $K - AB$ contains two arcs abutting on AB from opposite sides.*

PROOF. Suppose, on the contrary, that there exists a component H of $K - AB$ containing two arcs XY and ZW abutting on AB from opposite sides at Y and W respectively. By Theorem 42 of the present chapter and Theorem 13 of Chapter II, there exists an arc XZ lying in H. The point set $XY + ZW + XZ$ contains an arc YW. Let C and F denote points lying on YW in the order $YCFW$ and let CY and FW denote intervals of YW with endpoints as indicated. The arcs CY and FW abut on AB from opposite sides. Let CUF denote the interval of YW whose endpoints are C and F and let CVF denote the arc composed of $CY + FW$ together with either the point Y or the interval YW of AB according as W is or is not identical with Y. Let J denote the simple closed curve formed by the arcs CUF and CVF. By definition, the arc CVF crosses AB at the point Y or at the interval YW of AB according as Y is or is not identical with W. Hence, by Theorem 35, AB crosses CVF at that point or that interval. Therefore J separates A from B, contrary to the fact that no subset of the boundary of a connected domain separates two points of that boundary from each other.

THEOREM 53. *If A and B are two points of the compact continuum M and T is a totally disconnected closed point set that separates A from B in M but no proper subset of T does so, then there exists a simple closed curve containing T, but no point of $M - T$, and separating A from B.*

PROOF. By hypothesis, $M - T$ is the sum of two mutually separated point sets H and K containing A and B respectively. By Theorem 61 of Chapter I, the closed point sets $H + T$ and $K + T$ are connected. Let D denote the complementary domain of $K + T$ that contains A. Let W denote the boundary of D. By Theorem 24, W is a continuum. The point sets $H \cdot D$ and $(M - W \cdot T) - H \cdot D$ are mutually separated, they contain A and B respectively, and their sum is $M - W \cdot T$. Hence, by hypothesis, $W \cdot T = T$ and, by Theorem 61 of Chapter I, $H \cdot D + T$ is connected. Since the continuum W separates S, therefore, by Axiom 3, it is nondegenerate. Hence, since T is totally disconnected, there exists a point O belonging to W but not to T. By Theorem 26, $S - T$ is a connected domain. Hence, by Theorem 1 of Chapter II, it contains an arc α from B to O. The arc α contains an arc BF having only the point F in common with W. By Theorem 50, there exists a simple closed curve β, separating A from F, such that $\beta - \beta \cdot T$ is a subset of $D - (H - T)$. Since β separates A from F and contains no point of the arc BF therefore β separates A from B. Since $\beta \cdot T$ separates A from B in M, it is identical with T.

THEOREM 54. *If D is a connected point set and the simple closed curve J separates two points of $\bar{D}$ from each other and contains a point O of $\bar{D} - D$ but contains no other point of the boundary of D, then $J - O$ is a subset of D.*

PROOF. The point set $\bar{D} - O$ is connected and it contains two points that are separated from each other by J. Hence $\bar{D} - O$ and J have a point A in common. By hypothesis, A belongs to D. Hence if $J - O$ were not a subset of D it would contain both a point of D and a point not belonging to D and therefore, since it is connected, it would contain a point of the boundary of D, contrary to hypothesis.

THEOREM 55. *If D is a connected point set and the simple closed curve J separates two points of $\bar{D}$ from each other and contains the two points A and B of $\bar{D} - D$ but contains no other point of the boundary of D then D contains at least one segment lying on J and with endpoints at A and B.*

PROOF. The point set $\bar{D} - (A + B)$ is connected and it contains two points that are separated from each other by J. Hence $\bar{D} - (A + B)$ and

J have a point C in common. By hypothesis, C belongs to D. Let ACB denote the arc of J from A to B that contains C. If $ACB - (A + B)$ were not a subset of D it would contain both a point of D and a point not belonging to D and therefore, since it is connected, it would contain a point of the boundary of D, contrary to hypothesis.

THEOREM 56. *If the compact continuous curve M is the boundary of a connected domain and the simple closed curve J is a subset of M and A and B are two points of J, then either A or B is a cut point of M or $M - (A + B)$ is the sum of two mutually separated connected point sets one containing $s(AX_1B)$ and the other containing $s(AX_2B)$, where AX_1B and AX_2B are the arcs of J whose endpoints are A and B.*

PROOF. Suppose neither A nor B is a cut point of M. Then, by Theorems 49 and 45, neither A nor B is a limit point of a component of $M - J$. Hence, by Theorem 45, if H is a component of $M - J$, $J \cdot \bar{H}$ is a subset either of AX_1B or of AX_2B. If there is no component H of $M - J$ such that $J \cdot \bar{H}$ is a subset of AX_iB $(i = 1,2)$, let N_i denote AX_iB. If there exists at least one such component let N_i denote the point set obtained by adding all of them to AX_iB. The point sets N_1 and N_2 are connected and mutually separated and they contain $s(AX_1B)$ and $s(AX_2B)$ respectively. Furthermore $M - (A + B) = N_1 + N_2$.

THEOREM 57. *If K is the boundary of a complementary domain D of a compact continuum M and A is a cut point of K then A is a cut point of M and A is accessible from D.*

PROOF. By Theorem 53, there exists a simple closed curve J having only the point A in common with K and separating two points of K from each other. By Theorem 54, $J - A$ is a subset of D. Let I and E denote the complementary domains of J. The point sets $M \cdot I$ and $M \cdot E$ are mutually separated. But $M - A = M \cdot I + M \cdot E$. Hence A is a cut point of M. If PA is an arc of J with one endpoint at A, $PA - A$ is a subset of D and therefore A is accessible from D.

THEOREM 58. *If K is the boundary of a complementary domain D of a compact continuum M and A and B are two non-cut points of K but $A + B$ separates K then A and B are accessible from D.*

PROOF. By Theorem 24, the compact point set K is a continuum. By Theorem 53, there exists a simple closed curve J containing $A + B$, but no point of $K - (A + B)$, and separating two points of K from each other. By Theorem 55, D contains at least one segment lying on J and with endpoints at A and B. Thus A and B are accessible from D.

THEOREM 59. *Every point of the boundary of a complementary domain of a compact continuous curve is accessible from that domain.*

PROOF. Suppose A is a point of the boundary K of the complementary domain D of the compact continuous curve M. If A is a cut point of K then, by Theorem 57, it is accessible from D. If A is a non-cut point of K belonging to some simple closed curve J which is a subset of K then, by Theorems 22, 105, 53 and 108 of Chapter I, J contains another point B which is a non-cut point of K and, by Theorems 24 and 56, $A + B$ separates K and therefore, by Theorem 58, A is accessible from D.

Suppose now that A is a non-cut point of K that belongs to no simple closed curve which is a subset of K. If M is degenerate, D is $M - A$ and A is accessible from $M - A$. Suppose M is nondegenerate. If X is a point of M distinct from A and O is a point of D then, by Axiom 3 and Theorem 1 of Chapter II, there exists an arc OX from O to X lying in $S - A$ and the first point of M on this arc in the order from O to X is a point of K and therefore K is nondegenerate. Hence, by Theorem 93 of Chapter I, there exists at least one non-cut point of K distinct from A. Let B denote one such point. The continuous curve K contains an arc AB with endpoints at A and B. Suppose H is a component of $K - AB$. By Theorems 12 and 2 of Chapter II, A and B are accessible from H. Hence there exists an arc ATB lying, except for its endpoints, wholly in H. The arcs ATB and AB form a simple closed curve containing A and B and lying wholly in K. Thus the supposition that $\bar{H}\cdot(AB) = A + B$ leads to a contradiction. Hence $\bar{H}\cdot(AB)$ contains a point of AB distinct from A and from B. It follows, with the aid of Theorem 2 of Chapter II, that there exists an arc abutting on AB and lying, except for the point of abutting, wholly in H. By Theorem 11, there exists an arc AXB having, in common with AB, only its endpoints, A and B. Let J denote the simple closed curve formed by the arcs AB and AXB and let D_1 and D_2 denote the complementary domains of J. For each i ($i = 1,2$) let W_i denote the set of all components H of $K - AB$ such that there exists an arc abutting on AB and lying, except for one endpoint, wholly in $D_i\cdot H$. Let K_i denote the sum of all the point sets of the set W_i. By Theorem 52, K_1 and K_2 are mutually exclusive. Suppose that $s(AB)$ contains a limit point of $D_1\cdot K_2$. Then there exists a sequence of points $P_1, P_2, P_3,\cdots$ belonging to $D_1\cdot K_2$ and converging to a point O of AB and such that no two points of this sequence belong to the same component of $K - AB$. There exists (Figure 27) an arc A_1XB_1 lying wholly in D_1 except for the points A_1 and B_1 which lie on AB in the order AA_1OB_1B. Of the two complementary domains of the simple closed curve formed by $A_1X_1B_1$ and the interval A_1OB_1 of AB, let I denote the one which is a subset of D_1. There exists a number m such that the points $P_m, P_{m+1},\cdots$ are all in I. But, for each n, there exists an arc P_nY_n abutting on AB at some point Y_n and lying, except for Y_n, wholly in some one point set of the collection W_2. For each n greater than m, the arc P_nY_n contains both a point P_n of I and a point of $S - \bar{D}_1$. It follows that, for each n greater than m, there exists an interval C_nF_n of P_nY_n lying wholly in the domain $D_1 - (I_1 + s(A_1X_1B_1))$

except for the points C_n and F_n which belong to AXB and $A_1X_1B_1$ respectively. It follows that K is not a continuous curve. Thus the supposition that AB contains a limit point of $D_1 \cdot K_2$ leads to a contradiction. It follows that the compact and connected point set $AXB + D_1 \cdot K$ is closed. Hence, by Theorem 11, there exists an arc AZ_1B lying, except for its endpoints, wholly in D_1 and containing no point, other than A and B, of this point set and therefore no point of K_2. Similarly there exists an arc AZ_2B lying,

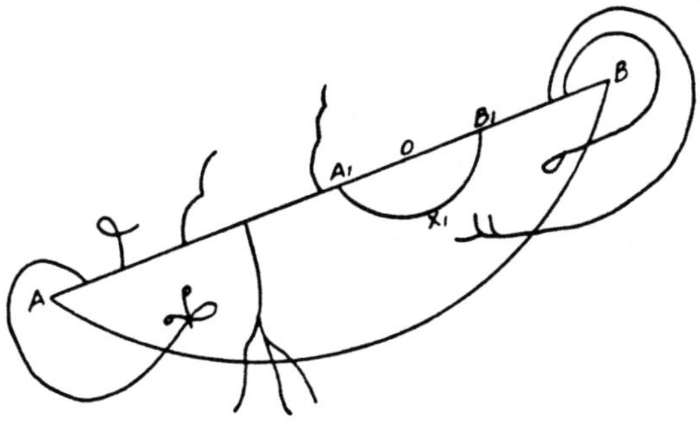

FIGURE 27

except for its endpoints, wholly in D_2 and containing no point of K_1. The arcs AZ_1B and AZ_2B form a simple closed curve J. Let E denote the complementary domain of J that does not contain $s(AB)$. Let N_1 and N_2 denote $AZ_1B + E \cdot K_1$ and $AZ_2B + E \cdot K_2$ respectively. The point sets N_1 and N_2 are compact continua, $N_1 - (A + B)$ and $N_2 - (A + B)$ are connected and $N_1 \cdot N_2 = A + B$. Hence, by Theorem 44, there exists a simple closed curve Q containing A and B and separating $N_1 - (A + B)$ from $N_2 - (A + B)$. The curve J contains an arc ALB (Figure 28) lying,

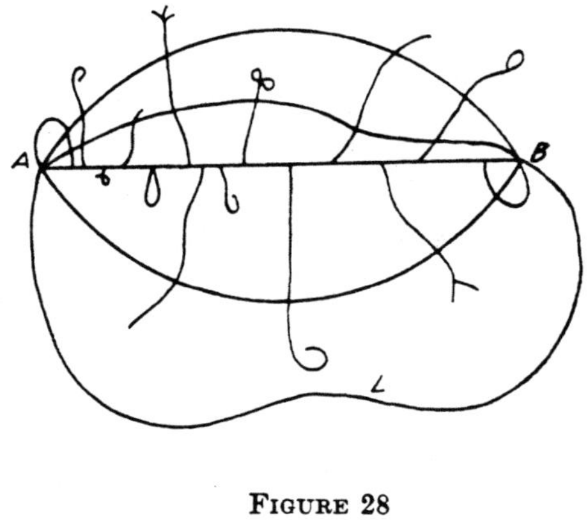

FIGURE 28

except for its endpoints, wholly in E and therefore containing no point of K except A and B. The segment $s(ALB)$ lies wholly in some component V of $S - K$. The point A belongs to the boundary of V. Hence, if V is not identical with D, A belongs to the outer boundary of D with respect to V and therefore, by Theorem 42, A belongs to a simple closed curve lying in K, contrary to supposition. Hence V is identical with D and A is accessible from D.

THEOREM 60 (GEHMAN). *If K is the sum of a finite number of continuous curves, M is a compact continuous curve and N is a component of $M - K \cdot M$ then $\bar{N}$ is a continuous curve.*

PROOF. Suppose $\bar{N}$ is not a continuous curve. Then there exist two simple domains D_1 and D_2, an arc AB and an infinite sequence of mutually exclusive continua M', M_1, M_2, $M_3, \cdots$ such that $\bar{D}_2$ is a subset of D_1 and if D denotes the domain $D_1 - \bar{D}_2$ then (1) for each n, M_n is a component of $\bar{N} \cdot \bar{D}$ containing both a point of J_1, the boundary of D_1 and a point of J_2, the boundary of D_2, (2) the sequence M_1, M_2, $M_3, \cdots$ converges to M', (3) the arc AB has no point in common with any of the continua M', M_1, M_2, $M_3, \cdots$ and lies wholly in D except for its endpoints A and B which lie on J_1 and J_2 respectively, (4) the point set $D \cdot M_1 + D \cdot M_2 + D \cdot M_3 + \cdots$ is a subset of a component U of $D - D \cdot (M' + AB)$. Clearly $\bar{N}$ is connected im kleinen at every point of N. It follows that M' is a subset of K. Suppose O is a point of $M' \cdot D$ and R is a region containing O. Since M is a continuous curve there exists a region R' containing O such that every point of $M \cdot U \cdot R'$ can be joined to O by an arc t lying wholly in $M \cdot R \cdot D$. The arc t intersects infinitely many of the continua M_1, M_2, $M_3, \cdots$. If X and Y are points of t belonging to two continua of this sequence then they belong to $\bar{N}$ but, since it lies in D and joins two components of $D \cdot \bar{N}$, the interval XY of t is not a subset of $\bar{N}$ and therefore it contains a point of K. It follows that every region containing O contains a point of K belonging to U. With the help of this fact and the fact that K is connected im kleinen at every point of M', it may be seen that there exist a number k and three mutually exclusive arcs A_1B_1, A_2B_2, A_3B_3 (Figure 29) lying in K and such that if U' denotes the component of $U - M_k$ whose boundary contains M' then (1) A_1B_1, A_2B_2 and A_3B_3 lie wholly in U' except that A_1, A_2 and A_3 belong to M_k and B_1, B_2 and B_3 belong to M', (2) if U_1 denotes the component of $U' - U' \cdot (A_1B_1 + A_2B_2)$ whose boundary contains $A_1B_1 + A_2B_2$, U_2 denotes the component of $U' - U' \cdot (A_2B_2 + A_3B_3)$ whose boundary contains $A_2B_2 + A_3B_3$ and U_3 denotes the component of $U' - U' \cdot (A_1B_1 + A_3B_3)$ whose boundary contains $A_1B_1 + A_3B_3$ then $U_3 - A_2B_2 = U_1 + U_2$ and U_1 and U_2 are mutually separated. There exists a number m such that, for each n equal to or greater than m, M_n intersects U_1. But the point set $M_m \cdot U_1$ is a subset of $\bar{N}$. Therefore U_1 contains a point P_1 belonging to N. Similarly U_2 contains a point P_2 belonging to N. There exists an arc P_1P_2

lying wholly in N. Since P_1 belongs to U_1 and P_2 does not, the arc P_1P_2 intersects β, the boundary of U_1. But β is a subset of $K + M_k$; and N, and therefore P_1P_2, contains no point of K. Hence P_1P_2 intersects M_k. The point set $M_k + P_1P_2$ contains a connected subset of $\bar{N}\cdot\bar{D}$ containing M_k and a point P_1 not belonging to M_k. But this involves a contradiction since M_k is a maximal connected subset of $\bar{N}\cdot\bar{D}$.

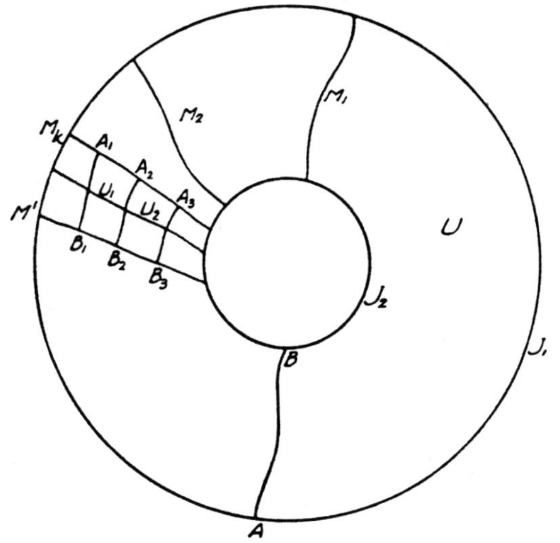

FIGURE 29

THEOREM 61 (WILDER). *If M and K are compact continuous curves and K is a subset of M and D is a component of $M - K$ then every point of the boundary of D with respect to M is accessible from D.*

PROOF. Suppose O is a point of the boundary of D with respect to M. Let P denote a point of D and let $I_1, I_2, I_3, \cdots$ denote a sequence of simple domains closing down on O such that I_1 does not contain P. For each n let O_n denote a point of $D\cdot I_{n+1}$ and let PO_n denote an arc from P to O_n lying wholly in D. For every n, PO_n contains a point O_n of I_1. For every n, let $P_{1n}O_n$ denote an interval of PO_n lying wholly in I_1 except for the point P_{1n} which belongs to J_1, the boundary of I_1. For every n equal to or greater than 2, $P_{1n}O_n$ contains a point O_n of I_2. For each such n, let $P_{2n}O_n$ denote an interval of $P_{1n}O_n$ lying wholly in I_2 except for the point P_{2n} which belongs to J_2. This process may be continued. Thus there exists an infinite sequence $\alpha_1, \alpha_2, \alpha_3, \cdots$ such that (1) for each m, α_m is an infinite sequence of arcs $P_{mm}O_m, P_{m,m+1}O_{m+1}, P_{m,m+2}O_{m+2}, \cdots$, (2) for each m and n, the arc $P_{mn}O_n$ lies wholly in I_m except for the point P_{mn} which belongs to J_m, the boundary of I_m, (3) for each m and n such that $n \geqq m$, the arc $P_{m+1,n}O_n$ is an interval of the arc $P_{mn}O_n$. For each n let H_n denote the domain $I_n - \bar{I}_{n+1}$. There exists a sequence $\beta_1, \beta_2, \beta_3, \cdots$ such that (1) for each n, β_n is a convergent sequence of arcs $b_{n1}, b_{n2}, b_{n3}, \cdots$, (2) each arc of β_n is an interval of some arc of α_n and lies wholly in H_n except for its endpoints which belong to J_n and J_{n+1} respectively, (3) for each n, each arc of β_{n+1} is an interval of some arc of α_n that contains an arc of β_n.

Case I. Suppose there exists an infinite increasing sequence of positive integers $n_1, n_2, n_3, \cdots$ and, for each i, an infinite increasing sequence of positive integers $m_{i1}, m_{i2}, m_{i3}, \cdots$ such that if γ_i denotes the sequence $b_{n_i m_{i1}}$, $b_{n_i m_{i2}}$, $b_{n_i m_{i3}}, \cdots$ then (1) the arc $b_{n_i m_{i1}}$ intersects every other arc of the sequence γ_i, (2) each arc of the sequence γ_{i+1} is an interval of an arc of α_{n_i} that contains an arc of γ_i. There exists a number j such that PO_j contains $b_{n_1 m_{11}}$. For each i there exists an arc r_i belonging to α_{n_i} and containing $b_{n_{i+1} m_{i+1,1}}$ and some arc q_i of the set γ_i. The point set $O + PO_j + r_1 + r_2 + r_3 + \cdots + q_1 + q_2 + q_3 + \cdots$ is a continuous curve T containing O and P. By Theorem 13 of Chapter II, there exists an arc PO lying wholly in T. The arc PO lies, except for O, wholly in D.

Case II. Suppose it is not true that there exist an infinite sequence $n_1, n_2, n_3, \cdots$ and, for each i, an infinite sequence $m_{i1}, m_{i2}, m_{i3}, \cdots$ satisfying the conditions described under Case I. In this case there exist an infinite increasing sequence of positive integers $n_1, n_2, n_3, \cdots$ and, for each i, a sequence γ_i of arcs $t_{i1}, t_{i2}, t_{i3}, \cdots$ such that if t_i denotes the continuum to which the sequence γ_i converges then (1) the sequence γ_i is a subsequence of β_{n_i}, (2) no two arcs of γ_i have a point in common, (3) for each j, t_{ij} and $t_{i,j+2}$ compose a pair that separates $t_{i,j+1}$ from t_i in $\bar{H}_{n_i}$, (4) every arc of γ_{i+1} is an interval of an arc of α_{n_i} that contains some arc of γ_i. For each i let A_i denote a definite point of $t_i \cdot H_i$. For each i there exists in H_i a region R_i containing A_i such that every point of $M \cdot R_i$ can be joined to A_i by an arc lying wholly in $M \cdot (H_i - H_i \cdot t_{i1})$. There exists a number j_i such that t_{ij_i} contains a point P_i of R_i. There exists an arc P_iA_i lying in $M \cdot (H_i - H_i \cdot t_{i1})$. The arc P_iA_i contains an arc $P_i'A_i'$ having in common with t_{ij_i} only the point P_i' and having in common with t_i only the point A_i'. For each m greater than j_i, $t_{i1} + t_{im}$ separates t_{ij_i} from t_i in $\bar{H}_{n_i}$ and therefore the arc $P_i'A_i'$ intersects t_{im}. Every arc in H_{n_i} from A_i' to a point of $s(P_i'A_i')$ contains a point of some arc of the sequence $t_{ij_i}, t_{ij_i+1}, t_{ij_i+2}, \cdots$ and therefore of D. But since K is a continuous curve there exists a region R_i', containing A_i', such that if there is any point of K in R_i' then every such point can be joined to A_i' by an arc lying wholly in $K \cdot H_{n_i}$ and therefore containing no point of D. There exists, on $P_i'A_i'$, a point P_i'' such that the interval $P_i''A_i'$ of $P_i'A_i'$ is a subset of R_i'. The arc $P_i''A_i'$ contains no point of K except possibly A_i'.

There exists a number j such that PO_j contains a point B of $P_1''A_1'$. For each positive integer i there exists an arc q_i belonging to α_{n_i} and containing both a point X_i of $P_i''A_i'$ and a point Y_i of $P_{i+1}''A_{i+1}'$. Let BX_1 denote B or the interval BX_1 of $P_1''A_1'$ according as B is or is not identical with X_1 and, for each i, let $X_{i+1}Y_i$ denote X_{i+1} or the interval $X_{i+1}Y_i$ of $P_{i+1}''A_{i+1}'$ according as X_{i+1} is, or is not, identical with Y_i. The point set $O + PO_j + BX_1 + X_2Y_1 + X_3Y_2 + X_4Y_3 + \cdots + q_1 + q_2 + q_3 + \cdots$ is a compact continuous curve N containing O and lying, except for O, wholly in D. The curve N contains an arc PO. Therefore O is accessible from D.

THEOREM 62. *If G is a collection of complementary domains of the continuous curve M such that the boundary of G^* is compact and A is a point of M such that every region that contains it intersects infinitely many domains of the collection G then every region that contains A contains the boundary of some domain of the collection G.*

PROOF. There exist an infinite sequence $g_1, g_2, g_3, \cdots$ of distinct domains of the set G and an infinite sequence of points $A_1, A_2, A_3, \cdots$ converging to A such that, for each n, A_n is a point of g_n. Suppose R is a region containing A. Suppose that, for each n, β_n, the boundary of g_n, contains a point B_n of $S - R$. There exists an ascending sequence of positive integers $n_1, n_2, n_3, \cdots$ such that $B_{n_1}, B_{n_2}, \cdots$ converges to some point B of $\beta\cdot(S - R)$ where β is the boundary of G^*. There exist two simple domains I_1 and I_2 containing A such that (1) $\bar{I}_2$ is a subset of I_1, (2) $\bar{I}_1$ does not contain B. There exists a number m such that, for every k greater than m, A_{n_k} belongs to I_2 and B_{n_k} belongs to $S - \bar{I}_1$. For each k, the domain $g_{n_{m+k}}$ contains an arc t_k lying wholly in the domain $I_1 - \bar{I}_2$ except that one of its endpoints belongs to J_1, the boundary of I_1 and the other one belongs to J_2, the boundary of I_2. There exists an infinite subsequence $t_1', t_2', t_3', \cdots$ of the sequence $t_1, t_2, t_3, \cdots$ such that if, for each n, C_n and F_n denote the points $J_1\cdot t_n'$ and $J_2\cdot t_n'$ respectively and C_2C_1 denotes the arc of J_1 from C_2 to C_1 that contains C_3, and F_2F_1 denotes the arc of J_2 from F_2 to F_1 that contains F_3 then the points $C_2, C_3, C_4, C_5, \cdots$ lie on C_2C_1 in the order $C_2C_3C_4C_5\cdots$ and $F_2, F_3, F_4, F_5, \cdots$ lie on F_2F_1 in the order $F_2F_3F_4F_5\cdots$. Let J denote the simple closed curve formed by the arcs C_2C_1, F_2F_1, t_1' and t_2' and let I denote the complementary domain of J which is a subset of $I_1 - \bar{I}_2$. For each n equal to or greater than 2, let α_n denote the simple closed curve formed by t_n', t_{n+1}' and the intervals C_nC_{n+1} and F_nF_{n+1} of C_2C_1 and F_2F_1 respectively. Let D_n denote the complementary domain of α_n which is a subset of $I_1 - \bar{I}_2$. By Theorem 44 of Chapter I and Theorem 19 of the present chapter, for each n, the point set $\beta\cdot\bar{D}_n$ contains a continuum M_n that intersects both of the intervals C_nC_{n+1} and F_nF_{n+1}. For each n, let M_n' denote the component of $M\cdot\bar{D}_n$ that contains M_n. The continuum M_n' is a component of $M\cdot(\bar{I} - I_2)$. The domain $I_1 - \bar{I}_2$ contains a point O of the limiting set of the sequence $M_1', M_2', M_3', \cdots$. By Theorem 14 of Chapter II, the continuum M is not connected im kleinen at the point O. Thus the supposition that Theorem 62 is false leads to a contradiction.

THEOREM 63. *No compact continuous curve has uncountably many complementary domains.*

PROOF. Suppose, on the contrary, that there exists a compact continuous curve M such that G, the set of all its complementary domains, is uncountable. For each domain g of the set G, let B_g denote the boundary of g. By Theorem 119(a) of Chapter I there exist two mutually exclusive closed subsets K_1 and K_2 of M and an uncountable subcollection H of G such that if h is an element of H then B_h intersects both K_1 and K_2. There exists a point A of K_1 such that every region that contains A intersects infinitely many elements of H. There exists a point A of K_1 such that every region that contains A intersects infinitely many elements of H. There exists a region R containing A but no point of K_2. By Theorem 62, R contains B_h for some domain h of H. Since B_h intersects K_2 this involves a contradiction.

THEOREM 64. *If the boundary M of a connected domain D is connected and compact then in order that M be a continuous curve it is necessary and sufficient that for every closed subset K of M there exist a compact continuum which contains K and is a subset of $K + D$.*

PROOF. Suppose the boundary of the connected domain D is a compact continuous curve M. Let K denote a closed subset of M. By Theorem 15, for each point P of M there exists a simple domain I_P containing P such that $\bar{I}_P$ does not contain all of D and such that if C_P denotes the boundary of I_P then $M \cdot I_P + C_P$ is a subset of some region of the collection G_1. Let Q_P denote the collection of all components of $D \cdot I_P$ whose boundaries contain P. By Theorem 43 of Chapter II, the compact continuum $M + C_P$ is a continuous curve T. Each component of $D \cdot I_P$ is a complementary domain of T distinct from D and therefore has on its boundary at least one point of C_P. Hence, by Theorem 62, the collection Q_P is finite. There exists a simple domain I'_P, containing P, such that $\bar{I}'_P$ is a subset of I_P. There are not more than a finite number of components of $D \cdot I_P$ which contain points of I'_P and points of the boundary of I_P but whose boundaries do not contain P. Hence there exists a simple domain R_P, containing P and lying in I_P, such that every component of $D \cdot I_P$ that contains a point of R_P belongs to the collection Q_P. It follows that every point of M in R_P is on the boundary of some domain of the finite collection Q_P. By Theorems 37 and 4 of Chapter I, there exists a finite set of points $P^1, P^2, \cdots, P^n$, belonging to K, such that every point of K belongs to some domain of the collection $R_{P^1}, R_{P^2}, \cdots, R_{P^n}$. Let W_1 denote the collection of all the domains of the collections $Q_{P^1}, Q_{P^2}, \cdots, Q_{P^n}$. If w is a domain of the collection W_1, there exists a region of the collection G_1, containing both $w \cdot M$ and the boundary of w, and each point of K belongs to the boundary of some domain of W_1. Let O denote some definite point of D. There exists a compact continuum N_1 containing O, lying wholly in D, and such that each domain of the finite collection W_1

contains at least one point of N_1. With the help of the fact that each domain of W_1 is itself bounded by a continuous curve, it follows that there exists a finite collection W_2 of domains such that (1) each domain of W_2 is a subset of some domain of W_1 and each domain of W_1 contains at least one domain of W_2, (2) if w is a domain of W_2 there exists a region of the collection G_2 containing both $w \cdot M$ and the boundary of w, and w is bounded by a compact continuous curve that contains a point of K, (3) each point of K is on the boundary of some domain of W_2. There exists a finite collection H_2 of compact continua such that for each continuum h of H_2 there exists a domain w of W_1 (and for each domain w of W_1 there exists a continuum h of H_2) such that h is a subset of w that intersects both N_1 and every domain of W_2 that lies in w. Let N_2 denote the closed and compact point set obtained by adding together all the continua of the collection H_2. This process may be continued. Thus there exist a sequence $W_1, W_2, W_3, \cdots$ of finite collections of domains, a sequence $H_1, H_2, H_3, \cdots$ of finite collections of compact continua and a sequence $N_1, N_2, N_3, \cdots$ of closed and compact point sets, such that (1) for every n, each domain of W_{n+1} is a subset of some domain of W_n and each domain of W_n contains at least one domain of W_{n+1}, (2) if w is a domain of W_n there exists a region of the collection G_n containing both $w \cdot M$ and the boundary of w, and w is bounded by a compact continuous curve that contains a point of K, (3) each point of K belongs to the boundary of some domain of W_n, (4) for each continuum h of H_{n+1} there exists a domain w of W_n (and for each domain w of W_n there exists a continuum h of H_{n+1}) such that h is a subset of w that intersects both N_n and each domain of W_{n+1} that lies in w, (5) N_n is the point set obtained by adding together all the continua of the collection H_n. Let T denote the sum of K and all the closed sets $N_1, N_2, N_3, \cdots$. The point set T is a continuum. It contains K and it is a subset of $D + L$. The necessity of the condition of Theorem 64 is therefore established.

It will be shown that this condition is also sufficient. Suppose that the boundary of a connected domain D is a compact continuum M which is not a continuous curve. Then there exist four simple domains R_1, R_2, R_3 and R_4, two points A and B lying on the boundaries of R_1 and R_4 respectively, a subcontinuum M' of M, an infinite sequence $A_1B_1, A_2B_2, \cdots$ of mutually exclusive arcs, and an infinite sequence $M_1, M_2, M_3, \cdots$ of subcontinua of M such that (1) for each n $(1 < n \leqq 4)$, $\bar{R}_n$ is a subset of R_{n-1}, (2) the points $A_1, A_2, \cdots$ are in the order $A_1A_2 \cdots A$ on C_1, the boundary of R_1, and $B_1, B_2, \cdots$ are in the order $B_1B_2 \cdots B$ on C_4, the boundary of R_4, (3) A is the sequential limit point of $A_1, A_2, \cdots$ and B is the sequential limit point of $B_1, B_2, \cdots$ and M' is the sequential limiting set of $A_1B_1, A_2B_2, \cdots$ (4) no arc of the set $A_1B_1, A_2B_2, \cdots$ intersects M, (5) for each n, the continuum M_n contains a point between A_n and A_{n+1} on the arc A_1A_2A of C_1 and a point between B_n and B_{n+1} on the arc B_1B_2B of C_4 but contains no point without R_1 or in R_4. For each n, let J_n denote the simple closed curve formed by

A_nB_n, $A_{n+1}B_{n+1}$ and the intervals A_nA_{n+1} and B_nB_{n+1} of the arcs A_1A_2A and B_1B_2B and let I_n denote the component of $S - J_n$ that lies in $R_1 - \bar{R}_4$. Let T_n denote the set of all points of $M \cdot I_n$ that lie on C_2 or C_3 according as n is even or odd. Let L denote $T_2 + T_4 + T_6 + \cdots$ and let K denote $\bar{L}$. The point set K is a closed subset of M but there exists no continuum which contains K and lies in $K + D$.

THEOREM 65. *If M and N are two mutually exclusive closed and compact point sets, H and K are components of M and N respectively and G is a set of simple closed curves such that, for every two points that belong to M and N respectively, there exists at least one curve of the set G that separates these two points from each other, then there exists a simple closed curve J such that* (1) *J separates H from K,* (2) *each point of J belongs to some curve of the set G,* (3) *J contains no point of $M + N$.*

PROOF. If P is a point of M then, for each point X of N, let J_{PX} denote a curve of the set G that separates P from X and let R_{PX} denote the complementary domain of J_{PX} that contains X. Let G_P denote the set of all the domains R_{PX} for all points X of the set N. By Theorems 37 and 4 of Chapter I, there exists a finite subset L_P of the set G_P such that every point of N belongs to some domain of the set L_P. Let D denote the sum of all the domains of L_P and let D_X denote the component of D that contains K. The boundary of the complementary domain of $\bar{D}_K$ that contains P is a simple closed curve J_P. This curve separates P from K and it contains no point of N. For each point P of M let R_P denote the complementary domain of J_P that contains P. Let G denote the set of all the domains R_P for all points P of M. There exists a finite subset Q of the set G such that every point of M belongs to some domain of the set Q. Let W denote the sum of all the domains of the set Q and let D_H denote the component of W that contains H. The boundary of the complementary domain of $\bar{D}_H$ that contains K is a simple closed curve J that satisfies all the requirements of the conclusion of Theorem 65.

THEOREM 66 (CLEVELAND). *Suppose M is a continuous curve with a compact boundary. If A is a point of M and B is a point distinct from A such that every arc from A to B contains at least one point of M distinct from A and from B then there exists a simple closed curve which is a subset of M and which separates A from B.*

PROOF. By Theorem 59, A and B do not belong to the boundary of the same complementary domain of M. Let I denote a simple domain containing A such that $\bar{I}$ does not contain B. There exists a region R, containing A, such that $\bar{R}$ is a subset of I and $M \cdot \bar{R}$ is a subset of the component of $I \cdot M$ that contains A. By Theorem 62, there are not more than a finite number of complementary domains of M having A on their boundaries and containing points of $S - \bar{R}$. If there be any, let $D_1, D_2, \cdots, D_n$ denote them, let $T_1, T_2, \cdots, T_n$ denote their respective boundaries and let H denote the set $T_1 + T_2 + \cdots + T_n$. Let K denote the component of $M \cdot \bar{I}$ that contains A and let E denote the sum of the sets H and K. Let D denote the complementary domain of E which contains B and let L denote the boundary of D. With the aid of Theorem 45 of Chapter II and Theorem 24 of the present chapter and the fact that no simple closed curve has a continuum of condensation it may be shown that K is a continuous curve. Furthermore, by Theorem 42, for each i, T_i is a continuous curve. It follows that E is a continuous curve. Hence so is L.

The point A does not belong to L. For suppose it does. By Theorem 59, every point of L is accessible from D. Hence there exists an arc AB lying, except for A, wholly in D. Let P denote the first point that the arc AB has in common with the boundary of R. The interval AP of AB has only the point A in common with M and therefore $AP - A$ lies in one of the domains $D_1, D_2, \cdots, D_n$. But this is impossible since the interval BP of AB contains no point of the boundary of any one of these domains. Hence A lies in one complementary domain of the continuous curve L and B lies in another one. Hence, by Theorem 43, there exists a simple closed curve which separates A from B and is a subset of L and therefore of M.

THEOREM 67. *If O is a point of a locally compact closed point set M there exists a sequence of simple domains $D_1, D_2, D_3, \cdots$ such that, for each n, $\bar{D}_{n+1}$ is a subset of D_n and the point sets $M \cdot D_1, M \cdot D_2, M \cdot D_3, \cdots$ have only the point O in common.*

The truth of Theorem 67 may be established with the aid of Theorem 7 of Chapter I and Theorem 15 of the present chapter.

THEOREM 68. *If M is a continuous curve with a compact boundary, O is a point of M not lying on the boundary of any complementary domain of M, K is a nondegenerate closed and compact point set containing O, and A is a point distinct from O then there exists a sequence of simple closed curves $J_1, J_2, J_3, \cdots$, converging to O, such that* (1) *for each n, J_n lies in M and separates O from A, and J_{n+1} separates O from J_n,* (2) *if D_n denotes the complementary domain of J_n that contains O then $\bar{D}_{n+1}$ is a subset of D_n, and $K \cdot D_1, K \cdot D_2, K \cdot D_3, \cdots$ have only the point O in common.*

PROOF. Let A' denote A or some definite point on the boundary of the component of $S - M$ that contains A according as A is or is not a point of M. With the help of Theorem 15, it may be shown that there exists a sequence of simple domains $D_1, D_2, D_3, \cdots$ such that $K \cdot D_1, K \cdot D_2, K \cdot D_3, \cdots$ have only the point O in common and such that if, for each n, C_n denotes the boundary of D_n then $D_{n+1} + C_{n+1}$ is a subset of D_n. By Theorem 66, each point of the closed and compact point set $A' + C_1 + K \cdot (S - D_1)$ is separated from O by a simple closed curve lying in M. Therefore, by Theorem 17 of Chapter III and Theorem 65 of the present chapter, there exists a simple closed curve J_1 lying in M and separating O from $A' + C_1 + K \cdot (S - D_1)$. The curve

J_1 is necessarily a subset of D_1 and it separates O from A. By a similar argument it may be shown that there exists a simple closed curve J_2 lying wholly in $M \cdot D_2$ and separating O from $A + C_2 + J_1 + K \cdot (S - D_2)$. This process may be continued indefinitely. It follows that there exists a sequence of simple closed curves $J_1, J_2, J_3, \cdots$ such that J_{n+1} lies wholly in $M \cdot D_{n+1}$ and separates O from $A + C_{n+1} + J_n + K \cdot (S - D_{n+1})$. This sequence satisfies all the requirements of Theorem 68.

THEOREM 69. *If M is a compact continuous curve, K is a closed subset of M and H is a subset of K and of some connected subset of $K + (S - M)$, but no point set of which H is a proper subset satisfies these conditions, then if the continuum E is a subset of $K - H$ there exists a simple closed curve lying wholly in $M - K$ and separating H from E.*

PROOF. For each complementary domain D of M whose boundary intersects K, let $C(D)$ denote the common part of K and the boundary of D. By Theorems 42 and 64, there exists a compact continuum which contains $C(D)$ and which is a subset of $D + C(D)$. For each D, let $T(D)$ denote a definite continuum having this property. Let $D_1, D_2, D_3, \cdots$ denote the set of all complementary domains of M whose boundaries intersect K and let N denote the point set $T(D_1) + T(D_2) + T(D_3) + \cdots$. It is easy to see that a given subset of K is contained in some connected subset of $K + (S - M)$ if and only if it is contained in some connected subset of $K + N$. Suppose that P is a limit point of N that does not belong to N and that R is a region containing P. Let R_1 denote a region, containing P, such that $\bar{R}_1$ is a subset of R. Infinitely many of the point sets $T(D_1), T(D_2), \cdots$ contain points of R_1. But, by Theorem 62, not more than a finite number of the domains $D_1, D_2, D_3, \cdots$ intersect both $\bar{R}_1$ and $S - R$. Hence there exists a positive integer m such that $T(D_m)$ is a subset of R. But $T(D_m)$ intersects K. Hence R intersects K. It follows that P is a limit point of K. But K is closed. Hence P belongs to K. Thus K contains every limit point of N that does not belong to N. It follows that $K + N$ is closed. Let W denote the component of $K + N$ that contains H. The point set W contains no point of $K - H$. If E is a continuum lying in $K - H$ then, by Theorem 44 of Chapter I, $K + N$ is the sum of two mutually exclusive closed sets Q and Z containing W and E respectively. Let U denote the common part of K and Q and let V denote the common part of K and Z. Every arc from a point of U to a point of V contains a point of $M - K$. Hence, by Theorem 66, for every two points that belong to U and V respectively there exists a simple closed curve that lies wholly in M and separates these two points from each other. It follows, by Theorem 65, that there exists a simple closed curve that lies wholly in $M - K$ and separates H from E.

THEOREM 70. *If K is a closed subset of a compact continuous curve M and L is a subset of M lying in a connected subset of $K + (S - M)$ then L lies in a closed and connected subset of $K + (S - M)$.*

The truth of Theorem 70 follows easily with the aid of results established in the course of the proof of Theorem 69.

THEOREM 71. *If K is a closed subset of a compact continuous curve M and A and B are two points of M that lie in a connected subset of $K + (S - M)$ then A and B lie in a closed and connected subset of $K + (S - M)$.*

Theorem 71 is a special case of Theorem 70. It may be seen, with the aid of the following example, that it does not remain true in the plane if the phrase "lie in a closed and connected subset of" is replaced by "are the extremities of an arc lying in".

EXAMPLE 2. In a Cartesian plane, let M_1 denote the point set obtained by adding to the straight line interval from $(0, -1)$ to $(0,1)$ that part of the graph of $y = \sin(1/x)$ consisting of all points of it whose abscissae are positive numbers not more than 1. Let N denote the set of all points of the graph of $y = x + \sin(1/x)$ with such abscissae. Let H denote the sum of M_1, N and the straight line interval whose endpoints are the point of M_1 whose abscissa is 1 and the point of N with the same abscissa. The continuum H is the boundary of only one compact domain. Let D denote this domain. For each pair of integers m and n where n is positive and m is numerically less than n, let A_{mn} denote the point $(0,m/n)$, let B_{mn} denote the point $(1/n,m/n)$ and let $A_{mn}B_{mn}$ denote the straight line interval whose extremities are A_{mn} and B_{mn}. Let G denote the set of all such intervals $A_{mn}B_{mn}$ and let T denote the point set obtained by adding together all the intervals of this set. Let M denote the continuous curve $M_1 + N + T$. For each interval t whose endpoints are on M and which lies, except for its endpoints, in D and in some interval of the set G, let P_t denote the midpoint of t. Let L denote the set of all such points P_t and let K denote L plus all the points of the straight line interval from $(0, -1)$ to $(0,1)$. Let A denote some definite point of L and let B denote the origin $(0,0)$. Clearly $K + (S - M)$ contains a continuum that contains both A and B. But it does not contain an arc that does so.

THEOREM 72. *In order that the compact continuum M be a continuous curve it is necessary and sufficient that it be true that if t is an arc and H is a domain with respect to $M + t$ and X and Y are points of $S - H$ which do not lie together in any continuum which is a subset of $S - H$ then H contains a simple closed curve that separates X from Y.*

PROOF. It will first be shown that this condition is necessary. Suppose that M is a continuous curve. Suppose t is an arc and H is a domain with respect to $M + t$. If $M + t$ is connected it is a continuous curve. If it is not connected and H intersects both M and t and the points X and Y belong to $S - H$ but no closed and connected subset of $S - H$ contains both X and Y then neither does any closed and connected subset of $S - H \cdot M$. In each case the situation is reduced to one where a domain with respect to a compact continuous curve intersects every continuum that contains both of the points X and Y. Suppose then that K is a closed subset of a compact continuous curve M and that X and Y are points of $K + (S - M)$ such that every continuum that contains both X and Y intersects $M - K$.

Case 1. Suppose X and Y both belong to M and therefore to K. In this case it follows, by Theorems 69 and 70, that $M - K$ contains a simple closed curve that separates X from Y.

Case 2. Suppose neither X nor Y belongs to M. Let D_X denote the complementary domain of M which contains X and let C_X denote the boundary of D_X. Let D_Y and C_Y denote sets determined in a similar manner by Y. If either C_X or C_Y contains no point of K then either the outer boundary of D_X with respect to Y or that of D_Y with respect to X is a simple closed curve lying in $M - K$ and separating X from Y. If C_X contains a point E of K and C_Y contains a point F of K then there exist arcs XE and YF that have only E and F, respectively, in common with M and clearly every continuum that contains both E and F intersects $M - K$. Hence $M - K$ contains a simple closed curve J that separates E from F. Clearly J separates E from Y.

The case where one of the points X and Y belongs to K and the other one belongs to $S - M$ may be treated in a similar manner.

That this condition is sufficient may be shown as follows. Suppose that M is a compact continuum which is not a continuous curve. There exist four simple domains R_1, R_2, R_3, R_4, two points A and B lying on the boundaries of R_1 and R_4 respectively, a subcontinuum M' of M, an infinite sequence A_1B_1, $A_2B_2, \cdots$ of arcs, an infinite sequence M_1, $M_2, \cdots$ of subcontinua of M and an infinite sequence of point sets T_1, $T_2, \cdots$ satisfying all the conditions (1)–(5) described in the proof of Theorem 64. Let K denote the point set $T_1 + T_2 + \cdots$ together with all of its limit points. Let t denote an arc with A and B as endpoints and composed of the arc A_1B_1 together with the arcs A_1A_2A of C_1 and B_1B_2B of C_4. Let X denote a point of T_2 and let Y denote a point of K belonging to M'. Let H denote the point set $(M + t) - K$. Every continuum that contains both X and Y contains a point of H. But clearly there exists no simple closed curve that lies wholly in H and separates X from Y. Hence the condition is sufficient.

The condition of Theorem 72 does not remain sufficient if the phrase "do not lie together in any continuum which is a subset of $S - H$" is replaced by the phrase "are separated from each other by H". The truth of this statement may be seen with the aid of the following example.

EXAMPLE 3. In a Cartesian plane let O and A denote the origin and the point (2,0) respectively and, for every positive integer n, let B_n denote the point $(1,1/n)$. Let M denote the sum of all the straight line intervals OA, OB_1, $OB_2, \cdots$ and AB_1, AB_2, $AB_3, \cdots$. If t is an arc, H is a domain with respect to $M + t$ and X and Y are points of $S - H$ such that $S - H$ is the sum of two mutually separated point sets containing X and Y respectively, then H contains a simple closed curve that separates X from Y. But M is not a continuous curve.

THEOREM 73. *If G is an uncountable set of compact continua each having a proper subcontinuum that separates S and $\bar{G}^*$ is compact then there exists an uncountable subset of G such that every two continua of this subset have a point in common.*

PROOF. For each continuum g of the set G let B_g denote a proper subcontinuum of g that separates S, let P_g denote some definite point of $g - B_g$ and let q_g denote some definite domain containing P_g but no point of B_g. By Theorem 92 of Chapter I, there exist a region R, a point A of R, and an uncountable subset H of G such that (1) if Q is any region whatsoever containing A there exist uncountably many continua h of the set H such that P_h belongs to Q, (2) if h is any continuum of the set H then q_h contains R. By Theorems 12 and 15, there exists a simple domain E containing A and bounded by a simple closed curve J such that $J + E \cdot H^*$ is a subset of R. There exists an uncountable subset H' of H such that if h is any continuum of the set H' then P_h belongs to E. Let D denote the simple domain $S - \bar{E}$. For each continuum g belonging to H' there exists a connected domain D_g lying in D and bounded by a subcontinuum β_g of B_g. By Theorem 92 of Chapter I, there exist a point O and an uncountable subset H'' of H' such that if g is a continuum of the set H'' then D_g contains O. Every two continua of the set H'' have a point in common. For suppose, on the contrary, that there exist two, h and k, that do not. The connected domains D_h and D_k have the point O in common, their exteriors have $S - D$ in common and their boundaries are mutually exclusive continua. Hence, by Theorem 2 of Chapter III, one of these domains is a subset of the other one. Suppose, for example, that D_h is a subset of D_k. Then β_h is a subset of $D_k + k$. But β_h has no point in common with k. Hence it is a subset of D_k. Thus h contains a point of D_k. But h is connected and it contains no point of the boundary of D_k. Hence it is a subset of D_k and therefore of D. But P_h belongs to h and lies in $S - D$. This involves a contradiction.

DEFINITION. A domain is said to be *simply connected* if and only if it is connected and contains one of the complementary domains of every simple closed curve that lies wholly in it.

THEOREM 74. *If the boundary of the connected domain D is a compact point set β then in order that D be simply connected it is necessary and sufficient that β be a continuum.*

PROOF. The condition of Theorem 74 is necessary. For suppose, on the contrary, that D is simply connected and β is not a continuum. Then β is the sum of two mutually exclusive closed point sets H and K and, by Theorem

13, there exists a simple closed curve J containing no point of β but separating some point of H from some point of K. Since each of the two components of $S - J$ is a domain containing at least one point of the boundary of D therefore each of them intersects D. But D is connected. Therefore it contains a point of J. Hence, since J contains no point of the boundary of D, J is a subset of D. But neither of its complementary domains is a subset of D. This involves a contradiction.

The condition of Theorem 74 is also sufficient. For suppose that β is a continuum and that J is a simple closed curve lying in D. Let P denote a point of β. Of the complementary domains of J, let E denote the one that contains P and let I denote the other one. Since β is a connected subset of $S - J$ it must lie wholly in E. Hence I contains no point of β. But, since the boundary of I is a subset of D, I contains a point of D. Therefore, since it is connected, I is a subset of D. Thus D is simply connected.

DEFINITION. The continuum M is said to be a *triod* if and only if it is the sum of three continua such that the common part of all three of them is a proper subcontinuum of each of them and is also the common part of every two of them.

A continuum is said to be *atriodic* if and only if it contains no triod.

A triod is said to be *simple* if and only if it is the sum of three arcs OX, OY and OZ such that no two of them have any point in common except O. The point O is called its *emanation* point.

THEOREM 75. *If a compact nondegenerate continuous curve contains no simple triod then it is either an arc or a simple closed curve.*

PROOF. Suppose M is a compact continuous curve containing no triod. By Theorem 22 of Chapter I, M contains a countable set N of points X_1, X_2, $X_3, \cdots$ such that every point of M is a limit point of N. For each n, X_n can be joined to X_{n+1} by an arc X_nX_{n+1} lying wholly in M. In view of the fact that M contains no simple triod it is clear that, for every n, $X_1X_2 + X_2X_3 + \cdots + X_{n-1}X_n$ is either a simple closed curve or an arc. It is clear that if, for some n, it is a simple closed curve then this curve is identical with M. Suppose, on the other hand, that, for every n, it is an arc A_nB_n, the notation being so assigned that, in the order from A_n to B_n on the arc A_nB_n, for every i and j less than or equal to n, A_i precedes B_j. Let N' denote the point set constituted by the sum of all the arcs A_1B_1, A_2B_2, $A_3B_3, \cdots$. If P_1, P_2, $P_3, \cdots$ is a set of points such that, for every n, P_n follows B_n on some arc of the set A_1B_1, A_2B_2, $A_3B_3, \cdots$ then the set P_1, P_2, $P_3, \cdots$ does not have more than one limit point. For suppose there exists such a set with two limit points O_1 and O_2. Then if D_1 and D_2 are two simple domains, containing O_1, such that $\bar{D}_2$ is a subset of D_1 and such that $\bar{D}_1$ does not contain O_2, there exists an infinite set G of arcs (Figure 30) such that (1) each arc of G is, for some n, i and j (with i and j less than or equal to n) a subinterval of the interval B_iB_j of the arc A_nB_n, (2) no two arcs of G have a point in common, (3) each arc of G lies entirely in the domain $D_1 - \bar{D}_2$ except that one of its endpoints is on the boundary of D_1 and the other one is on the boundary of D_2, (4) if X and Y are points of two different arcs of G then no connected subset of $M \cdot (D_1 - \bar{D}_2)$ contains both X and Y. But, since M is a compact continuous curve, there exists no set of arcs having these properties.

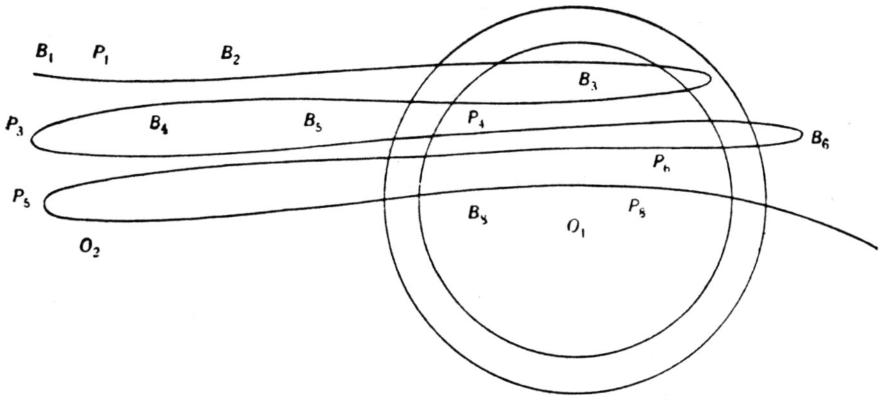

FIGURE 30

Similarly if $P_1, P_2, P_3, \cdots$ is a set of points such that, for every n, P_n precedes A_n on some arc of the set $A_1B_1, A_2B_2, A_3B_3, \cdots$, then the point set $P_1, P_2, P_3, \cdots$ cannot have more than one limit point. If neither of the sets $A_1, A_2, A_3, \cdots$ and $B_1, B_2, B_3, \cdots$ is infinite then N', and therefore M, is an arc. There remain two conceivable cases. Either (1) the sets $A_1, A_2, A_3, \cdots$ and $B_1, B_2, B_3, \cdots$ are both infinite or (2) one is finite and the other infinite. Suppose they are both infinite. Consider the fact that M contains no simple triod and that no set with the properties indicated above for $P_1, P_2, P_3, \cdots$ has more than one limit point. In view of these facts it is clear that either M is an arc or $A_1, A_2, A_3, \cdots$ and $B_1, B_2, B_3, \cdots$ have a common limit point O in which case the point set $N' + O$ is a simple closed curve identical with M.

Suppose the set $A_1, A_2, A_3, \cdots$ is infinite while $B_1, B_2, B_3, \cdots$ is finite. With the aid of the same considerations it is clear that, in this case, if O denotes the limit point of $A_1, A_2, A_3, \cdots$ and m denotes a positive integer such that, for every n greater than m, B_n coincides with B_m, then either $N' + O$ is an arc from O to B_m or O is identical with B_m and N' is a simple closed curve. Of course a parallel argument applies in case $B_1, B_2, B_3, \cdots$ is infinite. Thus, in every case, M is either an arc or a simple closed curve.

Theorem 76. *If a, b and c are three mutually exclusive arcs of the boundary of a simple domain D there do not exist two mutually exclusive compact subcontinua of $\bar{D}$ each intersecting each of the arcs a, b and c.*

Proof. Suppose there do exist two such mutually exclusive subcontinua of $\bar{D}$. Call them x and y and let J denote the boundary of D. With the help of Theorem 13 of the present chapter and Theorem 16 of Chapter III, it may be seen that there exists an arc AOB with endpoints on J, lying, except for its endpoints, wholly in D and separating x from y in $\bar{D}$. Let AXB and AYB denote the two arcs of J from A to B. Of the arcs a, b and c there is one, say b, that contains neither A nor B. The arc b is a subset of one of the segments AXB and AYB. But b contains both a point of x and a point of y and the segments AXB and AYB are connected subsets of $\bar{D} - AOB$. Thus there is a connected subset of $\bar{D} - ACB$ intersecting both x and y. But this is impossible since AOB separates x from y in $\bar{D}$.

Theorem 77. *Suppose* (1) *D is a domain bounded by a simple closed curve J,* (2) *a, b and c are three mutually exclusive arcs of J,* (3) *H_1, H_2, λ_1 and λ_2 are compact continua,* (4) *for each i ($i = 1,2$), H_i is the sum of three continua α_i, β_i and γ_i such that $\alpha_i \cdot \beta_i = \beta_i \cdot \gamma_i = \gamma_i \cdot \alpha_i = \lambda_i$,* (5) *$\lambda_i$ is a subset of D,* (6) *$\alpha_i \cdot J$, $\beta_i \cdot J$ and $\gamma_i \cdot J$ are subsets of a, b and c respectively. Then $H_1 \cdot H_2 \cdot D$ exists.*

Proof. There exist three mutually exclusive arcs A_1A_2, B_1B_2 and C_1C_2 each lying wholly in D except for its endpoints and such that A_1A_2 separates a from $\lambda_1 + \lambda_2 + b + c + C_1C_2 + B_1B_2$, B_1B_2 separates b from $\lambda_1 + \lambda_2 + a + c + A_1A_2 + C_1C_2$, and C_1C_2 separates c from $\lambda_1 + \lambda_2 + a + b + A_1A_2 + B_1B_2$, in $\bar{D}$, and the points A_1, A_2, B_1, B_2, C_1 and C_2 lie on J in the order $A_1A_2B_1B_2C_1C_2A_1$. Let A_2B_1, B_2C_1 and C_2A_1 denote the arcs of J, with endpoints as indicated, that contain no point of $a + b + c$. Let J' denote the simple closed curve $A_1A_2 + A_2B_1 + B_1B_2 + B_2C_1 + C_1C_2 + C_2A_1$ and let D' denote the component of $S - J'$ which lies wholly in D. Let H_i' denote the component of $H_i \cdot (D' + J')$ that contains λ_i. For each i, H_i' intersects each of the arcs A_1A_2, B_1B_2 and C_1C_2. It follows, by Theorem 76, that H_1' and H_2' have at least one point in common. But they are subsets of D. Hence $H_1 \cdot H_2 \cdot D$ exists.

Theorem 78. *If D is a simple domain and G is an uncountable collection of compact subcontinua of $\bar{D}$ each containing three points of J, the boundary of D, then there exists an uncountable subcollection H of G such that every two continua of the collection H have a point in common.*

Proof. For each continuum g of G let A_g, B_g and C_g denote three distinct points of $g \cdot J$. There exist three mutually exclusive arcs t_a, t_b and t_c of J and an uncountable subcollection H of G such that if h is any continuum of the collection H then the points A_h, B_h and C_h lie on the arcs

t_a, t_b and t_c respectively. By Theorem 76, every two continua of the collection H have at least one point in common.

THEOREM 79. *Suppose m is a positive integer and G is an uncountable collection of point sets such that* (1) $(\overline{G^*})$ *is compact and* (2) *each element g of G is the sum of m distinct points $A_{g1}, A_{g2}, A_{g3}, \cdots, A_{gm}$ and a continuum B_g which contains no one of these points and which does not separate S. Then there exist an uncountable subcollection H of G and two simple domains D_1 and D_2 such that* (1) *$\bar{D}_2$ is a subset of D_1 and* (2) *if g is any element of H then B_g is a subset of D_2 and the points $A_{g1}, A_{g2}, A_{g3}, \cdots, A_{gm}$ are all in the exterior of D_1.*

PROOF. By Theorem 118 of Chapter I, there exists an uncountable subcollection G' of G such that if g is any element of G' and U is any set of domains properly covering g then there are uncountably many elements of G' each properly covered by U. Let h denote a definite element of G'. By Theorem 18, there exists a simple domain I_0 containing B_h and bounded by a simple closed curve J_1 such that $I_0 + J_1$ contains no point of the closed and compact point set $A_{h1}, A_{h2}, A_{h3}, \cdots, A_{hm}$. By Theorem 15, there exists a simple domain I'_0, containing B_h, such that $\bar{I}'_0$ contains no point of J_1. The point set $\bar{I}'_0$ is a subset of I_0. With the aid of Axiom 3 and repeated applications of Theorem 15 it may be seen that there exist two sets of simple domains $I_1, I_2, I_3, \cdots, I_m$ and $I'_1, I'_2, I'_3, \cdots, I'_m$ such that $I_0, I_1, I_2, I_3, \cdots, I_m$ are mutually exclusive and, for each n, not greater than m, I'_n contains A_{hn} and $\bar{I}'_n$ is a subset of I_n. Let V denote the set of domains $I'_0, I'_1, I'_2, \cdots, I'_m$. The point set h is properly covered by V. Hence there exists an uncountable subcollection G' of G such that every element of G' is properly covered by V. With the help of the fact that V is a finite collection and that G' is uncountable, it may be seen that there exist an uncountable subcollection H of G' and a number j belonging to the set $0, 1, 2, \cdots, m$ such that if g is any element of H then B_g is a subset of I'_j. Let D_1 and D_2 denote I_j and I'_j respectively. The domains D_1 and D_2 satisfy all the requirements of Theorem 79.

THEOREM 80. *The compact continuum M is a triod if and only if it contains a continuum K such that $M - K$ has more than two components.*

PROOF. Clearly if M is a triod it contains such a continuum. Suppose M is a compact continuum containing such a continuum K. By Theorem 35 of Chapter I, $M - K$ is the sum of three mutually separated point sets T_1, T_2 and T_3. The point sets $K + T_1$, $K + T_2$ and $K + T_3$ are continua and K is the common part of every two of them.

THEOREM 81. *The compact continuum M is a triod if it contains three continua α, β and γ such that* (1) *α and β are mutually exclusive,* (2) *α separates M, and* (3) *γ intersects α and separates some two points of β from each other in M.*

PROOF. By hypothesis, $M - \gamma$ is the sum of two mutually separated point sets T_1 and T_2 such that $T_1 \cdot \beta$ and $T_2 \cdot \beta$ exist and $M - \alpha$ is the sum of two mutually separated point sets H and K such that H contains β. The point sets $H \cdot T_1 + \alpha + \gamma$, $H \cdot T_2 + \alpha + \gamma$ and $K + \alpha + \gamma$ are continua, their sum is M and the common part of every two of them is the continuum $\alpha + \gamma$.

It is not true in the plane that the compact continuum M is a triod if it contains two continua β and γ such that β separates two points of γ from each other in M and γ separates two points of β from each other in M. If the continuum M is the sum of three arcs AXB, AYB and AZB such that the common part of every two of them is $A + B$ then M contains two such continua β and γ. But it is not a triod.

DEFINITION. The continuum M is said to be *unicoherent* if and only if it is true that if it is the sum of two continua their common part is a continuum. A continuum M is said to be *hereditarily unicoherent* if and only if every subcontinuum of it is unicoherent.

THEOREM 82. *If M is an atriodic hereditarily unicoherent continuum, there do not exist in M three points A, B and C and three continua M_{AB}, M_{BC} and M_{AC} such that M_{AB} contains A and B but not C, M_{BC} contains B and C but not A, and M_{AC} contains A and C but not B.*

PROOF. Suppose there do exist such points and continua. Since M_{AB} and M_{AC} have A in common and their sum is unicoherent therefore $M_{AB} \cdot M_{AC}$ is a continuum T_A containing A. The point set $M_{BC} \cdot (M_{AB} + M_{AC})$ is a continuum L_{BC} containing B and C. The point sets $L_{BC} \cdot M_{AB}$ and $L_{BC} \cdot M_{AC}$ are continua T_B and T_C containing B and C respectively. Since the continuum L_{BC} is the sum of the continua T_B and T_C therefore $T_B \cdot T_C$ exists. It follows that $T_B \cdot T_C$ is a continuum. But $T_B \cdot T_C = T_B \cdot T_A = T_A \cdot T_C$ and $T_B \cdot T_C$ contains no one of the points A, B and C. Hence M is triodic, contrary to hypothesis.

THEOREM 83. *If M is a hereditarily unicoherent continuum, then M is triodic if and only if it contains three points such that if X and Y are any two of them there exists a subcontinuum of M containing X and Y but not the third one.*

THEOREM 84. *If G is an uncountable set of compact triodic continua and the sum of all the continua of G is a subset of a closed and compact point set then there exists an uncountable subset H of G such that every two continua of H have a point in common.*

PROOF. For each continuum g of the set G let k_g denote some definite triod which is a subset of g. Let G' denote the set of all continua k_g for all elements g of G. Every continuum of the set G' is the sum of two continua a and b such that $a \cdot b$ is a proper subcontinuum both of a and of b. Hence, by Theorem 20, if a continuum of the set G' contains a continuum that separates S, then some *proper* subcontinuum of it separates S. It follows, by Theorem 73, that if uncountably many continua of G' separate S then there exists an uncountable subset H of G such that every two continua of the set H have a point in common.

Suppose, on the other hand, that there do not exist uncountably many continua of the collection G' each containing a continuum that separates S. By Theorem 79 there exist an uncountable subset G_1 of G' and two domains D_1 and D_2 such that (1) $\bar{D}_2$ is a subset of D_1, (2) each continuum g of the set G_1 contains a continuum B_g and three continua a_g, b_g and c_g such that (a) $a_g \cdot b_g = b_g \cdot c_g = c_g \cdot a_g = B_g$, (b) each of the continua a_g, b_g and c_g contains a point of $S - D_1$, (c) B_g is a subset of D_2. Each continuum of the set G_1 contains a subcontinuum of $\bar{D}_2$ containing at least three points of the boundary of D_2. Therefore, by Theorem 78, there exists an uncountable subcollection H of G such that every two continua of the collection H have at least one point in common.

THEOREM 85. *If G is an uncountable set of mutually exclusive nondegenerate compact continuous curves such that G^* is compact, then all but countably many curves of the set G are either arcs or simple closed curves.*

Theorem 85 is a consequence of Theorems 75 and 84.

THEOREM 86. *If M is a compact continuous curve and ω is a point of $S - M$ there do not exist two complementary domains of M with the same outer boundary with respect to ω.*

PROOF. Suppose, on the contrary, that there exist a point set J and two distinct components D_1 and D_2 of $S - M$ such that J is the outer boundary, with respect to ω, both of D_1 and of D_2. By Theorem 43, J is a simple closed curve. Let I denote the complementary domain of J that does not contain ω. Let A and B denote two points of J. By Theorem 59, there exists an arc AOB lying, except for its endpoints, wholly in D_1. Let AXB and AYB denote the arcs of J whose endpoints are A and B. The point set $I - s(AOB)$ is the sum of two mutually separated domains I_1 and I_2 bounded respectively by the simple closed curves $AOB + AXB$ and $AOB + AYB$. The domain D_2 is a subset of $I - s(AOB)$ and therefore of $I_1 + I_2$. But X and Y are limit points of D_2. Therefore $D_2 \cdot I_1$ and $D_2 \cdot I_2$ exist. Thus the connected point set D_1 is the sum of two mutually separated point sets $D_2 \cdot I_1$ and $D_2 \cdot I_2$. This involves a contradiction.

If, in the statement of Theorem 86, "continuous curve" is replaced by "continuum", the resulting proposition does not hold true in the plane. Indeed there exists, in the plane, a compact continuum having infinitely many complementary domains all having the same outer boundary with respect to the same point. An example of such a continuum is indicated in Figure 31.

THEOREM 87. *If M is a compact continuous curve, ω is a point of $S - M$, and a and b are two complementary domains of M whose boundaries have a* *point P in common then P belongs either to the outer boundary of a with respect to ω or to the outer boundary of b with respect to ω.*

PROOF. Let J_a and J_b denote the outer boundaries of a and b respectively with respect to ω. Suppose P belongs neither to J_a nor to J_b. Let I_a and I_b denote the complementary domains of J_a and J_b respectively that contain P and let E_a and E_b denote $S - \bar{I}_a$ and $S - \bar{I}_b$ respectively. Suppose I_a contains a point X of E_b. Since E_b is a connected point set containing X and ω, E_b contains a point of J_a and therefore a point of a. Furthermore I_b contains the point P which is a limit point of a and therefore I_b intersects a. Thus a contains both a point of I_b and a point of E_b and therefore a point of J_b. But J_b is a subset of M and a is a complementary domain of M. This involves a contradiction. Hence I_a is a subset of $\bar{I}_b$ and therefore of I_b. Similarly I_b is a subset of I_a. Since I_a is identical with I_b, therefore J_a is identical with J_b. But this is contrary to Theorem 86.

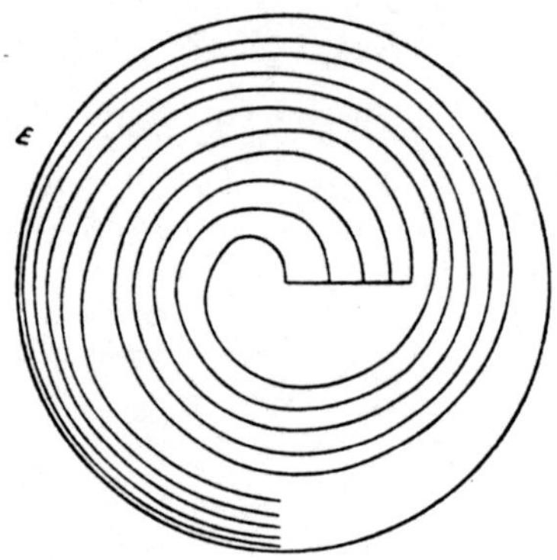

FIGURE 31

THEOREM 88. *If M is a compact continuous curve there do not exist three complementary domains of M all having the same boundary.*

PROOF. Suppose D_1, D_2 and D_3 are three complementary domains of M all having the same boundary β. Let ω denote some point of D_3. The domains D_1 and D_2 have the same outer boundary with respect to ω. But this is contrary to Theorem 86.

THEOREM 89. *If D_1 and D_2 are complementary domains of a compact continuous curve and B_1 and B_2 are their respective boundaries and B is the boundary of a complementary domain E of the point set $D_1 + D_2 + B_1 + B_2$ then the three point sets B, B_1 and B_2 do not have more than two points in common.*

Proof. Suppose that B_1, B_2 and B have, in common, three distinct points X, Y and Z. Let J_1 and J_2 denote the outer boundaries of D_1 and D_2 respectively with respect to E. The domain E is a component of $S - (J_1 + J_2)$, and J_1 and J_2 are simple closed curves having more than one point in common. It follows that B, the boundary of E, is a simple closed curve. Let I denote $S - (E + B)$. The connected domains D_1 and D_2 are mutually exclusive subsets of I and the points X, Y and Z belong to B, the boundary of I. With the help of Theorem 59, it may be seen that there exists an arc XO_1Y lying wholly in D_1 except for its endpoints. By Theorems 24 and 36, $D_1 - s(XO_1Y)$ is the sum of two mutually exclusive connected domains each having XO_1Y as a part of its boundary. Of these two domains let W denote one having Z on its boundary. The domain W is a complementary domain of the continuous curve $B_2 + XO_1Y$. Hence its boundary is a continuous curve and therefore there exists an arc O_1Z lying, except for its endpoints, wholly in W. The point set $XO_1Y + O_1Z$ is a simple triod t_1 lying, except for the points X, Y and Z, wholly in D_1. Similarly there exists a simple triod t_2 lying, except for X, Y and Z, wholly in D_2. These triods have no point in common except X, Y and Z. But they both lie wholly in I except that X, Y and Z belong to the boundary of I. But this is contrary to Theorem 77.

Theorem 90. *If M is a compact continuous curve and K is the set of all points X such that X is common to the boundaries B_1 and B_2 of two complementary domains D_1 and D_2 of M and is on the boundary of some complementary domain of the point set $D_1 + D_2 + B_1 + B_2$, then K is countable.*

Theorem 90 may be established with the aid of Theorems 89 and 63.

Theorem 91. *If D is a complementary domain of a compact continuous curve M, J is the outer boundary of D with respect to some point ω of $S - M$ and I is the complementary domain of J that contains D, then no point of I belongs to a simple closed curve that lies in M and separates ω from D.*

Theorem 91 may be proved with the assistance, in particular, of Theorem 45.

Theorem 92. *If the points O and ω are separated from each other by the compact continuous curve M and K is the sum of all the simple closed curves that lie in M and separate O from ω then K is closed.*

Proof. Suppose P is a limit point of K. Then if P does not belong to K it is a limit point either of N or of N_2 where N is the sum of the elements of the set G of all simple closed curves that lie in M and separate both O and P from ω, and N_2 is the sum of all those that separate O, but not P, from ω. Suppose that P is a limit point of N. The case where it is a limit point of N_2 may be treated in a similar manner. If P lies on the boundary of a complementary domain of M then, by Theorem 91, it must lie on the outer boundary of that domain with respect to ω and if the simple closed curve which constitutes this outer boundary should separate O from ω then P would belong to K. Suppose that P does not belong to the boundary of any complementary domain of M whose outer boundary with respect to ω separates O from ω. If there exist any complementary domains of M whose boundaries contain P, let H denote the set of all the outer boundaries of such domains with respect to ω, let U denote the set of all points X such that X is separated from ω by some simple closed curve of the set H and let T denote the point set consisting of U plus its boundary. By Theorem 55 of Chapter II, $M + T$ is a continuous curve. But P does not belong to the boundary of any complementary domain of $M + T$. Hence, by Theorem 68, there exists a sequence α of simple closed curves $J_1, J_2, J_3, \cdots$ converging to P such that (1) for each n, J_n lies in $M + T$ and separates P from $\omega + O$, and J_{n+1} separates P from J_n, (2) if D_n denotes the complementary domain of J_n that contains P then $\bar{D}_{n+1}$ is a subset of D_n, and $M \cdot D_1, M \cdot D_2, M \cdot D_3, \cdots$ have only the point P in common. If a simple closed curve β does not contain the point ω, the component of $S - \beta$ that does not contain ω will be called the interior of β, X will be said to be within β and β will be said to enclose X. If D is any domain such that $\bar{D}$ does not contain ω, then by the outer boundary of D will be meant the outer boundary of D with respect to ω. Suppose some curve J of the set G contains two points A and B in common with a curve Q of the set H. The point O is enclosed by the simple closed curve L consisting of a certain arc of J together with a certain interval, containing P, of the arc APB of Q. But L lies wholly in M, contains P and encloses O. Hence, in this case, P belongs to K. Suppose some curve J of the set G contains only one point A in common with one curve Q_1 of the set H and only one point B in common with a certain other curve Q_2 of the set H and suppose that the points A and B are distinct. If AP denotes one of the arcs of Q_1 from A to P, it may be shown that there is one arc of Q_2 from B to P which has no point except P in common with AP. Let BP denote such an arc of Q_2. Let AB denote the sum of the arcs AP and BP. There is an arc of J which together with AB forms a simple closed curve that lies wholly in M, encloses O and contains P. Hence, in this case also, P belongs to K. Suppose now that some curve of the set G has only one point E in common with T. In this case it may be seen that there exist other curves of G which pass through E and that indeed P is a limit point of the sum of all such curves. The point E lies on some curve J_E of the set H. There exists a region R_P containing P and such that every point of $M \cdot R$ can be joined to P by an arc which lies wholly in $M \cdot (S - O - E)$. But there exists a curve J of the set G which passes through E and contains a point X of R_P. Suppose J has no point except E in common with T. Let XP denote an arc which is a subset of $M \cdot (S - O - E)$. Let Z denote the last point that XP has in common with J and let Y denote the first point after Z that it has in common with T. Let ZY denote the interval of XP whose endpoints are Z and Y.

Let J_Y denote a curve of H which contains Y. There exists an arc YE which is a subset of $J_E + J_Y$. Let XYE denote the arc obtained by adding together the two arcs XY and YE. There exists on J an arc EWZ such that the simple closed curve formed by the arcs EWZ and XYE encloses O. But this curve is a subset of M and it has two points Y and E in common with T. It follows, by a result established above, that P belongs to K.

Suppose finally that no curve of the set G has a point in common with the point set T. If T exists, in other words if there exists at least one simple closed curve C belonging to the set H, then there exists a number n such that J_n contains at least one point within C and there exists a simple closed curve J of the set G such that J_n encloses at least one point of J and it may be shown that the point set $C + J_n + J$ contains a simple closed curve which encloses O and contains P and therefore P belongs to K. If, on the other hand, T does not exist then the following argument applies. For each n there exists a simple closed curve j_n which belongs to G and contains points within J_{n+1}. There exist two mutually exclusive arcs A_nB_n and C_nD_n lying on j_n and lying, except for their endpoints, wholly within J_n and wholly without J_{n+1}, the points A_n and C_n lying on J_n and the points B_n and D_n lying on J_{n+1}. If A_2 is B_1 let X_2 denote A_2 and let Y_2 denote C_2. If A_2 is not B_1 let X_2 and Y_2 denote C_2 and A_2 respectively of A_2 and C_2 respectively according as C_2 is or is not on the arc of J_{n+1} which has B_1 and A_2 as endpoints and which does not contain D_1. In any case, if B_1 is identical with X_2, let B_1X_2 denote the point B_1 and, if B_1 is distinct from X_2, let B_1X_2 denote the arc of J_{n+1} which has B_1 and X_2 as endpoints and which does not contain D_1. Similarly, if D_1 is Y_2 let D_1Y_2 denote D_1 and if D_1 is not Y_2 let D_1Y_2 denote the arc of J_{n+1} which has D_1 and Y_2 as endpoints and which does not contain B_1. Now let X_1, Y_1, Z_1 and W_1 denote the points A_1, C_1, B_1 and D_1 respectively and let Z_2 and W_2 denote B_2 and D_2 respectively or D_2 and B_2 respectively according as X_2 is identical with A_2 or with C_2. Let X_2Z_2 and Y_2W_2 denote the arcs A_2B_2 and C_2D_2 or the arcs C_2D_2 and A_2B_2 respectively, according as X_2 is identical with A_2 or with C_2. This process may be continued. Thus there exist four infinite sequences $X_1Z_1, X_2Z_2, X_3Z_3, \cdots$; $Y_1W_1, Y_2W_2, Y_3W_3, \cdots$; $Z_1X_2, Z_2X_3, Z_3X_4, \cdots$; and $W_1Y_2, W_2Y_3, W_3Y_4, \cdots$ such that, for every n, (a) X_n and Y_n lie on J_n, (b) X_nZ_n and Y_nW_n lie, except for their endpoints, wholly within J_n and wholly without J_{n+1} and indeed one of them coincides with A_nB_n and the other one with C_nD_n, (c) Z_nX_{n+1} and W_nY_{n+1} are arcs lying on J_{n+1} and with no point in common. Let a denote the point set $P + X_1Z_1 + Z_1X_2 + X_2Z_2 + Z_2X_3 + \cdots$ and let b denote the point set $P + Y_1W_1 + W_1Y_2 + Y_2W_2 + W_2Y_3 + \cdots$. It is easy to see that a and b are arcs. These arcs lie in M and have P as a common endpoint but they have no other point in common. The point set $j_1 + a + b$ contains a simple closed curve which passes through P and encloses O. Hence P belongs to K. Thus K contains all of its limit points.

THEOREM 93. *If the points O and ω lie in different complementary domains of the compact continuous curve M, and K is the sum of all the simple closed curves that lie in M and separate O from ω, then K is the sum of a finite number of continuous curves.*

PROOF. It has been shown that K is closed. Suppose it is not the sum of a finite number of continuous curves. Then, with the help of the fact that every component of K contains a simple closed curve that encloses O and that O does not belong to K, it may be seen that either there is at least one component of K which is not a continuous curve or there exist two mutually exclusive closed point sets such that there are infinitely many components of K each intersecting both of them. In either case there exist two mutually exclusive simple closed curves k_1 and k_2 and an infinite sequence of mutually exclusive continua $M_1, M_2, M_3, \cdots$ such that (1) k_1 separates k_2 from $\omega + O$, (2) each of the continua $M_1, M_2, M_3, \cdots$ contains at least one point of k_1 and at least one point of k_2 and is a component of $K \cdot \bar{H}$ where H is the set of all points that lie within k_1 and without k_2. For each n, M_n has a point P_n in common with k_2 and K contains a simple closed curve J_n which encloses O and contains P_n. Since, for each n, J_n contains a point of k_2 and encloses O, and O is without k_1, therefore J_n intersects k_1. For each n, J_n contains, in common with M_n, an arc t_n which has one endpoint on k_1 and the other one on k_2 and which lies, except for its endpoints, wholly in H. The arcs of the sequence $t_1, t_2, t_3, \cdots$ are mutually exclusive and no two of them can be joined by a connected subset of K that lies wholly in H. With the help of the fact that M is locally connected it may be seen that there exist two numbers i and j such that t_i and t_j can be joined by two mutually exclusive arcs E_1F_1 and E_2F_2 which are subsets of M and which lie wholly in H, the points E_1 and E_2 belonging to t_i and the points F_1 and F_2 belonging to t_j. For each n ($n = 1,2$), let A_n denote the last point that E_nF_n has in common with M_i and let B_n denote the first point following A_n that it has in common with M_j. Suppose that, either when $n = 1$ or when $n = 2$, A_nB_n contains at least one point of J_i distinct from A_n. Then, for that n, let C_n denote the first such point in the order from A_n to B_n on the arc A_nB_n and let L_n denote the last point of J_i, other than C_n, in the order from E_n to C_n on the interval E_nC_n of the arc E_nF_n. The interval L_nC_n of the arc E_nF_n has its endpoints L_n and C_n on J_i but it has no other point in common with J_i. Hence the point O is enclosed by a simple closed curve formed by the arc L_nC_n together with one of the arcs of J_i whose endpoints are L_n and C_n. Therefore every point of L_nC_n belongs to K. But this is impossible since L_nC_n lies wholly in H and contains a point A_n of M_n and a point C_n which does not belong to M_n. Suppose, on the other hand, that J_i contains no point of A_1B_1 distinct from A_1 and no point of A_2B_2 distinct from A_2. In this case, for each n ($n = 1,2$), the arc E_nF_n contains an interval L_nC_n which has only the point C_n in common with J_i and only the point L_n in common with J_j. By Theorem 35 of Chapter III, the point set

$J_i + J_j + L_1C_1 + L_2C_2$ contains a simple closed curve enclosing O and containing either L_1C_1 or L_2C_2. Hence L_1C_1 or L_2C_2 is a subset of K. Hence P_i can be joined to P_j by a connected subset of K lying in H. Thus again we have a contradiction. The truth of Theorem 93 is therefore established.

THEOREM 94. *If O and ω are points belonging to different complementary domains of the compact continuous curve M and K is the sum of all the simple closed curves that lie in M and separate O from ω, then* (1) *the boundary of every complementary domain of a component of K is either a simple closed curve or the sum of two simple closed curves each separating O from ω and* (2) *if T is a component of $M - K$, no component of K contains more than one limit point of T.*

Theorem 94 may be proved with the help of Theorems 93, 91 and 45.

THEOREM 95. *If D_1 and D_2 are two simple domains, with J_1 and J_2 as their respective boundaries, such that $\bar{D}_2$ is a subset of D_1, D is the domain $D_1 - \bar{D}_2$, AB is an arc lying wholly in D except for the points A and B which belong to J_1 and J_2 respectively, XA_1 and XB_1 are two arcs lying in D, having only X in common with each other and having, in common with AB, only the points A_1 and B_1 respectively at which they abut on AB from opposite sides, and A_1B_1 is the interval of AB whose endpoints are A_1 and B_1, then the simple closed curve $XA_1 + A_1B_1 + XB_1$ separates J_1 from J_2.*

THEOREM 96. *Suppose that* (1) *D_1 and D_2 are two simple domains such that $\bar{D}_2$ is a subset of D_1 and K is a compact subcontinuum of $\bar{D}_1 - D_2$ containing at least one point of J_1, the boundary of D_1 and at least one point of J_2, the boundary of D_2,* (2) *AB is an arc containing no point of K and lying wholly in $D_1 - \bar{D}_2$ except for the points A and B which belong to J_1 and J_2 respectively,* (3) *CF is an arc having in common with K only the points C and F which belong to J_1 and J_2 respectively,* (4) *U and V are the two components of $(D_1 - \bar{D}_2) - (AB + K)\cdot(D_1 - \bar{D}_2)$ whose boundaries contain AB and* (5) *W denotes one of the components of $S - (K + CF)$ whose boundaries contain CF. Then no point of $K\cdot(D_1 - \bar{D}_2)$ is accessible both from $U\cdot W$ and from $V\cdot W$.*

PROOF. Suppose, on the contrary, that there exist two arcs A_UO and A_VO lying wholly in $U\cdot W$ and $V\cdot W$ respectively except for the point O which belongs to $K\cdot D$ where D is the domain $D_1 - \bar{D}_2$. There exist arcs A_UA_1 and A_VB_1 lying wholly in U and V respectively except for the points A_1 and B_1 which belong to $D\cdot(AB)$. The sum of A_UO, A_VO, A_UA_1, A_VB_1 and the interval A_1B_1 of the arc AB contains a simple closed curve α containing A_1B_1 and an interval A'_VO of the arc A_VO. With the help of Theorem 95 it may be seen that the simple closed curve α separates J_1 from J_2. The domain W contains an arc A_UXA_V from A_U to A_V. Since W is a component of $S - (K + CF)$, this arc does not intersect $K + CF$. In particular it does not contain the point O of K. Hence the sum of A_UXA_V and the arc

$A_UO + A_VO$ contains a simple closed curve γ containing an interval B_UOB_V of $A'_UO + A'_VO$. The arc B_UOB_V is a subset both of α and of γ and γ and K have only the point O in common. But α separates the points C and F of K from each other. It follows with the help of Theorem 28 that γ separates C from F. Therefore the arc CF intersects γ. But CF does not intersect A_UXA_V or $A_UO + A_VO$, and γ is a subset of the sum of these arcs. This involves a contradiction.

Theorem 96 does not remain true even in the plane if, in the conclusion, "no point of $K \cdot (D_1 - \bar{D}_2)$ is accessible both from $U \cdot W$ and from $V \cdot W$" is replaced by "$K \cdot (D_1 - \bar{D}_2)$ does not contain both a limit point of $U \cdot W$ and a limit point of $V \cdot W$".

Theorem 97. *Suppose* (1) *the compact continuum M is the boundary of the connected domain D,* (2) *XY is an arc lying wholly in D except for its endpoints which belong to M,* (3) *D_1 and D_2 are the two mutually exclusive connected domains whose sum is $D - s(XY)$ and* (4) *M_1, the boundary of D_1 and M_2, the boundary of D_2, have in common a point P of $S - XY$ which is accessible from $D_1 + D_2$. Then P separates X from Y in M.*

Proof. Let O denote a point of $s(XY)$. There exist arcs OA_1P and OA_2P lying, except for their endpoints, wholly in D_1 and D_2 respectively. Let J denote the simple closed curve $OA_1P + OA_2P$. The arc A_1OA_2 of J crosses the arc XY at the point O. Hence XY crosses A_1OA_2 at O. It follows that X and Y belong to different complementary domains of J. Let I denote the one that contains X and let E denote the one that contains Y. The point set $M - P$ is the sum of the two mutually separated point sets $M \cdot I$ and $M \cdot E$. Hence P separates X from Y in M.

Definition (Schoenflies). The point P is said to be *accessible from all sides* from the domain D if P is accessible from every subset of D which has P on its boundary and which is a complementary domain of the point set consisting of the boundary of D together with an arc lying wholly in D except for its endpoints which belong to the boundary of D and which are both distinct from P.

Example 4. In a Cartesian plane, let P denote the origin and let A, C, F, L, B, T and N denote the points $(0,-1)$, $(0,1)$, $(-1,1)$, $(-1,-2)$, $(0,-2)$, $(1/\pi,-2)$ and $(1/\pi,0)$ respectively. Let M denote the continuum composed of the straight line intervals AC, CF, FL, LT and TN together with the portion of the graph of $y = \sin(1/x)$ which lies between the lines $x = 0$ and $x = 1/\pi$. Let AB denote a straight line interval with endpoints as indicated (Figure 32). Let D denote the complementary domain of M which contains the segment $s(AB)$. The point set $D - s(AB)$ is the sum of two mutually exclusive connected domains D_1 and D_2, where D_1 is the interior of the rectangle $CFLB$. The point P is accessible from D_1 but not from D_2. Hence it is accessible from D but not accessible from all sides from D.

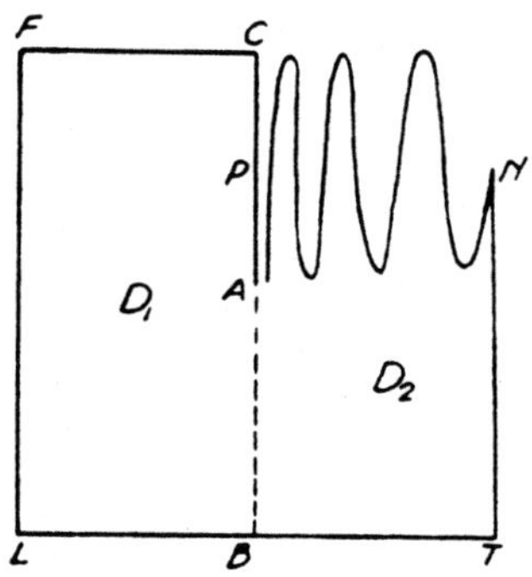

FIGURE 32

THEOREM 98. *If the compact continuum M is the boundary of the connected domain D and every point of M is accessible from all sides from D then M is a continuous curve.*

PROOF. Supose M is not a continuous curve. Then, by Theorem 73 of Chapter II, there exists a nondegenerate subcontinuum K of M such that no finite point set separates two points of K from each other in M. But if X and Y are two points of K there exists an arc XY lying, except for X and Y, wholly in D and $D - s(XY)$ is the sum of two mutually separated connected domains D_1 and D_2. If M_1 and M_2, the boundaries of D_1 and D_2, had in common a point P of $S - XY$ then, by Theorem 97, P would separate X from Y in M. Hence $M_1 - XY$ and $M_2 - XY$ are mutually separated. But their sum is $M - (X + Y)$. Therefore every pair of points of K disconnects M. But there are uncountably many mutually exclusive pairs of points of K and no one of them disconnects K. This is contrary to Theorem 105 of Chapter I.

THEOREM 99 (SCHOENFLIES). *In order that the compact continuum M be a continuous curve it is necessary and sufficient that* (1) *there do not exist two mutually exclusive closed and compact point sets such that infinitely many complementary domains of M intersect both of them and* (2) *every boundary point of a complementary domain of M be accessible from all sides from that domain.*

PROOF. This condition is necessary. For suppose M is a compact continuous curve. Suppose condition (1) is not fulfilled. Then there exist two closed and compact point sets H and K and an infinite sequence D_1, D_2, $D_3, \cdots$ of distinct complementary domains of M such that, for each n, D_n contains both a point of H and a point of K. For each n there exists an arc A_nB_n, lying wholly in D_n, such that its endpoints A_n and B_2 belong to H and K respectively. There exists an ascending sequence of numbers

$n_1, n_2, n_3, \cdots$ such that the sequences $A_{n_1}, A_{n_2}, A_{n_3}, \cdots$ and $B_{n_1}, B_{n_2}, B_{n_3}, \cdots$ are convergent. Let A and B denote the respective sequential limit points of these sequences. There exist two simple closed curves J_1 and J_2 such that J_1 separates A from J_2 and J_2 separates both A and J_1 from B. Let D denote the complementary domain of $J_1 + J_2$ which contains neither A nor B. There exists a number j such that, for each i greater than j, $A_{n_i}B_{n_i}$ contains an arc $A'_{n_i}B'_{n_i}$ lying wholly in D except that its endpoints A'_{n_i} and B'_{n_i} belong to J_1 and J_2 respectively. There exist a component U of $D - (A'_{n_{j+1}}B'_{n_{j+1}} + A'_{n_{j+2}}B'_{n_{j+2}})$ and a subsequence $m_1, m_2, m_3, \cdots$ of $n_{j+1}, n_{j+2}, \cdots$ such that, for every i, $A_{m_i}B_{m_i}$ is a subset of U such that the points $A_{m_1}, A_{m_2}, A_{m_3}, \cdots$ lie in the order indicated on J_1. For every i, $M \cdot \overline{U}$ contains a continuum N_i separating $A'_{m_i}B'_{m_i}$ from $A'_{m_{i+1}}B'_{m_{i+1}}$ in $\overline{U}$. For each i, let M_i denote the component of $M \cdot \overline{U}$ that contains N_i. The continua $M_1, M_2, M_3, \cdots$ are mutually exclusive and each of them intersects both J_1 and J_2 and D intersects the limiting set of the sequence M_1, M_2, $M_3, \cdots$. Therefore, by Theorem 14 of Chapter II, M is not a continuous curve. Thus the supposition that condition (1) is not fulfilled leads to a contradiction. By Theorems 42 and 59 of the present chapter and Theorems 21 and 43 of Chapter II, every point of the boundary of a complementary domain of M is accessible from all sides from that domain.

This condition is also sufficient. For suppose M is a compact continuum which is not a continuous curve. Then there exist a domain D, two mutually exclusive simple closed curves J_1 and J_2, and an infinite sequence of mutually exclusive continua $M_1, M_2, M_3, \cdots$ such that (1) D is bounded by $J_1 + J_2$ and (2) for each n, M_n contains a point of J_1 and a point of J_2 and is a component of $M \cdot \overline{D}$. There exist two mutually exclusive arcs A_1B_1 and A_2B_2 lying wholly in D, except that A_1 and A_2 belong to J_1 and B_1 and B_2 belong to J_2, and such that neither A_1B_1 nor A_2B_2 intersects any one of the continua $M_1, M_2, M_3, \cdots$. There exist an infinite ascending sequence n_1, n_2, $n_3, \cdots$ and a component U of $D - (s(A_1B_1) + s(A_2B_2))$ such that, for each i, M_{n_i} is a subset of $\overline{U}$ and such that if, for each i, C_i denotes a point of $J_1 \cdot M_{n_i}$ then the points $C_1, C_2, C_3, \cdots$ lie on J_1 in the order $C_1C_2C_3\cdots$. For each i there exists an arc t_i separating M_{n_i} from $M_{n_{i+1}}$ in $\overline{U}$, containing no point of M and lying wholly in U except that its endpoints belong to J_1 and J_2 respectively. For each i, let D_i denote the component of $S - M$ that contains t_i. If there are infinitely many domains in the sequence D_1, D_2, $D_3, \cdots$ then condition (1) is not fulfilled. If there are not infinitely many domains in this sequence then there exists a component W of $S - M$ containing infinitely many arcs of the sequence $t_1, t_2, t_3, \cdots$. In this case, if N denotes the boundary of W, there exist infinitely many components of $N \cdot \overline{D}$ each intersecting both J_1 and J_2 and N is not a continuous curve and therefore, by Theorem 98, some point of N is not accessible from all sides from W.

THEOREM 100. *If K is a point set such that $\bar{K}$ is compact then in order that K be a simple closed curve it is necessary and sufficient that $S - K$ be the sum of two mutually separated connected point sets such that every point of K is accessible from each of them.*

PROOF. The necessity of this condition is a consequence of Axiom 4, Theorem 21 of Chapter II and Theorem 99 of the present chapter. It will be shown to be sufficient. Suppose $S - K$ is the sum of two mutually separated point sets S_1 and S_2 such that every point of K is accessible both from S_1 and from S_2. Since every point of K is a limit point of S_2 and no point of S_1 is a limit point of S_2 therefore no point of S_1 is a limit point of K. It follows that S_1 is a domain. Similarly S_2 is a domain. Since K is the outer boundary of S_1 with respect to S_2 therefore, by Theorem 25, K is a continuum. By Axiom 3, K contains at least two points A and B. There exist arcs AX_1B and AX_2B lying, except for their endpoints, wholly in S_1 and S_2 respectively. Let I and E denote the complementary domains of the simple closed curve J formed by the arcs AX_1B and AX_2B. The point set $I + (X_1 + X_2)$ is a connected set joining the point X_1 of S_1 to the point X_2 of S_2. Hence I contains a point of K. Similarly E contains a point of K. The point sets $M \cdot I$ and $M \cdot E$ are mutually separated and $K \cdot I + K \cdot E = K - (A + B)$. Thus the compact continuum K is disconnected by the omission of $A + B$. Hence it is a simple closed curve.

DEFINITION. The point set M is said to be *uniformly connected im kleinen* if for every set G of regions covering $\bar{M}$ there exists a set H of regions covering $\bar{M}$ such that if h is any region of H then $h \cdot M$ is a subset of a component of the common part of M and some region of G.

DEFINITION. The point set M is said to *have the property S* if for every set G of regions covering $\bar{M}$ there exists a finite set H of closed and connected subsets of M covering M such that every point set of the collection H is a subset of some region of the set G.

The point set M is said to *have property S′* if for every set G of domains covering $\bar{M}$ there exists a finite set H of connected subsets of M covering M such that every point set of the collection H is a subset of some domain of the set G.

THEOREM 101. *If the point set M is uniformly connected im kleinen and P is a point of $\bar{M}$ and R is a region containing P then there exists a region R' containing P such that $M \cdot R'$ is a subset of some connected subset of $M \cdot R$.*

PROOF. There exists a region R_1 containing P such that $\bar{R}_1$ is a subset of R. Let G denote the set of regions consisting of R and all regions that contain no point of $\bar{R}_1$. The set G covers $\bar{M}$. Hence, since M is uniformly connected im kleinen, there exists a set H of regions covering $\bar{M}$ such that if h is any region of H then $h \cdot M$ is a subset of a component of the common part of M and some region of G. Let h_P denote a region of H containing P. Since $h_P \cdot M$ contains a point of R_1 it is not a subset of any region of G except R. Hence it is a subset of some connected subset of $M \cdot R$.

THEOREM 102. *Every compact continuous curve is uniformly connected im kleinen.*

It is to be noted that the only difference between properties S and S' is that in the statement of property S it is required that every set of the collection H be *closed*. It follows that every point set which has property S also has property S'. But the converse does not hold true. For instance every compact simple domain in a Euclidean plane has property S' but no simple domain, and indeed no point set which is not closed, can have property S. However, it may be seen that, *as applied to compact continuua*, properties S, S', connectedness im kleinen and uniform connectedness im kleinen are all equivalent.

THEOREM 103. *If the point set M is uniformly connected im kleinen then $\bar{M}$ is connected im kleinen provided it is compact.*

PROOF. Let P denote a point belonging to $\bar{M}$. Let K denote a region containing P and let L denote a region containing P such that $\bar{L}$ is a subset of K. Since M is uniformly connected im kleinen therefore, by Theorem 101, there exists a region R containing P such that if X and Y are two points of $M \cdot R$ then X and Y lie together in a connected subset of M that lies wholly in L. Suppose now that A is any point of $\bar{M} \cdot R$. There exist two sequences of points $A_1, A_2, A_3, \cdots$ and $P_1, P_2, P_3, \cdots$ belonging to $M \cdot R$ and converging to A and P respectively. For each n there exists a connected point set t_n containing A_n and P_n and lying wholly in $M \cdot L$. Let t denote the limiting set of the sequence $t_1, t_2, t_3, \cdots$. The point set t is a closed and connected subset of $\bar{M}$ containing A and P and lying wholly in K. Thus $\bar{M}$ is connected im kleinen at every point P that belongs to it.

It does not follow that if a point set is uniformly connected im kleinen then its boundary is connected im kleinen. Consider the following example.

EXAMPLE 5 (ROSENTHAL). In a Cartesian plane, let O denote the origin and let A, B and C denote the points (2,0), (2,2) and (0,2). Let F denote the point (1,0) and, for each n, let O_n and F_n denote the points $(0,1/n)$ and $(1,1/n)$ respectively and let O_nF_n denote the straight line interval whose endpoints are O_n and F_n. For each n, subdivide the interval O_nF_n into a finite number of congruent subdivisions each of length less than the distance from O_n to O_{n+1} and let U_n denote the set of circles having these subdivisions as diameters (Figure 33). Let U denote the set of all circles J such that, for some n, J belongs to U_n. Let K denote the point set obtained by adding together the interiors of the circles of the set U and let N denote the square $OABC$ plus its interior. Let β denote the point set obtained by adding together the straight line intervals OA, AB, BC and OC and all the circles of the set U. Let M denote $N - K$. The point set M is a compact continuous curve and therefore, by Theorem 42, the boundary of every one of its complementary domains is a continuous curve. Indeed every one of them is bounded either by a circle or a square. But the boundary of M itself is the continuum β and this continuum is not connected im kleinen at any point of the interval OF except the point O.

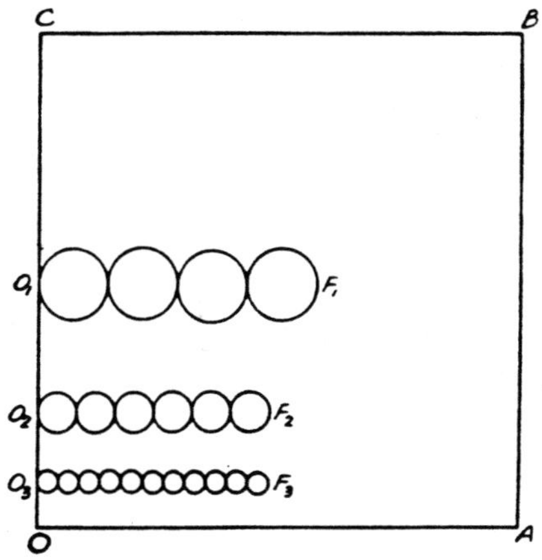

FIGURE 33

THEOREM 104. *Every point set having property S' is connected im kleinen.*

PROOF. Suppose, on the contrary, that there exists a point set M which has property S' but which is not connected im kleinen. Then, for some point P of M and some region R_1 containing P, there exists an infinite sequence α of distinct points $P_1, P_2, P_3, \cdots$, all belonging to $M \cdot R$, such that (1) α converges to P and (2) no point of α lies with P in a connected subset of $M \cdot R$. There exists a region R_2 containing P such that $\bar{R}_2$ is a subset of R_1. Let G denote the pair of domains R_1 and $S - \bar{R}_2$. The set G covers $\bar{M}$. Hence, since it has property S', M is the sum of the elements of a finite set β of connected point sets each lying wholly either in R_1 or in $S - \bar{R}_2$. Since each point of the infinite sequence α belongs to some point set of the finite set β therefore there exists a point set N belonging to β and containing infinitely many points of α. Since N contains infinitely many points of α it intersects R_2. Hence it is not a subset of $S - \bar{R}_2$. Therefore it is a subset of R_1. Thus $N + P$ is a connected subset of $M \cdot R_1$ containing P and points of α. But this involves a contradiction.

An example, in a Cartesian plane, of a compact and connected point set which is connected im kleinen but which does not have property S' is the set of all points of the curve $y = \sin (1/x)$ whose abscissae lie between 0 and 1. From

Theorem 104 and the existence of this example it follows that, as applied to compact and connected point sets, condition S' is stronger than that of connectivity im kleinen.

THEOREM 105. *If the point set M is uniformly connected im kleinen then it has property S'.*

PROOF. Suppose G is a set of regions covering $\bar{M}$. Then, by hypothesis, there exists a finite set H of regions covering $\bar{M}$ such that if h is any region of H then there exists a region g_h of the set G such that $M \cdot h$ is a subset of some component C_h of $M \cdot g_h$. Let $h_1, h_2, h_3, \cdots, h_n$ denote the regions of the set H. The point set M is the sum of the finite number of connected point sets $C_{h_1}, C_{h_2}, C_{h_3}, \cdots, C_{h_n}$ and each of these point sets is a subset of some region of the set G. This establishes Theorem 105.

It is not, however, true that every compact point set which has property S' is uniformly connected im kleinen. Consider, for example, the set of points $K - P$ where, in a Euclidean plane, K is a circle and P is a point of K.

It follows from the above results that condition S', as applied to compact point sets, is stronger than that of connectedness im kleinen but weaker than that of uniform connectedness im kleinen.

THEOREM 106. *If the boundary of the compact connected domain D is a nondegenerate compact continuum M then, in order that M be a simple closed curve, it is necessary and sufficient that D be uniformly connected im kleinen.*

PROOF. Suppose D is uniformly connected im kleinen. It will first be shown that M cannot contain two arcs that have in common only one point, that point being an endpoint of only one of them. Suppose it does contain two such arcs EFT and FK with no point in common except F. Let α denote a simple domain containing F such that $\bar{\alpha}$ contains no one of the points E, K and T. It may be shown that there exists a simple domain γ containing F such that $\gamma \cdot D$ is a subset of a component of $\alpha \cdot D$. There exist, in γ, points E', T' and K' and arcs $E'K'$ and $K'T'$ such that (1) E' and T' are on the arc EFT in the order $EE'FT'T$ and K' is on the arc FK, (2) the interval $E'T'$ of the arc EFT lies wholly in γ, (3) $E'K'$ has, in common with EFT, only the point E' and, in common with FK, only the point K', (4) $K'T'$ has, in common with EFT, only the point T' and, in common with FK, only the point K', (5) if J denotes the simple closed curve formed by the arcs $E'K'$ and $K'T'$ and the interval $E'FT'$ of the arc EFT then the segment FK' of the arc FK lies wholly in the complementary domain of J which is a subset of α. The arc $E'K'$ together with the interval $E'F$ of EFT and the interval FK' of FK form a simple closed curve J_1. The arc $K'T'$, the interval FT' of EFT and the interval FK' of FH form a simple closed curve J_2. Let I, I_1 and I_2 denote the complementary domains of J, J_1 and J_2 respectively that are subsets of α. The domain I is the sum of I_1, I_2 and the segment FK' of the arc FK. Let P, P_1 and P_2 denote points on the segments FK', $E'F$ and FT' respectively. The point P is a boundary point of D and the domain I contains P. Hence there are points of D in I and therefore in I_1 or in I_2. There exist regions R_1 and R_2, containing P_1 and P_2 respectively, such that neither $\bar{R}_1$ nor $\bar{R}_2$ contains any point of ET or of J that does not belong to $E'T'$. The point P_1 is on the boundary of D. It follows that there exist, in $\gamma \cdot R_1$, points of D lying either in $S - \bar{I}$ (and therefore both in $S - I_1$ and in $S - I_2$) or in I_1. Similarly either I_2 or $S - \bar{I}$ intersects $D \cdot \gamma \cdot R_2$. Thus either both I_1 and I_2 intersect D or only one of them intersects D in which latter case there exist points of D lying in $\gamma \cdot (S - \bar{I}) \cdot (R_1 + R_2)$. But if a connected point set lies entirely in α and contains either (1) a point of I_1 and a point of I_2 or (2) a point of I_1 or I_2 and a point of $(S - \bar{I}) \cdot (R_1 + R_2)$; then it intersects either EFT or FK. Therefore it is not a subset of D. Thus the supposition that M contains a simple triod leads to a contradiction.

It will be shown that M is connected im kleinen. Suppose that it is not. Then there exist three simple domains D_1, D_2 and D_3 and an infinite sequence of mutually exclusive continua $t_1, t_2, t_3, \cdots$ such that if J_1, J_2 and J_3 denote the boundaries of D_1, D_2 and D_3 respectively then (1) $D_2 + J_2$ is a subset of D_1 and $D_3 + J_3$ is a subset of D_2, (2) for each n, t_n is a component of $M \cdot (\bar{D}_1 - D_3)$ intersecting both J_1 and J_3. There exist, on J_2, a point O and a sequence of points $O_1, O_2, O_3, \cdots$, converging to O, such that O_m belongs to t_m. There exists a region R containing O such that $D \cdot R$ is a subset of a component of $D \cdot (D_1 - \bar{D}_3)$. There exist, in R, four points O_{m_1}, O_{m_2}, O_{m_3} and O_{m_4} of the sequence $O_1, O_2, O_3, \cdots$. Of these four points two, A_1 and A_3, are separated from each other on J_2 by the sum of the other two, A_2 and A_4. For each i $(1 \leqq i \leqq 4)$ let α_i denote that one of the continua t_{m_1}, t_{m_2}, t_{m_3} and t_{m_4} to which the point A_i belongs. For each such i there exists, in R, a region R_i containing A_i but no point of any one of the sets α_1, α_2, α_3 and α_4 except α_i. The regions R_1 and R_3 contain points B_1 and B_3 respectively belonging to D. Every continuum lying in $D_1 - D_3$ and containing both B_1 and B_3 intersects either α_2 or α_4. But there exists a connected subset of $D \cdot (D_1 - D_3)$ containing both B_1 and B_2. Thus the supposition that M is not a continuous curve has led to a contradiction. But it is compact and it contains no simple triod. Hence, by Theorem 75, M is either an arc or a simple closed curve. Suppose it is an arc ABC. Then, since no arc separates S, D is $S - ABC$. There exists a simple domain U_1 containing B but neither A nor C. There exists a simple domain U_2, containing B, such that $D \cdot U_2$ is a subset of a component of $D \cdot U_1$. But there exist two arcs XB and YB having only the point B in common with ABC, abutting on ABC at B from opposite sides and lying wholly in U_2. The points X and Y do not lie together in any connected subset of $D \cdot U_1$. Thus the supposition that M is an arc leads to a contradiction. Hence it is a simple closed curve. Thus the condition of Theorem 106 is sufficient.

This condition is also necessary. For suppose, on the contrary, that M is a simple closed curve but D is not uniformly connected im kleinen. Then there exist a point O belonging to M, a simple domain I containing O, and two infinite sequences of points $X_1, X_2, X_3, \cdots$ and $Y_1, Y_2, Y_3, \cdots$ lying in D and converging to O such that for no n is X_n joined to Y_n by a connected subset of D that lies entirely in I. But there exists, in D, a simple domain D' such that the point O belongs to J', the boundary of D', and every point of J' belongs either to M or to the boundary of I. For some n, the points X_n and Y_n are both in D'. But D' is a connected subset of $D \cdot I$. Thus the supposition that D is not uniformly connected im kleinen leads to a contradiction.

It is not, however, true that, in every space satisfying Axioms 0 and 1–5, every compact simply connected domain which has property S' is bounded by a simple closed curve. Consider, for example, in a Euclidean plane, the point set $(K + A) - OA$ where K is the interior of a circle with center at O, A is a point on this circle and OA is the straight line interval whose endpoints are O and A. This point set has property S' but it is not uniformly connected im kleinen.

THEOREM 107. *If the boundary of the compact and connected domain D is a compact continuum M then in order that M be a continuous curve it is necessary and sufficient that D have property S'.*

PROOF. This condition is sufficient. For suppose, on the contrary, that D has property S' but M is not connected im kleinen. Then there exist five simple domains D_1, D_2, D_3, D_4 and D_5 and an infinite sequence of mutually exclusive continua $t_1, t_2, t_3, \cdots$ such that if J_1, J_2, J_3, J_4 and J_5 denote the boundaries of D_1, D_2, D_3, D_4 and D_5 respectively then, for each n less than 5, (1) $D_{n+1} + J_{n+1}$ is a subset of D_n, (2) for each n, t_n is a component of $M \cdot (\bar{D}_1 - D_5)$ containing a point of J_1 and a point of J_5. For each m, the point set t_m contains a point A_m of J_3. There exists an infinite sequence $B_1, B_2, B_3, \cdots$ of distinct points of the sequence $A_1, A_2, A_3, \cdots$ lying in the order $B_1 B_2 B_3 \cdots B_n B_{n+1} \cdots$ on the curve J_3. For each n, let k_n denote the point set of the sequence $t_1, t_2, t_3, \cdots$ which contains B_n. For each even number n let R_n denote a connected domain containing B_n but no point of $J_2 + J_4 + k_{n-1} + k_{n+1}$. Since B_n is a point of the boundary of D there exists, in R_n, a point C_n belonging to D. Let G denote the set of domains $D_1 - \bar{D}_5$, D_4 and $S - \bar{D}_2$. The set G covers $\bar{D}$. But D is not the sum of a finite number of connected point sets each lying wholly in some domain of the set G. For consider the points $C_2, C_4, C_6, \cdots$. If m and n are even numbers such that $m < n$ then every connected point set lying wholly in some one domain of the set G and containing both C_m and C_n must contain a point either of k_{n-1} or of k_{n+1} and therefore of M. Thus there exists no connected subset of D lying wholly in some one domain of the set G and containing two distinct points of the sequence $C_2, C_4, C_6, \cdots$. It follows that D does not have property S'. Thus the supposition that the condition of Theorem 107 is not sufficient has led to a contradiction.

This condition is also necessary. For suppose, on the contrary, that M is a continuous curve but D does not have property S'. There exists a set G of simple domains covering $\bar{D}$ such that D is not the sum of a finite number of connected point sets each lying in some domain of the set G. For each point P of $\bar{D}$ there exist three simple domains x_P, y_P and z_P, all containing P, such that $\bar{x}_P$ is a subset of y_P, $\bar{y}_P$ is a subset of z_P, and z_P belongs to G. Let H denote the set of all domains x_P for all points P of $\bar{D}$. There exists a finite subcollection H' of H covering $\bar{D}$. For each domain h of the collection H' there exists a point P_h of $h \cdot \bar{D}$ such that $h = x_{P_h}$. Let y_h denote y_{P_h} and let z_h denote z_{P_h}. If h is any domain of the collection H' and P is any point of D belonging to h, let T_{hP} denote the component of $D \cdot z_h$ that contains P. There exists a domain q of the collection H' containing infinitely many points $P_1, P_2, P_3, \cdots$, belonging to D, such that the sets $T_{qP_1}, T_{qP_2}, T_{qP_3}, \cdots$ are mutually exclusive. For each i, the boundary of z_q contains a limit point of T_{qP_i}. Hence T_{qP_i} intersects both q and $S - y_q$. Hence it contains a point X_i belonging to β, the boundary of y_q. The infinite sequence $X_1, X_2, X_3, \cdots$ has an infinite subsequence $Y_1, Y_2, Y_3, \cdots$ of points lying in the order $Y_1 Y_2 Y_3 \cdots$ on β. For each n, let K_n denote the set of the sequence $T_{qP_1}, T_{qP_2}, T_{qP_3}, \cdots$ to which Y_n belongs. For each n, the boundary of K_n contains a continuum L_n that lies wholly in $\bar{y}_q - q$ and intersects both β and α, the boundary of q. No two of the continua $L_2, L_4, L_6, \cdots, L_{2n}, L_{2n+2}, \cdots$ have a point in common and no two of them lie in the same component of $M \cdot (\bar{y}_q - q)$. It follows, by Theorem 14 of Chapter II, that M is not a continuous curve. Thus the supposition that the condition of Theorem 107 is not necessary has led to a contradiction.

As will be shown by examples, in order that a compact continuum in Euclidean space of three dimensions be a continuous curve it is not necessary that it satisfy either of the conditions of Theorem 99 and it is not sufficient that it satisfy the condition of Theorem 100. It will be shown however that a compact continuum, in any space having a certain property A, is a continuous curve provided it is the common boundary of two mutually exclusive connected domains both of which are uniformly connected im kleinen.

There exists, in three-dimensional Euclidean space, a compact continuous curve M which separates S, the set of all points, into just two connected domains, S_1 and S_2, such that (a) *the boundary of S_1 contains a point which is not accessible from S_1,* (b) *the domain S_2 is uniformly connected im kleinen,* (c) *the boundary of S_1 is not a continuous curve.* Consider the following example.

EXAMPLE 6. Let K denote a cube so situated that b, one of its bases is horizontal. Let k and f denote two opposite lateral faces of K. Let $ABCD$ and $EFGH$ denote two squares concentric with k and lying within it and such that $EFGH$ is within $ABCD$, AB and EF being horizontal. Let I and J denote points in the face f such that the planes IAB and JCD are parallel to b and I is equidistant from A and B, while J is equidistant from C and D. We thus have a right prism whose triangular bases are ABI and DCJ. Let P denote this prism and let Q denote the right pyramid whose base is the square $EFGH$ and whose apex is at O, a point half-way between I and J. For each positive integer n, let α_n denote a plane separating k from f, parallel to f and at a distance from f equal to a/n, where a is one half the length of a side of K. For each n, let β_n denote a plane parallel to, and half-way between, the planes α_n and α_{n+1}. Let I_n denote the portion of the interior of P which is intercepted between the planes α_n and β_n. Let R denote the interior of the pyramid Q. Let S_1 denote the point set composed of R together with the sets $I_1, I_2, I_3, I_4, \cdots$. Let N denote the point set composed of the surface K together with its interior and let M denote $N - S_1$. The set M is a continuous curve, S_1 is one of its complementary domains, and the straight line interval JI is a portion of the boundary of S_1. But no point of JI, except O, is accessible from S_1. The boundary of S_1 is not connected im kleinen at any point of JI except O. Clearly S_2, the exterior of the cube K, is uniformly connected im kleinen.

If ϵ is a positive number there exists, in Euclidean space of three dimensions, a compact continuous curve M such that there are infinitely many complementary domains of M all of diameter greater than ϵ. Consider the following example.

EXAMPLE 7. Consider again the cube K and the domains $I_1, I_2, I_3, \cdots$ of Example 6, it being understood that the distance between the planes IAB and JCD of that example is 2ϵ. Let M denote the point set $N - (I_1 + I_2 + I_3 + \cdots)$ where N has the same meaning as in Example 6. Here M is a continuous curve and each I_n is one of its complementary domains. Though M is a compact continuous curve its complementary domains do not form a contracting sequence.

There exists, in three-dimensional Euclidean space, a compact continuous curve M which separates S, the set of all points, into two mutually exclusive connected domains, S_1 and S_2, such that (a) *every point of M is a limit point both of S_1 and of S_2,* (b) *not every point of M is accessible from S_2.*

EXAMPLE 8. Let K denote a square whose vertices A, B, C and D are the points $(0,-1,-1)$, $(0,1,-1)$, $(0,1,1)$ and $(0,-1,1)$ respectively. For each integer n (positive or negative) let K_n denote a square whose vertices A_n, B_n, C_n and D_n are the points $(1/n,-1,-1)$, $(1/n,1,-1)$, $(1/n,1,1)$ and $(1/n,-1,1)$ respectively. For each n let K'_n denote a square whose vertices A'_n, B'_n, C'_n and D'_n are so situated that A'_n is half-way between A_n and A_{n_1}, B'_n is half-way between B_n and B_{n_1}, C'_n is half-way between C_n and C_{n_1}, and D'_n is half-way between D_n and D_{n_1}, where n_1 denotes $n + 1$ or $n - 1$ according as n is positive or negative. For every n and every pair of integers i and j which are numerically less than $4^{|n|}$, let P'_{ijn} denote that point within the square K'_n whose y and z are $i/4^{|n|}$ and $j/4^{|n|}$, respectively. For each n, let T_n denote the set of all points P'_{ijn} for which i and j are even numbers numerically less than $4^{|n|}$ and let U_n denote the set of all points P'_{ijn} such that i and j are odd numbers numerically less than $4^{|n|}$. For every point P'_{ijn} belonging to T_n or to U_n let K'_{ijn} denote a circle lying in the plane of K'_n, with center at P'_{ijn} and with a radius equal to $1/4^{|n|+2}$. Thus with each point of the set $T_n + U_n$ there is associated a definite circle having that point as center, and every one of these circles lies wholly without every other one. Now let P''_{ijn} and K''_{ijn} denote the orthogonal projections of P'_{ijn} and K'_{ijn} respectively onto the plane of the square K_{n1} and let P_{ijn} denote the orthogonal projection of P'_{ijn} onto the plane of K_n. For each point P'_{ijn} belonging to the set T_n, let S_{ijn} denote the right circular cylinder whose bases have the circles K'_{ijn} and K''_{ijn} as their perimeters, and for each point P'_{ijn} belonging to U_n let C_{ijn} denote the right circular cone whose base has K'_{ijn} as its perimeter and whose apex is the point P_{ijn}. It is to be noted that S_{ijn} exists only when i and j are even and that C_{ijn} exists only when i and j are odd. For each n, let P_n denote the right parallelepiped which has K_n and K'_n as opposite lateral faces and let R_n denote the interior of P_n. Let I_n denote the portion of R_n remaining after the removal of every point of R_n which lies either within, or on, the surface of any C_{ijn} (i and j odd). For each n let L_n denote the point set obtained by adding to I_n the interiors of all the cylinders S_{ijn} (i and j even) together with the bases of these cylinders minus their perimeters. Let Z denote a rectangular disk lying in the plane $z = 2$ and with sides congruent to, and parallel to, the respective sides of the upper base of P_1. Let P' denote a parallelepiped whose upper base is Z and whose lower base is the upper base of P_1. Let P'' denote a parallelepiped whose upper base is Z and whose lower base is the upper base of P_{-1}. Let W denote the sum of the interiors of P' and P'' and the upper bases of P_1 and P_{-1} minus their perimeters. Let H denote the point set $W + L_1 + L_2 + L_3 + \cdots$ and let M denote its boundary. The point set H is a domain and M is a continuous curve which separates S into two connected domains S_1 and S_2 where S_1 is H. Every point of the disk Z belongs to M. But no point of Z minus its perimeter is accessible from S_2, though every point of M is a limit point both of S_1 and of S_2. Neither S_1 nor S_2 is uniformly connected im kleinen.

There exists, in Euclidean space of three dimensions, a compact continuum M which separates S, the set of all points, into two mutually exclusive connected domains such that (a) *both of these domains are uniformly connected im kleinen and every point of M is accessible from one of these domains, but* (b) *M is not a continuous curve.*

EXAMPLE 9. Let K denote the cube whose faces lie in the planes $x = 0$, $x = 2$, $y = 0$, $y = 2$, $z = 0$ and $z = 2$. For each positive integer n, let A_n denote the point $(1/n,1/n,0)$ and let B_n denote the point $(1/n,1/n,2)$. Let M denote the point set composed of the surface of K together with all the straight line intervals A_1B_1, A_2B_2, $A_3B_3, \cdots$.

There exists, in three-dimensional Euclidean space, a compact continuum M

which is not a continuous curve but which separates S, *the set of all points, into two mutually exclusive connected domains*, S_1 *and* S_2 *such that* (a) *every point of* M *is accessible both from* S_1 *and from* S_2, (b) *the domain* S_1 *is uniformly connected im kleinen*.

EXAMPLE 10. Returning to Example 8, let H denote a cube whose upper base lies in the plane $z = 1$, whose center is at the point $(0,0,-1)$, and one of whose faces is in the plane $x = 2$. For every positive integer n, the upper base of the parallelepiped P_n of Example 8 is a part of the upper base of H but the remainder of P_n is wholly within H. Let H denote the set of points composed of (a) the exterior of H, (b) the upper bases of the parallelepipeds $P_1, P_2, P_3, \cdots$ minus their perimeters, (c) the interiors of $P_1, P_2, P_3, \cdots$ and (d) the interiors of those cylinders S_{ijn} of Example 8 for which n is positive, together with the bases of those cylinders minus their perimeters. Let M denote the boundary of R. The point set M is a compact continuum and it separates S into two connected domains S_1 and S_2, where S_1 is R, and furthermore every point of M is accessible both from S_1 and from S_2. But M is not a continuous curve. It fails to be connected im kleinen at the point $(0,0,0)$. The domain S_1 is uniformly connected im kleinen, but S_2 is not.

DEFINITION. A space Σ satisfying Axiom 0 is said to have property A if and only if it is true that if the continua H and K are separated from each other in Σ by the compact continuum M and D is a region containing both H and K then H and K are separated from each other in $\bar{D}$ by a continuum which is a subset of $M \cdot \bar{D}$.

THEOREM A. *If, in a space satisfying Axioms* 0, 1 *and* 2 *and having property* A, *the compact continuum* M *separates* S, *the set of all points, into two mutually separated connected domains* S_1 *and* S_2 *such that* (1) *every point of* M *is a limit point both of* S_1 *and of* S_2, *and both* S_1 *and* S_2 *are uniformly connected im kleinen; then* M *is a continuous curve.*

PROOF. Suppose that M is not connected im kleinen. Then there exist a point A of M and a region R_1 containing A such that, if R is any region whatsoever containing A, there exists a point of $M \cdot R$ which does not belong to T, the component of $M \cdot \bar{R}_1$ which contains A. But, since S_1 and S_2 are uniformly connected im kleinen therefore, by Theorem 101, in the proof of which no use was made of any of the Axioms 3, 4 and 5, there exists a region R_2, containing A, such that if X and Y are two points belonging to $S_1 \cdot R_2$ or both belonging to $S_2 \cdot R_2$ then X and Y lie together in some connected subset of $S_1 \cdot R_1$ or of $S_2 \cdot R_1$. Let B denote a point of $M \cdot R_2$ not belonging to T. Let $\alpha_1, \alpha_2, \alpha_3, \cdots$ and $\beta_1, \beta_2, \beta_3, \cdots$ denote two sequences of regions closing down on A and B respectively. For each n there exist, in α_n, points A_n and A_n' belonging to $S_1 \cdot R_2$ and $S_2 \cdot R_2$ respectively and, in β_n, points B_n and B_n' belonging to $S_1 \cdot R_2$ and $S_2 \cdot R_2$ respectively. There exist arcs A_nB_n and $A_n'B_n'$ lying in $S_1 \cdot R_1$ and $S_2 \cdot R_1$ respectively. By hypothesis there exists, in $M \cdot \bar{R}_1$, a continuum M_n that separates A_nB_n from $A'_nB'_n$ in $\bar{R}_1$. Since α_n and β_n are connected, therefore, for each n, M_n contains both a point of α_n and a point of β_n. Therefore, by Theorem 58 of Chapter I, the limiting set of the sequence $M_1, M_2, M_3, \cdots$ is a closed and connected subset of $M \cdot \bar{R}_1$ containing both A and B. Thus the supposition that M is not connected im kleinen leads to a contradiction.

THEOREM 108. *If neither of the two intersecting continua H and K separates S, each of them has a compact boundary, and $H + K$ is not identical with S, then every component of $H \cdot K$ is bounded by a compact continuum.*

PROOF. Suppose, on the contrary, that there exists a component T of $H \cdot K$ whose boundary β is the sum of two mutually exclusive closed point sets β_1 and β_2. Clearly β is a subset of the sum of the boundaries of H and K. Hence it is compact. Let A_1 and A_2 denote points belonging to β_1 and β_2 respectively. By Theorem 13, there exists a simple closed curve J separating A_1 from A_2 and containing no point of β. The continuum T contains both A_1 and A_2. Hence it contains a point of J. But J contains no point of the boundary of T. Hence, since it is connected, J is a subset of T. Let ω denote some definite point of $S - (H + K)$. Of the two components of $S - J$, let E denote the one that contains ω and let I denote the other one. Since I contains a point of β, it contains a point X not belonging to $H \cdot K$. But J is a subset of $H \cdot K$ and it separates X from the point ω which belongs neither to H nor to K. Therefore either H or K separates ω from X according as X belongs to $S - H$ or to $S - K$. But this is contrary to hypothesis.

THEOREM 109. *If H and K are two compact continua whose common part T is not connected then there exist two complementary domains of $H + K$ such that the boundary of each of them intersects both $H - T$ and $K - T$.*

PROOF. Let G_H denote the set of all complementary domains of H which do not intersect K and let G_K denote the set of all complementary domains of K which do not intersect H. Let H' denote the point set obtained by adding to H all the domains of the set G_H and let K' denote the set obtained by adding to K all the domains of the set G_K. Let T' denote the common part of H' and K'. Suppose P is a point of $T' - T$. Let h denote the domain of G_H that contains P and let k denote the domain of G_K that contains P. Since h is a complementary domain of H containing no point of K therefore it is a complementary domain of $H + K$. Similarly k is a complementary domain of $H + K$. But h has the point P in common with k. It follows that h is identical with k and therefore h is a subset of T'. Furthermore the boundary of h is a subset of H and that of k is a subset of K. Therefore the boundary of h is a subset of T. By Theorem 24, it is a continuum. Let G denote the set of all domains h of G_H that intersect $T' - T$. By hypothesis, T is the sum of two mutually exclusive closed point sets T_1 and T_2. Suppose g is a domain of the collection G. Since β_g, the boundary of g, is a connected subset of $T_1 + T_2$, therefore β_g is a subset either of T_1 or of T_2. For each i less than 3, let T'_i denote the sum of T_i and all domains g of G such that β_g is a subset of T_i. By Theorem 4 of Chapter II, T'_1 and T'_2 are mutually exclusive closed point sets. But $T' = T'_1 + T'_2$. Hence T' is not connected. But α, the boundary of T', is a subset of T. Hence α is compact. Therefore, by Theorem 22, $H' + K'$ separates S. Hence $S - (H' + K')$ has at least two components D and E. Let γ denote the boundary of D. If γ were a subset of H, since D contains no point of K, D would be a subset of H', contrary to the fact that it is a component of $S - (H' + K')$. Similarly γ is not a subset of K. But it is easy to see that it is a subset of $H + K$. Hence γ intersects both $H - T$ and $K - T$. Similarly the boundary of E intersects both of these point sets.

THEOREM 110. *If A and B are two points and H, K, L and M are four compact continua such that the common part of every two of them is $A + B$ and no one of them is disconnected by the omission of $A + B$, and $H + K$ separates $L - (A + B)$ from $M - (A + B)$, then $L + M$ separates $H - (A + B)$ from $K - (A + B)$.*

PROOF. Suppose, on the contrary, that $H - (A + B)$ and $K - (A + B)$ lie in the same component of $S - (L + M)$. Then, by Theorem 109, there exists a component D of $S - (L + M)$ having a point X of $L - (A + B)$ and a point Y of $M - (A + B)$ on its boundary and containing no point of $H + K$. The point set $D + X + Y$ is a connected subset of $S - (H + K)$ intersecting both L and M, contrary to the hypothesis that $H + K$ separates L from M. Thus the supposition that Theorem 110 is not true leads to a contradiction.

THEOREM 111. *If H and K are two compact continua having in common only the points A and B, and $H - (A + B)$ and $K - (A + B)$ are connected and D is a component of $S - (H + K)$ and M is a compact continuum containing A and B and lying, except for A and B, wholly in D then M separates $H - (A + B)$ from $K - (A + B)$ in $D + (H + K)$.*

PROOF. By Theorem 43 of Chapter I there exists a subcontinuum M' of M which is irreducible from A to B. By Theorem 47 of that chapter, $M' - (A + B)$ is connected. By Theorem 109, there exists a component E of $S - (H + K + D)$ whose boundary contains a point C of $H - (A + B)$ and a point F of $K - (A + B)$. By Theorem 44, there exists a simple closed curve J which separates $H - (A + B)$ from $K - (A + B)$ and has only the points A and B in common with $H + K$. The point set $C + E + F$ is connected and it contains two points, C and F, which are separated from each other by J. Hence J contains a point X of E. The arc AXB of J has only A and B in common with $H + K + D$. Hence $AXB - (A + B)$ is a subset of E, the complementary domain of $H + K$ which contains X. By Theorem 42, there exists an arc AXB lying, except for its endpoints, wholly in E. The point set $H + K$ separates $M' - (A + B)$ from the segment $s(AXB)$. Therefore, by Theorem 110, $S - (M' + AXB)$ is the sum of two mutually separated point sets U and V containing $H - (A + B)$ and $K - (A + B)$ respectively. The point sets $U \cdot (D + H + K)$ and $V \cdot (D + H + K)$ are mutually separated. But their sum is $(D + H + K) - M'$ and they contain $H - (A + B)$ and $K - (A + B)$ respectively. Thus M separates $H - (A + B)$ from $K - (A + B)$ in $D + H + K$.

THEOREM 112. *If A and B are two points and H and K are two compact continua, having only A and B in common, such that $H - (A + B)$ and $K - (A + B)$ are connected, then there exist only two components of $S - (H + K)$ that are not components either of $S - H$ or of $S - K$.*

PROOF. By Theorem 44, there exists a simple closed curve J containing A and B and separating $H - (A + B)$ from $K - (A + B)$. Let I and E denote the complementary domains of J that contain $H - (A + B)$ and $K - (A + B)$ respectively. Let AXB and AYB denote the arcs of J from A to B. By Theorem 30 of Chapter III, the segments $s(AXB)$ and $s(AYB)$ are separated from each other in $\bar{I}$ by H and, in $\bar{E}$, by K. It follows that they are separated in S by $H + K$. Let D_1 and D_2 denote the components of $S - (H + K)$ that contain $s(AXB)$ and $s(AYB)$ respectively. Since $I + X + Y$ is a connected point set containing no point of K and $E + X + Y$ is one containing no point of H, therefore X and Y lie in the same component of $S - H$ and in the same component of $S - K$. Hence neither the boundary of D_1 nor the boundary of D_2 is a subset of H or of K. Suppose D is a complementary domain of $H + K$ whose boundary β is not a subset of H or of K. Then β contains a point B_1 belonging to $H - (A + B)$, and therefore to I, and a point B_2 belonging to $K - (A + B)$ and therefore to E. By Theorem 29 of Chapter I, $D + B_1 + B_2$ is connected. Hence it contains a point belonging to J and therefore either to AXB or to AYB. Hence D is identical either with D_1 or with D_2. Thus D_1 and D_2 are the only complementary domains of $H + K$ that are not complementary domains either of H or of K.

THEOREM 113. *If A and B are two points and H and K are two compact continua, with only A and B in common, such that $H - (A + B)$ and $K - (A + B)$ are connected and D is a component of $S - (H + K)$ not bounded by a subset of H or a subset of K, and J is a simple closed curve separating A from B, then there exists an arc of J lying wholly in D except for its endpoints which belong to H and K respectively.*

PROOF. Suppose there is no such arc of J. Let H_1 denote the sum of all segments on J that lie wholly in D and have both ends in H. Similarly let K_1 denote the sum of all segments on J that lie in D and have both ends on K. The point set $J \cdot D = H_1 + K_1$ and, with the help of Theorem 110 of Chapter I, it may be seen that $H + H_1$ and $K + K_1$ are closed and mutually exclusive. It follows, by Theorem 44, that there exists a simple closed curve J' containing A and B and separating $(H + H_1) - (A + B)$ from $(K + K_1) - (A + B)$. By an argument entirely analogous to that employed in a similar connection in the proof of Theorem 112, it may be shown that each arc of J' with endpoints at A and B lies, except for its endpoints, in a complementary domain of $H + K$ whose boundary is not a subset of H or of K. But, by Theorem 112, there are only two such domains. It follows that one of them is identical with D. Hence there exists an arc AXB of J' lying, except for its endpoints, wholly in D and containing no point of H_1 or of K_1 and therefore no point of J. But A and B are separated by J. Thus the supposition that Theorem 113 is false has led to a contradiction.

THEOREM 114. *If A and B are two points, H and K are compact continua having only A and B in common, D is a complementary domain of $H + K$, but not of H or of K, and M is a closed and compact subset of $D + H + K$ which has only A and B in common with $H + K$ and which separates $H - (A + B)$ from $K - (A + B)$ in $D + H + K$, and $H - (A + B)$ and $K - (A + B)$ are connected; then $H - (A + B)$ is separated from $K - (A + B)$ in $D + H + K$ by a continuum which contains A and B and which is a subset of M.*

PROOF. Suppose M contains no continuum containing A and B. Then, by Theorem 44 of Chapter I, M is the sum of two mutually exclusive closed point sets M_A and M_B containing A and B respectively. By Theorem 13, there exists a simple closed curve J separating A from B and containing no point of M. By Theorem 113, there exists an arc XY of J lying wholly in D except for the points X and Y which belong to H and K respectively. Therefore M does not separate $H - (A + B)$ from $K - (A + B)$ in $D + H + K$. But this is contrary to hypothesis. So M contains a continuum N containing A and B. By Theorem 111, N separates $H - (A + B)$ from $K - (A + B)$ in $D + H + K$.

THEOREM 115. *If A and B are points, H and K are compact continua having only A and B in common, $H - (A + B)$ and $K - (A + B)$ are connected and M is a closed and compact point set containing A and B and separating $H - (A + B)$ from $K - (A + B)$, then $M - (A + B)$ is not connected and M contains two continua M_1 and M_2 having only A and B in common and such that $M_1 - (A + B)$ and $M_2 - (A + B)$ are separated from each other by $H + K$.*

PROOF. By Theorem 112, there exist two components D_1 and D_2 of $S - (H + K)$ such that neither of them is bounded by a subset of H or a subset of K. The point set $M \cdot D_1 + A + B$ separates $H - (A + B)$ from $K - (A + B)$ in $D_1 + H + K$. Hence, by Theorem 114, $M \cdot D_1 + A + B$ contains a continuum M_1 containing A and B and separating $H - (A + B)$ from $K - (A + B)$ in $D_1 + H + K$. Similarly $M \cdot D_2 + A + B$ contains a continuum M_2 containing A and B and separating $H - (A + B)$ from $K - (A + B)$ in $D_2 + H + K$. The point sets $D_1 \cdot (M - A - B)$ and $(S - D_1) \cdot (M - A - B)$ are mutually separated. Therefore $M - (A + B)$ is not connected.

THEOREM 116. *If A and B are two points and each of the compact continua H, K and L contains A and B, and no two of them have in common any point except A and B and no one of them is separated by the omission of $A + B$, then there is no component of $S - (H + K + L)$ whose boundary intersects each of the point sets $H - (A + B)$, $K - (A + B)$ and $L - (A + B)$.*

PROOF. Suppose there exists a component D of $S - (H + K + L)$ whose boundary intersects each of these point sets. There exist only two complementary domains of $H + K$ whose boundaries are not subsets either of H or of K. Of these, let E denote one that contains no point of L and let I denote the other one. The domain I contains $L - (A + B)$. For if it did not then $L - (A + B)$ would be a subset of a complementary domain of $H + K$ that is bounded by a subset of H or a subset of K and thus one of the sets H and K would separate $L - (A + B)$ from the point set obtained by omitting $A + B$ from the other one and, by Theorem 115, one of the sets $H - (A + B)$ and $K - (A + B)$ would not be connected, contrary to hypothesis. The domain E is a component of $S - (H + K + L)$ and there is no point of $L - (A + B)$ on its boundary. Therefore D and E have no point in common. Suppose D is a subset of a complementary domain U of $H + K$ whose boundary is a subset of H. Since the connected point set $L - (A + B)$ contains a limit point of D and therefore of U and the boundary of U contains no point of $L - (A + B)$, therefore U contains $L - (A + B)$. But $\bar{U}$ contains no point of $K - (A + B)$. Thus H separates $L - (A + B)$ from $K - (A + B)$. Therefore, by Theorem 115, $H - (A + B)$ is not connected, contrary to hypothesis. Similarly the supposition that D is a subset of a complementary domain of $H + K$, bounded by a subset of K, leads to a contradiction. Therefore D is a subset of I. Let X and Y denote points, of $H - (A + B)$ and $K - (A + B)$ respectively, belonging to the boundary of D. The point set $D + X + Y$ is a connected subset of $I - I \cdot L$ and therefore of $(I + H + K) - L$. But, by Theorem 111, L separates X from Y in $I + H + K$. This involves a contradiction.

THEOREM 117. *If A and B are two points and H, K, L and M are compact continua, each containing $A + B$, such that $H - (A + B)$, $K - (A + B)$, $L - (A + B)$ and $M - (A + B)$ are mutually exclusive connected point sets, then* (1) *there are two of these continua whose sum separates from each other the two point sets obtained by subtracting $A + B$ from each of the remaining ones and* (2) *there exist four arcs AX_1B, AX_2B, AX_3B and AX_4B such that no two of them have any point in common except A and B, such that $AX_1B + AX_2B$ separates $AX_3B - (A + B)$ from $AX_4B - (A + B)$ and such that each of* *the four complementary domains of the continuum* $AX_1B + AX_2B + AX_3B + AX_4B$ *contains one of the point sets* $H - (A + B)$, $K - (A + B)$, $L - (A + B)$ *and* $M - (A + B)$.

PROOF. There exists a simple closed curve J separating A from B. The curve J contains a point C of H and a point O of K. Let CXO denote one of the arcs of J whose endpoints are C and O. Let F denote the first point that the arc CXO has in common with the closed point set $K + L + M$. Of the point sets K, L and M, let K' denote the one that contains F and let L' and M' denote the other ones. The interval AB of the arc AXO contains an arc $C'F'$ having in common with $H + K + L + M$ only the points C' and F' which belong to H and K' respectively. The point set $(H + K' + C'F') - AB$ is connected and it contains no point of L' or of M'. By Theorem 112, there are only two complementary domains of $L' + M'$ that are not complementary domains of L' or of M'. Since $(H + K' + C'F') - AB$ is connected, there is one of these domains that contains no point of this point set. Let D denote one that does not. By Theorem 113, there is an arc NT of J lying wholly in D except that its endpoints N and T belong to L' and M' respectively. By Theorem 44, there exists a simple closed curve J_1 containing A and B and separating the connected point sets $(H + K' + C'F') - (A + B)$ and $(L' + M' + NT) - (A + B)$ from each other. Let AX_1B and AX_2B denote the arcs of J_1 from A to B. There exists a simple closed curve J_2 containing A and B and separating $H - (A + B)$ from $K' - (A + B)$. One of the arcs of J_2 from A to B contains a point of $C'F'$. Let AX_3B denote this arc. Similarly there exists an arc AX_4B lying, except for its endpoints A and B, in the component of $S - (H + K + L + M)$ that contains $NT - (N + T)$. The point set $AX_1B + AX_2B$ separates $AX_3B - (A + B)$ from $AX_4B - (A + B)$.

Let I denote the complementary domain of J that contains $H + K' + C'F'$ and let D denote the complementary domain of $H + K'$ that contains $C'F'$. The arc AX_3B lies, except for its endpoints, wholly in I and separates $H - (A + B)$ from $K' - (A + B)$ in $\bar{I}$. Similarly if E denotes the complementary domain of J that contains AX_4B then AX_4B lies, except for its endpoints, wholly in E and separates $L' - (A + B)$ from $M' - (A + B)$ in $\bar{E}$. Of the point sets H and K', let H'' denote the one which lies, except for the points A and B, wholly in the component of $I - s(AX_3B)$ which is bounded by the simple closed curve $AX_1B + AX_3B$ and let K'' denote the one which lies, except for the points A and B, wholly in the component of $I - s(AX_3B)$ which is bounded by $AX_2B + AX_3B$. Of the point sets L' and M', let L'' denote the one which lies, except for A and B, in the component of $E - s(AX_4B)$ which is bounded by $AX_1B + AX_4B$, and let M'' denote the one which lies, except for A and B, in the component of $E - s(AX_4B)$ which is bounded by $AX_2B + AX_4B$. The point set $H'' + M''$ separates $K'' - (A + B)$ from $L'' - (A + B)$; and $K'' + L''$ separates $H'' - (A + B)$ from $M'' - (A + B)$. But $H'' + K''$ does not separate

$L'' - (A + B)$ from $M'' - (A + B)$, $L'' + M''$ does not separate $H'' - (A + B)$ from $K'' - (A + B)$, $K'' + M''$ does not separate $H'' - (A + B)$ from $L'' - (A + B)$, and $H'' + L''$ does not separate $K'' - (A + B)$ from $M'' - (A + B)$. Furthermore each of the four complementary domains of the continuum $AX_1B + AX_2B + AX_3B + AX_4B$ contains one of the point sets $H - (A + B)$, $K - (A + B)$, $L - (A + B)$ and $M - (A + B)$.

Theorem 118. *Suppose A and B are two points and H, K, L and M are compact continua, all containing A and B, such that $H - (A + B)$, $K - (A + B)$, $L - (A + B)$ and $M - (A + B)$ are mutually exclusive connected point sets. Suppose I is a domain containing A and bounded by a simple closed curve J such that $I + J$ does not contain B. If X is any one of the point sets H, K, L and M, let T_x denote the component of $X \cdot (I + J)$ that contains A and let P_X denote a point of $J \cdot T_X$. Then in order that $H + L$ separate $K - (A + B)$ from $M - (A + B)$ it is necessary and sufficient that $P_H + P_L$ separate P_K from P_M on the simple closed curve J.*

Proof. By Theorem 117, there exist four arcs AX_1B, AX_2B, AX_3B and AX_4B such that no two of them have any point except A and B in common and such that (1) each of the four complementary domains of their sum contains one of the point sets $H - (A + B)$, $K - (A + B)$, $L - (A + B)$ and $M - (A + B)$, (2) $AX_1B + AX_2B$ separates $AX_3B - (A + B)$ from $AX_4B - (A + B)$. For each n, less than 5, let Y_n denote the first point of J on the arc AX_nB, in the order from A to B, and let AY_n denote the interval of AX_nB whose endpoints are A and Y_n. Let N denote the point set $AX_1B + AX_2B + AX_3B + AX_4B$. Of the continua H, K, L and M, let x denote the one which lies, except for A and B, in the component of $S - N$ which is bounded by $AX_1B + AX_3B$, let y denote the one lying, except for A and B, in the component of $S - N$ which is bounded by $AX_3B + AX_2B$, let z denote the one lying, except for A and B, in the component of $S - N$ which is bounded by $AX_2B + AX_4B$, and let w denote the one which lies, except for A and B, in the component of $S - N$ which is bounded by $AX_1B + AX_4B$. With the help of Theorem 28 of Chapter III, it may be shown that no two of the arcs $Y_1P_xY_3$, $Y_3P_yY_2$, $Y_2P_zY_4$ and $Y_4P_wY_1$ of J have in common any point other than a common endpoint, and that $x \cdot J$, $y \cdot J$, $z \cdot J$ and $w \cdot J$ are subsets of $Y_1P_xY_3$, $Y_3P_yY_2$, $Y_2P_zY_4$ and $Y_4P_wY_1$ respectively. The point set $x + z$ separates $y - (A + B)$ from $w - (A + B)$ in S and the point set $P_x + P_z$ separates P_y from P_w on J.

Theorem 119. *Suppose O is a point, O_1 is a point distinct from O, O_2 is a point distinct from O, and, for each n less than 3, a_n, b_n, c_n and d_n are four compact continua, each containing O and O_n, such that $a_n - (O + O_n)$, $b_n - (O + O_n)$, $c_n - (O + O_n)$ and $d_n - (O + O_n)$ are mutually exclusive connected point sets. Suppose furthermore that there exists a region R containing O such that $R \cdot a_1 = R \cdot a_2$, $R \cdot b_1 = R \cdot b_2$, $R \cdot c_1 = R \cdot c_2$ and $R \cdot d_1 =$*

$R \cdot d_2$. *Suppose that* $a_1 + b_1$ *separates* $c_1 - (O + O_1)$ *from* $d_1 - (O + O_1)$. *Then* $a_2 + b_2$ *separates* $c_2 - (O + O_2)$ *from* $d_2 - (O + O_2)$.

PROOF. Let N denote the point set $a_1 + b_1 + c_1 + d_1 + a_2 + b_2 + c_2 + d_2$. By Theorem 15, there exists a domain I bounded by a simple closed curve J such that $N \cdot I + J$ is a subset of R not containing O_1 or O_2. For each point set X of the set $a_1, b_1, c_1, d_1, a_2, b_2, c_2, d_2$, let T_X denote the component of $X \cdot (J + I)$ that contains A. Let A, B, C and D denote points belonging to $J \cdot T_{a_1}$, $J \cdot T_{b_1}$, $J \cdot T_{c_1}$ and $J \cdot T_{d_1}$ respectively. By hypothesis and Theorem 118, $A + B$ separates C from D on J. But $T_{a_1} = T_{a_2}$, $T_{b_1} = T_{b_2}$, $T_{c_1} = T_{c_2}$ and $T_{d_1} = T_{d_2}$. Therefore, by Theorem 118, $a_2 + b_2$ separates $c_2 - (O_1 + O_2)$ from $d_2 - (O_1 + O_2)$.

DEFINITION. Suppose O is a point and a, b, c and d are four compact continua containing O such that $a - O$, $b - O$, $c - O$ and $d - O$ are mutually exclusive connected point sets. Then a and c are said to *ordinally separate* b from d if and only if there exist a point O', distinct from O, and four compact continua a', b', c' and d' all containing O', but none containing O, such that (1) $(a + a') - (O + O')$, $(b + b') - (O + O')$, $(c + c') - (O + O')$, $(d + d') - (O + O')$ are mutually exclusive connected point sets and (2) $(a + a') + (c + c')$ separates $(b + b') - (O + O')$ from $(d + d') - (O + O')$.

THEOREM 120. *If* O *is a point,* a, b, c *and* d *are four compact continua, containing* O, *such that* $a - O$, $b - O$, $c - O$ *and* $d - O$ *are mutually exclusive connected point sets and* a *and* c *ordinally separate* b *from* d, *then* (1) b *and* d *ordinally separate* a *from* c *and* (2) a *and* d *do not ordinally separate* b *from* c.

Theorem 120 may be proved with the use of Theorems 118 and 119.

THEOREM 121. *Suppose* O *is a point and* a, b, c, d *and* e *are five compact continua containing* O *such that* (1) $a - O$, $b - O$, $c - O$, $d - O$ *and* $e - O$ *are mutually exclusive connected point sets,* (2) *of any four of these continua, there are two that ordinally separate the other two. Suppose that* a *and* c *ordinally separate* b *from* e *and that* a *and* d *ordinally separate* c *from* e. *Then* a *and* d *ordinally separate* b *from* e.

Theorem 121 may be established with the use of Theorem 118 and the fact that if A, B, C, D and E are five points on a simple closed curve J, and A and C separate B from E on J, and A and D separate C from E on J, then A and D separate B from E on J.

THEOREM 122 (N. E. RUTT AND J. H. ROBERTS). *If* O *is a continuum and* G *is a collection of continua, all containing* O, *such that no two of them have in common any point of* $S - O$ *and the sum of all the continua of* G *is a compact continuum* M *and* A *and* B *are two points of* $S - M$ *which are not separated from each other by any one continuum of the collection* G *then* A *and* B *are not separated from each other by* M.

PROOF. Suppose, on the contrary, that M separates A from B. Let G' denote the collection of all continua x such that x is O plus a component of the point set obtained by omitting O from some continuum of the collection G. The collection G' satisfies all the conditions imposed on G in the hypothesis of the present theorem together with the additional condition that no continuum of G' is disconnected by the omission of O. Let H denote the collection of all subcontinua h of M such that (1) h separates A from B, (2) each continuum of the collection G' either is a subset of h or has no point of $S - O$ in common with h. Since, by supposition, M itself satisfies these two conditions, therefore H exists. Let N denote any continuum of the collection H. There exists a simple closed curve J separating A from N. Let D denote the complementary domain of J that contains N. The point B lies in D and N separates B from J. There exists, in $S - O$, an arc from B to some point of J. This arc contains an arc B_1A_1 lying wholly in $D - N$ except for the points B_1 and A_1 which belong to N and J respectively. Let g_1 denote the continuum of the collection G' that contains B_1. Since the compact continua g_1 and B_1A_1 have only the point B_1 in common and neither of them separates B from A therefore, by Theorem 20, their sum does not separate B from A. Hence there exists an arc from B to A containing no point of $g_1 + B_1A_1$. This arc contains an arc B_2A_2 lying in $D - N$ except for the points A_2 and B_2 which belong respectively to J and to some continuum g_2 of the collection G' which is a subset of N. The compact continua $g_1 + B_1A_1$ and $g_2 + B_2A_2$ have only O in common and neither of them separates B from A. Hence their sum does not separate B from A. It follows that there exists an arc B_3A_3 containing no point of either of these two continua and lying wholly in $D - N$ except that A_3 belongs to J and B_3 belongs to some continuum g_3 of the collection G' which is a subset of N. Since the compact continua $g_1 + B_1A_1 + g_2 + B_2A_2$ and $g_3 + B_3A_3$ have only O in common and neither of them separates B from A, there exists an arc B_4A_4 containing no point of either of these continua and lying wholly in $D - N$ except that A_4 belongs to J and B_4 belongs to a continuum g_4 of the collection G' which is a subset of N. Two of the points A_1, A_2, A_3 and A_4, say A_1 and A_3, separate the other two on J. It follows, with the help of Theorem 30 of Chapter III, that the continuum $g_1 + B_1A_1 + g_3 + B_3A_3$ separates A_2 from A_4 in $\bar{D}$. But $B_2A_2 + g_2 - O$ and $B_4A_4 + g_4 - O$ are connected subsets of $\bar{D} - (g_1 + B_1A_1 + g_3 + B_3A_3)$ containing B_2 and B_4 respectively. Therefore $g_1 + g_2$ separates $g_2 - O$ from $g_4 - O$ in N. Thus $N - (g_1 + g_2)$ is the sum of two mutually separated point sets K_2 and K_4 containing $g_2 - O$ and $g_4 - O$ respectively. The common part of the compact continua $K_2 + g_1 + g_3$ and $K_4 + g_1 + g_3$ is the continuum $g_1 + g_3$ and their sum N separates B from A. Hence, by Theorem 20, one of these continua separates B from A. But clearly both of these continua belong to the collection H.

Thus every continuum of the collection H has a proper subcontinuum also belonging to H. But, with the aid of Theorem 38 of Chapter I, it may be seen that if H' is any monotonic subcollection of H then the common part of all the continua of H' exists and belongs to H. Therefore, by Theorem 39 of Chapter I, there exists a continuum of the collection H that has no proper subset belonging to H. Thus the supposition that Theorem 122 is false leads to a contradiction.

That Theorem 122 does not remain true in the plane on the *removal* of the supposition that all the continua of G contain O may be seen from the example, in the plane, in which M is a circle, G is the collection of all degenerate subsets of M, A is a point within M and B is a point without M. However this supposition may be *replaced* by the supposition that the collection G is countable. Indeed the following theorem holds true.

THEOREM 123. *If O is a point and G is a countable collection of closed and compact point sets such that no two of them have in common any point of $S - O$ and the sum of all the point sets of the collection G is a closed and compact point set M, and A and B are two points of $S - M$ which are not separated from each other by any one point set of the collection G, then A and B are not separated from each other by M.*

PROOF. Suppose, on the contrary, that A and B are separated from each other by M. By Theorem 26, they are separated from each other by some closed and connected subset K of M. Let G' denote the collection of all point sets g such that g is the common part of K and some point set of the collection G. Suppose there exists a point set h of the collection G' such that there is a component t of h not containing O. By Theorem 52 of Chapter I, there exists a domain D, containing t, such that (1) $\bar{D}$ does not contain O and does not contain all of K and (2) the boundary of D contains no point of h. By Theorems 29 and 50 of Chapter I, there exists a closed and connected subset N of K containing t and a point of the boundary of D and lying wholly in $\bar{D}$. Let W denote the collection of all point sets w such that w is the common part of N and some point set of the collection G'. There is more than one element in W since one of its elements contains t and one contains a point of the boundary of D and no one does both. Thus W is a nondegenerate countable collection of mutually exclusive closed point sets and their sum is the compact continuum N. But this is contrary to Theorem 56 of Chapter I. Therefore every component of a point set of the collection G' contains O and consequently every point set of G' is a continuum containing O. Therefore, by Theorem 122, A and B are not separated from each other by K. Thus the supposition that they are not separated from each other by M leads to a contradiction.

THEOREM 124. *If O is a point and G is a countable nondegenerate collection of continua containing O such that no two of them have in common any point except O and the sum of all the continua of G is a continuum M and the boundary of M is compact and the boundary of every continuum of the collection G is* *compact and no continuum of G is disconnected by the omission of O and no continuum of G separates S, then M − O is not connected.*

PROOF. Suppose, on the contrary, that $M - O$ is connected. The continuum M is not identical with S. For suppose it is. Let X and Y denote points of $M - O$ belonging to different continua of the collection G. There exists, in $S - O$, an arc XY. Let H denote the collection of all point sets h such that h is the common part of XY and some continuum of the collection G. The collection H is a countable nondegenerate collection of mutually exclusive closed and compact point sets filling up the compact continuum XY. But this is contrary to Theorem 56 of Chapter I. Let B denote a point not belonging to M. There exists a simple closed curve J separating B from M. Let D denote the complementary domain of J that contains M. Let $M_1, M_2, M_3, \cdots$ denote the continua of the collection G. A point set M_n of the collection G will be said to be of class 1 if there exists an arc XY which has only the point X in common with M and only the point Y in common with J, the point X being a point of M_n distinct from O. Under these conditions the arc XY will be said to *join* M_n *simply to* J. If a, b, c and d are four sets of the collection G and a and c are simply joined to J by the mutually exclusive arcs X_1Y_1 and X_3Y_3 respectively and also by the mutually exclusive arcs $X_1'Y_1'$ and $X_3'Y_3'$ respectively and, furthermore, of the two domains into which the point set $(a + X_1Y_1 + c + X_3Y_3) - (Y_1 + Y_3)$ separates the domain D, one contains $b - O$ and the other contains $d - O$, then of the two domains into which $(a + X_1'Y_1' + c + X_3'Y_3') - (Y_1' + Y_3')$ separates D, one contains $b - O$ and the other contains $d - O$. If a, b, c and d are four distinct point sets of the collection G and there exist two arcs X_1Y_2 and X_3Y_3 satisfying, with respect to a, b, c and d, the conditions stipulated above, then a and c are said to ordinally separate b from d. It is to be noted that this does not imply that b and d ordinally separate a from c. The point sets b and d may conceivably not be of class 1. It can be shown that (1) if a, b, c and d are four point sets of class 1 then two of them ordinally separate the other two from each other and if a and c ordinally separate b from d then b and d ordinally separate a from c, but a and d do not ordinally separate b from c, (2) if a and c are continua of class 1 and b is a continuum of the collection G (whether of class 1 or not) and there exists some continuum of G which is ordinally separated from b by a and c then there exists a continuum of class 1 which is ordinally separated from b by a and c.

With the aid of the fact that $M - O$ is connected it can be proved that there exist four continua a_1, a_2, a_3 and a_4, all of class 1, such that a_1 and a_3 are ordinally separated from each other by a_2 and a_4. Let us now confine our attention to the collection G' consisting of a_1 and a_3 together with the continua of G which are ordinally separated from a_4 by a_1 and a_3. It can be seen that if x and y are two continua of G' distinct from a_1 and a_3, and x is of class 1 then y is ordinally separated either from a_1 by x and a_3 or from

a_3 by x and a_1. In the first case, y will be said to follow x. In the second case it will be said to precede x. The continuum a_1 is said to precede every other continuum of G' of class 1 and every other continuum of G' is said to follow a_1. The continuum a_3 is said to follow every other continuum of G' of class 1 and every other continuum of G' is said to precede a_3. It is to be observed that no meaning is here attached either to the statement that x precedes y, or to the statement that x follows y, in case y is not of class 1. It is clear that if x and y are two continua of the collection G', both of class 1, then (a) either x follows y or y follows x, and (b) if x follows y then y precedes x and does not follow x. Also, if x, y and z are continua of G' and y follows x and x follows z then y follows z. If x, y and z are continua of G', the statement that z is *between* x and y means that (a) x and y are of class 1, and (b) z either follows x and precedes y or follows y and precedes x. If x and y are two continua of G' of class 1, then by the *segment* xy is meant the collection of all continua of the collection G' which are between x and y, while by the *interval* xy is meant the collection of continua consisting of x and y and all continua of the segment xy. It is easy to see that if U is the point set obtained by adding together all the continua of the segment xy, then no point of $U - O$ is a limit point of $M - U$. It is also clear that if a segment exists, that is to say that it contains at least one continuum of the collection G', then it contains at least one continuum of class 1. There does not exist more than one pair (x,y) of continua of class 1 such that there is no continuum of G' between x and y. For suppose there are two such pairs (x,y) and (z,w). Let x' and y' denote x and y respectively or y and x respectively according as x precedes y or y precedes x. Let z' and w' denote z and w respectively or w and z respectively according as z precedes w or w precedes z.

The point sets x' and z' must be distinct. For suppose x' is z'. Then since (x',y') and (z',w') are distinct pairs, either y' precedes w' or w' precedes y'. In the first case y' is between z and w and, in the second case, w' is between x and y, both of which are contrary to hypothesis. Since x' and z' are distinct one of them precedes the other one. If x' precedes z' let a and b denote x' and y' respectively and let c and d denote z' and w' respectively. If z' precedes x' let a and b denote z' and w' respectively and let c and d denote x' and y' respectively. Then b either precedes c or is identical with it. For otherwise c would be between a and b, contrary to hypothesis. Let t denote either b or the point set obtained by adding together all the point sets of the interval bc, according as b is, or is not, identical with c. Then the point sets $t - O$ and $M - O$ are mutually separated and therefore $M - O$ is not connected. Thus the supposition that there exists more than one pair (x,y) of continua of class 1 such that there is no continuum of G' between x and y has led to a contradiction. It follows that there exist two continua a and b of class 1, belonging to G', such that a precedes b and such that if x and y are two continua of class 1 belonging to G' and to the interval ab then there is a continuum of G' of class 1 between x and y.

Let e denote a continuum of G', of class 1, lying between a and b. Let x_1 denote the point set, of lowest subscript in the sequence $M_1, M_2, M_3, \cdots$, which belongs to the segment ae, is of class 1 and is distinct from M_1. The intervals ax_1 and ex_1 do not both contain M_1. If ax_1 contains M_1, let y_1 and z_1 denote x_1 and e respectively. If ax_1 does not contain M_1 let y_1 and z_1 denote a and x_1 respectively. In either case y_1 precedes z_1, the interval y_1z_1 is a subset of the interval ae and the interval y_1z_1 does not contain M_1. Similarly, the interval y_1z_1 contains two continua y_2 and z_2, of class 1, such that y_2 precedes z_2 and the interval y_2z_2 does not contain M_2. This process may be continued. Thus there exist two infinite sequences $y_1, y_2, y_3, \cdots$ and $z_1, z_2, z_3, \cdots$ of continua of class 1, all belonging to the interval ae, such that, for every n, (1) y_n precedes z_n, (2) y_n either precedes, or is identical with, y_{n+1} and z_{n+1} either precedes, or is identical with, z_n, (3) the interval y_nz_n does not contain M_n. In a similar way it may be shown that there exist two sequences $u_1, u_2, u_3, \cdots$ and $v_1, v_2, v_3, \cdots$ of continua, of class 1, all belonging to the interval eb such that, for every n, (1) u_n precedes v_n, (2) u_n either precedes, or is identical with, u_{n+1} and v_{n+1} either precedes, or is identical with, v_n, (3) the interval u_nv_n does not contain M_n.

Let L denote the sum of all the intervals $z_1u_1, z_2u_2, z_3u_3, \cdots$. The point sets $L - O$ and $M - L$ are mutually separated and therefore $M - O$ is not connected. Thus the supposition that Theorem 124 is false has led to a contradiction.

THEOREM 125. *If D_1 and D_2 are two domains and ω is a point not belonging to $\bar{D}_1 + \bar{D}_2$ then there does not exist a closed point set K, not containing ω, such that* (1) *there is a point of $D_1 + D_2$ which does not belong to K and which is not separated from ω by K,* (2) *each of the domains D_1 and D_2 contains every point of the boundary of the other one which does not belong to K.*

PROOF. Suppose, on the contrary, that there does exist such a point set K. Let O denote a point of $D_1 + D_2$, not belonging to K, that is not separated from ω by K. Let B_1 and B_2 denote the boundaries of D_1 and D_2 respectively. There exists an arc $O\omega$ containing no point of K. In the order from O to ω on $O\omega$ there exists a point Z which is the last point that $O\omega$ has in common with $B_1 + B_2$. Since Z belongs either to B_1 or to B_2, it therefore belongs either to D_1 or to D_2. Hence, since D_1 and D_2 are domains, the boundary of one of them contains a point which lies on $O\omega$ between Z and ω. Hence Z is not the last point of $B_1 + B_2$ on the arc $O\omega$. Thus the supposition that Theorem 125 is false leads to a contradiction.

THEOREM 126. *If D is a domain and ω is a point not belonging to $\bar{D}$, then the boundary of D does not contain two closed point sets B_1 and B_2 such that* (a) *either B_1 and B_2 have no point in common or $B_1 \cdot B_2$ is a point set which does not separate D from ω,* (b) *each of the point sets B_1 and B_2 separates D from ω.*

PROOF. Suppose, on the contrary, that the boundary of D contains two such closed point sets B_1 and B_2. Let D_1 and D_2 denote the complementary domains of B_1 and B_2 respectively which contain D. Let T_1 and T_2 denote the boundaries of D_1 and D_2 respectively. Clearly $T_1 - T_1 \cdot T_2$ is a subset of D_2 and $T_2 - T_1 \cdot T_2$ is a subset of D_1, and $T_1 \cdot T_2$ does not separate D from ω. But this is contrary to Theorem 125.

THEOREM 127. *If O is a point and G is a countable nondegenerate collection of continua such that no two of them have in common any point except O and the sum of all the continua of G is a compact continuum M and no continuum of the collection G is disconnected by the omission of O then $M - O$ is not connected.*

PROOF. Suppose, on the contrary, that $M - O$ is connected. That every continuum of the collection G contains O may be proved by an argument similar to that used in an analogous situation in the proof of Theorem 123. That the continuum M is not identical with S may be shown as in the proof of Theorem 124. Let ω denote a point not belonging to M. There exists a simple closed curve J separating ω from M. Let D denote the complementary domain of J that contains M and let E denote the complementary domain of M that contains J. Let $M_1, M_2, M_3, \cdots$ denote the point sets of the collection G. If, for some n, M_n contains a point set β which is the boundary of some component d, other than E, of $S - M$, then add all such domains d to M_n and let M'_n denote the point set constituted by the sum so obtained. In case M_n contains no such point set β then let M'_n denote M_n. We now have a collection G' of point sets $M'_1, M'_2, M'_3, \cdots$ each of which either is identical with, or contains, some point set of the collection G. It is clear that every point set of the collection G' is a continuum with a compact boundary, that every two of them have in common only the point O, that their sum M' is a continuum lying wholly in D and bounded by a compact point set and finally that $M' - O$ is connected and, for every n, M'_n is connected. No complementary domain of M', except E, is bounded by a subset of any one continuum of the collection G'. It is conceivable, however, that there may be a domain, not containing E, which is *complementary to some point set of the collection* G' and which is therefore bounded by a subset of that point set. If such a domain exists it must contain points of M.

Case 1. Suppose E is the only component of $S - M'$. Suppose that a continuum M'_n of the collection G' separates S. Then there exists a component H of $S - M'_n$ containing no point of E. The domain H must be a subset of M. Let X denote a point of H. There exists an arc XY not containing O and lying wholly in H, except for the point Y which belongs to the boundary of H. Every point of the compact continuum XY belongs to some point set of the countable collection G' and XY contains the point Y of M'_n and the point X which does not belong to M'_n. This is contrary to Theorem 56 of Chapter I. Thus the supposition that S is separated by some point set of the collection G' leads to a contradiction. Hence, in Case 1, the collection G' satisfies all the conditions imposed on G by the hypothesis of Theorem 124. Thus, in Case 1, we have a contradiction.

Case 2. Suppose now that there exists a component D' of $S - M'$ distinct from E. The boundary k of the domain D' is not a subset of any one point set of the collection G'. It follows that k is the sum of the elements of a collection g of two or more closed point sets $k_1, k_2, \cdots$ each of which is a subset of some point set of the collection G and no two of which are subsets of the same point set of this collection. If k_n is a point set of the collection g, one complementary domain d_n of k_n contains D'. By Theorem 126, there does not exist more than one n such that d_n fails to contain J. By Theorem 123 there does exist one such n. Call it m. For no n does any complementary domain of k_n, except d_n, contain a point of k. The point set k is a continuum since it is the boundary of a complementary domain of the continuum M' and the boundary of M' is compact. That k contains O and that every k_n is a continuum containing O may be shown by a method similar to that used in a similar connection in the proof of Theorem 123. The point set $k - O$ is connected. For suppose it is not. Let w denote the outer boundary of D' with respect to E. Then $k - O$ is the sum of two mutually separated point sets k' and k'', where k' contains at least one point of $w - O$. It is clear that w is a subset of k_m. By Theorem 28, the point set $w - O$ is connected. Hence $w - O$ is a subset of k'. The point set w is the boundary of each of two mutually exclusive connected domains D_1 and D_2 where D_1 contains E and D_2 contains D'. The domain D_2 contains every point of $k - O$ which does not belong to w. If it contains any points of M which do not belong to k, let Z denote the set of all such points. If X is a point of Z then X belongs to a complementary domain of k. Call this domain d_X and call its boundary k_X. Then k_X is a subset of k. By Theorems 24 and 28, k_X is connected and if k_X contains O then $k_X - O$ is connected. Hence, in any case, the point set $(k_X + O) - O$ is a subset either of k' or of k''. Now let K' denote the point set obtained by adding together k', all points of $M - O$, if there are any, which do not belong to $w + D_2$, and all points X of Z such that $(k_X + O) - O$ is a subset of k'. Let K'' denote the set of points obtained by adding together k'' and all points X of Z such that $(k_X + O) - O$ is a subset of k''. With the help of Theorem 4 of Chapter II, and the fact that k' and k'' are mutually separated, it is easy to prove that K' and K'' are mutually separated. But clearly $K' + K'' = M - O$. Hence $M - O$ is not connected. Thus the supposition that $k - O$ is not connected has led to a contradiction. Since $k - O$ is connected it easily follows that there are infinitely many point sets in the collection g. Thus $k_1, k_2, \cdots$ is an *infinite* sequence.

If, for some n, k_n has a complementary domain other than d_n then let k'_n denote the point set obtained by adding to k_n all such domains. If k_n has no complementary domain except d_n, let k'_n denote k_n. Let g' denote the countable collection of point sets $k'_1, k'_2, k'_3, \cdots$ and let N denote their sum.

The point sets of the collection g' are all closed, no two of them have, in common, any point except O, their sum N is a continuum and, with the aid of the fact that $k - O$ is connected, it can be seen that $N - O$ is connected. Furthermore, no point set of the collection g' separates S. If a point P is separated from D' by a point set k_r of the collection g and P belongs to a point set of the collection G which contains some point set of the collection g, then P belongs to M_i, the point set of the collection G which contains k_r. For suppose that P belongs to M_j, where j is distinct from i. Let V denote the complementary domain of k_r to which P belongs. Suppose first that $M_j - O$ is not a subset of V. Then clearly $M_j - O$ is the sum of the mutually separated sets L_1 and L_2, where L_1 consists of those points of $M_j - O$ which belong to V and L_2 consists of all those which do not belong to V. Hence $M_j - O$ is not connected. Thus, in case $M_j - O$ is not a subset of V, a contradiction is reached. Suppose secondly that $M_j - O$ is a subset of V. Then, since by hypothesis M_j contains some point set of the collection g, therefore V contains a point Y which belongs to k. But Y is a limit point of D'. Hence V contains a point of D', contrary to the supposition that k_r separates P from D'. Thus again a contradiction is reached. It follows that P belongs to M_i. With the help of this fact it may be shown that, for every n, $k'_n - O$ is connected. For suppose, on the contrary, that there exists a positive integer j such that $k'_j - O$ is the sum of two mutually separated point sets H_1 and H_2. Let M_q denote the point set of the collection G which contains k_j. If B is the boundary of a complementary domain of k_j that contains no point of D', then, by Theorems 24 and 28, $(B + O) - O$ is connected. Hence $(B + O) - O$ is a subset either of H_1 or of H_2. For each i less than 3, let H'_i denote the point set obtained by adding to H_i every point of $M_q - O$ which lies in a complementary domain of k_j which contains no point of D' and whose boundary is, except for the point O, a subset of H_i. Every point of $M_q - O$ belongs either to H'_1 or to H'_2. For suppose this is not the case. Then d_j, the complementary domain of k_j which contains D', contains a point P belonging to $M_q - O$. Since the point P belongs neither to D' nor to its boundary k, it is separated from D' by k. It follows, with the help of Theorem 123, that P is separated from D' by some point set k_t of the collection g. Clearly k_t must be distinct from k_j. But it has been shown above that this is impossible. It follows that $M_q - O$ is the sum of the two point sets H'_1 and H'_2. But clearly these two point sets are mutually separated. Hence $M_q - O$ is not connected. Thus the supposition that $k'_j - O$ is not connected has led to a contradiction. So, for every n, $k'_n - O$ is connected. It follows, by Theorem 124, that $N - O$ is not connected. But this involves a contradiction.

As a special case of Theorem 123 we have the Theorem E that if M denotes S, the set of all points, A and B are two distinct points of M and G is a countable collection of mutually exclusive closed subsets of M no one of which separates A from B in M, and G^* is closed and compact, then A and

B are not separated from each other in M by G^*. This theorem does not remain true in the plane if the phrase "if M denotes S, the set of all points" is replaced by "if M is a compact continuum not separating S". Consider the following example.

EXAMPLE 11. In a Cartesian plane, let A denote the origin of coordinates, let B denote the point $(1,0)$ and, for each positive integer n, let B_n denote the point $(1,1/n)$. Let M denote the continuum obtained by adding together the intervals AB, AB_1, AB_2, $AB_3, \cdots$. Let C denote the point $(1/2,0)$ and, for each n, let C_n denote the point $(1/2,1/2n)$. Let G denote a countable collection of point sets such that g belongs to G if and only if, for some n, g is the set whose only element is C_n. In this case each point set of the collection G is degenerate. Clearly no point set of the collection G separates A from B in M. But A is separated from B in M by the sum of all the point sets of the collection G.

As a consequence of Theorem E we have the Theorem F that if M denotes S, G is a countable collection of mutually exclusive closed subsets of M no one of which separates M and G^* is closed and compact then M is not separated by the sum of all the point sets of the collection G. It is clear that this proposition does not remain true in the plane if "M denotes S" is replaced by "M is a compact continuum". If M is a simple closed curve, A and B are two points belonging to M and the point sets of the collection G are the degenerate sets whose elements are A and B respectively then no point set of the collection G separates M, but their sum does separate M. It is to be observed that in this case M *disconnects* S. Consider, however, the following example.

EXAMPLE 12. In a Cartesian plane, for each n, let d_n denote the arc composed of the four straight line intervals A_nB_n, B_nC_n, C_nD_n and D_nE_n, where A_n, B_n, C_n, D_n and E_n are the points $(1/2^{n-1},0)$, $(1/2^{n-1},1/2^{n-1})$, $(-1/2^{n-1},1/2^{n-1})$, $(-1/2^{n-1},-1/2^{n-1})$ and $(1,-1/2^{n-1})$ respectively. Let d denote the straight line interval from $(0,0)$ to $(1,0)$. Let G denote the collection of point sets d_1, d_2, $d_3, \cdots$ and let M denote the sum of d and all the point sets of the collection G. Clearly M is a compact continuum, no point set of the collection G separates M and M does not separate S. Nevertheless, the sum of all the point sets of the collection G does separate M.

There exists (Mazurkiewicz), in the plane, a countable collection of mutually exclusive closed point sets such that their sum is a continuum and all but one of them is a continuum. Indeed there exists, in the plane, a continuum M containing a closed point set K such that (1) $M - K$ has only countably many components, (2) K contains no limit point of any component of $M - K$, (3) each component of K is a straight line ray and (4) each component of $M - K$ is a ray which is the sum of a straight ray and four straight arcs.

EXAMPLE 13. Let AB denote an interval of length 1 on the X-axis such that A is to the left of B. Let H_1, H_2, $H_3, \cdots$ denote a sequence such that

(1) H_1 is the collection whose only element is the middle third segment of AB, (2) for each n, H_{n+1} is a collection such that h belongs to it if and only if h is the middle third segment of some component of $AB - (H_1^* + H_2^* + \cdots + H_n^*)$. Let H denote the collection $H_1 + H_2 + H_3 + \cdots$ and let C denote the totally disconnected closed point set $AB - H^*$. For each element D of H let A_D denote the left end, let B_D denote the right end, and let O_D denote the midpoint, of the segment D and let α_D and β_D denote two infinite simple sequences of arcs such that (1) the nth term of α_D is the sum of three sides of a rectangle so situated that its fourth side is a subinterval of the interval A_DO_D with center at the midpoint of A_DO_D and length equal to the length of A_DO_D multiplied by $n/(n + 1)$ and its other horizontal side lies at a distance equal to n *below* AB and (2) the nth term of β_D satisfies the same conditions except for the substitution of β_D in place of α_D and B_D instead of A_D throughout in this description.

If D is an element of H_k and E is a shorter element of H lying to the right of D and x is the nth term of β_D then by an F-ray from x through the center of E is meant a ray which starts at the left end of x and is the sum of (a) the arc x, (b) a vertical interval of length $1/(kn)$ with its lower end at the right end of x, (c) a horizontal interval with its left end at the upper end of this vertical interval and its right end at a point directly above the center of E and (d) the straight ray that starts downward at the right end of that horizontal interval.

If D is an element of H_k and E is a shorter element of H lying to the left of D and x is the nth term of α_D then by an F-ray from x through the center of E is meant a ray which starts at the right end of x and is the sum of (a) the arc x, (b) a vertical interval of length $1/(kn)$ with its lower end at the left end of x, (c) a horizontal interval with its right end at the upper end of this vertical interval and its left end at a point directly above the center of E and (d) the straight ray that starts downward at the left end of that horizontal interval.

There exists a sequence $Q_1, Q_2, Q_3, \cdots$ such that (1) q belongs to Q_1 if and only if it is true that if D denotes the only element of H_1 then, for some n, q is an F-ray either from the nth term of α_D through the center of the nearest element of H_{n+1} to the left of D or from the nth term of β_D through the center of the nearest element of H_{n+1} to the right of D, (2) for each k greater than 1, q belongs to Q_k if and only if either (a) there exists an element D of H_k such that there is no other element of H_k to the right of D but, for some n, either q is an F-ray from the nth term of β_D through the center of the nearest element of H_{k+n} to the right of D or q is an F-ray from the nth term of α_D through the center of the nearest element of H_{k+1+n} to the left of D, (b) there exists an element D of H_k such that there is no other element of H_k to the left of D but, for some n, either q is an F-ray from the nth term of α_D through the center of the nearest element of H_{k+n} to the left of D or q is an F-ray from the nth term of β_D through the center of the nearest element of H_{k+1+n} to the right of D or (c) there exists an element D of H_k such that there is an element of H_k to the right of D and also one to the left of D and, for some n, either q is an F-ray from the nth term of β_D through the center of the nearest element of H_{k+1+n} to the right of D or q is an F-ray from the nth term of α_D through the center of the nearest element of H_{k+1+n} to the left of D.

Let K denote the sum of all straight vertical rays that start downward either at a point of C or at the midpoint of the only element of H_1, and let M denote K plus the sum of all the point sets $Q_1^*, Q_2^*, Q_3^*, \cdots$.

THEOREM 128. *If β_1, the boundary of the connected domain D_1, is compact and D_2 is a connected domain containing β_1 then the common part of D_1 and D_2 is connected.*

PROOF. Let β_2 denote the boundary of D_2 and let O denote some definite point of $D_1 \cdot D_2$. Suppose X is a point of $D_1 \cdot D_2$ distinct from O. Since D_1 is a connected point set containing X and O but no point of β_1 therefore β_1 does not separate X from O. Similarly β_2 does not separate X from O. Furthermore, β_1 is compact and β_1 and β_2 are mutually exclusive. Therefore, by Theorem 20, $\beta_1 + \beta_2$ does not separate X from O. Hence, by Theorem 6 of Chapter II, X belongs to the component of $S - (\beta_1 + \beta_2)$ that contains O. Therefore $D_1 \cdot D_2$ is connected.

THEOREM 129. *If O is a point, K is a closed and compact point set and G is a countable collection of closed point sets containing O such that* (1) *if g is one of them, $g - O$ is connected and* (2) *no two of them have in common any point other than O and G^*, the sum of all the elements of G, has no point other than O in common with K, and $G^* + K$ is compact and closed and $(G^* + K) - O$ is connected then if g is an element of G, $g - O$ is a component of G^*.*

PROOF. Let M denote $G^* + K$. Suppose K has a component H not containing O. Then there exists a domain D, containing H, such that $\bar{D}$ does not contain O. Let L denote the component of $M \cdot \bar{D}$ that contains H. The boundary of D contains a point B of L. Thus the connected subset L of M contains a point B not belonging to H. Hence L is not a subset of K. Let W denote the collection whose elements are K and the elements of the set G. Let Q denote the set of all point sets q such that q is the common part of L and some point set of the collection W. The collection Q is a nondegenerate countable collection of mutually exclusive closed point sets filling up the compact continuum L. But this is contrary to Theorem 56 of Chapter I. It follows that every component of K contains O. Let g denote a point set of the collection G and let N denote the component of $M - K$ that contains the connected point set $g - O$. Suppose N is distinct from $g - O$. Then there exists an infinite subcollection G'' of G such that $N + O$ is the sum of all the point sets of the collection G''. Since every component of the closed and compact point set K contains O, K is a compact continuum. Let I denote the complementary domain of K that contains

N. The boundary of I is a subcontinuum β of K, O is a limit point of $g - O$ and therefore of N and $N + \beta$ is closed. It follows that $N + \beta$ is a compact continuum.

Let Σ denote a space such that (1) P is a point in Σ if and only if either P is β or P is a point, in the original sense, belonging to the domain I, (2) R is a region in the space Σ if and only if either R is a region, in the original sense, which is a subset of I or, for some connected domain D in the original sense which contains the continuum β, R is a set such that P belongs to it if and only if P is either β or a point of $D \cdot I$. By Theorem 29 of Chapter I and Theorem 128 of the present chapter, every region in Σ which is not a region in the original sense is connected. It may be seen that Axioms 0 and 1–5 hold true in the space Σ. In this space let O' denote the "point" β, let G' denote a collection such that g' belongs to it if and only if, for some point set g, in the original sense, belonging to G'', g' is the set whose "points" are β and the points, in the original sense, of g, and let M' denote the sum of all the point sets of the collection G'. Here G' is a countable collection of continua all containing the "point" O', each two of them have only O' in common, no one of them is disconnected by the omission of O', and neither is their sum. But this is contrary to Theorem 127.

THEOREM 130. *If M is a compact continuous curve which does not separate S, A and B are two points of M and G is a countable collection of mutually exclusive closed subsets of M no one of which separates A from B in M, then the sum of all the point sets of the collection G does not separate A from B in M.*

PROOF. Let G^* denote the sum of all the point sets of the collection G. Suppose that G^* separates A from B in M. Then, by Theorems 16, 154 and 149 of Chapter I and Theorems 12 and 84 of Chapter II, there exists a closed subset T of G^* which is irreducible with respect to the property of separating A from B in M. Let U denote the collection of all point sets u such that u is the common part of T and some point set of the collection G. Since G^* separates A from B in M, no element of G contains A or B. But every element of U is a subset of some element of G, and no element of G separates A from B in M. It follows that no element of the countable collection U separates A from B in M. Let H denote the component of $M - T$ that contains A and let K denote the component of $M - \bar{H}$ that contains B. Let N denote the common part of $\bar{H}$ and $\bar{K}$. The point set N is a subset of T. Suppose N is not a subset of any one point set of the set U. Then, by Theorem 56 of Chapter I, N is not connected. The point set $\bar{K} - N$ is identical with K. Hence it is connected. The point set $\bar{H} - N$ contains H and is a subset of $\bar{H}$. Hence, by Theorem 29 of Chapter I, $\bar{H} - N$ is connected. Therefore $\bar{K} - N$ is a subset of a component of $S - \bar{H}$ and $\bar{H} - N$ is a subset of a component of $S - \bar{K}$. Let H' denote the continuum obtained by adding to $\bar{H}$ all the components of $S - \bar{H}$ except the one that contains $\bar{K} - N$ and let K' denote the continuum obtained by adding to $\bar{K}$ all the components of $S - \bar{K}$ except the one that contains $\bar{H} - N$. Since M does not separate S there is not more than one component of $S - \bar{H}$, or of $S - \bar{K}$, which is not a subset of M. By Theorem 22, $S - (\bar{H}' + \bar{K}')$ has at least two components. Hence it has a component D which is a subset of M. Since D is a complementary domain of $\bar{H} + \bar{K}$ but not of $\bar{H}$, therefore the boundary of D contains a point P belonging to $\bar{K} - N$, that is to say to K. By Theorems 2 and 12 of Chapter II, there exists a region R containing P such that $M \cdot R$ is a subset of K. But R contains a point of D and D is a subset of $M - K$. Thus the supposition that N is not a subset of any one point set of the set U leads to a contradiction. Let g denote the point set of U that contains N. The point sets $\bar{K} - \bar{K} \cdot g$ and $(M - g) - (\bar{K} - \bar{K} \cdot g)$ are mutually separated and they contain A and B respectively. Thus g separates A from B in M. But this involves a contradiction.

Theorem 130 does not remain true even in the plane on the removal of the supposition that M is compact. Consider the following example.

EXAMPLE 14. In the Cartesian plane, let C denote a circle with center at the origin and with a radius of 1. Let M denote the continuum consisting of the circle C together with all points lying without it. Let A and B denote the points $(0,1)$ and $(0,-1)$ respectively, let g_1 denote the set of all points on OX whose abscissae are greater than or equal to 1, let g_2 denote the set of all those whose abscissae are less than or equal to -1 and let G denote the collection whose elements are the two rays g_1 and g_2.

THEOREM 131. *If H and K are two compact continua whose common part is the sum of two mutually exclusive continua L and N and neither H nor K separates S then $S - (H + K)$ is the sum of two connected domains.*

PROOF. Let G_H denote the set of all components of $H - (L + N)$. Suppose there are two point sets X and Y, belonging to the set G_H, such that each of the continua L and N contains both a limit point of X and a limit point of Y. Let X' denote the point set obtained by adding to $X + (L + N)$ all of the complementary domains of $X + (L + N)$ except the one that contains the connected point set Y. Let Y' denote the point set obtained by adding to $Y + (L + N)$ all of its complementary domains except the one that contains X. Let T denote the sum of the continua X' and Y'. By Theorem 22, T separates S. Since H does not separate S, one of the complementary domains of T is a point set D lying wholly in H. The point sets X, D and Y are connected subsets of $H - (L + N)$ and both X and Y contain at least one limit point of D. Therefore $X + D + Y$ is connected, contrary to the fact that X is a component of $H - (L + N)$. Therefore, by Theorems 43 and 47 of Chapter I, there exists only one point set H', belonging to G_H, such that each of the sets L and N contains a limit point of H'. Similarly there exists only one component K' of $K - (L + N)$ such that each of the sets L and N contains a limit point of K'. By a modification of Theorem 44, there exists a compact continuum J, intersecting both L and N, such that (1) J is connected im kleinen at every point of $J - J \cdot (L + N)$, (2) J is the outer boundary, with respect to $K - (L + N)$, of a domain I that contains $H - (L + N)$. Since J is connected im kleinen at every point of $J - J \cdot (L + N)$, therefore no component of $J - J \cdot (L + N)$ contains a limit point of the sum of all the remaining components of this point set. Suppose $J - J \cdot (L + N)$ has more than two components. Let M_1 and M_2 denote two of them and let U denote the sum of all the others. Since neither M_1 nor M_2 contains a limit point of the connected point set $U + L + N$ therefore $U + L + N$ is a continuum. Hence, by Theorem 28, $J - (U + L + N)$ is connected. But it is the sum of the two mutually separated point sets M_1 and M_2. This involves a contradiction. Hence $J - J \cdot (L + N)$ does not have more than two components. There exist two components D' and E' of $S - (H' + K')$ each bounded by a point set containing both a point of H' and a point of K'. Since J separates H' from K' therefore it intersects both D' and E'. The point sets $J \cdot D'$ and $J \cdot E'$ are mutually exclusive and every component of $J \cdot D'$ or of $J \cdot E'$ is a component of $J - J \cdot (L + N)$. It follows that $J \cdot D'$ and $J \cdot E'$ are the components of $J - J \cdot (L + N)$.

Suppose there exists a component Q of $S - (H' + K' + L + N)$ distinct from D' and E'. If the boundary of Q were a subset of $H' + L + N$ then either Q would contain a point of $S - H$ in which case H would separate S, or Q would be a subset of H in which case $Q + H'$ would be a connected subset of $H - (L + N)$ having H' as a proper subset in which case H' would not be a component of $H - (L + N)$. Therefore the boundary of Q is not a subset of $H' + L + N$. Similarly it is not a subset of $K' + L + N$. Therefore each of the point sets H' and K' contains a limit point of Q. But J separates H' from K' and Q is connected. Therefore either $J \cdot D'$ or $J \cdot E'$ intersects Q. But Q, D' and E' are mutually exclusive. It follows that D' and E' are the only components of $S - (H' + K' + L + N)$.

Let V denote the set of all continua x such that x is the point set obtained by adding $H' + K' + L + N$ to some component of $D' \cdot (H + K)$. Let D and E denote the domains $D' - D' \cdot (H + K)$ and $E' - E' \cdot (H + K)$ respectively. Suppose X and Y are two points of D. Suppose g is any continuum of the collection V. Then $g = (H' + K' + L + N) + t$ where t is some component of $D' \cdot (H + K)$. But either $L + t$ or $N + t$ is a continuum. In the first case g is the sum of the continua $H' + K' + L + N$ and $L + t$ and in the second case it is the sum of the continua $H' + K' + L + N$ and $N + t$. But $(H' + K + L + N) \cdot (L + t) = L$ and $(H' + K' + L + N) \cdot (N + t) = N$. Furthermore, since X and Y belong to the same complementary domain of $H' + K' + L + N$, therefore $H' + K' + L + N$ does not separate X from Y and, since each of the point sets $L + t$ and $N + t$ is a subset either of H or of K, neither of them separates X from Y.

It follows that g does not separate X from Y. But the sum of all the continua of V is a compact continuum and the common part of every two continua of V is the continuum $H' + K'$. It follows, by Theorem 122, that X is not separated from Y by V^*, the sum of all the continua of the collection V. Therefore the domain D is connected. Hence it is a component of $S - (H + K)$. Similarly E is a component of $S - (H + K)$. But $S = D' + E' + (H + K)$, $D = D' - D' \cdot (H + K)$ and $E = E' - E' \cdot (H + K)$. Therefore $S = D + E + (H + K)$. It follows that D and E are the only components of $S - (H + K)$.

THEOREM 132. *Suppose O is a point and G is a countable collection of (two or more) compact continua, all containing O, such that* (1) *no continuum of G separates S and no two continua of G have in common any point except O,* (2) *no continuum of G is separated by the omission of O,* (3) *the sum of all the continua of G is a compact continuum M. Suppose a and b are two continua of G such that there exists an arc which contains a point of $a - O$ but no other point of M and such that there also exists an arc which contains a point of $b - O$ but no other point of M. Then there exists a simple closed curve which separates $a - O$ from $b - O$ and contains no point of $M - O$.*

PROOF. By Theorem 127, $M - O$ is not connected. Hence, by Axiom 3, there exists a point ω which does not belong to M. By Theorem 12, there exists a simple closed curve C separating M from ω. As in the proof of Theorem 124, an element x of G will be said to be of class 1 if there exists an arc AB whose endpoints A and B belong to $x - O$ and C respectively and which has only the point A in common with M. By Theorem 123, M does not separate S. From this and the hypothesis it follows that a and b are of class 1. If there are only a finite number of continua in the set G the truth of Theorem 132 can be easily established with the aid of Theorem 44. Let us suppose that there are infinitely many continua in G. Then it is easy to see that there exists an element c, of class 1, distinct from a and from b. There are two cases.

Case 1. Suppose there exists no G-element which is ordinally separated from c by a and b. If x and y are two G-elements, distinct from each other and from a and b, and x is of class 1 then y is ordinally separated either from a by x and b or from b by x and a. In the first case y will be said to follow x. In the second case it will be said to precede x. If x, y and z are G-elements the statement that z is between x and y means that (a) x and y are both of class 1, (b) z either follows x and precedes y or follows y and precedes x. If x and y are two G-elements of class 1 then by the segment xy is meant the collection of all G-elements which are between x and y while by the interval xy is meant the collection whose elements are x and y and the G-elements of the segment xy. With the aid of results established in the course of the proof of Theorem 124, it may be seen that either (a) there exist two G-elements e and f, of class 1, such that e precedes f and there is no G-element between them, or (b) there exist two infinite sequences $y_1, y_2, y_3, \cdots$ and $z_1, z_2, z_3, \cdots$ of G-elements, of class 1, such that, for every m and n, (1) y_m precedes z_n, (2) y_n precedes, or is identical with, y_{n+1} and z_{n+1} either precedes, or is identical with, z_n, (3) there exists no G-element which is common to all the segments $y_1z_1, y_2z_2, y_3z_3, \cdots$. In Case 1 (a), let K denote the point set obtained by adding together a and e and all point sets of G (if there are any) that precede e and let H denote the point set obtained by adding together f and all point sets of G (if there are any) that follow f. In Case 1(b), let K denote the point set obtained by adding together all the point sets of the sequence $a, y_1, y_2, y_3, \cdots$ together with all point sets g of the collection G such that g precedes some point set of this sequence and let H denote the point set obtained by adding together all the point sets of the sequence $b, z_1, z_2, z_3, \cdots$ together with all point sets g of the collection G such that g follows some point set of this sequence. In either case H and K are compact continua with only the point O in common and $H + K = M$. By Theorem 123, neither H nor K separates S. Let I denote the connected domain $S - H$. Let H' denote the boundary of I. The point set H' is a compact continuum containing a point B of $b - O$. The point set K is a compact continuum and $K - O$ is a subset of I containing $a - O$. Let A denote a point of $a - O$. By Theorem 50, there exists a simple closed curve β, separating A from B, such that $\beta - O$ is a subset of $I - (K - O)$ and therefore of $S - (H + K)$. Since $a - O$ and $b - O$ are connected, β separates them from each other.

Case 2. Suppose there exists a G-element which is ordinally separated from c by a and b. Then there exists a G-element d of class 1 satisfying these conditions. By a modification of the argument given under Case 1 to prove the existence of H and K it may be shown that there exist two continua H_1 and K_1 containing a and b respectively, and two continua H_2 and K_2 containing a and b respectively, such that (1) the continua $H_1 + H_2$ and $K_1 + K_2$ have only O in common, (2) their sum is M and if the continuum g of the collection G intersects one of them then $g - O$ is a subset of that one, (3) if, in the argument given under Case 1 to show that there exists a simple closed curve β separating $a - O$ from $b - O$ but containing no point of $M - O$, H is replaced by $H_1 + H_2$ and K is replaced by $K_1 + K_2$ then that argument becomes valid for Case 2.

CONVENTION. Select some definite point and call it *the point at infinity* and denote it by the symbol ω. As in Chapter III, the point set M will be said to be *bounded* if and only if ω is neither a point of M nor a limit point of M; and if J is a simple closed curve not containing ω then the interior of J with respect to ω will be called the *interior* of J and the exterior of J with respect to ω will be called the *exterior* of J. If the point set K is a subset of the interior of the bounded simple closed curve J then K will be said to be *within* J and J will be said to *enclose* K. If K is a subset of the exterior of J it will be said to be *without* J.

THEOREM 133. *If G is a finite set of bounded simple closed curves and M is a closed and compact point set and each point of M is within some curve of the set G, but no point of M is on any curve of the set G, then there exists a finite set G' of simple closed curves such that* (1) *each point of M is within some curve of G',* (2) *every curve of the set G' lies entirely without every other curve of the set G',* (3) *each curve of the set G' is within some curve of the set G.*

PROOF. There exists a subset H of G such that (1) each point of M is within some curve of the set H, (2) no curve of the set H encloses the interior of any other curve of the set H. If P is a point of M there exists a simple closed curve J'_P, enclosing P, such that (1) every point of J'_P is on some curve of the set G and (2) every point within J'_P is within every curve of the set G that encloses P and without every curve of G that does not enclose P. The set M_P of all points of M that lie within J'_P is closed and compact. It follows that there exists a simple closed curve J_P lying within J'_P and enclosing M_P. Such a curve will be said to suitably enclose P. It is clear that if J_{P_1} and J_{P_2} are curves that suitably enclose the points P_1 and P_2 respectively then J_{P_1} either lies entirely without J_{P_2} or suitably encloses every point of M that is enclosed by J_{P_2}. For each point P of M select just one J_P and let K denote the set of curves thus selected. There is a finite subset G' of the set K such that every point of M is within some curve of the set G'. The set G' satisfies the conditions of Theorem 133.

THEOREM 134. *Suppose that M is a closed and compact set of points and G is a set of bounded simple closed curves such that* (1) *every point of M is either on or within some curve of G,* (2) *if a point of M is not within any curve of G then there exists a curve g of the set G such that P is on g and such that if P is a limit point of a subset of M then every such subset of M contains points within g. Then there is a finite set G' of curves such that every curve of G' is a curve of G and such that G' satisfies with respect to M the same conditions* (1) *and* (2) *that are assumed as being satisfied by G.*

PROOF. If P is a point of M that does not lie within any curve of the set G there exists a simple closed curve J_P belonging to G and containing P such that P is not a limit point of any subset of M that contains no points within J_P. There exists a simple closed curve C_P, enclosing P, such that every point of M within C_P, except the point P, is within J_P. Let K denote the set of all such curves C_P for all such points P. Every point of M is within a curve of the set $G + K$. There exist a finite set of curves $C_{P_1}, C_{P_2}, C_{P_3}, \cdots, C_{P_n}$, belonging to K, and a finite set $g_1, g_2, g_3, \cdots, g_m$ belonging to G, such that every point of M is within a curve of one or the other of these two sets. It is clear that the set G' of curves $g_1, g_2, g_3, \cdots, g_m, J_{P_1}, J_{P_2}, J_{P_3}, \cdots, J_{P_m}$ satisfies the requirements of Theorem 134.

DEFINITION. An *arc-curve chain* is a finite set K of simple closed curves $a_1, a_2, a_3, \cdots, a_n$ and arcs A'_1A_2, A'_2A_3, $A'_3A_4, \cdots$, $A'_{n-1}A_n$ such that (1) every simple closed curve of K is without every other simple closed curve of K, (2) no two arcs of K have a point in common, (3) for every i less than n, the arc $A_i'A_{i+1}$ lies entirely without every simple closed curve of K except that A_i' is on a_i and A_{i+1} is on a_{i+1}. The arc-curve chain K is said to *envelop* the point set M if every point of M is either within a simple closed curve of K or on an arc of K.

THEOREM 135. *In order that the closed and compact point set M be a subset of an arc it is necessary and sufficient that every closed and connected subset of M be either a degenerate point set or an arc t such that no point of t, except its endpoints, is a limit point of $M - t$.*

PROOF. In order that a closed and compact point set be an arc it is evidently necessary that it satisfy these conditions. It will be shown that it is also sufficient. Suppose that the point set M satisfies all of these conditions. There exists a sequence $H_1, H_2, H_3, \cdots$ such that (1) for each n, H_n is a collection of regions covering M and H_{n+1} is a subcollection of H_n and of G_n, (2) if K and L are two mutually exclusive closed subsets of M there exists a number m such that no region of H_m contains both a point of K and a point of L. Let the point at infinity be so selected that it does not belong to M. Every nondegenerate component of M is an arc. If A and B are the endpoints of such an arc then there exist two arcs AXB and AYB, with no point in common except A and B, such that the simple closed curve $AXBYA$ formed by these arcs encloses every point of the arc AB except A and B, contains no point of M except A and B and encloses no point of M that is not on the arc AB. There exists a region R_A containing A and belonging to the collection H_1. There exists, in R_A, a simple closed curve J_A enclosing A but not B, containing neither A nor any point of M that is not on the arc AB, and such that every point of M within J_A belongs to R_A and every point of $AXBYA$ that lies within J_A is a point of R_A. There exists a simple closed curve a, containing A, such that every point of a belongs either to $AXBYA$ or to J_A and every point within a is without $AXBYA$ and within J_A. The curve a contains the point A but no other point of M, encloses no point of AB and lies in R_A. Similarly there exists a simple closed curve b that lies entirely without a, contains B but no other point of M and lies in a region of H_1 that contains B, and encloses no point of M that does not lie in that region. Each point P of M which is the only point of some degenerate component of M can be enclosed by a simple closed curve p lying in a region of H_1 and containing no point of M and such that every point of M that it encloses lies in that region of H_1. The point set composed of all degenerate components of M together with all endpoints of nondegenerate components of M is closed. It follows, by Theorem 134, that there exists a finite set U of simple closed curves $a_1, a_2, a_3, \cdots, a_n, b_1, b_2, b_3, \cdots, b_n, p_1, p_2, p_3, \cdots, p_m$ and a set λ of nondegenerate components of M, $A_1B_1, A_2B_2, \cdots, A_nB_n$ such that every point of M is either within a curve of the set U or on an arc of the set λ and such that, for every i less than $n + 1$ and j less than $m + 1$, (1) each of the curves a_i, b_i, p_j lies in a region of H_1 that contains at least one point of M and that contains every point of M that is enclosed by that curve, (2) a_i contains A_i and b_i contains B_i but neither of them contains any other point of M, (3) a_i lies entirely without b_i, (4) the arc A_iB_i lies entirely without a_i and entirely without b_i except that A_i and B_i are on a_i and b_i respectively, (5) p_j contains no point of M.

For each i less than $n + 1$ there exist regions R_{A_i} and R_{B_i}, belonging to H_1, such that the common parts of M and the interiors of a_i and b_i are subsets of R_{A_i} and R_{B_i} respectively. There exists, in R_{A_i}, an arc $X_iY_iZ_i$ such that (1) $X_iY_iZ_i$ lies wholly within a_i except for its endpoints X_i and Z_i which belong to a_i, (2) the arc $X_iY_iZ_i$ and the arc $X_iA_iZ_i$ of a_i form a simple closed curve a_i' that contains no point of M except A_i and neither encloses nor contains any point of any curve of the set U except a_i. There exists, in R_{A_i}, an arc $X_i'Y_i'Z_i'$ lying entirely within a_i except that its endpoints X_i' and Z_i' are on a_i in the order $X_i'X_iA_iZ_iZ_i'$ and such that (1) $X_i'Y_i'Z_i'$ contains no point of M or of any curve of the set U, (2) every point of M within a_i is either within a_i' or within the simple closed curve a_i'' formed by $X_i'Y_i'Z_i'$ and the arc of a_i which has X_i' and Z_i' as its endpoints and which does not contain A_i. Similarly, for each i there exist, in R_{B_i}, two simple closed curves b_i' and b_i'' such that (1) every point of M that lies within b_i is within either b_i' or b_i'', (2) the curve b_i' contains B_i but no other point of M, (3) the curve b_i'' contains no point of M, (4) no curve, except b_i, of the set U contains any point on or within b_i', (5) the interiors of b_i' and b_i'' are subsets of the interior of b_i. No member of the set U' of curves a_1', a_2', $\cdots$, a_n', b_1', b_2', $\cdots$, b_n' contains or encloses any point of any other curve of the set U' or of the set V'' of curves a_1'', a_2'', $\cdots$, a_n'', b_1'', b_2'', $\cdots$, b_n'', p_1, p_2, $\cdots$, p_m. If a curve of V'' encloses a point of a_i or of b_i it must enclose a_i, b_i and A_iB_i. The set U' has, as a subset, a set V' of simple closed curves a_{m_1}', a_{m_2}', $\cdots$, $a_{m_{n^*}}'$, b_{m_1}', b_{m_2}', $\cdots$, $b_{m_{n^*}}'$ such that every point of M is either within some curve of V' or V'' or on some arc of the set $A_{m_1}B_{m_1}$, $A_{m_2}B_{m_2}$, $\cdots$, $A_{m_{n^*}}B_{m_{n^*}}$ and such that no curve of the set V'' either contains or encloses a point of any curve of the set V'. No curve of the set V'' contains a point of M. It follows with the aid of Theorem 134 that there exists a finite set of simple closed curves p_1^*, p_2^*, $\cdots$, $p_{m^*}^*$ such that (1) every point of M that is within a curve of the set V'' is also within a curve of the set p_1^*, p_2^*, $\cdots$, $p_{m^*}^*$, (2) every curve of the set p_1^*, p_2^*, $\cdots$, $p_{m^*}^*$ is without every other curve of this set and within some curve of the set V'' and therefore without every curve of the set V'. Replace the symbols a_{m_i}' by a_i^*, b_{m_i}' by b_i^*, A_{m_i} by A_i^* and B_{m_i} by B_i^*. We now have a set V of simple closed curves a_1^*, a_2^*, $\cdots$, $a_{n^*}^*$, b_1^*, b_2^*, $\cdots$, $b_{n^*}^*$, p_1^*, p_2^*, $\cdots$, $p_{m^*}^*$ and a set Q of arcs $A_1^*B_1^*$, $A_2^*B_2^*$, $\cdots$, $A_{n^*}^*B_{n^*}^*$ such that (1) every point of M is either within a curve of the set V or on an arc of the set Q, (2) no two arcs of the set Q have a point in common, (3) for every i, less than $n^* + 1$, the arc $A_i^*B_i^*$ lies entirely without every curve of the set V except that A_i^* is on a_i^* and B_i^* is on b_i^*, (4) if g is a curve of the set V there exists a region of H_1 containing all points of M that are within g, (5) every curve of the set V is entirely without every other curve of the set V. There exists a set Q' of arcs $C_1D_2, C_2D_3, C_3D_4, \cdots, C_{n^*-1}D_{n^*}, C_{n^*}E_1, F_1E_2, F_2E_3, F_3E_4, \cdots, F_{m^*-1}E_{m^*}$ such that (1) for every i less than $n^* + 1$, C_i is a point of b_i^* distinct from B_i^*; for every i greater than 1 and less than $n^* + 1$, D_i is a point of a_i^* distinct from A_i^*; for every i less than $m^* + 1$, E_i is a point of p_i^*; for every i less than m^*, F_i is a point of p_i^* distinct from E_i, (2) no arc of the set Q' has a point in common with any other arc of the set Q' or with any arc of the set Q, (3) every arc of Q' lies, except for its endpoints, entirely without every curve of the set V. The arcs of Q and of Q' and the curves of V form an arc-curve chain K_1 enveloping M. For every i less than $n^* + 1$, there exists an arc-curve chain α_i such that (1) α_i envelops the set of all points of M that lie within a_i^*, (2) the last curve of α_i contains the point A_i^*, but lies, except for this point, entirely within a_i^*, (3) every arc of α_i, and every curve of α_i except the last one, lies entirely within a_i, (4) if g is a curve of α_i there exists a region of H_2 containing all points of M that are within g. Also, for every i less than $n^* + 1$, there exists an arc-curve chain β_i such that (1) β_i envelops the set of all points of M that lie within b_i^*, (2) the first curve of β_i contains the point B_i^*, but lies, except for this point, entirely within b_i^*, (3) every arc of β_i, and every curve of β_i except the first one, lies entirely within b_i, (4) if g is a curve of β_i there exists a region of H_2 containing all points of M that are within g. For every i less than $m^* + 1$, there exists, within p_i^*, an arc-curve chain λ_i enveloping all points of M that are within p_i^* and such that if g is a curve of λ_i then there exists a region of H_2 containing all points of M that are within g. For every i, less than $n^* + 1$ but greater than 1, there exists, on the first curve of α_i, a point X_i^* not lying on the first arc of α_i. There exists an arc $X_i^*D_i$ that lies, except for the point D_i, entirely within α_i and has no point except X_i^* in common with any arc or curve of α_i. For every i less than $n^* + 1$ there exists, on the last curve of β_i, a point Y_i^* not lying on the last arc of β_i. There exists an arc $Y_i^*C_i$ that lies, except for the point C_i, entirely within b_i and has no point except Y_i^* in common with any arc or curve of β_i. For every i less than $m^* + 1$ there exists, on the first curve of λ_i, a point W_i that does not lie on the first arc of λ_i. There exists an arc W_iE_i that lies entirely within p_i^* except for the point E_i and has no point, except W_i, in common with any arc or curve of λ_i. For every i less than m^* there exists, on the last curve of λ_i, a point Z_i^* that does not lie on the last arc of λ_i. There exists an arc $Z_i^*F_i$ that lies entirely within p_i^*, except for the point F_i, and has no point in common with the arc W_iE_i or any arc of λ_i and no point, except Z_i^*, in common with any curve of λ_i. We now have a new chain K_2 whose simple closed curves are the curves of the chains $\alpha_1, \alpha_2, \cdots, \alpha_{n^*}, \beta_1, \beta_2, \cdots, \beta_{n^*}, \lambda_1, \lambda_2, \cdots, \lambda_{m^*}$ and whose arcs are the arcs of these chains together with the arcs $X_2^*D_2, X_3^*D_3, \cdots, X_{n^*}^*D_{n^*}, Y_1C_1, Y_2C_2, \cdots$

$Y_{n^*}^*C_{n^*}$, W_1E_1, $W_2E_2, \cdots$, $W_{m^*}E_{m^*}$, $Z_1^*F_1$, $Z_2^*F_2, \cdots$, $Z_{m^*-1}^*F_{m^*-1}$. Every point of M is on an arc or within a curve of K_2. The common part of M and the interior of any one curve of K_1 is a subset of some region of H_2. Every point of a curve of K_2 is within or on a curve of K_1 and every arc of K_2 is either within a curve or on an arc of K_1. There exists a chain K_3 having a relation to K_2 similar to the above described relation of K_2 to K_1 and such that if g is a curve of K_3 then there is a region of H_3 containing all points of M that lie within g. This process may be continued. Thus there exists an infinite sequence of arc-curve chains K_1, K_2, $K_3, \cdots$ such that K_1, K_2 and K_3 are as described and, for each n, (1) each point of M is on an arc or within a curve of K_n, (2) if g is a simple closed curve of K_n there exists a region of H_n containing all points of M that lie within g, (3) each curve of K_{n+1} either lies wholly within some one curve of K_n or has a point of M in common with some curve of K_n and lies, except for that point, wholly within that curve, (4) each arc of K_n whose endpoints belong to M is an arc of K_{n+1}, (5) each arc of K_n whose endpoints do not belong to M is a subset of an arc of K_{n+1}, (6) no arc of K_{n+1} contains more than two points each lying on some curve of K_n, (7) if t is an arc of K_{n+1} which is not identical with any arc of K_n and which does not lie entirely within any curve of K_n then t contains an arc of K_n and each endpoint of t lies within or on a curve of K_n.

For each n, let t_n denote the point set obtained by adding together all the arcs of K_n. The point set $M + t_1 + t_2 + t_3 + \cdots$ is an arc containing M.

EXAMPLE 15. Consider the point set M (Figure 34) composed of the

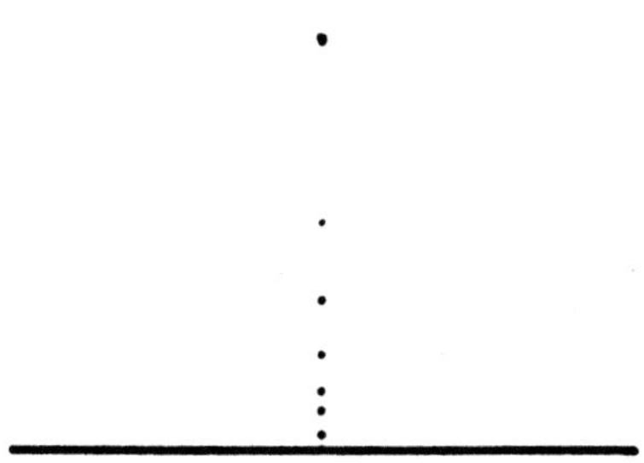

FIGURE 34

straight line interval t from (0,0) to (2,0) together with the infinite set of points (1,1), (1,1/2), (1,1/3), (1,1/4), $\cdots$. Every closed and connected subset of M is either a degenerate point set or an arc. But the point (1,0) of the arc t, though not an endpoint of t, is a limit point of the set $M - t$. It is accordingly impossible to pass an arc through M.

EXAMPLE 16. Consider, on the other hand, the point set M (Figure 35) composed of straight line intervals from $(-1,1)$ to $(1,1)$, from $(-1,-1)$ to

$(1,-1)$ and from $(0,0)$ to $(0,-1/2)$ respectively, together with the eight infinite point sets: $(1/2,1/2)$, $(1/3,1/3)$, $(1/4,1/4),\cdots$; $(1/2,1/2)$, $(2/3,2/3)$, $(3/4,3/4),\cdots$; $(-1/2,1/2)$, $(-1/3,1/3)$, $(-1/4,1/4),\cdots$; $(-1/2,1/2)$, $(-2/3,2/3)$, $(-3/4,3/4),\cdots$; $(-1/2,-1/2)$, $(-1/3,-1/3)$, $(-1/4,-1/4),\cdots$; $(-1/2,-1/2)$, $(-2/3,-2/3)$, $(-3/4,-3/4),\cdots$; $(1/2,-1/2)$, $(1/3,-1/3)$, $(1/4,-1/4),\cdots$; and $(1/2,-1/2)$, $(2/3,-2/3)$, $(3/4,-3/4),\cdots$. This point set satisfies the conditions of Theorem 135. It is accordingly a subset of an arc.

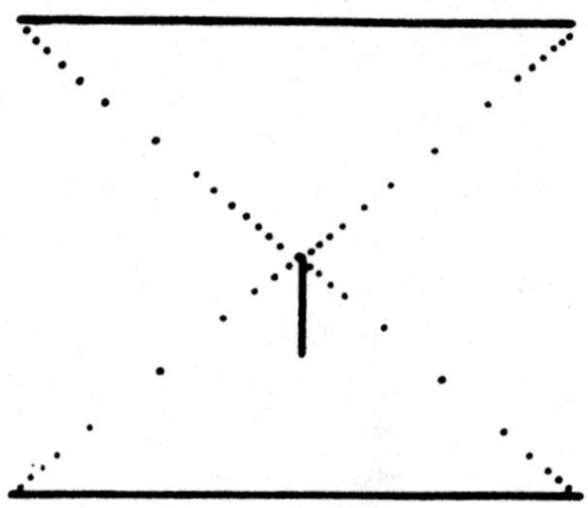

FIGURE 35

CHAPTER V

UPPER SEMI-CONTINUOUS COLLECTIONS

1. On the basis of Axioms 0, 1′ and C

AXIOM 1′. *There exists a sequence* $G_1, G_2, G_3, \cdots$ *such that* (1) *for each* n, G_n *is a collection of regions covering* S, (2) *for each* n, G_{n+1} *is a subcollection of* G_n, (3) *if* R *is a region,* X *is a point of* R *and* Y *is a point of* R *either identical with* X *or not, then there exists a positive integer* m *such that if* g *is any region belonging to the collection* G_m *and containing* X *then* $\bar{g}$ *is a subset of* $(R - Y) + X$, (4) *if* H *and* K *are two mutually exclusive closed point sets and* H *is compact then there exists a positive integer* m *such that no region of the collection* G_m *contains both a point of* H *and a point of* K, (5) *if* $M_1, M_2, M_3, \cdots$ *is a sequence of closed point sets such that, for each* n, M_n *contains* M_{n+1} *and, for each* n, *there exists a region* g_n *of the collection* G_n *such that* M_n *is a subset of* $\bar{g}_n$, *then there is at least one point common to all the point sets of the sequence* $M_1, M_2, M_3, \cdots$.

AXIOM C. *If* P *is a point, every region containing* P *contains a compact region containing* P.

DEFINITION. A collection G of mutually exclusive closed point sets is said to be *upper semi-continuous* if and only if it is true that if g is a point set of the collection G and $g_1, g_2, g_3, \cdots$ is a sequence of point sets of this collection and, for each n, A_n and B_n are points of g_n and the sequence $A_1, A_2, A_3, \cdots$ has a sequential limit point lying in g, then every infinite subsequence of $B_1, B_2, B_3, \cdots$ has a subsequence having a sequential limit point that lies in g.

THEOREM 1. *Every subcollection of an upper semi-continuous collection of mutually exclusive closed point sets is itself upper semi-continuous.*

DEFINITION. The collection G of point sets is said to *fill up* the point set M if and only if M is the sum of all the point sets of the collection G.

DEFINITION. The subset K of the point set M is said to be *closed relatively to* M if and only if M contains no limit point of K that does not belong to K.

THEOREM 2. *If* G *is an upper semi-continuous collection of mutually exclusive closed point sets,* K *is a subset of* G^* *which is closed relatively to* G^* *and* H *is the collection of all point sets of the collection* G *that intersect* K, *then* H^* *is closed relatively to* G^* *and if the elements of* H *are compact and* K *is closed and compact then* H^* *is closed and compact.*

PROOF. Suppose that P is a limit point of H^*. Then P is the sequential limit point of some sequence α of distinct points $P_1, P_2, P_3, \cdots$ all belonging to H^*. For each n, let h_n denote the point set of the collection H that contains P_n. If any one element of H contains infinitely many points of α then that element of H contains P and therefore P belongs to H^*. Suppose no element of H contains infinitely many points of α. Then there exists an infinite sequence of distinct numbers $n_1, n_2, n_3, \cdots$ such that the point sets $h_{n_1}, h_{n_2}, h_{n_3}, \cdots$ are all distinct. For each j, the point set h_{n_j} contains a point O_{n_j} belonging to K.

Suppose first that P is a point of G^*. Let g denote the element of G that contains P. The sequence $O_{n_1}, O_{n_2}, O_{n_3}, \cdots$ has a subsequence converging to a point O belonging to g. Since K is closed relatively to G^*, O belongs to K and therefore g is an element of H. Hence P belongs to H^*. Thus H^* contains every limit point of H^* which belongs to G^*. In other words, H^* is closed relatively to G^*.

Suppose now that K is closed and compact. Then the sequence O_{n_1}, $O_{n_2}, O_{n_3}, \cdots$ has a subsequence $O_{m_1}, O_{m_2}, O_{m_3}, \cdots$ converging to some point O of K. Let g denote the element of G that contains O. The sequence $P_{m_1}, P_{m_2}, P_{m_3}, \cdots$ has a subsequence converging to a point of g. But every subsequence of $P_1, P_2, P_3, \cdots$ converges to P. Hence P belongs to g and therefore to G^* and therefore, as has been shown, to H^*. Hence if K is closed and compact then H^* is closed. It is easy to see that, in this case, it is also compact.

Definition. If G is an upper semi-continuous collection of mutually exclusive closed point sets then by a *region with respect to G* is meant a subcollection R of G such that R^* contains no limit point of $G^* - R^*$.

Suppose G is a given upper semi-continuous collection of mutually exclusive closed point sets.

In Chapter I, and subsequent chapters, certain words and phrases were defined in terms of point and region. In the present chapter the same words and phrases, suitably qualified either explicitly or contextually, will be employed with the understanding that they may be defined in the same way as that in which the corresponding words and phrases were defined in the earlier chapters except that *point* (of S) is to be replaced by *element of G* and *region* is to be replaced by *region with respect to G*. A few definitions will be explicitly given, largely to exemplify the meaning of this statement.

Definition. The element g of G will be said to be a *limit element* of the subcollection H of G if and only if it is true that every region with respect to G that contains g (that is to say, has g as an element) contains at least one element of H distinct from g.

Definition. The subcollection H of G is said to be *closed* if and only if it is true that there is no limit element of H that does not belong to H.

Definition. The subcollections H and K are said to be *mutually separated* if and only if they have no element in common and neither of them contains a limit element of the other one.

DEFINITION. The subcollection H of G is said to be *connected* if and only if it is not the sum of two mutually separated collections.

DEFINITION. If H is a subcollection of G, the subcollection D of H is said to be a *domain* (of elements of G) *with respect to* H if and only if either D is H or no element of D is a limit element of $H - D$.

THEOREM 3. *In order that the element g of G be a limit element of the subcollection H of G it is necessary and sufficient that the point set g contain a limit point of the point set $(H^* + g) - g$.*

PROOF. This condition is sufficient. For suppose g contains a point P which is a limit point of $H^* - g$. Let R denote a region with respect to G containing g. By definition, R^* contains no limit point of $G^* - R^*$. But P belongs to R^*. Hence $H^* - g$ is not a subset of $G^* - R^*$. Therefore R^* contains a point X belonging to $H^* - g$. The element of G that contains X belongs both to H and to R and it is distinct from g. Thus R contains an element of H distinct from g and, by definition, g is a limit element of H.

This condition is also necessary. For suppose that no point of g is a limit point of $H^* - g$. Then each point of g belongs to a region containing no point of $H^* - g$ and the sum of all such regions for all points of g is a domain D containing g but no point of $H^* - g$. The point set $G^* - D \cdot G^*$ is closed relatively to G^* and therefore, by Theorem 2, the sum of all the elements of G that intersect $G^* - D \cdot G^*$ is closed relatively to G^*. Hence, by definition, the set of all elements of G that are subsets of D is a region with respect to G. This region with respect to G contains g but no element of $H - g$. Hence g is not a limit element of H.

THEOREM 4. *If H is a subcollection of G and H^* is closed then H is closed and if H and G^* are closed then H^* is closed.*

PROOF. This condition is sufficient. For suppose H^* is closed. Suppose g is a limit element of H. By Theorem 3, g contains a point P which is a limit point of H^*. Since H^* is closed, P belongs to H^*. Hence g belongs to H.

It is also necessary. For suppose H is closed. Suppose P is a limit point of H^*. Let g_P denote the element of G that contains P. By Theorem 3, g_P is a limit element of H. But H is closed. Hence g_P belongs to H. Therefore P belongs to H^*.

THEOREM 5. *If H is a subcollection of G then in order that H be connected it is sufficient that the point set H^* be connected and, in case the elements of H are all continua, this condition is also necessary.*

PROOF. Suppose H is the sum of two mutually separated collections H_1 and H_2. By Theorem 3, H_1^* and H_2^* are mutually separated and H^* is not connected. Therefore the condition of Theorem 5 is sufficient.

Suppose the elements of H are all continua and H^* is the sum of two mutually separated point sets K_1 and K_2. Then, by Theorem 28 of Chapter I, each point set of the collection H is a subset either of K_1 or of K_2. Let H_1 denote the collection of all of them that are subsets of K_1 and let H_2 denote the collection of all those that are subsets of K_2. Since H_1^* and H_2^* are identical with K_1 and K_2 respectively therefore they are mutually separated point sets. Hence, by Theorem 3, H_1 and H_2 are mutually separated collections. Therefore, in case the elements of H are continua the condition of Theorem 5 is necessary as well as sufficient.

DEFINITION. The subcollection H of G is said to be *compact* if and only if there is no infinite subcollection of H that has no limit element.

THEOREM 6. *If H is a subcollection of G then* (1) *if G^* is closed and H^* is compact, H is compact,* (2) *if all the point sets of the collection H are compact and H is compact then H^* is compact.*

PROOF. Suppose G^* is closed and H^* is compact and K is an infinite subcollection of H. Let $g_1, g_2, g_3, \cdots$ denote an infinite sequence of distinct point sets of the collection K. For each n, let P_n denote a definite point of g_n. Since H^* is compact there exists an infinite sequence of distinct numbers $n_1, n_2, n_3, \cdots$ such that the sequence $P_{n_1}, P_{n_2}, P_{n_3}, \cdots$ converges to a point P. Since G^* is closed, P belongs to G^* and therefore there exists a point set g containing P and belonging to the collection G. By Theorem 3, g is a limit element of the collection K. Thus every infinite subcollection of H has a limit element. Therefore, by definition, H is compact.

Suppose all the point sets of the collection H are compact and H is compact and N is an infinite subset of H^*. If there exists a point set g belonging to the collection G and containing infinitely many points of N, let L denote the set of all points of N that belong to g. Since g is compact, L has a limit point. Hence, in this case, N has a limit point. Suppose no point set of the collection G contains infinitely many points of N. Then there exists a subset L of N such that no two points of L belong to the same point set of the collection G. Let W denote the set of all elements of G that intersect L. By hypothesis, W has a limit element g. By Theorem 3, g contains a point P which is a limit point of $W^* - g$. By Theorem 10 of Chapter I, P is the sequential limit point of a sequence α of distinct points of $W^* - g$. There exists a subsequence $P_1, P_2, P_3, \cdots$ of α such that if i is distinct from j then P_i and P_j belong to different point sets of the collection W. For each n, let g_n denote the point set of the collection W that contains P_n and let X_n denote the point of L belonging to g_n. Since G is upper semi-continuous there exists a subsequence of $X_1, X_2, X_3, \cdots$ converging to a point of g. Since the points $X_1, X_2, X_3, \cdots$ all belong to N, therefore N has a limit point. Thus every infinite subset of H^* has a limit point. Therefore H^* is compact.

THEOREM 7. *If H is a closed and compact point set and K is a subset of $S - H$ such that, for every n, there is a region belonging to G_n and containing both a point of H and a point of K then H contains a limit point of K.*

PROOF. If H contained no limit point of K, the closed point set $\bar{K}$ would have no point in common with the closed and compact point set H and therefore, by Axiom 1′, there would exist a number n such that no region of G_n contains both a point of H and a point of $\bar{K}$. But this would be contrary to hypothesis.

THEOREM 8. *If H and K are mutually exclusive closed point sets and H is compact then there exists a compact domain D containing H such that $\bar{D}$ contains no point of K.*

PROOF. By Axioms 1 and C, for each point P of H there exists a compact region R_P, containing P, such that $\bar{R}_P$ contains no point of K. Let Q denote the set of all such regions R_P for all points P of H. Since H is compact therefore, by Theorems 37 and 4 of Chapter I, it is covered by some finite subcollection U of the collection Q. Let D denote the sum of all the regions of the collection U. Since every region of U is compact and U is a finite collection, therefore D is compact and $\bar{D}$ contains no point of K.

THEOREM 9. *If D is a domain containing the compact element g of the collection G there exists a domain W, containing g, such that every point set of the collection G that contains a point of W is a subset of D.*

PROOF. For each n, let W_n denote the point set obtained by adding together all regions of the set G_n that contain one or more points of g. Suppose that, for each n, there exists a point set g_n belonging to G and containing both a point A_n of W_n and a point P_n that does not belong to D. It follows, with the help of Theorem 7, that there exists an infinite sequence of numbers $n_1, n_2, n_3, \cdots$ such that the sequence $A_{n_1}, A_{n_2}, A_{n_3}, \cdots$ converges to a point of g. Since G is upper semi-continuous, some infinite subsequence of the sequence $P_{n_1}, P_{n_2}, P_{n_3}, \cdots$ converges to a point of g. But g is a subset of D and no point of the sequence $P_{n_1}, P_{n_2}, P_{n_3}, \cdots$ belongs to D. This involves a contradiction. Hence there exists a number m such that every point set of the collection G that intersects W_m is a subset of D.

THEOREM 10. *If the elements of the collection G are all compact and D is a domain containing at least one point set of the collection G, and R is the collection of all elements of G that lie wholly in D, then R is a region with respect to G.*

PROOF. Suppose P is a point of R^*. Let g denote the point set of the collection G that contains P. By Theorem 9, there exists a domain W, containing g, such that every point set of the collection G that intersects W is a subset of D and therefore belongs to R. Thus $W \cdot G^*$ is a subset of R^*. But, since W is a domain, no point of $W \cdot G^*$ is a limit point of $G^* - W \cdot G^*$. Hence P is not a limit point of $G^* - W \cdot G^*$. Hence it is not a limit point of $G^* - R^*$. Therefore, by definition, R is a region with respect to G.

THEOREM 11. *If the elements of the collection G are all compact and D is a domain containing the element g of G, then there exists a region R with respect to G, containing g, such that every point set of the collection $\bar{R}$ is a subset of D.*

PROOF. By Theorem 9, there exists a domain W containing g such that every point set of the collection G that contains a point of W is a subset of D. By Theorem 16 of Chapter I, there exists a domain V containing g such that $\overline{V}$ is a subset of W. Let R denote the collection of all elements of G that lie wholly in V. By Theorem 10, R is a region with respect to G. Suppose q is a limit element of R. Since R^* is a subset of V, q contains a point belonging to $\overline{V}$ and therefore to W. Hence q is a subset of D.

DEFINITION. The element g of G is said to be a *sequential limit element* of the sequence $g_1, g_2, g_3, \cdots$ of elements of G if, and only if, for every region R with respect to G that contains g, there exists a number δ such that if $n > \delta$ then g_n is an element of R.

THEOREM 12. *In order that the element g of G be a sequential limit element of the sequence $g_1, g_2, g_3, \cdots$ of elements of G it is necessary and sufficient that, for every subsequence $g'_1, g'_2, g'_3, \cdots$ of this sequence, there exist an increasing sequence of numbers $n_1, n_2, n_3, \cdots$ and a sequence of points $P_1, P_2, P_3, \cdots$, converging to some point of g, such that, for each i, P_i belongs to g'_{n_i}.*

PROOF. This condition is necessary. For suppose g is a sequential limit element of the sequence $g_1, g_2, g_3, \cdots$. Suppose $g'_1, g'_2, g'_3, \cdots$ is an infinite subsequence of this sequence. Then clearly g is a sequential limit element of the sequence $g'_1, g'_2, g'_3, \cdots$. Either there exist an increasing sequence of positive integers $n_1, n_2, n_3, \cdots$ and an element g' of G such that, for each i, g'_{n_i} is identical with g', or there exists an increasing sequence of numbers $m_1, m_2, m_3, \cdots$ such that the point sets $g'_{m_1}, g'_{m_2}, g'_{m_3}, \cdots$ are distinct from each other and from g. In the first case g' is identical with g. For if it were not, $G - g'$ would be a region with respect to G containing g but no element of the sequence $g'_{n_1}, g'_{n_2}, g'_{n_3}, \cdots$, contrary to hypothesis. Let P denote some definite point of g and, for each i, let P_i denote P. Then P is the sequential limit point of the sequence $P_1, P_2, P_3, \cdots$.

In the second case, by Theorem 3, g contains a point P which is a limit point of the point set $g'_{m_1} + g'_{m_2} + g'_{m_3} + \cdots$. By Theorem 10 of Chapter I, P is the sequential limit point of a sequence of distinct points $A_1, A_2, A_3, \cdots$ all distinct from P and all belonging to the set $g'_{m_1} + g'_{m_2} + g'_{m_3} + \cdots$. Since P is not a limit point of any one point set of the sequence $g'_{m_1}, g'_{m_2}, g'_{m_3}, \cdots$, it follows that there exists a subsequence $n_1, n_2, n_3, \cdots$ of the sequence $m_1, m_2, m_3, \cdots$ such that, for each i, A_{n_i} belongs to g'_{n_i}. For each i, let P_i denote A_{n_i}.

This condition is also sufficient. For suppose g is not a sequential limit element of the sequence $g_1, g_2, g_3, \cdots$. Then there exist an infinite increasing sequence of positive integers $m_1, m_2, m_3, \cdots$ and a region R with respect to G, containing g, such that no point set of the sequence $g_{m_1}, g_{m_2}, g_{m_3}, \cdots$ lies wholly in R. For each i, let B_i denote a point of g_{m_i} not belonging to R^*. The point set R^* contains no limit point of the point set $B_1 + B_2 + B_3 + \cdots$

and no point of this set belongs to g. Hence no subsequence of the sequence $B_1, B_2, B_3, \cdots$ converges to a point of g. Hence, since G is upper semi-continuous, there do not exist a subsequence $n_1, n_2, n_3, \cdots$ of the sequence $m_1, m_2, m_3, \cdots$ and a sequence $P_1, P_2, P_3, \cdots$ of points, converging to some point of g, such that, for each i, P_i belongs to g_{n_i}.

THEOREM 13. *If M is a connected point set and G is an upper semi-continuous collection of mutually exclusive closed point sets such that each point of M belongs to some point set of the collection G and each point set of the collection G contains at least one point of M then G is connected.*

PROOF. Suppose, on the contrary, that G is the sum of two mutually separated collections G_1 and G_2. Let M_1 denote the set of all points of M that belong to some element of G_1 and let M_2 denote the set of all those that belong to some element of G_2. One of the sets M_1 and M_2 contains a limit point of the other one. Suppose that M_1 contains a point O which is a limit point of M_2. Let g_O denote the point set of G that contains O. By Theorem 3, g_O is a limit element of G_2. But g_O belongs to G_1. This involves a contradiction.

THEOREM 14. *If M is a closed point set and G is an upper semi-continuous collection of mutually exclusive closed point sets such that every point set of the collection G contains a point of M and every point of M belongs to some point set of the collection G and K is a subset of M and W is the collection of all point sets of G that intersect K and no point set of W intersects $M - K$ then W is a domain with respect to G if and only if K is a domain with respect to M.*

PROOF. Suppose first that K is a domain with respect to M. Let g denote an element of W. Suppose g is a limit element of the set H of elements of G. By Theorem 3, g contains a limit point X of the point set $(H + g)^* - g$. Hence there exist a sequence $h_1', h_2', h_3', \cdots$ of distinct elements of H and a sequence of points $X_1, X_2, X_3, \cdots$, converging to X, such that, for each n, X_n belongs to h_n'. For each n, h_n' contains a point Y_n of M. The sequence $Y_1, Y_2, Y_3, \cdots$ has a subsequence $P_1, P_2, P_3, \cdots$ converging to some point P of g. For each n let h_n denote the point set of the sequence $h_1', h_2', h_3', \cdots$ to which P_n belongs. By Theorem 12, the sequence $h_1, h_2, h_3, \cdots$ of elements of G has h as a sequential limit element. Since M is closed, P belongs to M as well as to g. Hence, since g belongs to W, P belongs to K. But K is a domain with respect to M. Hence, for some n, P_n belongs to K and therefore h_n belongs to W. Thus W contains an element of every subset of G of which some element of W is a limit element. Therefore W is a domain with respect to G.

Suppose secondly that W is a domain with respect to G. Let P denote some point of K and let g denote the point set of G that contains it. Suppose P is a limit point of a subset N of M. Then it is the sequential limit point of a sequence of points $P_1, P_2, P_3, \cdots$ all belonging to N. For each n, P_n

belongs to a point set g_n of the set G. The element g of M is a sequential limit element of the sequence $g_1, g_2, g_3, \cdots$. Since W is a domain with respect to G it contains an element g_m of this sequence. The point P_m belongs to K. Thus K contains a point of every subset of M of which some point of K is a limit point. Therefore K is a domain with respect to M.

THEOREM 15. *Suppose G is an upper semi-continuous collection of mutually exclusive closed and compact point sets filling up S. Then if the elements of G are called "points" and every region with respect to G is called a "region", Axioms* 0, 1$'$ *and* C *hold true for the "points" and "regions" so obtained.*

PROOF. It is clear that Axiom 0 holds true for this interpretation of "point" and "region". Axiom C holds true for this interpretation. For suppose g is an element of G and R is a region with respect to G containing g. By Theorem 8, there exists a compact domain D containing g. By Theorem 11, there exists a region L with respect to G such that every point set of the collection L is a subset of D. But D is compact. Hence L is compact.

It will be shown that Axiom 1$'$ holds true for this interpretation. If n is a positive integer, g is an element of G and F is a finite subcollection of G_n that properly covers g, let R_{gnF} denote the "region" whose elements are the point sets of the collection G which are subsets of a point set obtained by adding together all the point sets of the collection F. For each n, let G_n' denote the collection of all such "regions" R_{gnF}.

It is clear that the sequence $G_1', G_2', G_3', \cdots$ satisfies, with respect to "point" and "region", all the conditions required of $G_1, G_2, G_3, \cdots$ under (1) and (2) in the statement of Axiom 1$'$. It will be shown that it also satisfies those required under (3). Suppose R is a region with respect to G, x is an element of R and y is an element of R either identical with x or not. By Theorem 8, there exists a compact domain I, containing the point set x, such that the point set $\bar{I}$ is a subset of $(R^* - y) + x$. There exists a domain D, containing x, such that $\bar{D}$ is a subset of I. There exists a region K with respect to G whose elements are the point sets of the collection G which are subsets of the point set $(R^* - y) + x$. Suppose that, for each n, there exists a region R_n with respect to G belonging to G_n', containing x and such that $\bar{R}_n$ is not a subset of K. For each n, there exist a finite subcollection H_n of G_n and a point set t_n belonging to G such that t_n is properly covered by H_n, x is covered by H_n and R_n is the collection of all point sets of the collection G which are subsets of H_n^*. There exists a number m such that no region of the set G_m intersects both $\bar{D}$ and $S - I$. For each n greater than m, some region of H_n contains a point of $S - I$ and therefore lies wholly in $S - \bar{D}$ and therefore t_n contains a point P_n of $S - \bar{D}$. But, as may be seen with the help of Theorem 7, there exist an ascending sequence of numbers $n_1, n_2, n_3, \cdots$ and a sequence of points $X_{n_1}, X_{n_2}, X_{n_3}, \cdots$, converging to some point of x, such that, for each i, X_{n_i} belongs to t_{n_i}. Since G is upper semi-continuous, some subsequence of $P_{n_1}, P_{n_2}, P_{n_3}, \cdots$ converges to a point of x.

But this is impossible since x is a subset of D and, for every i, P_{n_i} belongs to $S - D$. Therefore there exists a number m such that if W is a region with respect to G belonging to G'_m and containing x then $\overline{W}$ is a subcollection of K. Thus the sequence $G'_1, G'_2, G'_3, \cdots$ satisfies, with respect to "point" and "region", the conditions required of $G_1, G_2, G_3, \cdots$ under (3).

It will be shown that condition (4) is also satisfied. Suppose H is a closed and compact subcollection of G and K is a closed subcollection of G having no element ("point") in common with H. By Theorem 4 and 6, H^* and K^* are mutually exclusive closed point sets and H^* is compact. Suppose that for each n there exists a "region" R_n belonging to the collection G'_n and containing both a "point" (element) h_n of H and a "point" (element) k_n of K. Then, for each n, there exists an element t_n of the collection G such that if x is any element of R_n and P is a point of x, there exists a region belonging to the collection G_n and containing both P and some point of t_n. There exist regions q_n and r_n, belonging to G_n, such that q_n contains both a point of t_n and a point belonging to h_n and therefore to H^* and r_n contains both a point of t_n and a point belonging to k_n and therefore to K^*. With the aid of Axiom C and Theorem 4 of Chapter I, it may be seen that there exists a domain D containing H^* such that $\overline{D}$ is a compact subset of $S - K^*$. There exists a number m such that, for each n greater than m, r_n contains no point of $\overline{D}$ and therefore t_n contains a point of the closed and compact point set $\overline{D} - D$. It follows, with the help of Theorem 7, that there exist an ascending sequence of positive integers $n_1, n_2, n_3, \cdots$ and two sequences of points $A_{n_1}, A_{n_2}, A_{n_3}, \cdots$ and $B_{n_1}, B_{n_2}, B_{n_3}, \cdots$, converging to points A and B belonging to H^* and $\overline{D} - D$ respectively, such that, for each i, A_{n_i} and B_{n_i} both belong to t_{n_i}. Let g_A and g_B denote the elements of G that contain A and B respectively. Since the collection G is upper semi-continuous, g_A is identical with g_B. But g_A belongs to H and B belongs to $S - H^*$. This involves a contradiction. Therefore there exists a number m such that no "region" of G'_m contains both a "point" of H and a "point" of K.

Part (5) is also satisfied. For suppose that $M_1, M_2, M_3, \cdots$ is a sequence of closed "point" sets such that, for each n, M_n is a subset of some "region" of G'_n and M_{n+1} is a subset of M_n. By Theorem 115 of Chapter I, the closed point sets $M_1^*, M_2^*, M_3^*, \cdots$ have a point O in common. The point set of the collection G that contains O is an element of each of the sets $M_1, M_2, M_3, \cdots$. Thus condition (5) is satisfied by the sequence $G'_1, G'_2, G'_3, \cdots$.

Theorem 16. *Suppose G is an upper semi-continuous collection of mutually exclusive compact continua filling up S. Then if it is true that every region that contains a point contains a connected domain that contains that point, it is also true that every region with respect to G that contains an element of G contains a connected region with respect to G containing that element.*

Proof. Suppose g is an element of a region R with respect to G. Let W denote the collection of all elements x of R such that there exists a connected subcollection of R containing both x and g. The collection W is a region with respect to G. For suppose y is an element of W. By Theorem 9, there exists a domain D containing y such that every point set of the collection G that contains a point of D is a subset of the domain R^* and therefore is an element of R. By hypothesis, for each point P of y there is a connected domain R_P containing P and lying wholly in the domain D. Let V denote the sum of all such domains R_P for all points P of y. The point set V is a connected domain. Let L denote the collection of all point sets of the collection G that intersect V. By Theorems 1 and 14, the collection L is connected. Suppose now that y is a limit element of a subcollection H of G. By Theorem 3, the domain V contains a point of some element of H. Hence H contains an element of L. But since L is a connected subcollection of R containing y and there is a connected subcollection of R containing y and x, therefore L is a subcollection of W. Thus if the element y of W is a limit element of the subcollection H of G then H contains an element of W. Therefore, by definition, W is a region with respect to G. But it is a subcollection of R and it contains g.

THEOREM 17. *Suppose that, in addition to Axioms* 0, 1′ *and* C, *Axioms* 2, 3, 4 *and* 5 *hold true. Then if the compact continuum M is a subset of the domain D and $S - M$ is connected, there exists a simple domain I, containing M, such that $\bar{I}$ is a subset of D.*

PROOF. Suppose P is a point of the domain D. By Axiom C, D contains a compact region R_1 containing P. There exists a region R_2, containing P, such that $\bar{R}_2$ is a subset of R_1. Since the boundary of $\bar{R}_2$ is compact therefore, by Theorem 17 of Chapter IV, there exists a simple domain L, containing P, such that $\bar{L}$ contains no point of the boundary of R_2. Since $\bar{L}$ is a connected point set containing the point P of R_2 but no point of the boundary of R_2, therefore $\bar{L}$ is a subset of R_2 and therefore of D. Therefore for each point of M there exists a simple domain containing that point and lying, with its boundary, in D. It follows, by Theorems 4 and 37 of Chapter I, that M is covered by a finite collection H of simple domains such that if h is a domain of H then $\bar{h}$ is a subset of D. Let N denote the point set H^*. Since the boundary of each domain of H is compact and H is a finite collection, therefore β, the boundary of N, is compact. Therefore, by Theorem 18 of Chapter IV, there exists a simple domain I, containing M, such that $\bar{I}$ contains no point of β. Since $\bar{I}$ is a connected point set containing a point of N but no point of the boundary of N, therefore $\bar{I}$ is a subset of N and therefore of D.

THEOREM 18. *Suppose that, in addition to Axioms* 0, 1′ *and* C, *Axioms* 2, 3, 4 *and* 5 *hold true. Suppose G is an upper semi-continuous collection of mutually exclusive compact continua filling up S such that no continuum of G separates S. Then the following propositions hold true.*

I. *No arc of elements of G separates G.*

II. *If J is a simple closed curve of elements of G then $G - J$ is the sum of two mutually exclusive connected regions I and E with respect to G such that each element of J is a limit element both of I and of E.*

III. *If g is an element of a region D with respect to G, there exists a simple domain I of elements of G such that* (1) *g is an element of I and* (2) *$\bar{I}$ is a subset of D.*

PROOF. Suppose there exist two elements C and F of G and an arc AB of elements of G that separates C and F from each other in G. Let T_1 denote the subcollection of G consisting of the element A together with all elements X of AB distinct from A, if there are any, such that the interval AX of AB does not separate C from F. Let T_2 denote the collection $AB - T_1$. It is easy to see that every element of T_1 precedes every element of T_2 on AB. It follows that there exists an element O which is either the last element of T_1 or the first element of T_2 in the order from A to B on AB. Suppose O is the first element of T_2. Then there exists an infinite sequence $O_1, O_2, O_3, \cdots$ of elements of AB converging to O and lying on AB in the order $AO_1O_2O_3\cdots$. For each n, the intervals AO_n and O_nO of AB have only the element O_n in common and their sum separates C from F in G. Hence the common part of the compact continua $(AO_n)^*$ and $(O_nO)^*$ is the continuum O_n and $(AO_n)^* + (O_nO)^*$ separates the continuum C from the continuum F in S. But, since O_n belongs to T_1, AO_n does not separate C from F in G and therefore $(AO_n)^*$ does not separate C from F in S. Hence, by Theorem 20 of Chapter IV, $(O_nO)^*$ separates C from F in S. But O is the common part of all the continua $(O_1O)^*, (O_2O)^*, (O_3O)^*, \cdots$. Hence, by Theorem 7 of Chapter II, O separates C from F in S.

Suppose, on the other hand, that O is the last point of T_1. There exists a sequence $O_1, O_2, O_3, \cdots$ of elements of G converging to O and lying between O and B on AB in the order $BO_1O_2O_3\cdots$. In this case, the intervals AO and OO_n have only the element O in common and $(AO)^* + (OO_n)^*$ separates C from F but $(AO)^*$ does not separate C from F. Therefore each of the point sets $(OO_1)^*, (OO_2)^*, (OO_3)^*, \cdots$ separates C from F and therefore O, their common part, separates C from F.

Thus, in either case, O separates C from F. But this is contrary to hypothesis. Therefore G is not separated by any arc of elements of G.

II. Suppose J is a simple closed curve of elements of G. By Theorem 107 of Chapter I and Theorem 15 of the present chapter, the collection J is the sum of two arcs AXB and AYB of elements of G having in common only their end elements A and B. The common part of the compact continua $(AXB)^*$ and $(AYB)^*$ is the sum of the two mutually exclusive continua A and B. Furthermore, by I, neither AXB nor AYB separates G and therefore neither $(AXB)^*$ nor $(AYB)^*$ separates S. Hence, by Theorem 131 of Chapter IV, $S - J^*$ is the sum of two mutually exclusive connected domains U and V. Let β denote the boundary of U. By Theorem 24 of Chapter IV, β is a continuum. Let γ denote the collection of all elements of G that intersect β. By Theorems 1, 2 and 14, γ is a continuum of elements of G. But it is a subcollection of J. But every continuum of elements of G which is a *proper* subcollection of J is an arc of elements of G and γ separates C from F. Hence, by I, γ is identical with J. Thus every element of J contains a limit point of U and therefore is a limit element of I, the collection of all elements of G that lie wholly in U. Similarly every element of J is a limit element of E, the collection of all elements of G that lie in V.

III. Suppose g is an element of a region D with respect to G. The continuum g is a subset of the domain D^*. By Theorem 17, there exists a simple domain U, containing g, such that $\bar{U}$ is a subset of D^*. Let J denote the boundary of U. Let T denote the collection of all continua of G that intersect J. Let V denote the complementary domain of T^* that contains g. Let β denote the boundary of V. Since T^* is a compact continuum therefore, by Theorem 24 of Chapter IV, β is a continuum. Let K denote the collection of all continua of G that intersect β. Let W denote the collection of all continua of G that are subsets of V. The collection K is a compact continuum of elements of G, W is a connected domain of elements of G and K is the boundary of W. If B is an element of K and AB is an arc of elements of G lying, except for B, wholly in W then, by Theorem 20 of Chapter IV and I of the present theorem, $W - (AB - B)$ is connected. Therefore, by II of the present theorem and Theorem 34 of Chapter III, there exists a simple closed curve α of elements of G lying in W and separating g from K. Let I denote the component of $H - K$ that contains g. Since the connected domain D of elements of G contains the element g of W and contains no element of K, the boundary of W, therefore I is a subset of W. But W is a subset of D. It follows that $\bar{I}$ is a subset of D.

THEOREM 19. *Suppose that, in addition to Axioms* 0, 1′ *and* C, *Axioms* 2, 3, 4 *and* 5 *are satisfied. Suppose G is an upper semi-continuous collection of mutually exclusive compact continua filling up S such that no continuum of G separates S. Then if the elements of G are called "points" and every region with respect to G is called a "region", Axioms* 0, 1′, C, 2, 3, 4 *and* 5 *all hold true for the "points" and "regions" so obtained.*

Theorem 19 is a consequence of Theorems 15–18.

THEOREM 20. *If M is a closed point set whose components are all compact and G is the collection of all the components of M then G is upper semi-continuous.*

PROOF. Suppose, on the contrary, that there exist a point set g_0 of G, a sequence of distinct point sets $g_1, g_2, g_3, \cdots$ and two sequences of points $A_1, A_2, A_3, \cdots$ and $B_1, B_2, B_3, \cdots$ such that (1) for each n, A_n and B_n belong to g_n and g_n belongs to the collection G and is distinct from g_0 and (2) the sequence $A_1, A_2, A_3, \cdots$ converges to a point A belonging to g_0 while $B_1, B_2, B_3, \cdots$ either has no convergent subsequence or converges to a point not belonging to g_0. Let β denote the sequence $B_1, B_2, B_3, \cdots$ and let K_1 denote the point set whose elements are the points of the sequence β. The point sets g_0 and $\bar{K}_1$ are closed and mutually exclusive and g_0 is compact. Hence, by Theorem 8, there exists a domain D, containing g_0, such that $\bar{D}$ is a compact point set containing no point of $\bar{K}_1$. There exists a number m such that every point of the sequence $A_{m+1}, A_{m+2}, A_{m+3}, \cdots$ belongs to D. For every n greater than m, the compact continuum g_n contains the point A_n of D and the point B_n without D. Hence, by Theorems 50 and 29 of Chapter I, for every n greater than m, $\bar{D} \cdot g_n$ contains a continuum h_n containing A_n and a point C_n on the boundary of D. By Theorem 58 of Chapter I, the limiting set of the sequence $h_{m+1}, h_{m+2}, h_{m+3}, \cdots$ is a continuum t. This continuum contains A and a point of the boundary of D to which some subsequence of $C_1, C_2, C_3, \cdots$ converges. Hence $g_0 + t$ is a connected subset of M containing a point of the boundary of D. But this is contrary to the fact that g_0 is a maximal connected subset of M lying wholly in D.

Definition. The collection G of point sets is said to be *topologically contracting* if and only if it is true that if $g_1, g_2, g_3, \cdots$ is a sequence of point sets of the collection G and $A_1, A_2, A_3, \cdots$ and $B_1, B_2, B_3, \cdots$ are two sequences of points such that, for each n, A_n and B_n belong to g_n, then either neither of the sequences $A_1, A_2, A_3, \cdots$ and $B_1, B_2, B_3, \cdots$ converges or they converge to the same point.

Theorem 21. *Every topologically contracting collection of mutually exclusive closed point sets is upper semi-continuous.*

Theorem 22. *If G is a countable collection of mutually exclusive continua such that G^* is closed and compact then G is upper semi-continuous and its elements are the components of G^*.*

Proof. Suppose $g_1, g_2, g_3, \cdots$ is a sequence of elements of G, for each n, A_n and B_n are points of g_n and $A_1, A_2, A_3, \cdots$ converges to a point A of the element g of G but $B_1, B_2, B_3, \cdots$ converges to a point B not belonging to g. Then some subsequence of $g_1, g_2, g_3, \cdots$ converges to a continuum M which contains A and B and therefore is not a subset of any element of G. It follows that M is the sum of countably many mutually exclusive closed point sets, each of them being the common part of M and some element of G. This is contrary to Theorem 56 of Chapter I. If the component K of G^* intersects more than one element of G it is the sum of all those that it intersects. This is contrary to the same theorem. Hence K is an element of G.

Notation. For each point P of the compact continuum M, let $M(P)$ denote the set of all points X of M such that there do not exist uncountably many points each separating P from X in M.

Theorem 23. *If P is a point of the compact continuum M then $M(P)$ is a continuum and if X is a point of $M(P)$ distinct from P then every irreducible subcontinuum of M from P to X is a subset of $M(P)$.*

PROOF. The point set $M(P)$ is closed. For suppose, on the contrary, that there exists a point O which is a limit point of $M(P)$ but which does not belong to $M(P)$. Then there exists a sequence α of points of $M(P)$ converging to O and there exists an uncountable subset K of M such that every point of K separates O from P in M. For each point Y of α, $M - Y$ is the sum of two mutually separated point sets L_Y and N_Y containing O and P respectively. For each Y, N_Y contains infinitely many points of α; let A_Y denote one of them. Since the point set K is uncountable and there are only countably many points in α therefore there exist a point A of α and an uncountable subset K' of K such that, for *each* point Y of K', $A_Y = A$. The point A is separated from P in M by each point of the uncountable point set K'. But A belongs to $M(P)$. Thus the supposition that $M(P)$ is not closed leads to a contradiction. By a method similar to that used in Case I of the argument employed to prove Theorem 123 of Chapter I, it may be shown that $M(P)$ contains every subcontinuum of M which is irreducible from P to a point of $M(P)$ distinct from P. Therefore $M(P)$ is connected. It follows that it is a continuum.

THEOREM 24. *If A and B are points of the compact continuum M and B belongs to $M(A)$ then $M(B)$ is identical with $M(A)$.*

PROOF. Clearly A belongs to $M(B)$. Suppose P is a point of $M(A)$ distinct from A. Suppose X is a point distinct from A separating P from B in M. Then $M - X$ is the sum of two mutually separated point sets H and L containing B and P respectively. The point A belongs either to H or to L. In the first case, X separates A from P in M and in the second case it separates A from B in M. Hence if there exists an uncountable subset K of M such that every point of K separates B from P in M then each point of $K - A$ either separates A from B in M or separates A from P in M and therefore there exists an uncountable subset K' of $K - A$ such that either every point of K' separates A from B in M or every point of K' separates A from P in M. Therefore P and B do not both belong to $M(A)$. Thus the supposition that P does not belong to $M(B)$ leads to a contradiction. Hence $M(A)$ is a subset of $M(B)$. Similarly $M(B)$ is a subset of $M(A)$.

THEOREM 25. *If M is a compact continuum and G is the collection of all point sets $M(P)$ for all points P of M, then G is an upper semi-continuous collection of mutually exclusive continua filling up M and G is an acyclic continuous curve with respect to its elements.*

PROOF. That the continua of the collection G are mutually exclusive is a consequence of Theorem 24. Suppose $g_1, g_2, g_3, \cdots$ is a sequence of continua of G and $A_1, A_2, A_3, \cdots$ and $B_1, B_2, B_3, \cdots$ are two sequences of points, converging to A and B respectively, such that, for each n, A_n and B_n belong to g_n. Suppose there exists an uncountable subset K of M such that every point of K separates A from B in M. If X is a point of K, $M - X$ is the sum of two mutually separated point sets H and L containing A and B respectively and there exists a number m_X such that A_{m_X} and B_{m_X} belong to H and L respectively and therefore X separates A_{m_X} from B_{m_X} in M. There exist a number k and an uncountable subset K' of K such that $k = m_X$ for every point X of K'. Thus each point of the uncountable point set K' separates A_k from B_k in M. Thus the supposition that A and B do not belong to the same element of G leads to a contradiction. It follows that G is upper semi-continuous.

Suppose x and y are two elements of G. Let X and Y denote points belonging to x and y respectively. Let K denote the set of all points that separate X from Y in M. By Theorems 22 and 90 of Chapter I, there exists a point O of K such that there are uncountably many points of K each separating O from X in M and also uncountably many points of K each separating O from Y in M. The point O does not belong either to x or to y. It follows that O separates x from y in M. Let z denote the element of G that contains O. The point set z separates x from y in M. Thus, in the continuum G of elements, z separates x from y. Therefore, by Theorem 74 of Chapter II, G is an acyclic continuous curve with respect to its elements.

THEOREM 26. *If P is a point of a compact continuum M, the continuum $M(P)$ does not contain uncountably many cut points of M.*

PROOF. Suppose $M(P)$ contains an uncountable set K of cut points of M. By Theorem 111 of Chapter I, K contains two points B and C which are separated from each other in M by uncountably many different points. But no point of $M(B)$ is separated from B in M by uncountably many points of M and, by Theorem 24, $M(B)$ is identical with $M(P)$ and therefore C belongs to $M(B)$. This involves a contradiction.

THEOREM 27. *If M is a compact continuum and G is an upper semi-continuous collection of mutually exclusive continua filling up M such that every simple link of M is a subset of some element of G then G is an acyclic continuous curve with respect to its elements.*

PROOF. Suppose x and y are two continua of the collection G. If X and Y are points of x and y respectively, no simple link of M contains both X and Y and therefore, by Theorem 132 of Chapter I, X and Y are separated from each other in M by some point Z. Let z denote the continuum of the collection G that contains Z. The continuum z separates the continua x and y from each other in M. Hence the elements x and y are separated from each other in G by the element z of G. It follows, with the help of Theorem 74 of Chapter II, that G is a dendron with respect to its elements.

DEFINITION. The subset K of the continuum M is said to be a *graph-atomic subset* of M if and only if there exists an upper semi-continuous collection G of mutually exclusive continua, filling up M, such that (1) K is an element of G, (2) G is a continuous curve with respect to its elements regarded as points, (3) if H is an upper semi-continuous collection of mutually exclusive continua, filling up M, such that H is a continuous curve with respect to its elements regarded as points, then every element of G is a subset of some element of H or, in other words, each element of H is either an element of G or the sum of two or more elements of G.

DEFINITION. The subset K of the continuum M is said to be a *dendratomic subset* of M if and only if there exists an upper semi-continuous collection G of mutually exclusive continua filling up M and fulfilling conditions (1), (2) and (3) of the immediately preceding definition except for the substitution of "dendron" in place of "continuous curve". The point set K is said to be an *arcatomic subset* of M if and only if the same conditions are satisfied except for the substitution of "arc" in place of "continuous curve".

If M is a compact continuous curve and G is an upper semi-continuous collection of mutually exclusive continua filling up M and no simple link of M intersects two continua of G then, by Theorem 27, G is a dendron with respect to its elements. But if X and Y are two points of a dendron T there is only one arc in T having X and Y as its endpoints and every point of this arc, except X and Y, separates X from Y in T. It follows that if A and B are points of different continua of the collection G, they are separated, in M, by uncountably many elements of G and therefore, by Theorem 98 of Chapter II, they are separated in M by uncountably many *points*. Thus if P is a point of M, the subset $M(P)$ of M is what may be termed *dendratomic relatively* to dendrons such that no two elements of one of them contains points of the same simple link of M; that is to say the following theorem holds true.

THEOREM 28. *If M is a compact continuous curve and G denotes the set of all point sets $M(P)$ for all points P of M then G is a dendron with respect to its elements and if H is an upper semi-continuous collection of mutually exclusive continua, filling up M, such that* (1) *H is a dendron with respect to its elements and* (2) *no two continua of H intersect the same simple link of M, then every element of G is a subset of some element of H.*

Theorem 28 does not remain true in the plane if the stipulation requiring that M be a compact continuous curve is replaced by the weaker requirement that it be a compact continuum. Consider the following example.

EXAMPLE 1. Let A, B, A' and B' denote the points (0,1), (1,1), (0,0) and (1,0) respectively and let AB denote the straight line interval whose endpoints are A and B. Let K denote a totally disconnected perfect subset of AB having A as its leftmost point and B as its rightmost point. Let $t_1, t_2, t_3, \cdots$ denote the components of $AB - K$. For each n, let A_n and B_n respectively denote the left end and the right end of the segment t_n, let A'_n

and B'_n denote the projections of A_n and B_n respectively upon $A'B'$ and let $A_nA'_n$, $B_nB'_n$, and A'_nB_n denote straight line intervals with endpoints as indicated. For each n, let L_n denote the continuum $A_nA'_n + B_nB'_n + A_nB'_n + A'_nB_n$. For each point P of K, let q_P denote the vertical interval whose endpoints are P and its projection upon $A'B'$. Let Q denote the collection of all continua q_P for all points P of K such that P is not an endpoint of any segment of the sequence $t_1, t_2, t_3, \cdots$. Let H denote the collection whose elements are the continua $L_1, L_2, L_3, \cdots$ and the continua of the collection Q. Let M denote the sum of all the continua of the collection H. In this example, for each point P of M, $M(P)$ is identical with M. The collection H is a dendron with respect to its elements and no two continua of H intersect the same simple link of M. But there is only one element of the collection G and it is not a subset of any element of H. The points A and B are separated, in M, by uncountably many simple links of M that are not degenerate but not by any one that is degenerate. Indeed all simple links of M are nondegenerate.

Theorem 28 does not remain true in the plane on the removal of the requirement that no two continua of G intersect the same simple link of M. Consider the following example.

EXAMPLE 2. Let A, B, A', B', $t_1, t_2, t_3, \cdots$, K, $A_1A'_1$, $A_2A'_2$, $A_3A'_3, \cdots$, $B_1B'_1$, $B_2B'_2$, $B_3B'_3, \cdots$, and Q be the same as in Example 1. For each n, let Z_n denote a rectangle having $A_nA'_n$ and $B_nB'_n$ as two of its sides. For each n, let U_n denote a set of n horizontal intervals such that (a) each of them has its endpoints on $A_nA'_n$ and $B_nB'_n$ respectively and (b) they divide the interior of Z_n into $n + 1$ congruent rectangular domains. Let L_n denote the continuum obtained by adding to Z_n all the intervals of the set U_n. Let H denote the collection whose elements are the continua L_1, L_2, $L_3, \cdots$ and the continua of the collection Q. Let M denote the sum of all the continua of the collection H. In this example, M is a continuous curve and here, as in Example 1, M is not only a dendron, but an arc, with respect to its elements. As in Example 1, M is the only element of G and it is not a subset of any element of H. But here M is the only simple link of M and it not only intersects, but contains, every element of H.

DEFINITION. A collection G of closed point sets is said to be *continuous* if and only if it is true that if g is a point set of the collection G and $g_1, g_2, g_3, \cdots$ is a sequence of point sets of this collection and, for each n, A_n and B_n are points of g_n and the sequence $A_1, A_2, A_3, \cdots$ has a sequential limit point lying in g, then every infinite subsequence of $B_1, B_2, B_3, \cdots$ has a subsequence having a sequential limit point that lies in g and furthermore g is the sequential limiting set of the sequence $g_1, g_2, g_3, \cdots$.

THEOREM 29. *If U is an uncountable upper semi-continuous collection of mutually exclusive continua such that U^* is closed and compact then there exists an uncountable continuous subcollection of U.*

PROOF. By Theorem 118 of Chapter I, there exists an uncountable subcollection G of U such that if g_0 is any element of G and Q is any finite collection of regions properly covering g_0 then there are uncountably many elements of G each properly covered by Q. Let M denote U^*. By Theorems 22 and 113 of Chapter I, there exists a sequence $H_1, H_2, H_3, \cdots$ such that (1) for each n, H_n is a collection of domains with respect to M, (2) for each n, every element of H_n is a subset of a region belonging to the collection G_n, (3) if L and N are two mutually exclusive closed subsets of M there exists a number m such that no element of H_m intersects both L and N. Let g_1 and g_2 denote two elements of the collection G. There exist two domains D_1 and D_2, with respect to M, containing g_1 and g_2 respectively. There exists a number m_1, greater than 1, such that if a continuum g of G contains a point of an element of H_{m_1} that intersects g_i then g is a subset of D_i $(i = 1,2)$. For each i $(i = 1,2)$ there exists a finite collection Q_i of elements of H_{m_1} such that g_i is properly covered by Q_i and $\bar{Q}_i^*$ is a subset of D_i. For each such i there exist two elements g_{i1} and g_{i2} of G lying in Q_i^*. There exist two mutually exclusive domains D_{i1} and D_{i2}, with respect to M, containing g_{i1} and g_{i2} respectively and lying in $\bar{Q}_i^*$. There exists a number m_2, greater than m_1, such that if an element g of G intersects an element of H_{m_2} that intersects g_{ij} then g is a subset of D_{ij} $(i = 1,2; j = 1,2)$. For each i and j less than 3, there exists a finite subcollection Q_{ij} of H_{m_i} such that g_{ij} is properly covered by Q_{ij} and $\bar{Q}_{ij}^*$ is a subset of D_{ij}. For each such i and j there exist two elements g_{ij1} and g_{ij2} of G lying in Q_{ij}^*. There exist two mutually exclusive domains D_{ij1} and D_{ij2}, with respect to M, containing g_{ij1} and g_{ij2} respectively and lying in Q_{ij}^*. There exists a number m_3, greater than m_2, such that if an element g of G intersects an element of H_{m_3} that intersects g_{ijk} then g is a subset of D_{ijk} $(k = 1,2)$. This process may be continued indefinitely. Thus there exist four infinite sequences $m_1, m_2, m_3, \cdots$; $D_1, D_2, D_{11}, D_{12}, D_{21}, D_{22}, \cdots$; $Q_1, Q_2, Q_{11}, Q_{12}, Q_{21}, Q_{22}, \cdots$ and $g_1, g_2, g_{11}, g_{12}, g_{21}, g_{22}, \cdots$ such that $m_1, m_2, m_3, \cdots$ is an ascending sequence of positive integers and such that if t is a positive integer and, for each i, n_i is either 1 or 2 then (1) $g_{n_1n_2n_3\cdots n_t1}$ and $g_{n_1n_2n_3\cdots n_t2}$ are distinct elements of G lying in $Q^*_{n_1n_2n_3\cdots n_t2}$ and (2) $D_{n_1n_2n_3\cdots n_t1}$ and $D_{n_1n_2n_3\cdots n_t2}$ are mutually exclusive domains with respect to M containing $g_{n_1n_2n_3\cdots n_t1}$ and $g_{n_1n_2n_3\cdots n_t2}$ respectively and lying in $\bar{Q}^*_{n_1n_2n_3\cdots n_t}$, (3) if an element g of G intersects an element of H_{m_t} that intersects $g_{n_1n_2n_3\cdots n_tk}$ then g is a subset of $D_{n_1n_2n_3\cdots n_tk}$ $(k = 1,2)$, (4) $Q_{n_1n_2n_3\cdots n_tk}$ $(k = 1,2)$ is a finite subcollection of $H_{m_{t+1}}$ covering $g_{n_1n_2n_3\cdots n_tk}$ properly and $\bar{Q}^*_{n_1n_2n_3\cdots n_tk}$ $(k = 1,2)$ is a subset of $D_{n_1n_2n_3\cdots n_tk}$. Let G' denote the collection of all continua x such that, for some infinite sequence $n_1, n_2, n_3, \cdots$ of positive integers, each equal to 1 or 2, x is the common part of all the domains, with respect to M, of the infinite sequence $D_{n_1}, D_{n_1n_2}, D_{n_1n_2n_3}, D_{n_1n_2n_3n_4}, \cdots$. The collection G' is an uncountable continuous subcollection of G.

THEOREM 30. *If G is an upper semi-continuous collection of mutually exclusive continua filling up a compact continuum M, there do not exist uncountably many nondegenerate continua of the collection G each containing a cut point of M.*

PROOF. Suppose there exists an uncountable collection H of nondegenerate elements of G each containing a cut point of M. By Theorems 1 and 29, there exists an uncountable continuous subcollection H' of H. There exists a set T of cut points of M such that each point of T belongs to some continuum of the collection H' and each continuum of the collection H' contains only one point of T. By Theorem 111 of Chapter I, there exist an uncountable subset T' of T and two points A and B of T such that every point of T' separates A from B in M. With the help, in particular, of Theorem 91 of Chapter I, it follows that there exists a point C of T' which is a limit point both of T_A, the set of all points of T' that precede it on AB, and of T_B, the set of all points of T' that follow it on AB. The point set $M - C$ is the sum of two mutually separated point sets M_A and M_B containing T_A and T_B respectively. Let T'_A denote the sum of all the continua of G that intersect T_A and let T'_B denote the sum of all those that intersect T_B. The point sets T'_A and T'_B are subsets of M_A and M_B respectively. Let g_C denote the continuum of the collection G that contains C. Let F denote a point of g_C distinct from C. Since H' is continuous, F is a limit point both of T'_A and of T'_B and therefore both of M_A and of M_B. But F belongs either to M_A or to M_B. This involves a contradiction.

DEFINITION. If O, A_1, A_2 and A_3 are four points and, for each n less than 4, r_n is an irreducible continuum from A_n to O and no two of the continua r_1, r_2 and r_3 have, in common, any point except O then the continuum $r_1 + r_2 + r_3$ is a *triod of type* 1, the point O is an *emanation point*, and the continua r_1, r_2 and r_3 are the *rays* of this triod.

It may be easily shown that no triod of type 1 has more than one emanation point or more than one set of three rays. It is, therefore, permissible to speak of *the* emanation point and *the* rays of a given triod of type 1. If O is a point and M_1, M_2 and M_3 are three compact continua which have O in common and no two of them have, in common, any point except O then $M_1 + M_2 + M_3$ contains at least one triod of type 1 whose emanation point is O.

DEFINITION. The point P of the continuum M is said to be a *junction point* of M if and only if it is the emanation point of some triod of type 1 which is a subset of M and there exists a domain D containing P such that P is a cut point of the component of $M \cdot \bar{D}$ that contains P.

THEOREM 31. *If P is a cut point of the locally compact continuum M and D is a domain containing P and H is a connected subset of M containing the component of $M \cdot D$ that contains P, then P is a cut point of H.*

The truth of Theorem 31 may be easily established. It does not remain true in the plane if "continuum M" is replaced by "connected point set M".

THEOREM 32. *No separable dendron has uncountably many junction points.*

PROOF. Suppose the separable dendron M has uncountably many junction points. No junction point of M is an endpoint of M. Hence, by Theorem 50 of Chapter II, every one is a cut point of M. Therefore, by Theorem 13 of Chapter II and Theorem 111 of Chapter I, there exists an arc AB lying in M and containing uncountably many junction points of M. But every junction point of M belonging to AB is the boundary, with respect to M, of a component of $M - AB$ and there do not exist uncountably many components of $M - AB$. This involves a contradiction.

DEFINITION. If k is a positive integer, the point P of the continuum M is said to be of *Menger order k* with respect to M if and only if it is true that (1) every domain with respect to M that contains P contains a domain with respect to M which contains P and whose boundary with respect to M contains only k points, (2) if n is a positive integer less than k, there exists a domain D with respect to M, containing P, such that if U is any domain with respect to M which contains P and which is a subset of D then the boundary of U with respect to M contains more than n points.

THEOREM 33 (G. T. WHYBURN). *No compact continuum has uncountably many cut points that are not of Menger order* 2 *with respect to M.*

PROOF. Suppose there exist a compact continuum M such that the set of all cut points of M that are not of Menger order 2 with respect to M is an uncountable point set K. For every point P of M let $M(P)$ denote the set of all points X of M such that there do not exist uncountably many points of M each separating P from X in M. Let G denote the collection of all such point sets $M(P)$ for all points P of M. By Theorem 25, G is an upper semi-continuous collection of mutually exclusive continua and it is a dendron with respect to its elements. By Theorem 111, there exist two points A and B and an uncountable subset K' of K such that every point of K' separates A from B in M. The continua M_A and M_B are distinct elements of G. Let α denote the arc of elements of G having them as its end elements. Every point of K' belongs to some element of α and no element of α contains uncountably many points of K'. Let Q denote the collection of all elements of α that intersect K' and that are not junction elements of G. The collection Q is uncountable. Hence, by Theorem 29, there exists an uncountable continuous collection Q' whose elements belong to Q. Suppose H is any uncountable subcollection of Q'. By Theorem 112 of Chapter I, there exists an element q of H such that every segment of α that contains q contains uncountably many elements of H preceding q in α and uncountably many elements of H following q in α. Let L_{Aq} denote the sum of all the continua of α that precede q in α and let L_{Bq} denote the sum of all those that follow q in α. The continuum q contains a point O belonging to K'. Suppose q contains a point C distinct from O. With the aid of the fact that H is continuous it may be seen that C is a limit point of each of the connected point sets L_{Aq} and L_{Bq}. Hence $L_{Aq} + C + L_{Bq}$ is a connected subset of $M - O$ containing A and B, contrary to the fact that O separates A from B in M. It follows that q is identical with O. Let W denote the collection of all degenerate elements q of Q' such that every segment of α that contains q contains uncountably many elements of α preceding q, and uncountably many elements of α following q, in α. It has been shown that if H is any uncountable subcollection of Q' there is an element of H belonging to W. Let Y denote a definite element of Q' belonging to W. Since Y belongs to W there exist two sequences of points $A_1, A_2, A_3, \cdots$ and $B_1, B_2, B_3, \cdots$ each converging to Y and such that, for each n, A_n precedes Y, A_n precedes A_{n+1}, Y precedes B_n, and B_{n+1} precedes B_n, in the order from M_A to M_B on α. For each n, let t_n denote the common part of the sects A_nB_n and B_nA_n of G. Suppose D is any domain with respect to M containing Y. Since Y is not a junction point of α, it is the common part of all the point sets $t_1, t_2, t_3, \cdots$ and there exists a number m such that $t_m - (A_m + B_m)$ is a subset of D. But $t_m - (A_m + B_m)$ is a domain with respect to M and its boundary with respect to M consists of the two points A_m and B_m. Hence Y is a point of M of Menger order 2. But Y belongs to K. Thus the supposition that Theorem 33 is false has led to a contradiction.

Definition. The continuum M is said to be *hereditarily decomposable* if every nondegenerate subcontinuum of M is decomposable, and to be *hereditarily indecomposable* if every one is indecomposable.

Theorem 34. *If G is an upper semi-continuous collection of mutually exclusive continua filling up the compact hereditarily decomposable continuum M then, with respect to its elements regarded as points, G is a decomposable continuum.*

Proof. Suppose G is an indecomposable continuum with respect to its elements regarded as points. Then there exist three continua x, y and z of the collection G such that, with respect to its elements regarded as points, G is an irreducible continuum between each two of them. Let X, Y and Z denote points belonging to x, y and z respectively. Since M is decomposable it is not irreducible between each two of these points. Let A and B denote two of them between which it is not irreducible and let g_A and g_B denote the continua of the collection G that contain A and B respectively. There exists a proper subcontinuum K of M which is irreducible from A to B. Since M is hereditarily decomposable, K is the sum of two continua both distinct from it. Neither of these continua contains both A and B. Let K_A denote the one containing A and let K_B denote the other one. Let G_A denote the collection of all continua of G that intersect K_A and let G_B denote the collection of all those that intersect K_B. The continuum g_A is one of the continua x, y and z and g_B is another one. Hence, with respect to its elements regarded as points, G is an irreducible continuum from g_A to g_B.

Therefore $G_A + G_B$ is G. But G_A and G_B are continua with respect to their elements and neither of them is G since g_A belongs to G_A but not to G_B and g_B belongs to G_B but not to G_A. Hence G is a decomposable continuum with respect to its elements.

THEOREM 35. *If G is an upper semi-continuous collection of mutually exclusive continua filling up a compact regular curve M then G is a regular curve with respect to its elements.*

PROOF. Suppose g_1 and g_2 are two elements of G. By Theorem 67 of Chapter I, there exists a finite subset T of M separating g_1 from g_2 in M. For each point P of T let g_P denote the continuum of G that contains P. The sum of all the continua g_P for all points P of T separates g_1 from g_2 in G and it consists of only a finite number of continua of G. Hence G is a regular curve with respect to its elements.

There exists, however, a compact continuum M having no continuum of condensation but such that there is an upper semi-continuous collection G of mutually exclusive continua filling up M such that the continuum G of continua does have a continuum of condensation with respect to its elements. Consider the following example.

EXAMPLE 3. In a Euclidean plane let AB denote a straight line interval and let K denote a totally disconnected perfect point set lying on AB and containing A and B. For each segment x of AB which is a component of $AB - K$, let t_x denote a straight line interval perpendicular to AB at the mid-point of x and of the same length as x. Let M denote the sum of AB and all such intervals t_x. Let G denote the collection whose elements are (1) the closures of the components of $AB - K$ and (2) the degenerate subsets of M that are not subsets of any interval which is the closure of a component of $AB - K$. The collection G is, with respect to its elements, a dendron containing an arc which is a continuum of condensation of G. But M has no continuum of condensation.

It is to be noted, however, that none of the continua of condensation of Example 3 is *essential* in the sense defined below.

A collection will be said to be of *class* 1 if and only if (1) it is an upper semi-continuous collection of mutually exclusive continua, and (2) with respect to its elements regarded as points, it is a continuum with no continuum of condensation.

DEFINITION. The subcontinuum K of the continuum M is said to be an *essential continuum of condensation of M* if and only if it is true that (1) K is a continuum of condensation of M, and (2) if G is a collection of class 1 filling up M, then some continuum of G contains K.

THEOREM 36. *If the nondegenerate subcontinuum K of the compact continuum M has no continuum of condensation and for every collection G of class 1 filling up M there is a continuum of G that contains K, then K is a continuum of condensation of M.*

PROOF. Suppose K is not a continuum of condensation of M. Then there exists a point set D which is a domain both with respect to M and with respect to K. Since K has no continuum of condensation, D contains a free segment with respect to K, that is to say, a segment t of some arc lying in K such that no point of t is a limit point of $K - t$, in other words, such that t is a domain with respect to K. The point set t is also a domain with respect to M. If $M - t$ is a continuum, let G denote the collection whose elements are $M - t$ and the degenerate subsets of t. If $M - t$ is not connected, it is the sum of two mutually exclusive continua g_A and g_B containing the endpoints A and B respectively of t. In this case, let G denote the collection whose elements are g_A, g_B and the degenerate subsets of t. In the first case G is a simple closed curve, and in the second case it is an arc, with respect to its elements. In each case it is a collection of class 1 and no continuum of G contains K. This involves a contradiction.

THEOREM 37. *If K is a subcontinuum of the compact continuum M, and for every collection G of class 1 filling up M there is a continuum of G that contains K, then K contains a continuum of condensation of M.*

PROOF. Suppose K contains no continuum of condensation of M. Then it contains no continuum of condensation of itself. Hence, by Theorem 36, K is itself a continuum of condensation of M.

THEOREM 38. *If the arc-wise connected continuum K is a subset of the compact continuum M, and for every collection G of class 1 filling up M there is a continuum of G that contains K, then every two points of K belong to some essential continuum of condensation of M lying in K.*

Theorem 38 is a consequence of Theorem 36 and the fact that no arc has a continuum of condensation. That it does not remain true even in the plane on the omission of the stipulation that K be arc-wise connected may be seen from the following example.

EXAMPLE 4. In the plane, let K denote a compact indecomposable continuum, let AB denote an arc having only the point B in common with K and let M denote $K + AB$. Let C denote a point of K such that K is irreducible from B to C. If G is a collection of class 1 filling up M, some continuum of G contains the continuum K. But there is no continuum of condensation of M containing the points B and C.

THEOREM 39. *If the compact continuum M has no essential continuum of condensation then for every nondegenerate subcontinuum K of M there exists an upper semi-continuous collection G of mutually exclusive continua filling up M such that G is, with respect to its elements, either an arc or a simple closed curve and no continuum of the collection G contains K.*

PROOF. By hypothesis and Theorem 38 there exists a collection H of class 1 filling up M such that no continuum of H contains K. Let W denote the collection whose elements are the continua of the collection H that intersect K. There exists a free segment t with respect to H such that every element of t belongs to W. By an argument analogous to that used in a similar connection in the above proof of Theorem 36, it may be shown that there exists an upper semi-continuous collection G of mutually exclusive continua filling up M such that G is either an arc or a simple closed curve with respect to its elements and every element of t is an element of G. The continuum K is not a subset of any continuum of the collection G.

There exist a compact continuum M and two upper semi-continuous collections G_1 and G_2 of mutually exclusive continua filling up M such that (1) neither G_1 nor G_2 has a continuum of condensation with respect to its elements, (2) each degenerate subset of M is the common part of some continuum of G_1 and some continuum of G_2, (3) M has a continuum of condensation. This may be seen as follows.

EXAMPLE 5. In a Cartesian plane let AB denote an interval of OX. Let $P_1, P_2, P_3, \cdots$ denote the set of all points of AB whose abscissae are rational. For each n, let t_n denote a vertical interval of length $1/n$ having P_n as its lowest point. Let M denote the continuum consisting of all points belonging either to AB or to some interval of the sequence $t_1, t_2, t_3, \cdots$. Let G_1 denote the collection whose elements are $t_1, t_2, t_3, \cdots$ and the points of M that belong to no interval of this sequence. Let G_2 denote the collection whose elements are the interval AB and the points of M that do not belong to AB.

There exists a compact continuous curve M having a continuum of condensation K such that (1) $M - K$ is connected, (2) there does not exist a noncontracting sequence of mutually exclusive nondegenerate subcontinua of M, (3) K is an arc. Consider the following example.

EXAMPLE 6. Let K be a straight line interval AB and let M denote the right half of a continuum indicated by Figure 36.

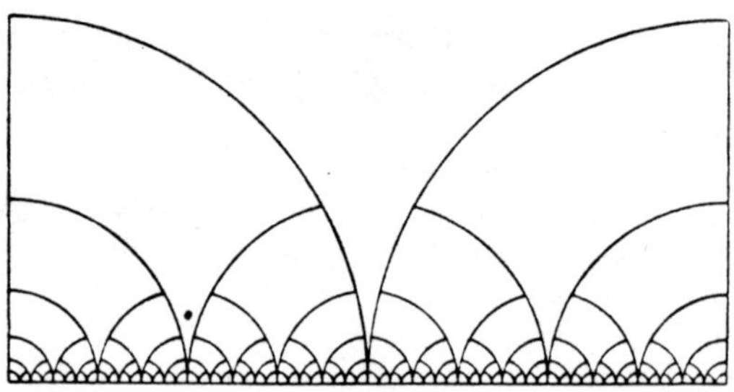

FIGURE 36

There exist a compact continuous curve M and a sequence $G_1, G_2, G_3, \cdots$ of upper semi-continuous collections of mutually exclusive continua filling up M such that (1) for each n, G_n is, with respect to its element, a continuum with no continuum of condensation, (2) for each n, every continuum of

G_{n+1} is a subset of some continuum of G_n, and (3) if G denotes the set of all continua g such that g is maximal with respect to the property of being, for each n, a subset of some continuum of G_n then every continuum of G is degenerate and thus G is the set of all degenerate subsets of M though M has a continuum of condensation.

EXAMPLE 7. Let M denote the same continuum as in Example 2. For each n, let G_n denote the collection whose elements are the continua t_n, t_{n+1}, $t_{n+2}, \cdots$ and the points of M that belong to no continuum of this sequence.

THEOREM 40. *If M is a compact continuum and $G_1, G_2, G_3, \cdots$ is a sequence of upper semi-continuous collections of mutually exclusive continua filling up M and, for each n, every continuum of G_{n+1} is a subset of some continuum of G_n and, for each n, G_n has no continuum of condensation with respect to its elements and, for each point P of M, M_P is the common part of $g_{P1}, g_{P2}, g_{P3}, \cdots$ where, for each n, g_{Pn} is the continuum of G_n that contains P, then the collection of all continua M_P, for all points P of M, is, with respect to its elements, a continuum with no essential continuum of condensation.*

DEFINITION. The compact continuum M is said to be a *web* if and only if there exist two upper semi-continuous collections H_1 and H_2 of mutually exclusive continua such that (1) each of these collections fills up M, (2) each of them is a dendron with respect to its elements and (3) there exists an uncountable subcollection W of the collection H_1 such that no element of W is a subset of any element of H_2. A compact continuum is said to be *webless* if and only if it contains no web.

THEOREM 41. *If the compact continuum M is a web there exist two upper semi-continuous collections Z_1 and Z_2 of mutually exclusive continua such that (1) for each i ($i = 1,2$), Z_i fills up M and is an arc with respect to its elements and (2) every continuum of Z_1 intersects every continuum of Z_2.*

PROOF. Let H_1, H_2 and W denote collections satisfying, with respect to M, all the conditions stated in the above definition of a web. There exist two mutually exclusive closed point sets L and T and an uncountable subcollection W_1 of W such that every continuum of the collection W_1 intersects both L and T. There exists a sequence γ of continua of the collection W_1 which converges to a continuum a_1 of that collection. There exist two points A and B belonging to a_1 but not to the same continuum of the collection H_2. Let α_2 and β_2 denote the continua of H_2 that contain A and B, respectively. Since H_2 is a continuous curve with respect to its elements there exists an arc $\alpha_2\beta_2$ of elements of H_2 having α_2 and β_2 as its end elements. Since H_2 fills up M and is a dendron with respect to its elements, the continua α_2 and β_2 are separated from each other in M by every continuum of the collection $\alpha_2\beta_2$ other than themselves. Let a_2 denote a continuum of the collection $\alpha_2\beta_2$ distinct from α_2 and β_2, let b_2 denote a continuum of this collection that separates a_2 from β_2 in M and let a_2b_2

denote the arc of elements of $\alpha_2\beta_2$ whose end elements are a_2 and b_2. Let D_A denote the set of all points of M that are separated from b_2 in M by a_2 and let D_B denote the set of all those that are separated from a_2 in M by b_2. The point sets D_A and D_B are domains with respect to M containing A and B respectively. Since γ converges to a_1 there exists a continuum b_1 belonging to γ, distinct from a_1 and intersecting both D_A and D_B. There exists an arc a_1b_1 of elements of H_1 having a_1 and b_1 as its end elements.

For each i less than 3, let Z_i denote the collection of all continua z such that, for some continuum x of a_ib_i, z is the sum of all continua y of H_i such that y is not separated from x in M by any continuum of the collection a_ib_i.

THEOREM 42. *If the compact continuum M has no essential continuum of condensation and there exists an infinite sequence α of mutually exclusive nondegenerate subcontinua of M converging to a nondegenerate continuum K that intersects no continuum of α then there exist two upper semi-continuous collections Z_1 and Z_2 of mutually exclusive continua such that* (1) *for each i* ($i = 1,2$), *Z_i fills up M and is an arc or a simple closed curve with respect to its elements, and* (2) *every continuum of Z_1 intersects every continuum of Z_2.*

PROOF. By hypothesis K is not an essential continuum of condensation of M. Hence there exists an upper semi-continuous collection G of mutually exclusive continua filling up M such that (1) with respect to its elements G is a continuum with no continuum of condensation, (2) K is not a subset of any one continuum of G. Let H denote the collection of all elements of G that intersect K. The collection H is a continuum of elements of G. Since H is not a continuum of condensation of G there exists an element h_0 of H which is not a limit element of $G - H$. There exists a domain D with respect to M containing h_0 such that (1) $\bar{D}$ does not contain every point of K and (2) no element of $G - H$ contains a point of $\bar{D}$. But h_0 contains a point P belonging to K. Therefore there exists a subsequence β of the sequence α containing all but a finite number of the elements of α and such that every element of β contains both points of D and points without D. There exist a subsequence $t_1, t_2, t_3, \cdots$ of the sequence β and a sequence of points $P_1, P_2, P_3, \cdots$ converging to P such that, for each n, P_n belongs to t_n. For each n, let q_n denote the component of $\bar{D}\cdot t_n$ that contains P_n. Since, for each n, q_n intersects the boundary of D therefore q, the limiting set of the sequence $q_1, q_2, q_3, \cdots$ contains a point belonging to the boundary of D and therefore not to h_0. Hence not every q_n is a subset of some element of H. For if that were the case then, since G is upper semi-continuous, q would be a subset of h_0. Let k denote a number such that q_k is not a subset of any continuum of the collection H. Since q_k is a subset of $\bar{D}$, each point of q_k belongs to some continuum of H. Therefore W, the set of all elements of H that intersect q_k, is a nondegenerate continuum of elements of G. Since each continuum of the set W intersects both of the mutually exclusive continua K and q_k, therefore every continuum of W is a nondegenerate point set. Since G has no continuum of condensation therefore, by Theorem 11 of Chapter II, W is a continuous curve. Hence there exists an arc ω of elements of W. Since ω is not a continuum of condensation of G therefore there exists an interval ω' of ω such that if L denotes the set of elements $G - \omega'$ then no element of ω', except its end elements, is a limit element of L. The set of elements $\bar{L}$ does not have more than two components. Let Z_1' denote the collection whose elements are the components of $\bar{L}$ (if there are any) and the elements of ω' that belong to no component of $\bar{L}$. Then Z_1' is an upper semi-continuous collection of mutually exclusive nondegenerate continua filling up M and, with respect to its elements, it is an arc or a simple closed curve. There exist a continuum g_0 of the collection Z_1' and a sequence α' of distinct, and therefore mutually exclusive, continua of Z_1' converging to some nondegenerate subcontinuum K' of g_0, the continua of α' being all distinct from g_0. There exists an upper semi-continuous collection G' of mutually exclusive continua filling up M such that (1) with respect to its elements, G' is a continuum with no continuum of condensation, (2) K' is not a subset of any continuum of G'. By an argument entirely parallel to that employed above to show the existence of Z_1', letting G', α' and K' play the role there played by G, α and K, it may be shown that there exist an upper semi-continuous collection Z_2 of mutually exclusive nondegenerate continua, filling up M, and an interval γ of Z_1' such that if L' denotes the set of elements $Z_1' - \gamma$ and Z_1 is the collection such that x belongs to it if and only if x is a component of L' or an element of Z_1' that belongs to no component of L' then (1) every continuum of the collection Z_2 intersects every continuum of the collection Z_1, (2) with respect to its elements, Z_2 is an arc or a simple closed curve, (3) every element of Z_2, with the possible exception of one or two, is an element of G'.

THEOREM 43. *If G and H are upper semi-continuous collections of mutually exclusive continua such that* (1) *every point set of the collection H is a subset of some one of the collection G and* (2) *every point belonging to a continuum of the collection G belongs to some one of the collection H, and for each element g of G, g' denotes the collection of all continua of H that are subsets of g and G' denotes the collection of all such collections g' for all elements g of G, and, indeed for every subcollection Q of G, Q' denotes the collection of all collections g' for all elements g of Q, then* (1) *if the elements of H are regarded as "points", G' is an upper semi-continuous collection of continua of such "points",* (2) *the element g of G is a limit element of the collection Q of elements of G if and only if the element g' of G' is a limit element of Q' and therefore, in particular,* (a) *if G is a continuous curve with respect to its elements, then G' is a continuous curve with respect to its elements and* (b) *if G is acyclic with respect to its elements, G' is acyclic with respect to its elements.*

THEOREM 44. *If M is a point set and F is a family of upper semi-continuous collections of mutually exclusive compact continua such that each collection of* *the family F fills up M and, for each point P of M, g_P denotes the component, containing P, of the common part of all the continua that contain P and belong to some collection of the family F, and G denotes the collection of all the continua g_P for all points P of M, then G is an upper semi-continuous collection of mutually exclusive compact continua filling up M.*

THEOREM 45. *If the compact continuum M is not a web and G_1 and G_2 are two upper semi-continuous collections of mutually exclusive continua filling up M such that each of them is a dendron with respect to its elements and G is the set of all point sets g such that, for some point P of M, g is the component containing P of the common part of the continuum of G_1 that contains P and the continuum of G_2 that contains P, then the collection G is a dendron with respect to its elements.*

PROOF. By Theorem 44, the collection G is upper semi-continuous. Suppose p is a continuum of G and suppose x is a continuum of G such that x and p are not subsets both of the same continuum of G_1 and of the same continuum of G_2. Suppose they belong to different continua q_x and q_p, respectively, of the collection G_1. Since G_1 is a dendron with respect to its elements, there exists an uncountable subcollection G_1' of G_1 such that every element of G_1' separates q_x from q_p in the continuum, of elements, G_1. Since M is not a web there is at least one continuum g of the collection G_1' which is a subset of some continuum t of the collection G_2 and which therefore belongs to the collection G. The continuum g separates q_x from q_p in G. A parallel argument applies in case p and x belong to different continua of G_2. Thus if H is the set of all components of the common part of the continuum of G_1 that contains p and the continuum of G_2 that contains p and x is any continuum of G not belonging to H then p is separated from x in G by some element of G. Suppose now that there exists an element h belonging to the collection H but distinct from p and not separated in G from p by any element of G. Then by Theorems 123 and 168 of Chapter I and Theorem 15 of the present chapter, the set T of all elements of G that are not separated from p in G by any element of G is a nondegenerate continuum. But T is a subset of H and H is a totally disconnected set of elements of G. This involves a contradiction. Therefore if x is any element whatsoever of G distinct from p then x and p are separated in G by some element of G. In other words every two elements of G are separated in G by some element of G. Therefore, by Theorem 123 of Chapter I, Theorem 15 of the present chapter and Theorem 74 of Chapter II, G is a dendron with respect to its elements.

THEOREM 46. *If M is a compact continuum with no essential continuum of condensation then M is a regular curve provided it is true that if K is a subcontinuum of M there do not exist two upper semi-continuous collections Z_1 and Z_2 of mutually exclusive continua such that* (1) *for each i ($i = 1,2$), Z_i fills up K and is an arc or simple closed curve with respect to its elements, and* (2) *every continuum of Z_1 intersects every continuum of Z_2.*

PROOF. Suppose there exists a compact continuum M which satisfies all of these conditions but is not a regular curve. By Theorem 129 of Chapter I, M contains a nondegenerate continuum K such that every point of K is a non-regular point of M. Since K is not an essential continuum of condensation of M there exists an upper semi-continuous collection G of mutually exclusive continua filling up M such that K is not a subset of any one continuum of the collection G and G has, with respect to its elements, no continuum of condensation. Let H denote the collection of all continua of G that intersect K. The collection H is a continuous curve with respect to its elements. Hence it contains an arc t of elements of G. Since t is not a continuum of condensation of G there exists a segment α of t such that no element of α is a limit element of the set of elements $G - \alpha$. But, by Theorem 42, all but countably many of the elements of α are points. Hence if O is any point which is an element of α and P is any point whatsoever of $M - O$ and g_P is the element of G that contains P then there exist two points X and Y, belonging to α, such that $X + Y$ separates O from g_P in G and therefore separates O from P in M. Thus the point O is a regular point of M. But O belongs to K. This involves a contradiction.

THEOREM 47. *If, under the hypothesis of Theorem 44, M is a compact continuum and, for every two collections G_1 and G_2 of the family F, either every element of G_1 is a subset of some element of G_2 or every element of G_2 is a subset of some element of G_1, then* (1) *if every collection of the family F is a continuous curve with respect to its elements, G is a continuous curve with respect to its elements and* (2) *if every collection of the family F is acyclic with respect to its elements, so is G.*

PROOF. If H is a collection of the family F, for each element h of H let h' denote the set of all continua of the collection G which are subsets of h and let H' denote the collection of all sets h' so obtained from the different elements h of H. Let F' denote the family of all such collections H' for all collections H of F. By Theorem 44, G is an upper semi-continuous collection. Let the continua of G be considered as "points" in the new sense. By Theorem 123 of Chapter I and Theorem 15 of the present chapter, the set G of "points", regarded as a space, satisfies Axioms 0, 1 and C. By Theorem 43, F' is a family of upper semi-continuous collections of continua and each collection of F' fills up the compact space G.

Suppose now that every collection of the family F is a continuous curve with respect to its elements. Then, by Theorem 43, every collection of F' is a continuous curve with respect to its elements. It will be shown that G is a continuous curve. Suppose that, *in the space* G, P is a point and D is a domain containing it. Let W denote the collection of all continua that contain P and belong to some collection of the family F'. Suppose that every continuum of the collection W contains a point not belonging to D. Let U denote the collection of all point sets $w \cdot (G - D)$ for all continua w of the collection W. By hypothesis, W is a monotonic collection of closed and compact point sets. Therefore so is U. Hence, by Theorems 37 and 38 of Chapter I, the point sets of the collection U have, in common, at least one point O. The point O belongs to $G - D$ and is therefore distinct from P. Thus t, the common part of the continua of the collection W contains at least two points O and P. But, by these two theorems, t is a continuum and therefore, by the definition of G, t contains only the point O. Thus the supposition that every continuum of W contains a point of $G - D$ leads to a contradiction. Therefore there exists a collection H' of the family F' such that the element of H' that contains P is a continuum K lying wholly in D. Since H' is a continuous curve with respect to its elements, therefore there exists a connected domain Q of elements of H' such that K belongs to Q and every element of Q is a subset of D. The point set Q^* is a connected domain containing P and lying wholly in D. Thus G is connected im kleinen at P. Hence it is a continuous curve.

Suppose now that every collection of the family F is a dendron with respect to its elements. Suppose A and B are any two points of G. There exists a collection H' of the family F' such that no continuum of H' contains both A and B. Let a and b denote the continua of H' that contain A and B respectively. There exists an arc ab of elements of H' having a and b as its end elements.

Suppose there are uncountably many nondegenerate elements of H'. Then, as may be seen with the aid of Theorem 118 of Chapter I, there exists an element k of H' containing two points C and T such that if, in the space G, D_C and D_T are two mutually exclusive domains containing C and T respectively then there are uncountably many elements of H' each containing both a point of D_C and a point of D_T. There exists a collection H'' of the family F' such that C and T do not belong to the same element of H''. Let c and t denote the elements of H'' that contain C and T respectively. The continua c and t are mutually exclusive subsets of k. There exist, in the space G, two mutually exclusive domains U_c and U_t containing c and t respectively such that no continuum of the collection H'' contains both a point of U_c and a point of U_t. But there exists an uncountable subcollection Q of H' such that every continuum of the collection Q contains both a point of U_c and a point of U_t. Each continuum of Q contains more than one continuum of H'' and is the sum of all the continua of H'' that have one or more points in common with it. For each continuum q of Q, let q' denote the set of all continua of H'' that are subsets of q. Let Q' denote the collection of all sets q' for all continua q of Q. The collection Q' is an uncountable collection of mutually exclusive nondegenerate continua of elements of H''. But H'' is a dendron with respect to its elements and, by Theorems 74 and 80 of Chapter II, no compact dendron contains uncountably many mutually exclusive nondegenerate continua. This involves a contradiction. Hence there do not exist uncountably many nondegenerate elements of H'. It follows that there is a point P which is an element of ab distinct from a and from b. The point P separates A from B in G. Thus every two points of the compact continuum G are separated in G by some one point. Hence, by Theorem 74 of Chapter II, G is an acyclic continuous curve.

THEOREM 48. *A compact continuum has a dendratomic subset if and only if it is not a web.*

PROOF. Suppose the compact continuum M is not a web. Let Q denote a family of collections such that a collection belongs to Q if and only if it is an upper semi-continuous collection of mutually exclusive continua which fills up M and is a dendron with respect to its elements. With the help of Theorem 45 it may be seen that there exists a subcollection F of Q such that (1) if G_1 and G_2 are two collections of the family F, either each element of G_1 is a subset of some element of G_2 or each element of G_2 is a subset of some element of G_1, and (2) for each collection U of the family Q there exists a collection V of the family F such that each element of the collection V is a subset of some element of the family U. Let G denote a collection related to F in the manner described in the statement of Theorem 44. By Theorem 47, G is a dendron with respect to its elements. Every element of G is a dendratomic subset of M.

Suppose now that the compact web M has a dendratomic subset. Let Z_1 and Z_2 denote collections satisfying, with respect to M, all of the conditions of Theorem 41. For each i ($i = 1,2$), let x_i and y_i denote two continua of the collection Z_i. By hypothesis there exists an upper semi-continuous collection G of mutually exclusive continua filling up M such that (1) G is a dendron with respect to its elements and (2) if a continuum of the collection G intersects a continuum which belongs either to Z_1 or to Z_2 then it lies wholly in that continuum. Let a and b denote continua of the collection G which are subsets of $x_1 \cdot x_2$ and $x_1 \cdot y_2$, respectively. Let U denote the collection of all continua of G that are subsets of x_1 and let V denote the collection of all those that are subsets of the continuum $x_2 + y_1 + y_2$. With respect to the elements of G, U and V are intersecting subcontinua of G whose common part is the set of all elements of G that are subsets either of $x_1 \cdot x_2$ or of $x_1 \cdot y_2$ and clearly this set is not a connected set of elements. But G is a dendron with respect to its elements and no dendron contains two intersecting continua whose common part is not connected. Thus the supposition that M has a dendratomic subset leads to a contradiction.

THEOREM 49. *Every web is a triod.*

PROOF. Suppose M is a web. Let Z_1 and Z_2 denote two upper semi-continuous collections satisfying, with respect to M, all of the conditions listed in the statement of Theorem 41. Let α and β denote two continua belonging to, but not end elements of, the collection Z_1 and let γ denote one belonging to Z_2 but distinct from its end elements. The continuum γ

intersects both α and β, and $M - \gamma$ is the sum of two mutually separated point sets H and K. Let X and Y denote points belonging to H and K respectively, and let h and k denote the continua of the collection Z_2 that contain X and Y respectively. The continua h and k are subsets of H and K respectively. But β intersects both of these continua. Thus $\beta \cdot H$ and $\beta \cdot K$ exist. The continuum γ separates $\beta \cdot H$ from $\beta \cdot K$ in M. Therefore, by Theorem 81 of Chapter IV, M is a triod.

THEOREM 50. *If the compact continuum M is not a triod it has a dendratomic subset and if it has more than one of them they are its arcatomic subsets.*

PROOF. That M has a dendratomic subset follows directly from Theorems 48 and 49. With the aid of the fact that every compact dendron which is not an arc is a triod it follows that if an upper semi-continuous collection of mutually exclusive continua fills up M and is a dendron with respect to its elements then it is an arc with respect to its elements. Hence if M has more than one dendratomic subset its dendratomic subsets are its arcatomic subsets.

THEOREM 51. *If A and B are two endpoints of the compact dendron M there exists an upper semi-continuous collection H of mutually exclusive continua filling up M such that, with respect to its elements, H is an arc whose end elements are A and B.*

PROOF. The dendron M contains an arc AB. For each point P of AB, let h_P denote P, or the continuum obtained by adding to P every component of $M - AB$ of which P is a limit point, according as there does not or does exist at least one such component. Let H denote the collection of all such continua h_P. This collection is upper semi-continuous, it is an arc with respect to its elements and its end elements are h_A and h_B. But $h_A = A$ and $h_B = B$.

THEOREM 52. *If a compact continuum M has arcatomic subsets then not only are they dendratomic subsets of M but every upper semi-continuous collection of mutually exclusive continua which fills up M and which is a nondegenerate dendron with respect to its elements is also an arc with respect to its elements.*

PROOF. Let G denote the collection of all arcatomic subsets of M. Suppose H is an upper semi-continuous collection of mutually exclusive continua filling up M such that H is a dendron with respect to its elements. Suppose this dendron has three end elements α, β and γ. With the help of Theorem 51 it may be seen that there exists an upper semi-continuous collection Q of mutually exclusive continua filling up M such that (1) every continuum of H is a subset of some continuum of Q, (2) α and β belong to Q, and (3) with respect to its elements, Q is an arc whose end elements are α and β. It follows that each of the continua α, β and γ contains every continuum of the collection G which it intersects. But if a, b and c denote the set of all the continua of G that lie in α, the set of all those that lie in β and the set of all those that lie in γ, respectively, then, *with respect to the elements of G regarded as points*, a, b and c are three mutually exclusive subcontinua of G and $G - a$, $G - b$ and $G - c$ are all connected. This is contrary to the fact that G is an arc with respect to its elements. Hence H has only two end elements. Therefore it is an arc with respect to its elements. It follows that every element of G is a dendratomic subset of M.

It is not true that if a compact continuum has arcatomic subsets then it is not a triod.

EXAMPLE 8. In a Euclidean plane let AB denote a straight line interval and let T denote a totally disconnected perfect subset of AB containing both A and B. Let $Q_1, Q_2, Q_3, \cdots$ denote the components of $AB - T$. For each n let A_n and B_n denote the end points of the segment Q_n, A_n being between A and B_n. For each point P of $T - A$ let J_P denote the circle with center at A which passes through P. For each positive integer n let α_n denote a set of n straight line intervals $A_{1n}B_{1n}, A_{2n}B_{2n}, \cdots, A_{nn}B_{nn}$ such that the points $A_{1n}, A_{2n}, \cdots, A_{nn}$ are equally spaced on J_{A_n} and $B_{1n}, B_{2n}, \cdots, B_{nn}$ are on J_{B_n} and, for each i ($1 \leqq i \leqq n$), A_{in} lies between B_{in} and A. Let K_n denote the continuum obtained by adding together J_{A_n}, J_{B_n} and all the intervals of the set α_n. Let K denote the point set $A + K_1 + K_2 + K_3 + \cdots$. Let G denote the collection of all continua g such that either g is A or, for some n, g is K_n or, for some point P of $T - K \cdot T$, g is J_P. Let M denote the sum of all the continua of the collection G. The point set M is a continuous curve and G is an arc with respect to its elements and its elements are arcatomic subsets of M. But M is a triod.

THEOREM 53. *Every compact irreducible continuum between two points has a dendratomic subset and if it has more than one of them then they are its arcatomic subsets.*

Theorem 53 is a consequence of Theorem 50 and the fact that no irreducible continuum between two points is a triod.

THEOREM 54. *If G is a collection of mutually exclusive compact continua filling up the compact continuum M and every two continua of G are separated from each other in M by some continuum of G then G is upper semi-continuous and G is a dendron with respect to its elements.*

PROOF. Suppose G is not upper semi-continuous. Then it may be seen that there exist two sequences of points $A_1, A_2, A_3, \cdots$ and $B_1, B_2, B_3, \cdots$ converging respectively to points A and B such that (1) for each n, there exists a continuum g_n of G containing both A_n and B_n, (2) A and B do not belong to the same continuum of the collection G. Let g_A and g_B denote the continua of G that contain A and B respectively. By hypothesis there exists a continuum g of the collection G such that $M - g$ is the sum of two mutually separated point sets M_A and M_B containing g_A and g_B respectively. Every continuum of the sequence $g_1, g_2, g_3, \cdots$ is a subset either of M_A or of M_B. Hence there exists an infinite sequence of distinct numbers $n_1, n_2, n_3, \cdots$ such that either all of the continua $g_{n_1}, g_{n_2}, g_{n_3}, \cdots$ are subsets of M_A or all of them are subsets of M_B. In the first case, the point B of M_B is the sequential limit point of the sequence of points $B_{n_1}, B_{n_2}, B_{n_3}, \cdots$ all belonging to M_A and, in the second case, the point A is the sequential limit point of a sequence of points all belonging to M_B. In either case, a contradiction is reached. Therefore G is an upper semi-continuous collection. That it is a dendron with respect to its elements may be proved with the help of Theorem 74 of Chapter II.

THEOREM 55 (HARLAN CROSS MILLER). *If the nondegenerate hereditarily decomposable compact point set M is an irreducible continuum from the point A to the point B there exists an upper semi-continuous collection G of mutually exclusive continua filling up M such that* (1) *the elements of G are the arcatomic subsets of M,* (2) *the element of G that contains B is the set of all points X of M such that M is irreducible from A to X, and the one that contains A is the set of all points X of M such that M is irreducible from B to X, and* (3) *if g is an element of G that contains neither A nor B then $M - g$ is the sum of two mutually separated connected point sets D_{gA} and D_{gB} containing A and B respectively and g is the sum of two intersecting continua α_g and β_g such that α_g is the boundary of D_{gA} with respect to M and β_g is the boundary of D_{gB} with respect to M.*

PROOF. By Theorem 145 of Chapter I, the set of all points X such that M is irreducible from A to X is a continuum M_B containing B and the set of all points X such that M is irreducible from B to X is a continuum M_A containing A.

Suppose P is a point of M belonging neither to M_A nor to M_B. There exists an irreducible subcontinuum p of M from A to P. The point set $M - p$ is connected and the set of all points X such that $(\overline{M - p})$ is an irreducible continuum from B to X is a continuum R_p that intersects p. The set $M - (\overline{M - p})$ is connected and the set of all points X such that the closure of $M - (\overline{M - p})$ is irreducible from A to X is a continuum L_p that intersects $(\overline{M - p})$. The continua R_p and L_p have at least one point in common and $R_p + L_p$ contains the point P.

The point set $M - (R_p + L_p)$ is the sum of two mutually separated connected point sets D_{pA} and D_{pB} that contain A and B respectively, L_p is the set of all points X such that $\overline{D}_{pA}$ is irreducible from A to X, and R_p is the set of all points X such that $\overline{D}_{pB}$ is irreducible from B to X. Every point of R_p is a limit point of D_{pB} and every point of L_p is a limit point of D_{pA}. So D_{pA} and D_{pB} are domains with respect to M but neither L_p nor R_p contains a domain with respect to M. Hence, since L_p and R_p are closed, $L_p + R_p$ contains no domain with respect to M.

Suppose now that x and y are two subsets of M such that each of them is an irreducible continuum from A to some point of $M - M_B$ and suppose that $R_x + L_x$ and $R_y + L_y$ are distinct but have a point O in common. Some point belongs to one of these two sets but not to the other one. Suppose C is a point belonging to $R_x + L_x$ but not to $R_y + L_y$. The point C belongs either to D_{yA} or to D_{yB}. Suppose it belongs to D_{yA}. Then there exists a continuum N lying in D_{yA} and containing A and C. The point set $N + R_y + L_y + D_{yB}$ is a continuum containing both A and B. Hence it is identical with M. But N and D_{yB} are mutually exclusive subcontinua of M and therefore $M - (N + D_{yB})$ is a domain with respect to M. But it is a subset of $R_y + L_y$ which contains no domain with respect to M. This involves a contradiction. It follows that if $R_x + L_x$ and $R_y + L_y$ are distinct they have no point in common. Let G denote a collection such that g belongs to it if and only if either g is M_A or M_B or, for some irreducible subcontinuum x of M from A to some point of $M - (M_A + M_B)$, g is $R_x + L_x$. The collection G fills up M, its elements are continua and every one of them separates M into two mutually separated connected point sets one containing A and the other containing B. It follows that every two elements of G are separated from each other in M by some element of G. Therefore, by Theorem 54, G is upper semi-continuous and, with respect to its elements, it is a dendron. It follows that, with respect to its elements, it is an arc from M_A to M_B. By Theorem 53, M has arcatomic subsets. Suppose there exists an element g of G which is not an arcatomic subset of M. Let t denote the set of all arcatomic subsets of M which are subsets of g. The set t is an arc of arcatomic subsets of M and t minus its end elements is a domain with respect to the set of all arcatomic subsets of M and therefore g contains a domain with respect to M. This involves a contradiction.

THEOREM 56. *If Axiom 2 holds true and G is an upper semi-continuous collection of mutually exclusive continua which fills up the compact continuum M and which is a nondendritic continuous curve with respect to its elements then there exists an upper semi-continuous collection H of mutually exclusive continua such that* (1) *each continuum of H is either a continuum of G or a point set obtained by adding, to some continuum g of G, one or more of the complementary domains of g,* (2) *no continuum of H separates S,* (3) *every point of M belongs to some continuum of H,* (4) *with respect to its elements, H is a nondendritic continuous curve.*

PROOF. By hypothesis, there exists a simple closed curve J of elements of G. Let T denote the collection whose elements are the continua of G and the points of $S - M$. The collection T fills up S and it is upper semi-continuous. By Theorems 15 and 16, if the elements of T are called "points" and every region with respect to G is called a "region", Axioms 0, 1′ and 2 hold true for the "points" and "regions" so obtained. In the space T, G is a compact continuum. Therefore, by Theorem 15 of Chapter II, since no "point" of G separates any two "points" of J from each other, for each "point" g of G which is separated, in T, by some other "point" of G from some "point" of the "simple closed curve" J, there exists a "point" t_g of G which separates g from every other "point" of J distinct from t_g but which is itself not separated from any "point" of J by any "point" of G. Thus for each continuum g of G which is separated, in S, by some other continuum of G from some continuum of the simple closed curve J of elements, there exists a continuum t_g of G which separates g from every other continuum of J distinct from t_g but which is itself not separated from any continuum of J by any continuum of G. Let Q denote the collection of all continua q of G such that q is not separated from any element of J by any element of G. For each element q of Q let h_q denote the continuum obtained by adding to q all the complementary domains of q that contain no element of J. Let H denote the collection of all such continua h_q for all continua q of the collection Q. The collection H satisfies all the conditions of Theorem 56.

THEOREM 57. *No closed and compact point set is the sum of countably many mutually exclusive continua such that each of them contains a limit point of the sum of all the others.*

PROOF. Suppose, on the contrary, that there exists a countable collection G of mutually exclusive continua $M_1, M_2, M_3, \cdots$ such that their sum is a closed and compact point set M and such that, for each n, M_n contains a limit point of $M - M_n$. By Theorem 56 of Chapter I, for each n, M_n is a component of M. Hence, by Theorem 20, G is an upper semi-continuous collection. With the help of Theorems 149 and 168 of Chapter I and Theorem 15 of the present chapter, it follows that if the elements of G are regarded as "points" and the term "region" is applied to every subcollection H of G such that H^* is a domain with respect to M, then Axioms 0 and 1 hold true for the "points" and "regions" so obtained. The countable "point" set G is compact and perfect. But this is contrary to Theorem 53 of Chapter I.

That Theorem 57 does not remain true, even in the plane, if the stipulation requiring that M be compact is omitted, may be seen with the aid of the following example.

EXAMPLE 9. For each positive integer n, let K_n denote the set of all decimal fractions in which there are only n digits equal to 1 and in which all the other digits are equal to 0. Let $d_{1n}, d_{2n}, d_{3n}, \cdots$ denote the fractions of the set K_n. For each n and m, let A_{mn} denote the point $(1/n, d_{mn}/n)$. Let K denote the point set consisting of the point (1,0) and all points A_{mn} for all values of the positive integers m and n. The points of the countable set K may be designated $P_1, P_2, P_3, \cdots$. For each n, let OP_n denote the straight line interval whose endpoints are O and P_n and let M_n denote the ray into which the point set $OP_n - O$ is thrown by an inversion of the plane about a circle of radius 2 with center at O. Let M denote the sum of all the point sets of the sequence $M_1, M_2, M_3, \cdots$. The point set M is closed and, for each n, M_n contains at least one limit point of $M - M_n$.

That Theorem 57 does not remain true in the plane if the requirement that M be closed is omitted, may be seen with the help of the following example.

EXAMPLE 10 (ANNA M. MULLIKIN). In a Cartesian plane, for each positive integer n, let A_n, B_n, C_n, D_n and E_n denote the points $(1/2^{n-1},0)$, $(1/2^{n-1},1/2^{n-1})$, $(-1/2^{n-1},1/2^{n-1})$, $(-1/2^{n-1},-1/2^{n-1})$ and $(1,-1/2^{n-1})$ respectively. For each n, let M_n denote the arc obtained by adding together the straight line intervals A_nB_n, B_nC_n, C_nD_n and D_nE_n. Let M denote the sum of all the arcs $M_1, M_2, M_3, \cdots$. For each n, M_n contains a limit point of $M - M_n$. Indeed for each n, the arc M_n contains a limit point of every point set which is the sum of infinitely many of the remaining arcs of this sequence and therefore M is connected.

That Theorem 57 does not remain true in the plane on the omission of the requirement that each point set of the countable set in question be closed, may be seen from the following example.

EXAMPLE 11. For each positive integer n, let P_n^n denote the point on OX whose abscissa is $1 - 1/n$. For each pair of positive integers m and n, such that $n > 1$ and $m > n$, let P_m^n denote the point on OX whose abscissa is $a_n + b_n/m$, where a_m and b_n denote the numbers $1 - 1/(n-1)$ and $1/(n-1) - 1/n$ respectively. Let M_1 denote the point set composed of the points P_1^1 and $(1,0)$. For each n greater than 1, let M_n denote the point set composed of the $n - 1$ points $P_n^2, P_n^3, \cdots, P_n^n$.

Theorem 47 does not remain true in the plane if "acyclic" is replaced by "a regular curve". Consider the following example.

EXAMPLE 12. In a Cartesian plane let O and C denote the points $(0,0)$ and $(1,0)$ respectively and for each n let O_n and C_n denote the points $(0,1/n)$ and $(1,1/n)$ respectively. Let OC denote a straight line interval with end-points as indicated and, for each n, let O_nC_n denote the straight line interval whose end points are O_n and C_n. For each positive integer n and *odd* positive integer i less than 2^n, let A_{in} denote the point $(i/2^n,1/n)$. Let $A_1, A_2, A_3, A_4, A_5, A_6, A_7, A_8, A_9, A_{10}, \cdots$ denote $A_{11}, A_{12}, A_{32}, A_{13}, A_{33}, A_{53}, A_{73}, A_{14}, A_{34}, A_{54}, \cdots$ respectively. For each n, let B_n denote the projection of A_n upon OC. Let M denote a point set such that P belongs to it if and only if either P belongs to OC or, for some n, P belongs to O_nC_n or to the straight line interval A_nB_n. For each n, let G_n denote a collection such that g belongs to it if and only if either (1) g is A_mB_m for some m greater than or equal to n or (2) g is a degenerate subset of M which is not a subset of A_mB_m for any m greater than or equal to n. For each n, G_n is a regular curve with respect to its elements. But G is the set of all degenerate subsets of M and M is not a regular curve. Hence G is not a regular curve with respect to its elements.

Definition. If H, L and N are subsets of the continuum M, N is said to *shield H from L in M* if, and only if, N has no point in common with L but intersects every connected subset of M which intersects both H and L.

Theorem 58. *If X and Y are two points of the compact irreducible continuum K from the point A to the point B then X does not shield $A + B$ from Y in K.*

Proof. If $K - X$ is connected it is itself a connected subset of K containing Y and intersecting $A + B$ but not X. Suppose $K - X$ is not connected. Then it is the sum of two mutually separated *connected* point sets K_A and K_B containing A and B respectively. If Y belongs to K_A, K_A is a connected subset of K that intersects $A + B$ but not X. If Y belongs to K_B then K_B is such a subset of K. So, in any case, X does not shield $A + B$ from Y in K.

Theorem 59. *If A and B are points of the continuum M, T is a compact subcontinuum of M containing A and B, H is a closed and compact subset of M and, for each point P of $T - (A + B + T \cdot H)$, there is a continuum lying in H and shielding $A + B$ from P in M, then A and B belong to the same component of H.*

Proof. Suppose A and B do not belong to the same component of H. Let G denote a collection such that g belongs to it if and only if either g is a component of H or g is a point of T that does not belong to H. The collection G is upper semi-continuous. With respect to its elements regarded as points, G is a compact continuum. If A' and B' are the components of H which contain A and B respectively there exists a subcollection Q of G which, with respect to its elements regarded as "points", is an irreducible continuum from the "point" A' to the "point" B'. By Theorem 58, no "point" of Q shields $A' + B'$, in Q, from any other "point" of Q. It follows that no component of H shields $A + B$, in M, from any point of $T - (A + B + T \cdot H)$.

Theorem 59 does not remain true, in the plane, if the supposition that, for each point P of $T - (A + B + T \cdot H)$ there is a continuum lying in H and shielding $A + B$ from P in M is replaced by the supposition that, for each such point P, there is a continuum lying in H and intersecting every subcontinuum of M that intersects both P and $A + B$. Consider the following example.

Example 13. Let A and B denote points belonging to different composants of a compact indecomposable plane continuum M. Let H denote $A + B$. If P is a point of $M - H$, M is an irreducible continuum either from P to A or from P to B. In the first case, B intersects every subcontinuum of M that intersects both P and $A + B$. In the second case, A intersects every such continuum. But no component of H intersects both A and B.

The following theorem may be easily proved with the assistance of Theorem 59.

THEOREM 60. *If the points A and B belong to different components of β, the boundary of the compact continuum M, there exists a component D of $M - \beta$ such that no component of β shields $A + B$ from D in M.*

THEOREM 61. *If β, the boundary of the compactly connected continuum M is compact and, for every component D* of *$M - \beta$ and component E of $S - \bar{D}$, the boundary of E is a subset of a component of β and K is a compact continuum intersecting M, then $M \cdot K$ is a subset of a component of $M \cdot K + \beta$.*

PROOF. Since $M \cdot K$ and β are closed and compact, the sum of all components of $M \cdot K + \beta$ that intersect $M \cdot K$ is closed. Suppose it is the sum of two mutually exclusive closed point sets H and L. These point sets both intersect $M \cdot K$. The continuum K contains a connected point set T, lying in M, such that $\bar{T}$ intersects both $H \cdot M \cdot K$ and $L \cdot M \cdot K$. Let A and B denote points belonging respectively to the subsets $\bar{T} \cdot H \cdot M \cdot K$ and $\bar{T} \cdot L \cdot M \cdot K$ of β. There exists a compact continuum M' lying in M and containing A and B. Suppose P is a point of $M' - M' \cdot \beta$. Let D_P denote the component of $M - \beta$ that contains P and let E_P denote the component of $S - \bar{D}_P$ that contains T. By hypothesis, the boundary of E_P is a subset of a component β_P of β. For each point P of $M' - M' \cdot \beta$, β_P shields $A + B$ from P in M. Hence, by Theorem 59, some component of β contains both A and B. This involves a contradiction.

If the supposition that M is compactly connected is omitted from the hypothesis of Theorem 61, the resulting proposition is not a consequence of these axioms. This may be seen from Example 20 of Chapter I.

2. On the basis of Axioms 0, 1′, C, 2, 3, 4 and 5

THEOREM 1. *Suppose G is an upper semi-continuous collection of mutually exclusive compact continua. For each continuum g of the collection G, let h_g denote either g or the continuum obtained by adding to g one or more of the complementary domains of g (if there are any) that contain no point of any continuum of G. Let H denote the collection of all continua h_g for all continua g of the collection G. Then the collection H is upper semi-continuous and the element g of G is a limit element of the subcollection Q of G if and only if the element h_g of H is a limit element of H_Q, the collection of all elements x of H such that x is identical with h_q for some element q of Q.*

Theorem 1 may be proved with the help, in particular, of Theorems 39 and 4 of Chapter II.

THEOREM 2. *Suppose G is an upper semi-continuous collection of mutually exclusive compact continua such that no continuum of G separates any two continua of G from each other and such that G^* is closed. For each continuum g of G that does not separate S, let h_g denote g and, for each continuum g of G that separates S, let h_g denote the continuum obtained by adding to g all the comple-*

mentary domains of g that contain no continuum of G. Let H denote the collection of all continua h_g for all continua g of the collection G. Let W denote the collection whose elements are the continua of H and the degenerate subsets of $S - H^*$. The collection W is an upper semi-continuous collection and if the elements of W are called "points" and every region with respect to W is called a "region", Axioms 1′, C, 2, 3, 4 *and* 5 *all hold true for the "points" and "regions" so obtained.*

Theorem 19 of Part 1 of the present chapter was established with the aid of a rather long sequence of propositions. Many of the propositions of this sequence may be modified by omitting, from their hypotheses, one or more stipulations requiring the compactness of certain closed point sets and replacing each stipulation thus omitted by a stipulation requiring the weaker condition that each of the point sets in question be either a closed and compact point set or the sum of such a point set and one or more of its complementary domains. With the aid of the new sequence of propositions thus obtained, together with Theorem 1, it is possible to establish Theorem 2.

THEOREM 3. *If M is a compact continuum which does not separate S and G is a countable upper semi-continuous collection of mutually exclusive subcontinua of M no one of which separates M, then the sum of all the continua of the collection G does not separate M.*

PROOF. Suppose, on the contrary, that M is separated by the sum of all the continua of the collection G. Let S' denote the collection whose elements are the continua of the collection G and the degenerate subsets of $S - G^*$. The collection S' is an upper semi-continuous collection of mutually exclusive compact continua and no continuum of S' separates any two continua of S' from each other. It follows, by Theorems 1 and 2, that if the elements of S' are regarded as "points" and every region with respect to S' is regarded as a "region", then Axioms 1′, C, 2, 3, 4 and 5 all hold true for the "points" and "regions" so obtained. In the space S', let M' denote the set of all "points" which are either elements of G or degenerate subsets of $M - G^*$. In this space, M' is a compact continuum not separating S' and G is a countable set of non-cut points of M' such that G separates M'. By Theorems 168, 149, 154, 16 and 37 of Chapter I, M' is separated by some closed subset T of G'. The point set $M' - T$ is the sum of two mutually separated point sets H and K. For each point X of H let H_X denote the component of $H + T$ which contains X. For each X, H_X contains at least one point of T. It follows that there exists, in H, a countable set of points $X_1, X_2, X_3, \cdots$ such that every point of H belongs to some point set of the collection $H_{X1}, H_{X2}, H_{X3}, \cdots$ and no two point sets of this collection have a point in common. There are two cases to be considered.

Case 1. Suppose that, for every n, H_{X_n} contains a limit point of $(H + T) - (H_{X_n} + T)$. Let U denote the collection of point sets H_{X_1}, $H_{X_2}, H_{X_3}, \cdots$ together with all points of T which do not belong to any H_X

but which are limit points of H. The sum of all the continua of the countable collection U is closed and compact and each continuum of this collection contains at least one limit point of the sum of all the others. But this is contrary to Theorem 57 of Part 1 of this chapter. Thus in Case 1 we have a contradiction.

Case 2. Suppose that there exists a point set E which belongs to the collection $H_{X_1}, H_{X_2}, H_{X_3}, \cdots$ and which contains no limit point of $(H + T) - (E + T)$. For every point Y of the set $M' - E$ let D_Y denote the component of $(M' - E) + T$ which contains Y. For every Y, D_Y is a continuum and D_Y and E have in common at least one point and their common part is a subset of T. They do not have more than one point in common. For suppose they do. Then their common part is a closed and countable nondegenerate point set N. Hence, by Theorems 27 and 53 of Chapter I, N is not connected. If there is any complementary domain of E which contains no point of D_Y let e denote the point set obtained by adding all such domains to E. If there is no such domain, let e denote E. Similarly if there is any complementary domain of D_Y which contains no point of E, let d denote the point set obtained by adding all such domains to E and if there is no such domain let d denote D_Y. Clearly N is a subset of $d \cdot e$. Suppose $d \cdot e$ contains a point P not belonging to N. Then P belongs both to a complementary domain U of E and to a complementary domain V of D_Y. Let β_U and β_V denote the boundaries of U and V respectively. Since β_U is a subset of E and β_V is a subset of D_Y therefore neither of the domains U and V contains a point of the boundary of the other one. But they are connected and they have a point in common. Hence U is identical with V. Therefore β_U is identical with β_V. Hence β_U is a subset of N. But N is totally disconnected and, by Theorem 24 of Chapter IV and Axiom 3, β_U is a nondegenerate continuum. Thus the supposition that $d \cdot e$ is not a subset of N leads to a contradiction. It follows that $d \cdot e$ is identical with N. Therefore, by Theorem 22 of Chapter IV, the point set $E + D_Y$ separates S'. But $E + D_Y$ is a subset of M' and M' does not separate S'. Therefore at least one of the complementary domains of $E + D_Y$ is a subset of M'. Let D denote one which is a subset of M' and let β denote its boundary. The point set β is a nondegenerate continuum containing a point P of $E - N$. There exists a region R containing P but no point of K. The region R contains a nondegenerate subcontinuum β_1 of β. Since T is totally disconnected, β_1 contains a point X not belonging to T. The point X belongs to $E - E \cdot T$. There exists a region R_1 containing X but no point of $K + T$. There exists in R_1 a point Y belonging to D. There exists an arc XY lying wholly in R_1. Let Y_1 denote the last point of β in the order from X to Y on the arc XY. Let YY_1 denote the interval of XY whose end points are Y and Y_1. The point set $E + YY_1$ is a connected subset of $M - (K + T)$, and therefore of H, containing the point Y_1 of E. Hence YY_1 is a subset of E. Thus the supposition that D_Y and E have more than one point in common has led to a contradiction. Hence the common part of D_Y and E is a single point of T. Also, for every two points Z and W of the point set $M' - E$, either D_Z and D_W are identical or they have no point in common. It follows that there exist, in $M' - E$, countably many points $Y_1, Y_2, Y_3, \cdots$ such that every point of $M' - E$ belongs to some point set of the sequence $D_{Y_1}, D_{Y_2}, D_{Y_3}, \cdots$. Let Q denote the collection of point sets consisting of $D_{Y_1}, D_{Y_2}, D_{Y_3}, \cdots$ and all points of T which belong to E and not to any D_Y but which are limit points of $M' - E$. Every point set of the collection Q which is not a D_Y is a limit point of the sum of all the point sets of the collection Q. Suppose that, for some Y, D_Y does not contain a limit point of the sum of all the other point sets of the collection Q. Let g_Y denote the point of T which is common to D_Y and E. No point of $D_Y - g_Y$ is a limit point of $M' - D_Y$. Furthermore no point of $M' - D_Y$ is a limit point of $D_Y - g_Y$. But $M' - g_Y = (M' - D_Y) + (D_Y - g_Y)$. Therefore $M' - g_Y$ is not connected. But this is contrary to hypothesis. Hence every point set of the collection Q contains a limit point of the sum of all the other sets of that collection. Furthermore Q^* is closed and compact. But this is contrary to Theorem 57 of Part 1.

THEOREM 4. *If A and B are two mutually exclusive continua and the compact continua H_1 and H_2 are both irreducible about $A + B$, and $H_1 \cdot H_2 = A + B$, then $H_1 + H_2$ is not the boundary of more than two connected domains.*

PROOF. Let F denote $H_1 + H_2$. By Theorem 48 of Chapter I, the point sets $H_1 - A$ and $H_2 - A$ are connected. But their common part is the connected point set B. Hence their sum is connected. Hence, if X_1 and X_2 are points belonging to $H_1 - (A + B)$ and $H_2 - (A + B)$ respectively, X_1 and X_2 are not separated from each other by A. Similarly they are not separated from each other by B. Therefore, by Theorem 20 of Chapter IV, they are not separated by $A + B$. Let D denote the complementary domain of $A + B$ that contains X_1 and X_2. By Theorem 48 of Chapter I, $H_1 - (A + B)$ and $H_2 - (A + B)$ are connected. But each of them contains a point of D and neither of them contains a point of the boundary of D. Hence $H_1 - (A + B)$ and $H_2 - (A + B)$ are subsets of D. Let G denote the collection whose elements are the continua A and B and the points of D. By Theorems 1 and 2, G is an upper semi-continuous collection and if its elements are regarded as points and every subcollection H of G such that H^* is a domain with respect to G^* is regarded as a region, then Axioms 1′, C, 2, 3, 4 and 5 are all satisfied. Hence, by Theorem 44 of Chapter IV, there exists a simple closed curve J of elements of G which separates $H_1 - (A + B)$ from $H_2 - (A + B)$. If t_1 and t_2 denote the components of $J - (A + B)$ and, for each i $(i = 1,2)$, D_i denotes the complementary domain of F that contains t_i then if F is the boundary of a connected domain that domain contains a point of t_1 or of t_2 and therefore must be identical with D_1 or with D_2.

THEOREM 5. *If α and β are two mutually exclusive closed point sets and each of the compact continua C_1 and C_2 is irreducible from α to β, and $C_1 \cdot C_2 = \alpha + \beta$, then $C_1 + C_2$ is the boundary of at least two distinct connected domains.*

PROOF. Let F denote the point set $C_1 + C_2$. By Theorem 47 of Chapter I, $C_1 - \alpha$, $C_2 - \alpha$, $C_1 - \beta$, $C_2 - \beta$, $C_1 - (\alpha + \beta)$ and $C_2 - (\alpha + \beta)$ are all connected. By an argument employed in a similar connection in the proof of Theorem 4, it follows that there exists a component E of $S - (\alpha + \beta)$ containing both $C_1 - (\alpha + \beta)$ and $C_2 - (\alpha + \beta)$. Let G_α denote the collection of all components of α and let G_β denote the collection of all components of β. Let U denote the collection of all components of $\alpha + \beta$. The collection U is the sum of the collections G_α and G_β and, by Theorem 20 of Part 1, each of the collections G_α, G_β and U is upper semi-continuous. Let G denote the upper semi-continuous collection whose elements are the elements of U and the points of E. Let H_1 denote the subcollection of G whose elements are the elements of G_α, the elements of G_β and the points of $C_1 - (\alpha + \beta)$ and let H_2 denote the collection whose elements are the elements of G_α, the elements of G_β and the points of $C_2 - (\alpha + \beta)$. The set of elements U is a totally disconnected set of elements and it consists of all elements common to the two closed, connected and compact sets H_1 and H_2. Let ω denote some definite point of E and let Q denote a simple closed curve lying in E and separating ω from $H_1 + H_2$. There exists an arc AB such that (1) A belongs to Q, (2) B belongs to $C_1 - (\alpha + \beta)$ and (3) B is the only element of AB that belongs to $H_1 + H_2$. Suppose B belongs to $C_1 - (\alpha + \beta)$. Then $C_2 - (\alpha + \beta)$ lies in a complementary domain of $Q + AB + H_1$ that does not contain ω. Call this domain D. By Theorem 50 of Chapter IV, there exists a simple closed curve J of elements of G such that (1) J separates some element of $C_2 - (\alpha + \beta)$ from ω, (2) $(J + U) - U$ is a subset of D and it contains no element of $C_2 - (\alpha + \beta)$. The simple closed curve J contains at least one element of U. With the help of the fact that $C_2 - (\alpha + \beta)$ is connected and that every element of U is a limit element of $C_2 - (\alpha + \beta)$ it easily follows that J contains U and separates the whole of $C_2 - (\alpha + \beta)$ from ω. Furthermore, no point of $C_1 - (\alpha + \beta)$ is separated from ω by J. It is easy to see that the curve J contains two segments t_1 and t_2 such that (1) for each i ($i = 1,2$), t_i contains no element of U, but the end elements of t_i belong to G_α and G_β respectively and (2) t_1 and t_2 are subsets of different complementary domains of $H_1 + H_2$. The segments t_1 and t_2 are point sets in the ordinary sense and they are connected subsets of $S - F$ and each of the sets α and β contains at least one limit point of t_1 and at least one limit point of t_2. In the space S, for each i let D_i denote the complementary domain of F that contains t_i. The continuum F is the boundary both of D_1 and of D_2. For let N denote the boundary of D_1. If N does not contain C_2 then $N \cdot C_2$ contains no connected point set containing both a point of some element of G_α and a point of some element of G_β. Hence, if I denotes the component of $S - J^*$ that does not contain ω, there exists an arc A_1A_2 from a point A_1 of t_1 to a point A_2 of t_2 and lying, except for these points, wholly in I and having no point in common with $N \cdot \bar{C}_2$. Since $C_1 - (\alpha + \beta)$ lies wholly in $S - (J + I)$, it has no point in common with N. But A_1 and A_2 are in different complementary domains of N. Thus the supposition that N does not contain C_2 has led to a contradiction. In a similar way it may be shown that N contains C_1. Therefore F is the boundary of D_1. By a similar argument it may be shown that F is the boundary of D_2. It follows that it is the boundary of at least two connected domains.

NOTE. In the above proof it is stated that the segment t_1 of an arc of elements of G is a point set in the ordinary sense. This does not imply that t_1 is necessarily a subset of an *arc* whose elements are all points in the ordinary sense. Consider the following example.

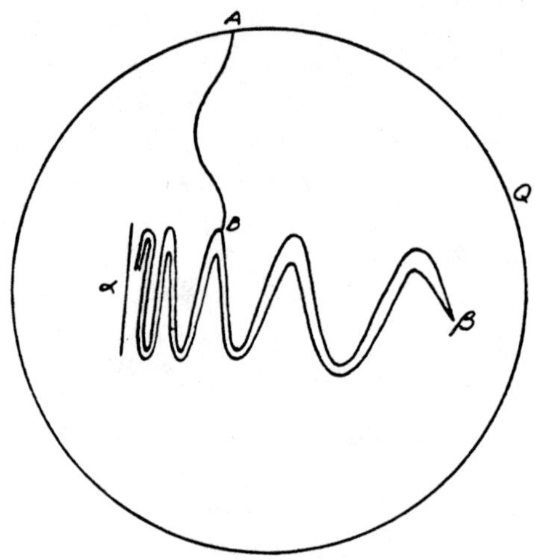

FIGURE 37

EXAMPLE 1. In a Cartesian plane, let α denote the straight line interval whose extremities are the points $(0,1)$ and $(0,-1)$ and let β denote the point $(1/\pi,0)$. Let C_1 denote the point set whose elements are the point β, the points of the interval α and the points of the graph of $y = \sin(1/x)$ that lie between $x = 0$ and $x = 1/\pi$. Let C_2 denote the point set whose elements are the point β, the points of α and the points of the graph of $y = x(x - 1/\pi) + \sin(1/x)$ that lie between $x = 0$ and $x = 1/\pi$. The point sets α, β, C_1 and C_2 (Figure 37) satisfy the conditions of the hypothesis of Theorem 5. But, in this case, neither t_1 nor t_2 is a subset of an ordinary arc. If the simple closed curve Q, employed in the proof of Theorem 5, should enclose $C_1 + C_2$, it would be possible, in this example, for the point set t_1 to be the set of all points of the graph of $y = x(x - 1/\pi)/2 + \sin(1/\pi)$ that lie between $x = 0$ and $x = 1/\pi$. This point set is not a subset of any arc whose elements are all points in the plane but it is a segment of an arc (of elements of G) of which the only element which is not a point (in the plane) is the straight line interval α.

THEOREM 6 (W. A. WILSON). *Suppose A, B, H_1, H_2, C_1 and C_2 are compact continua such that* (1) *A and B are mutually exclusive,* (2) *each of the continua H_1 and H_2 is irreducible about $A + B$,* (3) *C_1 and C_2 are subcontinua of H_1 and H_2 respectively,* (4) *$H_1 + H_2 = C_1 + C_2$ and* (5) *if α and β denote $C_1 \cdot C_2 \cdot A$ and $C_1 \cdot C_2 \cdot B$ respectively, then α and β exist and each of the continua C_1 and C_2 is irreducible from α to β. Then there are only two complementary domains of $H_1 + H_2$ that have $H_1 + H_2$ as their boundary.*

Theorem 6 may be established with the use of Theorems 4 and 5.

THEOREM 7. *If G_1 and G_2 are two upper semi-continuous collections of mutually exclusive continua filling up the compact continuum M and each of these collections is a continuous curve with respect to its elements and, for each point P of M, g_P is the component containing P of the common part of the continuum of G_1 that contains P and the continuum of G_2 that contains P, and G is the collection of all continua g_P for all points P of M, then the upper semi-continuous collection G is a continuous curve with respect to its elements.*

PROOF. Suppose that G is not a continuous curve with respect to its elements. Let V denote the collection whose elements are the continua of G and the points of $S - M$. The collection V fills up S and is upper semi-continuous. By Theorems 15 and 16 of Part 1, if the elements of V are called "points" and every region with respect to V is called a "region", Axioms 0, 1′ and 2 hold true for the "points" and "regions" so obtained. In the space V, G is a compact continuum which is not a continuous curve. Therefore, by Theorem 73 of Chapter II, there exists, *in that space*, a "subcontinuum" K of G such that K is not a "continuous curve" and no "point" separates any two "points" of K from each other in V. In other words, again considering, as space, the original space whose points are the points of S, there exists a subcollection G' of G which is a continuum, but not a continuous curve, with respect to its elements and which has the property that no one of its elements separates any two of its elements from each other in S. Let S' denote the collection whose elements are the continua of G' and the points of S that are not separated from any continuum of G' by any other continuum of G'. If the elements of S' are called "points" and a subcollection H of S' is called a "region" if and only if H^* is a domain with respect to $(S')^*$, then, as may be seen with the help of Theorems 1 and 2, Axioms 0, 1′, C, 2, 3, 4 and 5 all hold true for the "points" and "regions" so obtained. In the space Σ' so obtained, G' is a compact continuum which is not a continuous curve. Hence there exist, in Σ', two simple domains D_1

and D_2, a continuum T and an infinite sequence of mutually exclusive continua $M_1, M_2, M_3, \cdots$ such that (1) $\bar{D}_2$ is a subset of D_1, (2) for each n, M_n is a component of $(\bar{D}_1 - D_2)\cdot G'$ containing at least one point of β_1, the boundary of D_1, and at least one point of β_2, the boundary of D_2, (3) the sequence $M_1, M_2, M_3, \cdots$ converges to T and no one of its elements has any point in common with T. There exists an arc B_1B_2 containing no point of the continuum G' and lying wholly in $D_1 - \bar{D}_2$ except for its endpoints B_1 and B_2 which belong to β_1 and β_2 respectively. There exist a subdomain D of the domain $D_1 - \bar{D}_2$, an interval B_1A_1 of β_1 and an interval B_2A_2 of β_2 such that (1) $(B_1A_1)\cdot T = A_1$ and $(B_2A_2)\cdot T = A_2$, (2) D is bounded by $B_1B_2 + B_1A_1 + B_2A_2 + T$, (3) there exists an infinite subsequence $T_1, T_2, T_3, \cdots$ of the sequence $M_1, M_2, M_3, \cdots$ such that, for each n, T_n is a subset of $\bar{D}$. For each continuum g belonging either to G_1 or to G_2 and containing a continuum of G', let g' denote the set of all continua of G' which are subsets of g. For each i, let G_i' denote the set of all such continua g' for all continua g of G_i $(i = 1,2)$ that contain one or more continua of G'. For each i $(i = 1,2)$, every element of G_i' is a continuum of points of the space Σ' and, in the space Σ', G_i' is an upper semi-continuous collection of mutually exclusive continua filling up a compact continuum M'. It may be seen that G_i' is a continuous curve with respect to its elements. For each point P of G', let g_P denote the continuum of G_1' that contains P. Let O denote a point of T. Suppose that $T\cdot g_O$ contains no continuum containing both O and a point of $A_1 + A_2$. Then, by Theorems 16, 42 and 37 of Chapter I, there exist two domains W_1 and W_2, containing O and $A_1 + A_2$ respectively, such that (1) $T\cdot g_O$ contains no point of the boundary of W_1 or of the boundary of W_2, (2) $\bar{W}_1$ and $\bar{W}_2$ have no point in common. If $(W_1 + W_2)\cdot g_O$ is not identical with g_O there exists a domain W_3 containing the closed point set $g_O - (W_1 + W_2)\cdot g_O$ but containing no point of T. Let U denote $W_1 + W_2$ or $W_1 + W_2 + W_3$ according as $(W_1 + W_2)\cdot g_O$ is or is not identical with g_O. The domain U contains g_O but it contains no subcontinuum of T containing both O and a point of $A_1 + A_2$. There exists a sequence of points $P_1, P_2, P_3, \cdots$, converging to O, such that, for each n, P_n belongs to T_n. Since the upper semi-continuous collection G_1' is a continuous curve with respect to its elements, there exists a number δ such that if $n > \delta$ then g_P can be joined to g_O by a subcontinuum q_n of G_1' of which each element is a subset of U. Let q_n^* denote the sum of all the continua of the collection q_n. For each n greater than δ, q_n^* is a subcontinuum of G' containing P_n and O and lying in U. But P_n belongs to T_n and T_n is a component of $(\bar{D}_1 - D_2)\cdot G'$ not containing O. Hence, for each n greater than δ, $q_n^*\cdot T_n$ contains a continuum t_n containing both O and a point of $\beta_1 + \beta_2$. The limiting set of the sequence $t_{\delta+1}, t_{\delta+2}, t_{\delta+3}, \cdots$ is a continuum t containing O and a point of $A_1 + A_2$. But, for each n greater than δ, t_n is a subset of U. Hence t belongs to U. But this involves a contradiction. Hence, for every point O of T, g_O contains a subcontinuum of T containing both O and a point of $A_1 + A_2$.

Hence, for every point O of T, g_O is identical either with g_{A_1} or with g_{A_2}. Therefore T is a subset of $g_{A_1} + g_{A_2}$. It follows that g_{A_1} is identical with g_{A_2}. For, otherwise, since it contains both A_1 and A_2, T would be the sum of two mutually exclusive closed point sets $T \cdot g_{A_1}$ and $T \cdot g_{A_2}$, contrary to the fact that T is a continuum. Therefore T is a subset of g_{A_1}. Similarly T is a subset of x, the continuum of G_2' that contains A_1. Thus the common part of x and g_{A_1} contains the nondegenerate continuum T. This involves a contradiction.

THEOREM 8. *If α and β are two mutually exclusive arcs of the boundary of the simple domain D and G is a collection of mutually exclusive continua such that* (1) *each continuum of G is a subset of $\bar{D}$ containing both a point of α and a point of β,* (2) *the sum of all the continua of G is a closed and compact point set; then the collection G is upper semi-continuous.*

THEOREM 9. *If I is a domain bounded by a simple closed curve J and A, B, C and F are four points lying on J in the order $ABFC$ and AB, BF, CF and AC are arcs of J, with endpoints as indicated, such that no one of them contains any one of the points A, B, C and F except its endpoints and G_1 and G_2 are two uncountable collections of continua such that* (1) *for each i ($i = 1,2$), the continua of G_i are mutually exclusive and the sum of all the continua of G_i is a closed and compact subset of $\bar{I}$,* (2) *each continuum of G_1 contains a point of AB and a point of CF and* (3) *each continuum of G_2 contains a point of AC and a point of BF; then, if M denotes the sum of all the continua of G_1 and all the continua of G_2, there exists an upper semi-continuous collection G of mutually exclusive continua such that* (a) *with respect to its elements G is a space satisfying Axioms* 0, 1′, C, 2, 3, 4 *and* 5, (b) *every nondegenerate continuum of the collection G is a subset of the point set M,* (c) *in the space G there exists a domain D (of elements of G) such that every element of G that belongs to D is, in the original space, whose points are the points of S, a subset of M.*

PROOF. For each point P of the interval AB, let t_P denote the greatest connected subset of AB containing P and intersecting only countably many continua of the collection G_1. For each P of AB, t_P is either a point or an interval of AB. Let T denote the collection of all the continua t_P for all points P of AB. Suppose either X denotes A and Y denotes C or X denotes B and Y denotes F. If X belongs to T, let r_X denote XY plus the continuum of G_1 that contains X. If the arc t of the collection T has one endpoint at X, let r_t denote $t + x_t + y_t$ where x_t is the continuum of the collection T that contains the other endpoint of t and y_t is either XY or XY plus the continuum of G_1 that intersects XY according as XY and G_1^* are or are not mutually exclusive. For each point t of the collection T, distinct from A and from B, let r_t denote the continuum of the collection G_1 that contains t. For each arc t of the collection T, which contains neither A nor B, let r_t denote the continuum of the collection G_1 that contains the endpoints of T

if there exists such a continuum and, if there exists no such continuum, let r_t denote the point set $x_t + y_t$ where x_t is a continuum of G_1 containing one endpoint of t and y_t is the one containing the other one. For each continuum t of the collection T, let s_t denote the smallest connected subset of CF that contains the point set $r_t \cdot (CF)$ and let h_t denote the point set $t + r_t + s_t$. Let H_1 denote the collection of all continua h_t for all continua t of the collection T. Let H_2 denote the collection of continua obtained in an entirely analogous manner from the collection G_2, the arcs AC and BF now playing the role of AB and CF. For each continuum h of the collection H_1, let w_h denote the point set obtained by adding to h all the complementary domains of h that are subsets of I. Let W_1 denote the collection of all such continua w_h for all continua h of H_1. Let W_2 denote the collection of continua obtained in the same way from H_2. Let W denote the collection of all continua w such that w is a component of the common part of some continuum of W_1 and some continuum of W_2. Let Q denote the collection whose elements are the continua of W and the points that belong to no continuum of W. By Theorem 108 of Chapter IV, every continuum of the collection Q is bounded by a compact continuum. By Theorem 44 of Part 1, and Theorem 8 of Part 2, of the present chapter, W is an upper semi-continuous collection. The collection Q is an upper semi-continuous collection of mutually exclusive continua filling up S. Let G denote the collection of all point sets g such that g is the boundary of a continuum of the collection Q. As may be seen with the help of Theorems 1 and 2, G is an upper semi-continuous collection and if the elements of G are regarded as points and the term "region" is applied to every subcollection H of G such that H^* is a domain with respect to G^*, then Axioms 0, 1′, C, 2, 3, 4 and 5 all hold true for the "points" and "regions" so obtained. Let x_1 and x_2 denote the continua of W_1 that contain AC and BF respectively and let x_3 and x_4 denote the continua of W_2 that contain AB and CF respectively. Let D denote the set of all continua of G that contain no points of any one of the continua x_1, x_2, x_3 and x_4. In the space G, D is a connected domain.

THEOREM 10. *If D is a simple domain, then Axioms* 0, 1′, C, 2, 3, 4 *and* 5 *remain true if only the points of D are regarded as points and only those regions which are subsets of D are regarded as regions.*

THEOREM 11. *Suppose M is a compact continuum and G_1 and G_2 are two upper semi-continuous collections of mutually exclusive continua such that each of the collections G_1 and G_2 fills up M and each of them is a dendron with respect to its elements. For each point P of M, let g_P denote the component, containing P, of the common part of the continuum of G_1 that contains P and the continuum of G_2 that contains P. Let G denote the collection of all such point sets g_P for all points P of M. Then either G is a dendron with respect to its elements or* (1) *there is a subcollection of G which, regarded as a space, satisfies Axioms* 0, 1′, 2, 3, 4 *and* 5 *and* (2) *there is a collection X of continua such that* (a) *each* *continuum of the collection* X *is either a continuum of the collection* G *or the sum of some continua of the collection* G, (b) *regarded as a space*, X *satisfies Axioms* 0, 1′, C, 2, 3, 4 *and* 5.

PROOF. Suppose that G is not a dendron with respect to its elements. By Theorem 7, G is a continuous curve with respect to its elements. By Theorem 56 of Part 1, and Theorem 1 of Part 2, of the present chapter, there exists a subcollection H of G such that (1) each point of M either belongs to a continuum of the collection H or is separated by some continuum of H from every other continuum of H, (2) no continuum of H separates two continua of H from each other in S, (3) with respect to its elements H is a nondendritic continuous curve. For each continuum g of the collection G_1 that contains one or more continua of the collection H, let g' denote the collection of all continua of H that are subsets of g. Let G_1' denote the collection of all such collections g' for all continua g of the collection G_1. Let G_2' denote a collection of collections obtained in an analogous manner from the collection G_2. Let S' denote the collection of all continua P' such that P' is either a continuum of the collection H or a point which neither belongs to any continuum of the collection H nor is separated by any continuum of H from any other continuum of H. By Theorems 1 and 2, if the elements of S' are regarded as "points" and the subcollection Z of S' is regarded as a "region" if, and only if, Z^* is a domain with respect to $(S')^*$, then Axioms 0, 1′, C, 2, 3, 4 and 5 all hold true for the space Σ' so obtained. With the help of Theorem 43 of Part 1, it may be seen that, in the space Σ', G_1' and G_2' are dendrons with respect to their elements and H is a nondendritic continuous curve. In this space, let Q denote a simple closed curve lying in the continuous curve H. Since, in Σ', no continuum of the collection G_1' has a nondegenerate subcontinuum in common with any continuum of the collection G_2' therefore the point set Q intersects either two distinct continua of G_1' or two distinct continua of G_2' in a nondegenerate point set. Suppose it so intersects two continua a and b belonging to G_1'. There exist, on the curve Q, two mutually exclusive arcs t_1 and t_2 each containing both a point of a and a point of b. But G_1' is a dendron with respect to its elements. Therefore a and b are separated, in the continuous curve H, by uncountably many elements of G_1'. But every point set that separates a from b in H must intersect both t_1 and t_2. Hence if W denotes the collection of all continua of G_1' that intersect both t_1 and t_2, W is uncountable. It follows that there exists a sequence β of continua of the collection W that converges to a continuum w_0 of W. There exist two points X and Y belonging to w_0 but belonging to different continua of the collection G_2'. With the help of the fact that G_2' is upper semi-continuous it may be seen that there exist two mutually exclusive domains D_X and D_Y, containing X and Y respectively, such that no continuum of G_2' intersects both of them. Since H is a continuous curve there exist two connected domains I_X and I_Y, with respect to $H \cdot D_X$ and $H \cdot D_Y$ respectively, containing X and Y respectively. There exists a continuum w_1 belonging to the sequence β, but distinct from w_0, and containing both a point of I_X and a point of I_Y. There exist, in I_X and I_Y respectively, arcs AB and CF such that A and C belong to w_0, B and F belong to w_1 and neither AB nor CF has any point except its endpoints in common with w_0 or w_1. Let x and y denote two continua of G'_1 that separate w_0 from w_1 in G'_1. Let Q' denote the point set obtained by adding together x and y and all the continua of G'_1 that separate x from y in G'_1. There exist, on AB, two points A' and B', in the order $AA'B'B$ and, on CF, two points C' and F', in the order $CC'F'F$ such that Q' contains no point of the point set $AA' + BB' + CC' + FF'$, where AA' and BB' denote intervals of AB, and CC' and FF' denote intervals of CF. If AB' and CF' denote intervals of AB and CF respectively with endpoints as indicated then B and F lie in the connected subset w_1 of the point set $H - (w_0 + Q' + AB' + CF')$ and therefore, by Theorem 13 of Chapter II, there exists an arc BF lying wholly in this point set. The arc BF contains an arc $B''F''$ having in common with $BB' + FF'$ only the points B'' and F'', which belong to BB' and FF' respectively. Similarly if AB' and $C'F$ denote intervals of AB and CF respectively, the points A and C lie in the connected subset w_0 of the point set $H - (B''F'' + Q' + A''B'' + C'F'')$ and therefore there exists an arc $A''C''$ lying wholly in this point set and having in common with AA' only the points A'' and C'', which belong to AA' and CC' respectively. Let $A''B''$ and $C''F''$ denote intervals of AB and CF respectively. Of the continua of G'_1 that separate x from y in G'_1 each intersects both $A''B''$ and $C''F''$ and no one of them intersects either $A''C''$ or $B''F''$. Let D_1 and D_2 denote the complementary domains of the simple closed curve J formed by the arcs $A''B''$, $B''F''$, $C''F''$ and $A''C''$. Let α denote the collection of all continua h such that h contains a point of $A''B''$ and a point of $C''F''$ and is a component of the common part either of $\bar{D}_1$ or of $\bar{D}_2$ and some continuum of G'_1 that separates x from y in H. There exist an uncountable subcollection α' of α and a domain D such that (1) D is either D_1 or D_2, (2) every continuum of the collection α' is a subset of $\bar{D}$, (3) no two continua of the collection α' are subsets of the same continuum of the collection G'_1. By Theorem 78 of Chapter IV, there exists an uncountable subcollection α'' of α' such that each continuum of α'' has only one point in common with $A''B''$ and only one point in common with $C''F''$. There exist, between A'' and B'', and between C'' and F'', on the intervals $A''B''$ and $C''F''$ respectively, two points O and L such that, if OA'' and OB'' denote intervals of $A''B''$ and LC'' and LF'' denote intervals of $C''F''$ then (1) O and L belong to the same continuum of the collection α'', (2) if t_O and t_L are segments of $A''B''$ and $C''F''$ respectively containing O and L respectively, there are uncountably many continua of the collection α'' each containing both a point of $t_O \cdot (OB'')$ and a point of $t_L \cdot (LF'')$ and there are also uncountably many continua of α'' each containing both a point of $t_O \cdot (OA'')$ and a point of $t_L \cdot (LC'')$. Let q_O and q_L denote the continua of G'_2 that contain O and L respectively. There exists an uncountable subcollection U of the collection G_2' such that every continuum of U separates q_O from q_L in H and the sum of all the continua of U is a closed point set containing no point of q_O or of q_L. There exist points A_1, B_1, C_1 and F_1 such that (1) A_1 and B_1 lie on $A''B''$ in the order $A''A_1OB_1B''$ and C_1 and F_1 lie on $C''F''$ in the order $C''C_1LF_1F''$, (2) A_1 and C_1 belong to the same continuum of α'' and so do B_1 and F_1, (3) no continuum of the collection U intersects either the interval A_1B_1 of $A''B''$ or the interval C_1F_1 of $C''F''$, (4) if x and y denote either segments of A_1O and C_1L with endpoints at A_1 and C_1 respectively or segments of B_1O and F_1L with endpoints at B_1 and F_1 respectively, then there exist uncountably many continua of the collection α'' each intersecting both x and y. Let h_O denote the continuum of α'' that contains both O and L and let g_{1O} denote the continuum of G_1' that contains h_O. Let g_{1A} and g_{1B} denote the continua of G_1' that contain A_1 and B_1 respectively. By Theorems 43 and 47 of Chapter I, the point set $h_O - (O + L)$ contains a connected point set having O and L as limit points. This connected set is a subset of a component N of the point set $H - (g_{1A} + g_{1B} + A_1B_1 + C_1F_1)$. By Theorem 61 of Chapter IV, both O and L are accessible from N. It follows that there exists an arc OL lying, except for its endpoints, wholly in N. Let I_1 and I_2 denote the subsets of D which are bounded respectively by the simple closed curves $A''O + OL + LC'' + A''C''$ and $OB'' + B''F'' + LF'' + OL$. With the aid of the fact that OL contains no point either of g_{1A} or of g_{1B} and that G_1' is upper semi-continuous it may be seen that there exist two uncountable subcollections α_1'' and α_2'' such that every continuum of α_1'' is a subset of $I_1 + A_1O + C_1L$ and every continuum of α_2'' is a subset of $I_2 + B_1O + F_1L$. There exist, on the arc OL, two points O_1 and L_1, in the order OO_1L_1L, such that if O_1L_1 denotes the interval of OL whose endpoints are O_1 and L_1 then no continuum of U contains any point of $OL - O_1L_1$. Every continuum of U intersects both O_1L_1 and J, the boundary of D, and therefore contains a subcontinuum which is irreducible from O_1L_1 to J. For each continuum u of U select just one such subcontinuum of u and call it v_u. Let V denote the collection of all the continua v_u so obtained. No continuum of V contains a point of A_1OB_1 or of C_1LF_1 or of either of the intervals O_1O and L_1L of OL. Hence every continuum of V either lies wholly in $I_1 + O_1L_1 + A_1A''C''C_1$ and is irreducible from O_1L_1 to $A_1A''C''C_1$ or lies wholly in $I_2 + O_1L_1 + B_1B''F''F_1$ and is irreducible from O_1L_1 to $B_1B''F''F_1$. There exists an uncountable subcollection V' of V such that either every continuum of V' is a subset of $\bar{I}_1$ or every continuum of V' is a subset of $\bar{I}_2$. Suppose every one is a subset of $\bar{I}_1$. Let T denote the collection of all point sets t such that t contains both a point of the interval A_1O of $A''O$ and a point of the interval C_1L of $C''L$ and is a component of the common part of $\bar{I}_1$ and some continuum of the collection G_1'. With the help of the fact that G_1' is upper semi-continuous, it may be seen that the sum of all the continua of the collection T is a closed and compact point set. Let V'' denote the collection of all point sets w such that w intersects both the arc OL and the arc $A_1A''C''C_1$ of J and is a component of the common part of $\bar{I}_1$ and some continuum of the collection U. Let J_1 denote the boundary of I_1. The four points A_1, O, L and C_1 are on J_1 in the order A_1OLC_1, T is an uncountable collection of mutually exclusive continua each containing a point of A_1O and a point of C_1L, V'' is an uncountable collection of continua each intersecting OL and the interval $A_1A''C_1$ of J_1, every continuum of T is a subset of $\bar{I}_1$ and so is every continuum of V'' and, finally, the sum of all the continua of T is closed and compact and so is the sum of all the continua of V''. Therefore, by Theorem 9, if K denotes the sum of all the continua of T and all the continua of V'', there exists an upper semi-continuous collection Y' of mutually exclusive continua such that (a) with respect to its elements, Y' is a space satisfying Axioms 0, 1′, C, 2, 3, 4 and 5, (b) every nondegenerate continuum of the collection Y' is a subset of the point set K, (c) in the space Y' there exists a domain Z (of elements of Y') such that every element of Y' that belongs to Z is, in the space Σ', a subset of the point set K. By Theorem 17 of Part 1, there exists a simple domain X' of elements of Y' such that every element of X' belongs to Z. Each element of X' is a collection of elements of H. For each element x' of X', let x denote the continuum of points, in the original sense, of S obtained by adding together all the point sets (in the original sense of the word point) that are elements of x'. Let X denote the collection of all such continua (in the original sense) x for all elements x' of X'. The collection X is a collection of continua each of which is either an element of G or the sum of some elements of G and, by Theorem 43 of Part 1, with respect to its elements regarded as points, X satisfies Axioms 0, 1′, C, 2, 3, 4 and 5. All but countably many of the elements of X are elements of G. It follows that the set of all of those that are elements of G satisfies, with respect to its elements regarded as points, all of these axioms except C.

THEOREM 12. *Every compact continuum has graphatomic subsets.*

PROOF. Suppose M is a compact continuum. Let W denote the set of all upper semi-continuous collections Q of mutually exclusive continua such that Q fills up M and is a continuous curve with respect to its elements. There exists a well ordered sequence α whose terms are the elements of W. For each term w of the sequence α, let G_w denote the collection of all point sets x such that, for some point P of M, x is the component containing P of the common part of all continua y such that y contains P and belongs either to w or to some collection that precedes w in the well ordered sequence α. Suppose that, for some term w of α, the collection G_w is not a continuous curve with respect to its elements. There exists a first such term of α. Call it w_0. Let G'_{w_0} denote the collection of all point sets x such that, for some point P of M, x is the component containing P of the common part of all continua y such that y contains P and belongs to G_w for some collection w that precedes w_0 in the sequence α. If u follows v in α then every continuum of the collection G_u is a subset of some continuum of the collection G_v. Hence, by Theorem 47 of Part 1 of the present chapter, G'_{w_0} is a continuous curve with respect to its elements. But w_0 is a continuous curve with respect to its elements and it is easy to see that G_{w_0} is the collection of all point sets x such that, for some point P of M, x is the component containing P of the common part of the continuum of G'_{w_0} that contains P and the continuum of w_0 that contains P. Therefore, by Theorem 7, G_{w_0} is a continuous curve with respect to its elements. Thus, for every term w of the sequence α, G_w is a continuous curve with respect to its elements. Let G denote the set of all continua g such that, for some point P of M, g is the component containing P of the common part of all continua y such that y contains P and belongs to some collection of the set W. If g is a continuum of the collection G then, for every term w of α, g is a subset of some element of G_w and if h is a continuum of which g is a proper subset then there is a term t of α such that h is not a subset of any element of G_t. Therefore, by Theorem 47 of Part 1, every element of G is a graphatomic subset of M.

Examples may be given of compact continua, in Euclidean space of three dimensions, that do not have graphatomic subsets. Therefore Theorem 12 cannot be proved on the basis of Axioms 0, 1′ and C.

THEOREM 13. *If there exists no upper semi-continuous collection of mutually exclusive continua filling up the compact continuum M and such that some subcollection of that collection satisfies, with respect to its elements regarded as points, Axioms* 0, 1′, C, 2, 3, 4 *and* 5 *then M has a dendratomic subset.*

Theorem 13 may be proved with the help of Theorem 11 by an argument similar to that used to prove Theorem 12 with the help of Theorem 7.

THEOREM 14. *If M is a locally compact continuum which does not have uncountably many complementary domains and G is a collection of mutually exclusive compact continua filling up M and all but countably many of the elements of G separate S, then G is an upper semi-continuous collection and it is a dendron with respect to its elements regarded as points.*

PROOF. Suppose G is not upper semi-continuous. Then there exist a continuum g_0 of G, an infinite sequence $g_1, g_2, g_3, \cdots$ of continua of G and two infinite sequences of points $A_1, A_2, A_3, \cdots$ and $B_1, B_2, B_3, \cdots$ such that (1) for each n, A_n and B_n belong to g_n, and (2) the sequence $A_1, A_2, A_3, \cdots$ converges to a point A belonging to g_0 but no subsequence of the sequence $B_1, B_2, B_3, \cdots$ converges to a point of g_0. If some subsequence of the sequence $B_1, B_2, B_3, \cdots$ converges to a point B let D denote a domain, containing g_0, such that $\overline{D}\cdot M$ is a compact point set not containing B. If there exists no convergent subsequence of $B_1, B_2, B_3, \cdots$, let D denote a domain, containing g_0, such that $\overline{D}\cdot M$ is compact. In either case there exist a sequence of continua $h_1, h_2, h_3, \cdots$ and an infinite ascending sequence of positive integers $n_1, n_2, n_3, \cdots$ such that (1) for each m, h_m is a component of $\bar{D}\cdot g_{n_m}$, (2) the sequence h_1, h_2, h_3,$\cdots$ converges to a continuum K containing A and a point of the boundary of D. Let Q denote the collection of all continua of G that intersect K. By Theorem 56 of Chapter I, the collection Q is uncountable. If q is any continuum of the collection Q then since the sequential limiting set of the sequence h_1, h_2, h_3,$\cdots$ is not a subset of q therefore, by Theorem 4 of Chapter II, there do not exist infinitely many components of $S - q$ each containing a continuum of this sequence. Thus, for each q, there exists a component of $S - q$ which contains infinitely many continua of the sequence h_1, h_2, h_3,$\cdots$ and therefore contains $K - K\cdot q$. Clearly for each q there is only one such component of $S - q$. Call it E_q. Let Q' denote the set of all continua of Q that separate S. Since Q is uncountable and all but countably many of the continua of Q separate S, therefore Q' is uncountable. For each continuum q of the collection Q' select just one component of $S - q$ distinct from E_q and call it D_q. Suppose x and y are two elements of Q'. By Theorem 24 of Chapter IV and Theorem 2 of Chapter III, if D_x and D_y have a point in common then either D_x is a subset of D_y or D_y is a subset of D_x. Suppose D_x is a subset of D_y. Then x is a subset of D_y. But x contains a point of $K - K\cdot y$ and $K - K\cdot y$ is a subset of E_y. This involves a contradiction. Hence if x and y are two elements of the uncountable collection Q' then D_x and D_y have no point in common. But this is contrary to the hypothesis that K does not have uncountably many complementary domains. It follows that the collection G is upper semi-continuous.

Suppose there are uncountably many continua of G that do not separate M. Let G' denote the set of all those that separate S but not M. Then G' is uncountable. For each element g of G', let I_g denote a component of $S - g$ that contains no point of M. If x and y are two continua of G' and z is any continuum of G' distinct from x and y then $S - \bar{I}_x$ and $S - \bar{I}_y$ contain z. Therefore, by Theorem 24 of Chapter IV and Theorem 2 of Chapter III, if I_x and I_y had a point in common then one of the domains I_x and I_y would be a subset of the other one and therefore either x would be a subset of I_y or y would be a subset of I_x, contrary to the fact that neither I_x nor I_y contains a point of M. Hence, for every two continua x and y of the uncountable collection G', the domains I_x and I_y are mutually exclusive, contrary to the hypothesis that M does not have uncountably many complementary domains. Hence all but countably many of the continua of G separate M. Therefore the compact continuum G of elements does not have uncountably many non-cut elements. With the help of Theorem 104 of Chapter I it follows that G is a dendron with respect to its elements.

Theorem 14 does not remain true if the stipulation requiring that all but countably many of the elements of G separate S is replaced by one requiring that the sum of all the elements of G that do not separate S contains no connected point set containing points of two different elements of G.

EXAMPLE 2. Let A, B, A', B', t_1, t_2, t_3,$\cdots$, K, A_1A_1', A_2A_2', A_3A_3',$\cdots$,

$B_1B'_1$, $B_2B'_2$, $B_3B'_3, \cdots$, Q and Z_1, Z_2, $Z_3, \cdots$ be the same as in Example 2 of Part 1 of the present chapter. For each positive integer n, let T_n denote the collection whose elements are the center of Z_n and all rectangles concentric with, and similar to, the rectangle Z_n and lying within it. Let T_n denote the collection of all point sets t such that t belongs to T_n for some number n. Let N denote some definite rectangle enclosing the rectangle $ABB'A$, concentric with it, and similar to it. Let T' denote the collection whose elements are the rectangle N and all rectangles which are concentric with, and similar to, $ABB'A$ and which enclose $ABB'A$ and are enclosed by N. Let G denote the collection whose elements are the elements of T, the elements of T', the rectangles Z_1, Z_2, $Z_3, \cdots$ and the straight line intervals of the set Q. The only nondegenerate elements of G which do not separate S are those that belong to Q. Furthermore, the sum of all the elements of Q contains no connected point set that intersects more than one element of G. But G is not upper semi-continuous.

Theorem 15. *If the arc AB separates the compact continuum M but no proper subarc of AB does so, then the endpoints of AB belong to the boundary of the same complementary domain of M, that is to say there is a complementary domain of M whose boundary contains both A and B.*

Proof. Consider AB as an element and every point not on AB as an element. Let N denote the set of elements consisting of AB and all points of M not on AB. The arc AB is a cut element of N. Hence, by Axiom 3, Theorems 21 and 53 of Chapter IV, and Theorem 2, there exists a simple closed curve J of elements which contains AB but no other element of N and which separates N. Let C denote an element of $J - AB$. The element C is a point and $J - AB$ is a point set. The point set $J - AB$ is the sum of two connected point sets t_1 and t_2 having only the point C in common and such that $t_1 + AB$ and $t_2 + AB$ are continua and q_i, the set of all points of AB which are limit points of t_i $(i = 1,2)$ is either a point or an interval of AB. The set $J - AB$ is a connected subset of $S - M$ and therefore there exists a component D of $S - M$ which contains $J - AB$. The point set $q_1 + q_2$ is a subset of the boundary of D. Let W denote the smallest subarc of AB that contains $q_1 + q_2$. The arc W separates M. Hence it is identical with AB. It follows that A and B both belong to $q_1 + q_2$ and therefore that they are both limit points of D.

Theorem 16. *If S is normal no compact and countable point set separates S.*

Proof. Suppose, on the contrary, that there exists a compact and countable point set T such that $S - T$ is the sum of two mutually separated point sets H and K. Let A and B denote points belonging to H and K respectively. By Theorem 154 of Chapter I, there exists a closed subset T' of T which separates A from B. By Theorem 26 of Chapter IV, T' contains a continuum M which separates A from B. By Theorem 56 of Chapter I, M is degenerate. But this is contrary to Axiom 3.

THEOREM 17. *If S is normal no compact and countable point set separates a simple domain.*

PROOF. Suppose D is a simple domain. By Theorem 10, if only the points of D are regarded as points and only those regions which are subsets of D are regarded as regions, then Axioms 0, 1′, C, 2, 3, 4 and 5 hold true in the space so obtained. With the help of Theorem 153 of Chapter I, it may be seen that this space is also normal. Therefore, by Theorem 16, Theorem 17 is true.

THEOREM 18. *If S is normal and K is a compact and countable point set then $S - K$ is arcwise connected.*

PROOF. By Theorem 16, $S - K$ is connected. By Theorem 17, Axiom C and Theorem 15 of Chapter IV, it is also connected im kleinen. Furthermore, by Theorem 147(a) of Chapter I, $S - K$ is an inner limiting set. Hence, by Theorem 13 of Chapter II, $S - K$ is arcwise connected.

THEOREM 19. *If S is normal and A and B are distinct points not belonging to the compact continuous curve M, there exists an arc from A to B which does not disconnect M.*

PROOF. If A and B belong to the same complementary domain of M they can be joined by an arc lying in that domain and therefore containing no point of M. Suppose they lie in different complementary domains of M. A point set X will be said to be within a simple closed curve J or to be enclosed by J if and only if J separates X from B. If D is a domain and $\bar{D}$ does not contain B then by the outer boundary of D will be meant its outer boundary with respect to B. Let N_1 denote the set of all cut points X of M such that X lies on some simple closed curve which is a subset of M. Every point of N_1 belongs to a nondegenerate simple link of M. Since no simple link of M has a cut point, therefore, by Theorems 22 and 105 of Chapter I, no one of them contains uncountably many cut points of M. But, by Theorem 98 of Chapter II, M does not have uncountably many nondegenerate simple links. It follows that N_1 is countable. Let N_2 denote the set of all points X of M such that X is common to the boundaries γ_1 and γ_2 of two complementary domains D_1 and D_2 of M and is on the boundary of some complementary domain of the point set $D_1 + D_2 + \gamma_1 + \gamma_2$. By Theorem 90 of Chapter IV, N_2 is countable.

Let K denote the set of points obtained by adding together all simple closed curves that lie in M and enclose A. By Theorem 93 of Chapter IV, K is the sum of a finite number of continuous curves. For each point P of $M - K$ let M_P denote the component of $M - K$ which contains P and let T_P denote $\bar{M}_P$. Let H denote the set of all such point sets T_P for all points P of $M - K$. The set of all components of $M - K$ is topologically contracting. It can be seen that if x and y are two components of K then there do not exist infinitely many elements of H each intersecting both x and y. Furthermore no component of $M - K$ has limit points in each of three distinct components of K. It follows, with the help of Theorem 94, that the set H is the sum of two sets H_1 and H_2, where H_1 consists of all elements h of H such that h has only one point in common with K, and H_2 is a *finite* set consisting of all those elements h of H such that h has only two points in common with K, these points belonging to different components of K. Let K_1 denote the set of all points X of K such that X belongs to some point set of the set H_1, but not to any one of the set H_2. For each point X of K_1, let N_X denote the sum of all point sets of the set H that contain X. Let Q_1 denote the set of all point sets N_X for all points X of K_1. With the help of the fact that H is a topologically contracting collection it is easy to see that every element of Q_1 is a continuous curve and Q_1 is topologically contracting. It is also clear that no two elements of Q_1 have a point in common. Let K_2 denote the finite set consisting of all points X of K such that X belongs to some point set of the set H_2. Let L denote the point set obtained by adding together all point sets h of the set H that intersect K_2. Let Q_2 denote the set of all components of L. The set Q_2 cannot have more elements than K_2 has points. Hence Q_2 is a finite set. Every element of Q_2 is a continuous curve and no two of them have a point in common. No curve of the set Q_1 intersects one of the set Q_2. Let Q denote the set of all curves that belong either to Q_1 or to Q_2. If a curve of the countable set Q separates A from B it contains a simple closed curve that separates A from B and thus contains uncountably many points of K which is impossible. Hence no element of Q separates A from B. For each element q of Q let g_q denote either q or the point set obtained by adding to q all its bounded complementary domains, according as q does or does not fail to separate S. Let N_3 denote the set of all point sets g_q for all elements q of Q. Let N denote a set such that x belongs to it if and only if x is an element of N_1, N_2 or N_3. Let U denote the set of all continua x such that x is either an element of N or a point which does not belong to any element of N. The collection U is an upper semi-continuous collection of mutually exclusive continua and every point belongs to some one of them and no one of them separates S. Furthermore N is a countable subcollection of U. With the help of Theorems 2 and 18 it follows that there exists an arc t with endpoints at A and B and containing no point of any element of the set N. If a point P of the arc t belongs to some segment of t whose endpoints belong to the boundary of the same complementary domain of M then (1) there exists a segment A_PB_P which has this property and which contains every other segment of t that has this property with respect to this particular point P, and (2) since t contains no point of N_2, there exists *only one* domain D_P complementary to M and having both A_P and B_P on its boundary. Let W denote the set of all points P of t for which A_PB_P exists and for which A_PB_P does not lie in D_P. For each point P of W let $A_PZ_PB_P$ denote a definite arc lying wholly in D_P except for its endpoints A_P and B_P and let s_P denote the segment $A_PZ_PB_P - (A_P + B_P)$. Let AB denote the arc obtained by adding $M - W$ to the sum of all the segments s_P for all points P of W. The arc AB does not separate M. For suppose it does. Then there exist two points X and Y, belonging to M, such that every arc in M from X to Y contains at least one point of AB. It can be shown that AB contains a point set CD which is either a degenerate set or an arc, with C and D as endpoints, such that every arc in M from X to Y contains a point of CD but such that if u is any proper subinterval of CD then there exists, in M, an arc from X to Y that contains no point of u. If CD contained only one point that point would be a cut point of M, contrary to the fact that AB contains no point of N_1 and intersects no point set of the collection N_3. Hence CD is an interval. For each positive integer n, let C_n denote a point of CD such that the interval CC_n of CD is a subset of some region of the collection G_n. The curve M contains an arc α_n from X to Y that has no point in common with the interval C_nD of CD. But α_n contains at least one point of CD. Hence it contains a point of the interval CC_n. But every such point is in a region of G_n containing C. Therefore C is a limit point of the component of $(M + CD) - CD$ that contains X. But $M + CD$ is a compact continuous curve. Hence, by Theorem 61 of Chapter IV, M contains an arc XC which has no point except C in common with CD. By a similar argument it may be shown that M contains an arc YC that has only C in common with CD and two arcs XD and YD each having only D in common with CD. The point set $XC + XD$ contains an arc CX_1D, and $CY + YD$ contains an arc CY_1D. These two arcs have only their endpoints C and D in common and their sum is a simple closed curve J. The arc CD lies, except for its endpoints, wholly in one of the complementary domains of J. Let I denote this domain and let E denote the other complementary domain of J. There does not exist an arc with endpoints at C and D and lying, except for C and D, wholly in $(E + M) - M$. For if there were such an arc then, according to the method given for constructing AB, one such arc would be a part of AB which is impossible since CD is a part of AB. With the help of Theorem 66 of Chapter IV and the fact that $M + I$ is a continuous curve it follows that C is separated from D by a simple closed curve lying in $M + I$ and, therefore, that M contains an arc X_2Y_2 which lies wholly in E except for its endpoints X_2 and Y_2 which lie, respectively, on the segments CX_1D and CY_1D. If XX_1 denotes X or the interval XX_1 of XC according as X_1 is or is not X and YY_1 denotes Y or the interval YY_1 of YC according as Y_1 is or is not Y then the point set $s(CX_1D) + s(CY_1D) + XX_1 + YY_1 + X_2Y_2$ contains an arc from X_2 to Y_2 that contains no point of CD. Thus the supposition that Theorem 19 is false has led to a contradiction.

THEOREM 20. *No compact continuum has uncountably many junction points.*

PROOF. Suppose the set of all junction points of the compact continuum M is an uncountable point set K. For each point P of K there exists a domain D_P, containing P, such that P is a cut point of the component of $M \cdot \bar{D}_P$ that contains P. By Theorems 22 and 92 of Chapter I, there exist a domain D and an uncountable subset K_1 of K such that (1) K_1 is a subset of D and (2) if P is any point of K_1 then D_P contains D. For each point P of K_1, let N_P denote the component of $M \cdot \bar{D}$ that contains P and let T_P denote a triod lying in N_P and with P as its emanation point. By Theorem 84 of Chapter IV there exists an uncountable subset K_2 of K_1 such that if X and Y are points of K_2, T_X intersects T_Y. Let W denote the set of all triods T_P for all points P of K_2. The point set W^* is connected. Hence it is a subset of a component H of $M \cdot \bar{D}$. By Theorem 31 of Part 1 of this chapter, every point of K_2 is a cut point of H. But no point of K_2 is of Menger order 2 with respect to M. This is contrary to Theorem 33 of Part 1.

3. Equicontinuous collections on the basis of Axioms 0 and 1

Definition. A collection G of continua is said to be *equicontinuous* with respect to a given point set M if, for every set H of regions covering M, there exists a finite set H' of regions covering M such that if X and Y are two points of M lying together in some one region of H' and belonging to a continuum g of the collection G then there exists an arc XY lying both in g and in some region of the set H.

Definition. A collection G of point sets is said to be *self-compact* if every infinite sequence of point sets of the collection G has an infinite subsequence which converges to some point set of the collection G. The collection G of point sets is said to be *self-compact with respect to the point set* M if the collection of all point sets of G that contain one or more points of M is self-compact.

Theorem 1. *Suppose A, B, C and D are four distinct points, $ABCD$ is a simple closed curve and AB, BC, CD and AD are arcs of $ABCD$ no two of which have in common any point other than a common endpoint. Suppose G is a self-compact and equicontinuous collection of mutually exclusive arcs such that* (1) *each point of $AB + CD$ is an endpoint of one and only one arc of G,* (2) *BC and AD are arcs of G,* (3) *each arc of G has only one point in common with AB and only one point in common with CD. Let Q denote the point set which is the sum of all the arcs of the collection G. Then there exists a self-compact and equicontinuous collection G^1 of mutually exclusive arcs such that* (a) *Q is the sum of all the arcs of G^1,* (b) *AB and CD are arcs of G^1,* (c) *if g and g^1 are arcs belonging to G and G^1 respectively, g and g^1 have only one point in common.*

The truth of Theorem 1 will be established with the help of five lemmas. First certain terms will be defined.

Definition. If G is a set of arcs or curves, a *G-arc* or a *G-curve* is an arc or curve of the set G. If G is a set of arcs, a *G-interval* is an interval (and a

G-segment is a segment) of an arc of G. If G is a set of arcs satisfying the conditions stated in the hypothesis of Theorem 1, the G-arc g_1 is said to be *above* the G-arc g_2 if the endpoint of g_1 which lies on AB either is B or is between B and the endpoint of g_2 which lies on AB, and g_2 is said to be below g_1 if g_1 is above g_2. If P is a point of Q, g_P denotes the arc of G which contains P. If X and Y are points of Q, X will be said to lie *above* Y in case g_X is above g_Y. If x, y and z are arcs of G and y is above x and below z then y is said to be between x and z and every point of y is said to be between x and z.

DEFINITION. A connected point set K is said to be *simply related* to a collection G of arcs satisfying the conditions stated in the hypothesis of Theorem 1 if and only if (1) every point of K belongs to some arc of the set G, (2) K contains the whole of every G-interval whose endpoints are in K, (3) K is a domain with respect to Q and (4) there exist two G-arcs k_1 and k_2 such that (a) k_1 lies above k_2, every point of K is between k_1 and k_2 and both k_1 and k_2 have points in common with the boundary of K with respect to Q, (b) the common part of k_i and the boundary of K with respect to Q is an interval t_i of k_i ($i = 1,2$) and (c) no point of t_1 or of t_2, except their endpoints, is a limit point of any point set which lies between g_1 and g_2 and contains no point of K. The interval t_1 minus its endpoints will be called the upper base, and the interval t_2 minus its endpoints will be called the lower base, of the set K.

LEMMA 1. *If G is a collection of arcs satisfying the conditions stated in the hypothesis of Theorem 1, B_1C_1 and B_2C_2 are arcs of G, B_1 and B_2 are points of AB and C_1 and C_2 are points of CD, and B_1 is between B_2 and A, then C_1 is between C_2 and D.*

PROOF. There exists a point E between A and B on AB such that if X is either E or a point between A and E on AB and XY is an arc of G then Y is between C_2 and D. For if there exists no such point E then there exists on AB a sequence of distinct points $X_1, X_2, X_3, \cdots$ converging to A but such that, if Y_n denotes the other endpoint of the arc of G which has X_n as one endpoint, then Y_n is between C_2 and C. But since the set G is self-compact therefore the sequence of arcs X_1Y_1, X_2Y_2, $X_3Y_3, \cdots$ has an infinite subsequence $X_{n_1}Y_{n_1}, X_{n_2}Y_{n_2}, X_{n_3}Y_{n_3}, \cdots$ converging to an arc of G. Since the sequence $X_{n_1}, X_{n_2}, X_{n_3}, \cdots$ converges to A, this arc of G must contain A and must therefore be the arc AD. But this is impossible since, for every m, Y_{n_m} is between C_2 and C. Thus the supposition that there exists no point E having the property stipulated above leads to a contradiction. Now let M denote the set of all points E that have this property and let N denote the set of all the remaining points of the segment AB. There exists a point P which is either the highest point in the set M or the lowest point in the set N. That there is no highest point in M can be proved by an argument entirely analogous to that employed above to prove the existence of M.

Every point between P and A belongs to M. Hence, since G is self-compact, if PH denotes the arc of G which has P as an endpoint, there exists an infinite sequence of arcs P_1Z_1, P_2Z_2, $P_3Z_3, \cdots$ of G such that the points P_1, P_2, $P_3, \cdots$ belong to M and PH is the limiting arc of this sequence. The sequence Z_1, Z_2, $Z_3, \cdots$ converges to H. But either H is C_2 or it lies between C_2 and C; and the points Z_1, Z_2, $Z_3, \cdots$ are all between D and C_2. Therefore H is C_2. Hence P is B_2 and therefore B_1 belongs to M and thus C_1 is between C_2 and D. The truth of Lemma 1 is therefore established.

LEMMA 2. *Under the conditions stated in the hypothesis of Theorem* 1, *if P is a point, distinct from A_0 and D_0, belonging to an arc A_0D_0 of the set G, and R is a region containing P, then* (1) *if A_0D_0 is distinct from AD there exists, in R, a connected point set N such that* (a) *every point of N lies on a G-arc below A_0D_0,* (b) *the point P is a limit point of N,* (c) *there exists a region, containing P, such that every point below A_0D_0 and in this region belongs to the set N and* (2) *if A_0D_0 is distinct from BC then there exists, in R, a connected point set M such that, if in the statement of conditions* 1(a), 1(b) *and* 1(c), *N is replaced by M and "below" is replaced by "above", the conditions so obtained are fulfilled.*

PROOF. There exists, in R, a region R_0 containing P such that if X and Y are two points of R_0 lying on a G-arc g, then the interval of g whose endpoints are X and Y lies wholly in R. Let E and F denote two points, on A_0D_0 in the order A_0EPFD_0, such that the interval EPF of A_0D_0 lies wholly in R_0. Let R_P denote a region containing P such that $\bar{R}_P$ is a subset of R_0 containing no point of the arc A_0D_0 not lying on the segment EF of A_0D_0. With the help of the fact that the set G of arcs is equicontinuous and satisfies the other conditions of the hypothesis of Theorem 1, it can be proved that there exist two regions R_E and R_F containing E and F respectively and such that (1) $\bar{R}_E$ and $\bar{R}_F$ are subsets of R_0, (2) if Z is a point of R_P lying on a G-arc g which contains a point of R_E and a point of R_F then Z lies on an interval of g whose endpoints lie in R_E and R_F respectively and if a G-interval has both of its endpoints in R_E or both of its endpoints in R_F it contains no point of R_P, (3) if the arc g of G intersects both R_E and R_F every interval of g that intersects R_E and R_F contains $g \cdot R_P$.

Suppose that A_0D_0 is distinct from BC. With the aid of the conditions, other than those of equicontinuity, stipulated in the hypothesis of Theorem 1, it can be shown that there exists a G-arc A_1D_1 lying above A_0D_0 and such that every G-arc which lies between A_1D_1 and A_0D_0 intersects each of the regions R_E, R_P and R_F. Let M denote the point set composed of all points W such that E is on a G-segment which lies between A_1D_1 and A_0D_0 and whose endpoints are in R_E and R_F respectively. The set M is connected. For suppose that M is the sum of two mutually exclusive point sets M_1 and M_2. If a G-arc g intersects both M_1 and M_2, then, since the set $g \cdot M$ is clearly connected, it follows that one of the point sets $g \cdot M_1$ and $g \cdot M_2$ contains a limit point of the other one and therefore one of the sets M_1 and

M_2 contains a limit point of the other one. Suppose that no arc of G intersects both M_1 and M_2. Let M' denote the set of all points L of the segment A_0A_1 of AB such that the G-arc which has L as one of its endpoints intersects M_1 and let M_2' denote the set of all the remaining points of A_0A_1. Then every G-arc which has a point of M_2' as an endpoint intersects M_2. Since the segment A_0A_1 is a connected point set, one of the point sets M_1' and M_2' contains a limit point of the other one. Suppose, for instance, that M_1' contains a point X which is a limit point of M_2'. Then X is the sequential limit point of some sequence of points $X_1, X_2, X_3, \cdots$ belonging to M_2'. The arc g_X is the limiting set of the sequence of arcs $g_{X_1}, g_{X_2}, g_{X_3}, \cdots$. The arc g_X contains a point T belonging to the region R_P. The point T is the sequential limit point of a sequence of points $T_1, T_2, T_3, \cdots$ such that, for every n, T_n belongs to g_{X_n}. There exists a number m such that if $n > m$ then T_n is in R_P. The point T is the sequential limit point of the sequence $T_{m+1}, T_{m+2}, T_{m+3}, \cdots$. But every point of this sequence belongs to M_2. Thus M_1 contains a limit point of M_2.

So, no matter how M is divided into two mutually exclusive sets, one of them contains a limit point of the other one. In other words M is connected. Clearly P is a limit point of M. It is clear that no point between A_1D_1 and A_0D_0 is a limit point of the point set composed of all points of Q which are above A_1D_1 or below A_0D_0. It follows that there exists a region V containing P such that $\overline{V}$ is a subset of R_P containing no point of Q which lies above A_0D_0 but not between A_0D_0 and A_1D_1. Every point of $V \cdot Q$ above A_0D_0 belongs to M.

Lemma 3. *The point set Q is a compact continuous curve.*

Lemma 3 may be easily proved with the help of Lemma 2.

Lemma 4. *If x and y are G-arcs and P is a point of $s(x)$ then every region that contains P contains the closure of some connected subset of Q which is simply related to G and which has, as one of its bases, a segment of x and, as its other base, a segment of some G-arc lying between x and y.*

Lemma 4 may be established with the help of Lemma 3.

Lemma 5. *If G is a set of arcs satisfying the conditions of the hypothesis of Theorem 1, and the set K is simply related to G, then any point of the upper base of K can be joined to any point of its lower base by an arc that lies wholly in K and does not have more than one point in common with any one arc of the set G.*

Proof. Let Σ' denote the space obtained by interpreting the word point to mean point of Q and the word region to mean connected domain with respect to Q. By Lemma 3, and Theorem 12 of Chapter II, Axioms 0, 1 and 2 hold true in Σ' and the point P of Q is a limit point of the subset M of Q in the original sense if and only if it is in Σ'. From here on, in this argument, point and region will be understood to have this new meaning. There exists a sequence $G_1', G_2', G_3', \cdots$ satisfying all the requirements of Axiom 1 except for the substitution of G in place of S and of G_n' in place of G_n. There exists a sequence $H_1, H_2, H_3, \cdots$ satisfying, with respect to Q and $G_1', G_2', G_3', \cdots$, all of the conditions which, according to the statement of Theorem 117 of Chapter I, are satisfied by the sequence $W_1, W_2, W_3, \cdots$ with respect to M. If P is a point of Q and n is a positive integer, let R_{Pn} denote a region of H_n containing P and let D_{Pn} denote the set of all points X such that X lies on a G-interval ZW such that Z and W belong to R_{Pn}. That the set D_{Pn} contains points above g_P, unless P is on BC, and points below g_P, unless P is on AD, can be easily proved with the aid of Lemma 2. Clearly D_{Pn} is connected. If, for a given pair of positive integers h and k such that $h \geqq k$, the point set D_{Ph} intersects two G-arcs g_1 and g_2 then the set of all points of D_{Ph} that lie between g_1 and g_2 will be called an *elemental set of rank* k. According to this definition, if $n_1 > n_2$ then every elemental set of rank n_1 is also of rank n_2. It follows from Lemma 2 that if n is a positive integer, each point of K is in some elemental set of rank n which lies, together with its boundary, wholly in the point set K_* composed of K and its two bases. Such an elemental set will be called a *K-element of rank n*. If E and F are two points of K_* and E is above F, then by a *chain of K-elements from E to F*, or *joining E to F*, is meant a finite set of K-elements $K_1, K_2, K_3, \cdots, K_n$ such that (1) E belongs to the upper base of K_1 and F belongs to the lower base of K_n, (2) for each i less than n, the lower base of K_i and the upper base of K_{i+1} lie on the same arc of the set G and have points in common, and the set of all their common points is a segment t_i. The point set $K_1 + K_2 + \cdots + K_n + t_1 + t_2 + \cdots + t_{n-1}$ is simply related to G. It will be called *the set associated with the chain* $K_1K_2K_3 \cdots K_n$. Suppose that E is a point on the upper base of K, F is a point on its lower base and n is a positive integer. It will be shown that E can be joined to F by a chain of K-elements of rank n. Let K' denote the set of all points of K that can be joined to E by chains of K-elements of rank n. It can easily be seen, with the aid of Lemma 2, that there exists a K-element of rank n whose upper base contains the point E. It follows that the set K' exists.

Suppose that WZ is an arc of G that contains a point of K'. The common part of WZ and K is a segment W^1Z^1. Every point of W^1Z^1 belongs to K'. For suppose this is not the case. Then the segment W^1Z^1 is the sum of two mutually exclusive point sets S_1 and S_2 such that S_1 is a subset of K' but no point of S_2 belongs to K'. There exists a point P which either belongs to S_1 and is a limit point of S_2 or belongs to S_2 and is a limit point of S_1. In the first case there is a chain α_2 of K-elements of rank n from E to P. The lower base of the last element of this chain is a segment of W^1Z^1 containing P. Since P is a limit point of S_2 this segment must contain at least one point P_2 of S_2. Thus α_2 is a chain of K-elements of rank n from E to P_2. Thus the supposition that S_1 contains a limit point of S_2 leads to a contradiction. Suppose now that S_2 contains a point P which is a limit point of S_1. There exists a K-element e of rank n whose lower base W_2Z_2 contains P and is a segment of W^1Z^1. Since P is a limit point of S_1 there exists, on the segment W_2Z_2, a point P_1 belonging to S_1. There exists a chain $e_1e_2e_3 \cdots e_m$ of K-elements of rank n from E to P_1. The lower base of the last element e_m of this chain is a segment W_1Z_1 containing P_1. There exist a G-arc g', lying above WZ, and two segments $W_1'Z_1'$ and $W_2'Z_2'$ such that (1) $W_1'Z_1'$ is the common part of e_m and g', (2) $W_2'Z_2'$ is the common part of e and g', (3) $W_1'Z_1'$ and $W_2'Z_2'$ have a segment in common. Let e_m' denote the part of e_m which lies between g' and the arc of G that contains the upper base of e_m. Let e_{m+1}' denote the part of e which lies between g' and WZ. The set of elements $e_1, e_2, e_3, \cdots, e_{m-1}, e_m', e_{m+1}'$ is a chain of K-elements of rank n from E to P. It is thus established that if one point of W^1Z^1 belongs to K' then so does every other point of W^1Z^1. It has been shown that if a G-arc above g_F contains a point of K then so must some lower arc of G. It follows that if F does not belong to K' there exists an arc XY which is the uppermost arc of G that neither contains nor lies above any point of K'. Let P denote a point of K on the arc XY. There exists a K-element e of rank n whose lower base contains P. The set G contains an arc g that intersects e in a segment MN which contains a point P' belonging to K'. There exists a chain of K-elements of rank n from E to P'. If to this chain of elements there is added the portion of the K-element e which lies between g and XY, there is obtained a chain of K-elements of rank n from E to P. Thus the supposition that E cannot be joined to F by a chain of K-elements of rank n leads to a contradiction. It follows that there exists a chain, from E to F, whose links are all K-elements of rank 1. Let K_1 denote the set associated with this chain. There exists a simple chain of K_1-elements, of rank 2, from E to F. This process may be continued. It follows that there exists a sequence of chains $C_1, C_2, C_3, \cdots$ from E to F such that if, for each i, K_i denotes the set associated with C_i, then (1) every link of C_{i+1} is a K-element of rank i, (2) $\overline{K}_{i+1}$ is a subset of the point set composed of K_i plus its bases. Let t denote the set of all points X such that X belongs to every K_i. By a method analogous to that used in the proof of Theorem 1 of Chapter II, it may be shown that t is an arc from E to F and that it does not have more than one point in common with any given arc of the set G. The truth of Lemma 5 is thus established.

Proof of Theorem 1. Let $H_1, H_2, H_3, \cdots$ be as described in the above proof of Lemma 5.

Suppose $A'D'$ and $B'C'$ are two G-intervals not belonging to the same G-arc and $A'B'$ and $C'D'$ are two mutually exclusive arcs each of which lies in Q and has not more than one point in common with any one G-arc. Then by the *interior with respect to* Q of the simple closed curve $A'B'C'D'$ will be meant the set of all points X such that X is on a G-segment which has one endpoint on $A'B'$ and its other endpoint on $C'D'$. A point will be said to be *within* $A'B'C'D'$ if it belongs to the interior, with respect to Q, of $A'B'C'D'$.

If, on the simple closed curve $A'B'C'D'$, $X_1, X_2, X_3, \cdots, X_{n-1}, X_n, Y_1, Y_2, \cdots, Y_{n-1}, Y_n, X'_n, X'_{n-1}, \cdots, X'_2, X'_1, Y'_n, Y'_{n-1}, \cdots, Y'_2, Y'_1$ are points in the order $A'X_1X_2\cdots X_{n-1}X_nB'Y_1Y_2\cdots Y_{n-1}Y_nC'X'_nX'_{n-1}\cdots X'_2X'_1D'Y'_n Y'_{n-1}\cdots Y'_2Y'_1A'$ and $X_1X'_1, X_2X'_2, \cdots, X_nX'_n, Y_1Y'_1, Y_2Y'_2, \cdots, Y_nY'_n$ are arcs which, except for their endpoints, lie entirely within $A'B'C'D'$, and finally, for every i, j each less than $n+1$, $X_iX'_i$ has one and only one point in common with $Y_jY'_j$, and no point in common with $X_jX'_j$ unless $i = j$, then these two sets of arcs are said to constitute a *double ruling* of the interior of $A'B'C'D'$ (or merely a *double ruling* of $A'B'C'D'$). The arcs $X_1X'_1, X_2X'_2, \cdots, X_nX'_n$ are said to be parallel to $B'C'$ and to $A'D'$ and the arcs $Y_1Y'_1, Y_2Y'_2, \cdots, Y_nY'_n$ are said to be parallel to $A'B'$ and to $C'D'$. Each of these two sets of arcs is a *single ruling* of the double ruling formed by the two sets combined. The segments $s(A'B')$, $s(B'C')$, $s(C'D')$ and $s(A'D')$ are called the *sides* of $A'B'C'D'$.

If X is a point of $s(AB)$ and XY is the arc of G which has X as one of its endpoints, it may be easily proved, with the use of Theorem 4 of Chapter I, that there exists on XY a finite set of points $A_1, A_2, A_3, \cdots, A_n$ in the order $XA_1A_2\cdots A_{n-1}A_nY$, such that each of the intervals $XA_1, A_1A_2, \cdots, A_{n-1}A_n, A_nY$ of the arc XY lies wholly in some region of H_1. Let $C_1, C_2, \cdots, C_n$ denote n points in the order $BC_1C_2\cdots C_{n-1}C_nC$ on the arc BC and let $D_1, D_2, \cdots, D_n$ denote n points in the order $AD_1D_2\cdots D_{n-1}D_nD$ on the arc AD. With the help of Lemma 5 it is easily established that there exist two sets of arcs $A_1C_1, A_2C_2, \cdots, A_nC_n$ and $A_1D_1, A_2D_2, \cdots, A_nD_n$ such that no arc of either of these sets intersects any other arc of that set and such that, for every m less than $n+1$, (1) A_mC_m lies, except for its endpoints, entirely within $XBCY$, (2) A_mD_m lies, except for its endpoints, entirely within $AXYD$, (3) neither A_mC_m nor A_mD_m has more than one point in common with any one arc of the set G. With the help of the fact that the set G is equicontinuous, it is easy to show that there exist two points X^1 and X', in the order $AX^1XX'B$, and two arcs X^1Y^1 and $X'Y'$ belonging to G, such that if, for every i less than $n+1$, A^1_i is the point in which X^1Y^1 intersects A_iC_i and $A^1_0, A_0, A'_0, A^1_{n+1}, A_{n+1}$ and A'_{n+1} denote the points X^1, X, X', Y^1, Y and Y' respectively, then the subset of Q which forms the interior, with respect to Q, of the closed curve formed by the intervals $A^1_iA_i, A_iA'_i, A'_iA'_{i+1}, A'_{i+1}A_{i+1}, A_{i+1}A^1_{i+1}$ and $A^1_{i+1}A^1_i$ of the arcs $A_iD_i, A_iC_i, X'Y', A_{i+1}C_{i+1}, A_{i+1}D_{i+1}$ and X^1Y^1 respectively, lies entirely in some region of the collection H_1. For each point X of the interval AB make a similar construction except that when X is A, X^1 is X and when X is B, X' is X. With the help of Theorem 4 of Chapter I, it may be shown that there exists a finite set λ of intervals such that each point of AB belongs to some interval of the set λ and each interval of the set λ is the interval $X'X^1$ for some point X of AB. The intervals AA' and BB' clearly belong to λ. Of the intervals of λ that contain A' there is one, $X^1_1X'_1$, such that no other interval of λ that contains A' contains a point between X'_1 and B. Similarly, of the intervals of λ that contain X_1', there is one, $X_2^1 X_2'$, such that no other interval of λ that contains X_1' contains a point between X_2' and B. This process may be continued. It follows that there exists a finite set of points $X_1, X_2, \cdots, X_n$ such that the points $A, X_1^1, A', X_2^1, X_1', X_3^1, X_2', X_4^1, X_3', X_5^1, X_4', X_6^1, \cdots, X_{i+1}^1, X_i', X_{i+2}^1, \cdots, X_n^1, X_{n-1}', B', X_n', B$ lie on AB in the order indicated except that X_1^1 may be identical with A', X_n' may be identical with B' and, for one or more values of i, X_{i+1}^1 may be identical with X_i'. For each point X of AB, let t_{Xi} denote the arc $A_iA_i^1 + A_iA_i'$ and let U_X denote the set of arcs $t_{Xi}, t_{X2}, \cdots, t_{Xn}$. Let X_0 denote A and let X_{n+1} denote B. For each i less than $n + 1$, let U'_{X_i} denote the set of all arcs t such that t is an interval of some arc of U_{X_i} and has its endpoints on the arcs of G that contain X'_{i-1} and X_i' respectively and lies, except for its endpoints, wholly between these two G-arcs. Let V_1 denote the set of all arcs v such that, for some i less than $n + 1$, v belongs to U'_{X_i}. For each i less than $n + 2$, let h_i denote the arc of G that contains the point X_i'. It is easy to see that there exists a double ruling T_1 of $ABCD$ such that (1) $h_0, h_1, h_2, \cdots, h_{n+1}$ are the arcs of one of the single rulings of T_1, (2) if W_1 denotes the other single ruling of T_1, then each arc of V_1 is an interval of some arc of W_1 and each arc of W_1 has its endpoints on BC and AD respectively and has only one point in common with each arc of the set G. It is clear that each of the subdivisions into which T_1 divides Q lies in some region of H_1. In a similar manner each subdivision α of this set can itself be subdivided by a double ruling $T_{1\alpha}$ such that (1) each arc of one of its single rulings is an interval of an arc of G, (2) each arc of its other single ruling has its endpoints on the arcs which form respectively the upper base and the lower base of α and no arc of this ruling has more than one point in common with any one arc of G, (3) each of the subdivisions into which $T_{1\alpha}$ divides α is in some region of the collection H_2. It follows that there exists a double ruling T_2 satisfying the conditions (1) and (2) stated above as being satisfied by T_1 and also satisfying the additional condition that each of its subdivisions is in some region of H_2, for every α each arc of $T_{1\alpha}$ being an interval of an arc of one or the other of the rulings of T_2. This process may be continued. It follows that there exists an infinite sequence of double rulings $T_1, T_2, T_3, \cdots$ such that, for every n, (a) T_n satisfies the conditions (1) and (2) stated above for T_1, (b) each arc of T_n is an arc of T_{n+1}, (c) each subdivision of T_n is in some region of H_n. Let β denote the set of all arcs t such that, for some n, t belongs to one of the rulings of T_n and has its endpoints on AD and BC respectively. If P is a point on BC which is not an endpoint of an arc of the set β then there exists only one arc t_P that has one endpoint at P and the other one on AD, lies except for its endpoints entirely in $Q - ABCD$ and has no point in common with any arc of the set β. Let γ denote the set of all such arcs t_P for all such points P. Let G^1 denote the set of arcs composed of all the arcs of β together with all the arcs of γ and the arcs AB and CD. The set G^1 satisfies all the conditions required in the statement of Theorem 1.

Chapter VI

CONSEQUENCES OF AXIOMS 0, 1, 2, 3, 4, 5_1, 5_2, 6 AND 7

Axiom 5_1. *If P is a point of a region R, there exists a connected domain D containing P and bounded by a continuum T such that $D + T$ is a subset of R.*

Axiom 5_2. *If D is a connected domain whose boundary is a continuum and AB is an arc lying wholly in $D + B$, then $(D + B) - AB$ is connected.*

Axiom 6. *No compact continuum separates two noncompact point sets from each other.*

Axiom 7. *The set of all points is completely separable.*

Theorem 1. *If P is a point of a connected domain D, then $D - P$ is connected.*

Proof. Suppose, on the contrary, that $D - P$ is the sum of two mutually separated point sets H and K. Let A and B denote points belonging to H and K respectively. There exists a region R containing P, lying in D, and containing neither A nor B. By Axiom 5_1, there exists a connected domain L containing P and bounded by a continuum T such that $T + L$ is a subset of R. By Theorem 60 of Chapter I, $H + P$ is connected. But it contains the point P of L and the point A of $S - L$. Hence it contains at least one point of the boundary of L. But P does not belong to T. Hence $T \cdot H$ exists. Similarly $T \cdot K$ exists. Hence T is the sum of two mutually separated point sets $T \cdot H$ and $T \cdot K$. But T is connected. This involves a contradiction.

Theorem 2. *If P is a point of a region R, there exists a connected domain D containing P and bounded by a simple closed curve J such that $D + J$ is a subset of R.*

Proof. By Axiom 5_1, there exists a connected domain Q containing P and bounded by a continuum T such that $Q + T$ is a subset of R. By Axiom 5_2 and Theorem 34 of Chapter III, there exists a simple closed curve J lying in Q and separating P from some point ω of T. Let D and E denote the components of $S - J$ that contain P and ω respectively. Since T is connected it is a subset of E. Hence the connected point set D contains the point P of Q but no point of the boundary of Q. Hence D is a subset of Q and therefore of R.

Theorem 3. *If P is a point of a region R, there exists a compact simple domain D, containing P, such that $\bar{D}$ is a subset of R.*

PROOF. By Theorem 2, there exists a simple domain Q containing P such that $\bar{Q}$ is a subset of R. Suppose Q is not compact. Then it contains an infinite point set M that has no limit point. There exists a region R' containing P and lying in R but containing no point of M other than P. By Theorem 2, there exists a simple domain D containing P and bounded by a simple closed curve J such that $D + J$ is a subset of R', and therefore of R. The point set $(M + P) - P$ is a subset of E, the component of $S - J$ distinct from D. Thus J separates $(M + P) - P$ from D. But J is a compact continuum and $(M + P) - P$ is not compact. Hence, by Axiom 6, D is compact.

THEOREM 4. *Suppose M is a completely separable point set and Q is a set of domains having the property L that if D is a domain containing a point P of M then there exists a domain of the collection Q containing P and lying wholly in D. Then there exists a countable subset Q' of Q having property L.*

PROOF. By hypothesis, there exists a countable set U of domains having property L. Let W denote the set of all sensed pairs (x,y) such that x is a domain of the set U and y is a domain of the set U distinct from x and lying wholly in x. For each element w of the set W for which there exists a domain which belongs to Q, contains the second member of w and is contained by the first member of w, let q'_w denote a definite domain of Q having that relation to the domains of the pair w. Let Q' denote the set of all such domains q'_w for all elements w of the set W. The set Q' is countable and it fulfills all the requirements of Theorem 4.

THEOREM 5. *For each n, let G'_n denote the set of all compact simple domains g' such that g' is a subset of some region of the set G_n. Then Axiom 1 holds true if the word "region" is replaced by "compact simple domain" and G_n is replaced by G'_n.*

Theorem 5 may be easily established with the aid of Theorem 3.

THEOREM 6. *There exists a sequence of compact simple domains R_1, R_2, $R_3, \cdots$ such that* (1) *for each n, S is covered by the collection whose elements are the terms of the sequence R_n, R_{n+1}, $R_{n+2}, \cdots$ and* (2) *if P is a point of a domain D there exists a number δ such that if m and n are greater than δ, R_m and R_n have a point in common, and R_m contains P then $R_m + R_n$ is a subset of D.*

PROOF. With the help of Axiom 7 and Theorems 4 and 5 it may be easily seen that there exists a sequence of countable collections Q_1, Q_2, $Q_3, \cdots$ satisfying conditions (1) and (2) of Theorem 113 of Chapter I except that M is replaced by S and, for each n, G_n is replaced by G'_n as defined in the statement of Theorem 5. Let I_1, I_2, $I_3, \cdots$ denote the simple domains of the collection Q_1 and, for each n, let M_n denote the closed and compact point set $\bar{I}_n$. For each n, there exists a finite subcollection H_n of Q_n covering the closed and compact point set $M_1 + M_2 + M_3 + \cdots + M_n$. For each n let

$h_{n1}, h_{n2}, h_{n3}, \cdots, h_{nm_n}$ denote the point sets of the collection H_n. For each n, let R_n denote the nth term of the infinite sequence $h_{11}, h_{12}, \cdots, h_{1m_1}, h_{21}, h_{22}, \cdots, h_{2m_2}, h_{31}, h_{32}, \cdots, h_{3m_3}, \cdots$.

CONVENTIONS. Select a definite sequence of compact simple domains satisfying the conditions of Theorem 6. Hereafter in this chapter this particular sequence will be referred to as *the fundamental sequence* and the letters $R_1, R_2, R_3, \cdots$ will be reserved to denote its terms. For each n, J_n will denote the boundary of R_n.

Select a definite infinite sequence ω of distinct points such that (1) if S is compact, ω is convergent, (2) if S is not compact, no subsequence of ω converges. If S is compact, a point set M will be said to be bounded if $\bar{M}$ does not contain the point to which ω converges. If S is not compact, M will be said to be bounded if $\bar{M}$ is compact. If J is a bounded simple closed curve then, as may be seen with the help of Axiom 6, one of the complementary domains of J is bounded and the other one is unbounded. The bounded one will be called the interior of J and the unbounded one will be called the exterior of J. The interior of a bounded simple closed curve is compact and if S is noncompact the exterior of a bounded simple closed curve is noncompact. If the point set M is a subset of the interior of the simple closed curve J it is said to be within J, or to be enclosed by J. If it is a subset of the exterior of J it is said to be without J.

THEOREM 7. *If J and L are two bounded simple closed curves and A and B are two points of J each of which either is not on L or is on some segment that is common to J and L, then there exists an arc from A to B that lies, except for its endpoints, entirely within J and has not more than a finite number of points in common with L and crosses L at every point in which it intersects it.*

PROOF. By Theorem 58 of Chapter IV, the points A and B are accessible from I, the interior of J. It follows that there exists an arc AB lying, except for its endpoints, wholly in I. If the segment $s(AB)$ and the simple closed curve L intersect let M denote their common part. In view of the conditions of the hypothesis, it is clear that there exists a point A_1 which is the first point that $s(AB)$ has in common with L. If there exist any points of M that can be joined to A_1 by an arc of L that lies entirely in I then there must be a last such point. For otherwise there would exist on AB a point X which cannot be joined to A_1 by an arc of L lying in I but which is a limit point of a set K of points that lie on AB and can be so joined to A_1. But in this case there would exist on L a segment $s(XYZ)$ containing X and lying entirely within J. This segment would contain at least one point P of the set K. But P could be joined to A_1 by an arc A_1P of L that lies in I and the point set $A_1P + YXZ$ would be an arc of L lying in I. Thus the supposition that there exists no last point of AB which either is identical with A_1 or can be joined to A_1 by an arc of L that lies in I has led to a contradiction. If X' is any point of M the statement that the point Y' can be *properly joined* to

X' will signify that either Y' is identical with X' or Y' can be joined to X' by an arc of L that lies in I. Let B_1 denote the last point on AB that can be properly joined to A_1. By an argument similar to the above proof of the nonexistence of X it can be shown that B_1 cannot be a limit point of a set of points of M no one of which can be properly joined to A_1. But B_1 is the last point that can be properly joined to A_1. It follows that on AB there must be a point W, following B_1, such that the segment $s(B_1W)$ of the arc AB is entirely free of points of L. But L is a closed set of points. It follows that if there are, on $s(AB)$, any points of L that follow B_1 then there is a first such point A_2. There exists a point B_2 which is the last point that can be properly joined to A_2. If there are any points of M after B_2 there is a first such point A_3. Continue this process. There must exist only a finite number of the points $A_1, A_2, A_3, \cdots$. For suppose there are infinitely many of them. Then there must be infinitely many points $B_1, B_2, B_3, \cdots$. Suppose n is any positive integer. The points B_n and A_{n+1} are the extremities of two arcs of L. Let B_nA_{n+1} denote one of these arcs. The arc B_nA_{n+1} does not lie wholly in I. Hence there exists a point X_n which is the first point that it has in common with J. Let B_nX_n denote the arc from B_n to X_n on the arc B_nA_{n+1}. Let i and j denote two positive integers. If the segments $s(B_iX_i)$ and $s(B_{i+j}X_{i+j})$ had a point in common then B_{i+j} could be properly joined to B_i. But B_i is properly joined to A_i. Hence B_{i+j} could be properly joined to A_i. But B_i is the last point of M that can be properly joined to A_i, and B_{i+j} follows B_i. It follows that $s(B_iX_i)$ and $s(B_{i+j}X_{i+j})$ have no point in common. Thus if there are infinitely many of the points $A_1, A_2, A_3, \cdots$, then there are infinitely many arcs, all lying on L, such that no two of them have, in common, any point except possibly a common endpoint, and such that each of them has one endpoint on J and the other one on the interval A_1O of the arc AB where O is the last point before B that AB has in common with L. But, as may be seen with the help of Theorem 110 of Chapter I, no simple closed curve contains infinitely many arcs satisfying these conditions. It follows that there exists only a finite number m of the points $A_1, A_2, A_3, \cdots$. It is clear that O is either A_m or B_m. For every n, less than m, such that A_n is distinct from B_n, there exists an arc $A_nZ_nB_n$ of L lying wholly in I. With the aid of Theorem 19 of Chapter III and Theorem 58 of Chapter IV, it is easy to see that there exists an arc $A_nZ'_nB_n$ that lies entirely within J and, except for its endpoints, entirely within L. Let γ denote the point set $AA_1 + (A_1Z'B_1) + B_1A_2 + (A_2Z'_2B_2) + B_2A_3 + \cdots + B_{m-1}A_m + (A_mZ'_mB_m) + OB$, where AA_1, B_1A_2, $B_2A_3, \cdots$, $B_{m-1}A_m$ and OB are intervals of the arc AB and, for each i less than $m + 1$, $(A_iZ'_iB_i)$ denotes the arc $A_iZ'_iB_i$ or the point A_i according as B_i is or is not distinct from A_i. For every n less than m, the point set B_nA_{n+1} has no point except B_n and A_{n+1} in common with L. In view of these considerations, it is clear that the point set γ contains an arc from A to B that lies, except for its endpoints, entirely within J and has not more than

$2(m + 1)$ points in common with L. It follows that there exists an arc β from A to B lying, except for its endpoints, wholly within J and such that either (1) β does not intersect L or (2) $\beta \cdot L$ is finite and if β' is another arc satisfying these conditions, $\beta' \cdot L$ is at least as numerous as $\beta \cdot L$. Suppose there exist a point C of $\beta \cdot L$, a point A' between A and C on β, and a point B' between C and B on β, such that the segments $s(A'C)$ and $s(B'C)$ of β lie in the same complementary domain D of L. Then there exists, in $D \cdot I$, an arc $A''B''$ from some point A'' of $s(A'C)$ to a point B'' of $s(B'C)$. The sum of the arc $A''B''$ and the intervals AA'' and BB'' of β contains an arc δ from A to B. Since $\delta \cdot L$ is a subset of $\beta \cdot L$ that does not contain C, $\delta \cdot L$ is less numerous than $\beta \cdot L$. This involves a contradiction. Hence, if C is a point of $\beta \cdot L$, some interval of β crosses L at C. The truth of Theorem 7 is therefore established.

THEOREM 8. *If the bounded simple closed curve g has only a finite number of points in common with the bounded simple closed curve $ABCDA$ and does not contain A, B, C or D and, for every point that g and $ABCDA$ have in common, some arc of g crosses some arc of $ABCDA$ at that point, then the interior of $ABCDA$ can be divided by a double ruling into subdivisions each lying either wholly within or wholly without g, the arcs of one of the single rulings of this double ruling being parallel to AB and CD and those of the other one being parallel to AD and BC.*

INDICATION OF PROOF. If X and Y are any two of the points A, B, C and D such that one of the arcs of $ABCDA$ from X to Y contains no other one of these points then that arc will be called the *side* XY of $ABCDA$. There exists a finite set K of arcs of g such that each arc of the set K lies entirely within $ABCDA$ except that its endpoints are on $ABCDA$ and such that every point of g that is within $ABCDA$ lies on an arc of K. If t is an arc of the set K, either (1) t has both endpoints on the same side of $ABCDA$, (2) t has one endpoint on one side, and the other one on an adjacent side, of $ABCDA$, or (3) t has one endpoint on one side, and the other one on an opposite side, of $ABCDA$. Let us suppose, for example, that there are four arcs $A_1B_1C_1D_1E_1$, $A_2B_2C_2D_2E_2$, $A_3B_3C_3D_3E_3$ and $A_4B_4C_4D_4E_4$ of type (1), two arcs $A_5B_5E_5$ and $A_6B_6E_6$ of type (2) and one arc A_7E_7 of type (3), situated as indicated in Figure A. In this figure a double ruling γ, satisfying the conditions of Theorem 8, is represented. This indication of how a ruling may be constructed in the case of this example is given instead of a tedious formal proof covering all possible cases. Here the arcs of γ that are parallel to AB are $K_1D_1C_1B_1L_1$, $K_2D_2C_2B_2L_2$, $A_4B_4L_3$, $A_3B_3L_4$, $E_3D_3L_5$, $E_4D_4L_6$, E_7A_7, $K_3B_6A_6$ and $K_4B_5A_5$. Those parallel to BC are $H_1B_3C_3D_3M_6$, $H_2B_4C_4D_4M_5$, $E_2D_2M_4$, $E_1D_1M_3$, $A_1B_1M_2$, $A_2B_2M_1$, $H_3B_6E_6$ and $H_4B_5E_5$.

THEOREM 9. *If $ABCDA$ is a bounded simple closed curve and G is a set of bounded simple closed curves and each point on or within $ABCDA$ is within some curve of the set G, the interior of $ABCDA$ can be subdivided by a double* *ruling into subdivisions each lying within some curve of the set G, the arcs of one of the single rulings of this double ruling being parallel to AB and CD and those of the other one being parallel to AD and BC.*

PROOF. There exists a finite set G' of simple closed curves $g_1, g_2, g_3, \cdots, g_m$ such that every one of them belongs to the set G and such that every point on or within $ABCDA$ is within one of them. Let J denote $ABCDA$. Every point of the arc AB of J is within some curve of the set G'. It follows that for each point of AB there is a segment on J containing that point, lying wholly within some curve of G', and not containing any side of $ABCDA$. There exists a finite set Q of these segments such that every point of AB is on

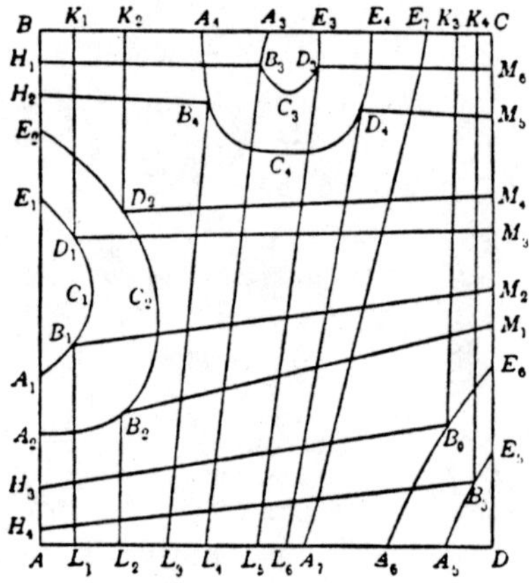

FIGURE A

some segment of Q. There exists a set of segments $A_1B_1, A_2B_2, \cdots, A_nB_n$ belonging to the set Q and with endpoints in the order $A_1AA_2B_1A_3B_2A_4B_3A_5B_4\cdots A_nB_{n-1}BB_n$. There exists a set of points $T_1, T_2, \cdots, T_{n-1}$ and arcs $T_1T_1', T_2T_2', \cdots, T_{n-1}T_{n-1}'$ such that, for every i and j less than n such that i is distinct from j, (1) T_i is on AB in the order $AA_{i+1}T_iB_iB$, (2) T_iT_i' lies within every region of the set G' that contains T_i, (3) every point of T_iT_i', except T_i, is within J, (4) T_iT_i' and T_jT_j' have no point in common, (5) not every point of the segment $s(T_iT_i')$ is on g_1. There exists a set of curves $g_1', g_2', \cdots, g_m'$ belonging to the set G' and such that, for every i less than $m + 1$, A_iB_i is entirely within g_i'. Within g_1' (and within J except for the point A_1) there is an arc A_1T_1' that has in common with T_1T_1' only the point T_1'. Within g_2' and J there is an arc $T_1'T_2'$ that has in common with T_1T_1' only the point T_1' and in common with A_1T_1' only the point T_1'. There exists a set of arcs $A_1T_1', T_1'T_2', T_2'T_3', \cdots, T_{n-1}'B_n$ such that (1) $T_i'T_{i+1}'$ is within g_{i+1}' and J if i is less than $n - 1$, $T_{n-1}'B_n$ is within g_n' and, except for the point B_n, within J, (2) no two nonconsecutive arcs of this set have any point in common and two consecutive ones have only one endpoint in common. It easily follows that there exist (Figure B) (1) on $ABCDA$, $4n$ points in the order $AE_1E_2\cdots E_nBF_1F_2\cdots F_nCL_1L_2\cdots L_nDM_1M_2\cdots M_nA$, (2) within $ABCDA$, a simple closed curve J' containing $4(n-1)$ points in the order $A'E_2'E_3'\cdots E_{n-1}'B'F_2'F_3'\cdots F_{n-1}'C'L_2'L_3'\cdots L_{n-1}'D'M_2'M_3'\cdots M_{n-1}'A'$, (3) arcs $A'\bar{E}_1E_1$, $E_2'\bar{E}_2E_2$, $E_3'\bar{E}_3E_3,\cdots$, $E_{n-1}'\bar{E}_{n-1}E_{n-1}$, $B'\bar{E}_nE_n$, $B'\bar{F}_1F_1$, $F'\bar{F}_2F_2$, $F_3'\bar{F}_3F_3,\cdots$, $F_{n-1}'\bar{F}_{n-1}F_{n-1}$, $C'\bar{F}_nF_n$, $C'\bar{L}_1L_1$, $L_2'\bar{L}_2L_2,\cdots$, $L_{n-1}'\bar{L}_{n-1}L_{n-1}$, $D'\bar{L}_nL_n$, $D'\bar{M}_1M_1$, $M_2'\bar{M}_2M_2,\cdots$, $M_{n-1}'\bar{M}_{n-1}M_{n-1}$, $A'\bar{M}_nM_n$ such that every one of these arcs lies, except for its endpoints, wholly between J' and J and no two of them have any point in common except that each of the points A', B', C' and D' is a common endpoint of two of them and such that no one of the points $\bar{E}_1$, $\bar{E}_2$, $\bar{E}_3,\cdots$, $\bar{E}_{n-1}$, $\bar{E}_n$, $\bar{F}_1$, $\bar{F}_2,\cdots$, $\bar{F}_{n-1}$, $\bar{F}_n$, $\bar{L}_1$, $\bar{L}_2,\cdots$, $\bar{L}_{n-1}$, $\bar{L}_n$, $\bar{M}_1$, $\bar{M}_2$, $\bar{M}_3,\cdots$, $\bar{M}_{n-1}$, $\bar{M}_n$ is on the curve g_1; the

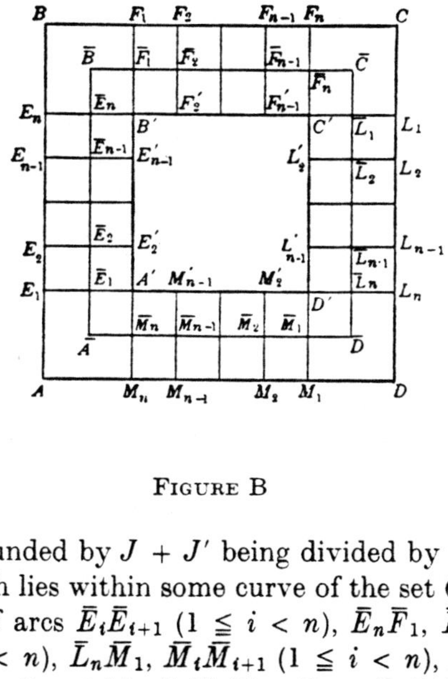

FIGURE B

connected domain bounded by $J+J'$ being divided by these arcs into subdivisions each of which lies within some curve of the set G'. By Theorem 7, there exists a set α of arcs $\bar{E}_i\bar{E}_{i+1}$ ($1\leqq i<n$), $\bar{E}_n\bar{F}_1$, $\bar{F}_i\bar{F}_{i+1}$ ($1\leqq i<n$), $\bar{F}_n\bar{L}_1$, $\bar{L}_i\bar{L}_{i+1}$ ($1\leqq i<n$), $\bar{L}_n\bar{M}_1$, $\bar{M}_i\bar{M}_{i+1}$ ($1\leqq i<n$), $\bar{M}_n\bar{E}_1$, lying except for their endpoints wholly within $E_iE_i'E_{i+1}'E_{i+1}$, $E_nB'F_1B$, $F_iF_i'F_{i+1}'F_{i+1}$, F_nCL_1C', $L_iL_i'L_{i+1}'L_{i+1}$, $L_nD'M_1D$, $M_iM_i'M_{i+1}'M_{i+1}$ and $M_nA'E_1A$ respectively (where $E_1'=M_n'=A'$, $E_n'=F_1'=B'$, $F_n'=L_1'=C'$ and $L_n'=M_1'=D'$), no one of these arcs having infinitely many points in common with g_1 or having, in common with g_1, any point at which it does not cross g_1. There exist, on the segments $\bar{E}_n\bar{F}_1$, $\bar{F}_n\bar{L}_1$, $\bar{L}_n\bar{M}_1$ and $\bar{M}_n\bar{E}_1$ respectively, points $\bar{B}$, $\bar{C}$, $\bar{D}$ and $\bar{A}$ that are not on g_1. The arcs of the set α form a simple closed curve $\bar{A}\bar{B}\bar{C}\bar{D}\bar{A}$ which will be called $\bar{J}$. This curve does not contain infinitely many points of g_1. Furthermore, the connected domain Q bounded

by $J + \bar{J}$ is subdivided by the arcs $E_i\bar{E}_i$, $F_i\bar{F}_i$, $L_i\bar{L}_i$, $M_i\bar{M}_i$ $(1 \leqq i < n + 1)$ into subdivisions each lying within some curve of the set G'. By Theorem 8, the interior of $\bar{J}$ may be divided by a double ruling λ into subdivisions each lying either entirely within or entirely without g_1, the arcs of one of the single rulings of λ being parallel to AB and CD and those of its other one being parallel to BC and AD. By the addition of certain arcs having their endpoints on J and $\bar{J}$ and lying, except for their endpoints, wholly in I, the interior of J, and wholly without $\bar{J}$, it is possible to continue the arcs of λ, together with suitably chosen additional arcs, in such a way that there will result a double ruling λ_1 of I such that every arc of the set α and every arc of the double ruling λ will be an interval of some arc of the double ruling λ_1. By a *side* of one of the subdivisions into which the double ruling λ_1 divides I will be meant a segment of the boundary of that subdivision which lies between two consecutive arcs of one of the single rulings of λ_1 and which has its endpoints on those arcs. Each subdivision into which I is divided by the ruling λ_1 is a subset either of one of the subdivisions into which the interior of J is divided by the arcs of λ or of one of the subdivisions into which the domain Q is divided by the arcs of the set α. It follows that each of the subdivisions into which the interior of J is divided by λ_1 is either wholly within some curve of the set G' or wholly without g_1. If any of them are wholly without g_1 then each such subdivision can itself be divided by a double ruling into subdivisions each lying either wholly within some curve of the set G' or wholly without g_2, the arcs of one single ruling of this double ruling being parallel to one side, and the arcs of the other one being parallel to an adjacent side, of this subdivision. The arcs of these rulings may be extended in such a way that there will result a new double ruling λ_2 of I such that each arc of λ_1 is also an arc of λ_2 and each subdivision formed by λ_2 is a subset of some subdivision formed by λ_1 and either lies wholly without g_1 and wholly without g_2 or lies wholly within some curve of the set G'. It can be shown in a similar way that there exists a double ruling λ_3 of I such that every subdivision formed by λ_3 either lies wholly without each of the curves g_1, g_2 and g_3 or lies wholly within some curve of the set G'. This process may be continued. It follows that there exists a double ruling λ_n of I such that every one of the subdivisions into which λ_n divides I is either wholly without every curve of the set $g_1, g_2, \cdots, g_m$ or wholly within one of them. But, by hypothesis, every point of I is within some curve of the set $g_1, g_2, \cdots, g_m$. It follows that every subdivision of I formed by the ruling λ_n is wholly within some curve of this set.

THEOREM 10. *If ABCDA is a bounded simple closed curve there exist two sets of arcs α_1 and α_2 such that* (1) *each arc of α_1 lies wholly within ABCDA except that its endpoints are on AB and CD respectively,* (2) *each arc of α_2 lies wholly within ABCDA except that its endpoints are on BC and DA respectively,* (3) *each point of ABCDA, with the exception of A, B, C and D, is an endpoint* *either of only one arc of* α_1 *or of only one arc of* α_2, (4) *through each point within* $ABCDA$ *there is only one arc of* α_1 *and only one arc of* α_2, (5) *if* x_1 *is an arc of* α_1 *and* x_2 *is an arc of* α_2, x_1 *has only one point in common with* x_2.

PROOF. Let J denote the simple closed curve $ABCDA$ and let I denote its interior. For each positive integer n, let W denote the set of all fundamental regions with subscript greater than n. By Theorem 9, the interior of J can be divided by a double ruling β_1 into subdivisions each lying within some region of the set W_1. Each of these subdivisions can be divided by a double ruling into subdivisions each lying in some region of the set W_2, the arcs of this ruling being so chosen that they can be extended in such a way that there will result a double ruling β_2 of I such that each arc of β_1 is an arc of β_2 and each subdivision of I formed by β_2 is in some region of the set W_2. This process may be continued. It follows that there exists an infinite sequence $\beta_1, \beta_2, \beta_3, \cdots$ of double rulings of I such that (1) for each n, every arc of the ruling β_n belongs also to the ruling β_{n+1}, (2) every subdivision of I formed by the ruling β_n is in some region of the fundamental sequence with subscript greater than n. Let γ_1 denote the set of all arcs t such that, for some n, t belongs to β_n and has one endpoint on AB and the other one on CD. Let γ_2 be defined in the same manner except that AB is replaced by BC and CD is replaced by DA. If P is a point of $J + I$, distinct from A, B, C and D and not on any arc of the set γ_1 it may be shown that there exists through P, and with endpoints on J, only one arc lying wholly in $J + I$ and not intersecting any arc of the set γ_1. Let λ_1 denote the set of all such arcs for all points P of $J + I$. Let λ_2 be defined in the same way except for the substitution of γ_2 in place of γ_1. Let α_1 denote the set of all arcs of the set $\gamma_1 + \lambda_1$ and let α_2 denote the set $\gamma_2 + \lambda_2$. The sets α_1 and α_2 satisfy the conditions of Theorem 10.

THEOREM 11. *If* S *is not compact, every closed and compact point set is a subset of a compact simple domain.*

PROOF. Suppose M is a closed and compact point set. By Theorem 29 of Chapter II, M is a subset of a compact continuum N. By Theorem 3 of the present chapter and Theorems 37 and 4 of Chapter I, there exists a finite collection Q of compact simple domains such that every point of N belongs to some domain of the collection Q and each domain of the collection Q contains at least one point of N. Let K denote the point set obtained by adding together all the domains of the set Q together with their boundaries. The closure of a simple domain is a continuous curve. Therefore, by Theorem 43 of Chapter II, the compact continuum K is a continuous curve. Suppose there exists an infinite sequence of distinct points $P_1, P_2, P_3, \cdots$, belonging to $S - K$, such that (1) no two points of this sequence belong to the same component of $S - K$, (2) no infinite subsequence of it converges. For each n, let D_n denote the component of $S - K$ that contains P_n and let O_n

denote a definite point belonging to the boundary of D_n. Since K is closed and compact, there exists an ascending sequence of positive integers $n_1, n_2, n_3, \cdots$ such that the sequence $O_{n_1}, O_{n_2}, O_{n_3}, \cdots$ converges to a point O of K. There exists a compact simple domain T_1 containing O. There exists a simple domain T_2, containing O, such that $\bar{T}_2$ is a subset of T_1. Since no infinite subsequence of the sequence $P_{n_1}, P_{n_2}, P_{n_3}, \cdots$ converges, there exists an infinite subsequence $m_1, m_2, m_3, \cdots$ of the sequence $n_1, n_2, n_3, \cdots$ such that no point of the sequence $P_{m_1}, P_{m_2}, P_{m_3}, \cdots$ belongs to the compact point set T_1. For each i, the domain D_{m_i} contains a point of each of the mutually exclusive compact continua $\bar{T}_2$ and J_1' where J_1' is the boundary of T_1. But this is contrary to Theorem 62 of Chapter IV. Therefore there exists a component E of $S - K$ containing infinitely many points of the sequence ω. Let β denote the boundary of E. Let J denote the outer boundary of E with respect to the component of $S - \beta$ that contains the connected point set N. By Theorem 42 of Chapter IV, J is a simple closed curve. It encloses N and therefore M.

THEOREM 12. *If S is not compact, there exists an infinite sequence of simple closed curves $Q_1, Q_2, Q_3, \cdots$ such that every point lies within some one of them and such that, for every n, Q_{n+1} encloses Q_n.*

PROOF. There exists a simple closed curve Q_1. By Theorem 11, there exists a simple closed curve Q_2 enclosing $Q_1 + \bar{R}_1$. By the same theorem there exists a simple closed curve Q_3 enclosing $Q_2 + \bar{R}_2$. This process may be continued. Thus there exists a sequence of simple closed curves $Q_1, Q_2, Q_3, \cdots$ such that, for each n, Q_{n+1} encloses $Q_n + R_n$. This sequence satisfies the requirements of Theorem 12.

THEOREM 13. *If k is an open curve then $S - k$ is the sum of two mutually separated connected domains each bounded by k.*

PROOF. Suppose O is a point of k. There exists on k an arc $A'OB'$. There exists a simple domain R' containing O but neither A' nor B'. There exists an arc $A'X'B'$ containing no point of R'. There exist points A and B such that A is the last point that $A'X'B'$ has in common with the ray OA' of k while B is the first point following A that $A'X'B'$ has in common with the ray OB' of k. The arc $A'X'B'$ contains a segment $s(AXB)$. This segment has no point except A and B in common with k. On k there exist points C' and D' in the order $C'AOBD'$. The open curve k is the sum of a segment $s(AOB)$ and two rays BD' and AC'. By Theorem 20 of Chapter II, no ray is compact. Hence the ray AC' contains at least one point without the curve $AOBXA$. Hence if it contains a point within $AOBXA$ it must contain a point, distinct from A, on $AOBXA$. But this involves a contradiction. Similarly there is no point of the ray BD' without $AOBXA$. It follows that the interior of $AOBXA$ contains no point of k. There exists a connected domain containing O but no point of AXB + ray AC' + ray BD'. By Theorem 30 of Chapter IV, there exists, in this domain, an arc CYD such that C is on the segment $s(OA)$, D is on the segment OB, and $s(CYD)$ is entirely without the curve $AOBXA$. There exists an arc CZD such that $s(CZD)$ is within $AOBXA$. The simple closed curve $CYDZC$ encloses no point of $k - s(COD)$. The interior of $CYDZC = s(COD)$ + the interior of $CODZC$ + the interior of $CODYC$. Let O_1 and O_2 denote definite points within $CODZC$ and $CODYC$ respectively. Let S_1 denote the component of $S - k$ containing O_1 and let S_2 denote the one containing O_2. Let O_1WO_2 denote an arc from O_1 to O_2. Suppose it contains no point of the interval COD of k. There exist points O_1' and O_2' such that O_1' is the last point that O_1WO_2 has in common with the segment $s(CZD)$ while O_2' is the first point following O_1' that O_1WO_2 has in common with the segment $s(CYD)$. The segment $s(O_1'W'O_2')$ of O_1WO_2 lies entirely without $CYDZC$. By Theorem 9 of Chapter III, the interior of $CYDZC$ lies either within $O_1'W'O_2'CO_1'$ or within $O_1'W'O_2'DO_1'$. Suppose it lies within the former. Then the ray OD of k contains a point D within $O_1'W'O_2'CO_1'$. Hence it contains a point on this curve. But the arc $O_2'CO_1'$ contains no point of the ray OD. Hence $O_1'W'O_2'$ must contain a point of this ray. Thus every arc from O_1 to O_2 contains a point of k. Therefore, by Theorem 13 of Chapter II, S_1 and S_2 are distinct components of $S - k$.

Suppose P is a point of $S - k$. There exists an arc PO. Let G denote the first point that PO has in common with k. There exist (Figure 38) arcs

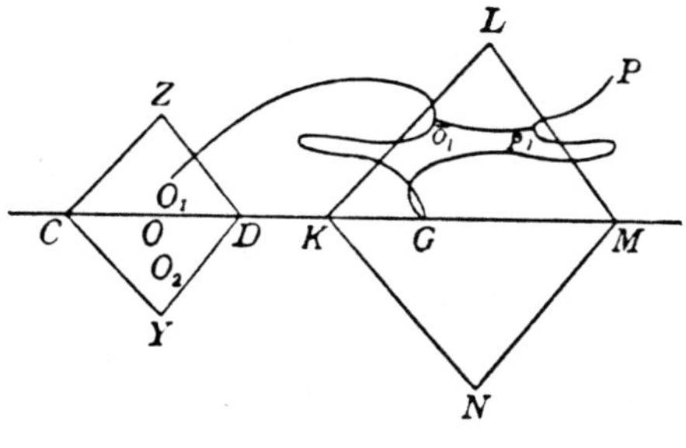

FIGURE 38

KLM and KNM having no point in common except K and M and such that K and M are the only points that the simple closed curve $KLMNK$ has in common with k, while the interior of $KLMNK = s(KGM)$ + the interior of $KGMLK$ + the interior of $KGMNK$. There exist arcs O_1O and O_2O lying, except for O, wholly within $CODZC$ and $CODYC$ respectively and therefore containing no point of k except O. Suppose G is distinct from O. Let F

denote a point such that O lies between F and G on k. Let r_1 and r_2 denote the rays of k from F and G respectively that do not contain O. Let Q denote the domain $S - (r_1 + r_2)$. The interval GOF of k is an arc lying, except for its endpoints, wholly in Q, and O_1O and O_2O are arcs lying wholly in Q and containing no point of GOF except O. Hence, by Theorem 10 of Chapter IV, there exist arcs O_1G and O_2G lying, except for G, wholly in Q and having only the point G in common with GOF and therefore having only G in common with k. The arc O_1G contains either a point within $KGMLK$ or a point within $KGMNK$. Suppose it contains a point $\bar{O}_1$ within $KGMLK$. Then O_2G cannot contain a point within $KGMLK$. For if it did then O_1 and O_2 could be joined by an arc containing no point of k. Hence O_2G must contain a point $\bar{O}_2$ within $KGMNK$. But PG contains a point within $KGMLK$ or a point within $KGMNK$. Suppose it contains a point $\bar{P}_1$ within $KGMLK$. Then there exists an arc $\bar{P}_1\bar{O}_1$ lying entirely within $KGMLK$ and therefore containing no point of k. The point set $PG + \bar{P}_1\bar{O}_1 + O_1G$ contains an arc PO_1 which contains no point of k. Hence in this case P belongs to S_1. If PG contains a point within $KGMNK$ then P belongs to S_2. Thus every point of $S - k$ belongs either to S_1 or to S_2. Thus if O is any given point of k, then $S - k$ is the sum of two mutually separated connected domains such that O is a limit point of each of them. But if $S - k$ is the sum of two mutually separated connected domains S_1' and S_2' and is also the sum of two mutually separated connected domains S_1'' and S_2'' then one of the domains S_1' and S_2' is S_1'' and the other one is S_2''. Hence every point of k is a limit point both of S_1 and of S_2.

THEOREM 14. *If S is noncompact there exist two sets W_1 and W_2 of open curves such that* (1) *through each point there is only one curve of W_1 and only one curve of W_2,* (2) *if x is a curve of W_1 and y is a curve of W_2, x and y have only one point in common.*

PROOF. Let $Q_1, Q_2, Q_3, \cdots$ denote an infinite sequence of simple closed curves satisfying the conditions of Theorem 12. For each n, let I_n denote the interior of Q_n. There exist (Figure 39) five infinite sequences $A_1, A_2, A_3, \cdots$; $B_1, B_2, B_3, \cdots$; $C_1, C_2, C_3, \cdots$; $D_1, D_2, D_3, \cdots$ and $V_2, V_3, \cdots$ such that (1) for each n, A_n, B_n, C_n and D_n are points lying on Q_n in the order $A_nB_nC_nD_n$, (2) for each n, V_{n+1} is a double ruling of I_{n+1} which subdivides I_{n+1} into nine subdivisions of which the central one is I_n, two of the arcs of V_{n+1} being parallel to $B_{n+1}C_{n+1}$ and the other two being parallel to $A_{n+1}B_{n+1}$ and $C_{n+1}D_{n+1}$. It follows that there exist two sequences of open curves $\cdots, k_{-3}, k_{-2}, k_{-1}, k_0, k_1, k_2, k_3, \cdots$ and $\cdots, t_{-3}, t_{-2}, t_{-1}, t_0, t_1, t_2, t_3, \cdots$ such that (1) if m is either 0 or an integer and so is n then k_m has only one point in common with t_n, (2) for every such n, t_n separates t_{n-1} from t_{n+1} and (3) if P is a point not lying on any open curve of one of these sequences then there exist m and n such that P is between k_m and k_{m+1} and also between t_n and t_{n+1}. For every m and n, let A_{mn} denote the intersection of k_m with t_n.

The intervals $A_{mn}A_{m(n+1)}$, $A_{m(n+1)}A_{(m+1)(n+1)}$, $A_{(m+1)(n+1)}A_{(m+1)n}$ and $A_{(m+1)n}A_{mn}$ of the open curves k_m, t_{n+1}, k_{m+1} and t_n form a simple closed curve Q_{mn}. By Theorem 10, there exist two sets of arcs W_{mn} and W'_{mn} such that (1) each arc of W_{mn} is entirely within Q_{mn} except that one of its endpoints is between A_{mn} and $A_{(m+1)n}$ on the curve t_n and the other one is between $A_{m(n+1)}$ and $A_{(m+1)(n+1)}$ on the curve t_{n+1}, (2) each arc of W'_{mn} is entirely within Q_{mn} except that one of its endpoints is between A_{mn} and $A_{m(n+1)}$ on k_n and the other one is between $A_{(m+1)n}$ and $A_{(m+1)(n+1)}$ on k_{m+1}, (3) through each point within Q_{mn} there is only one arc of the set W_{mn} and

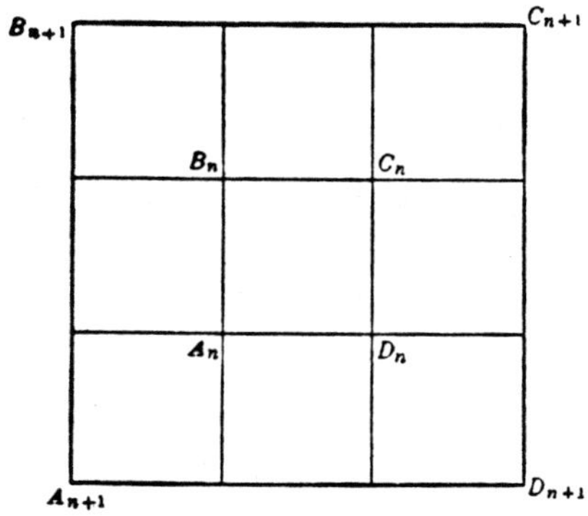

FIGURE 39

only one arc of the set W'_{mn}, (4) each point of Q_{mn}, with the exception of A_{mn}, $A_{m(n+1)}$, $A_{(m+1)n}$ and $A_{(m+1)(n+1)}$, is an endpoint of only one arc of W_{mn} or of only one arc of W'_{mn}, (5) each arc of W_{mn} has only one point in common with each arc of W'_{mn}. If P is a point on the arc t_0 and lying between the curves k_m and k_{m+1}, P is an endpoint of only one arc g_{m0} of the set W_{m0}. The other endpoint of g_{m0} is a point P_1 lying between k_m and k_{m+1} on the curve t_1. The point P_1 is an endpoint of only one arc g_{m1} of the set W_{m1}. The other endpoint of g_{m1} is a point P_2 lying between k_m and k_{m+1} on the curve t_2. This process may be continued. It follows that there exists an infinite sequence of arcs $\cdots, g_{m(-3)}, g_{m(-2)}, g_{m(-1)}, g_{m0}, g_{m1}, g_{m2}, g_{m3}, \cdots$ such that (1) P is an endpoint of g_{m0}, (2) for every n, g_{mn} is an arc of the set W_{mn} and the arcs g_{mn} and $g_{m(n+1)}$ have an endpoint in common. The sum of all the arcs of this sequence is clearly an open curve r_P containing the point P. Let W_1 denote the set of all the curves k_m, for all numbers m equal to 0 or a positive or negative integer, together with all the curves r_P for all points P of t_0. Similarly if P is a point of k_0 there exists an open curve s_P which passes through P and is the sum of infinitely many arcs each of which is, for some m and n, an arc of the set W'_{mn}. Let W_2 denote the set of all the curves t_n together with all the curves s_P for all points P of k_0. It is clear that every point belongs to only one curve of the set W_1 and to only one curve of the set W_2 and that if x is a curve of the set W_1 and y is a curve of the set W_2 then x and y have only one point in common.

In the next chapter we shall have occasion to refer to the two following axioms which, of course, are never both satisfied by the same space.

AXIOM 8. *Space is not compact.*

AXIOM 8′. *Space is compact.*

Chapter VII

CONCERNING TOPOLOGICAL EQUIVALENCE AND THE INTRODUCTION OF DISTANCE

Definitions. Suppose M is a set, N is a set, and T is a collection such that (1) each element of T is a sensed pair, (2) if x is a sensed pair of the collection T, the first term of x is an element of M and the second one is an element of N, (3) if P is an element of M there is only one sensed pair of the collection T whose first term is P, (4) if y is an element of N there is at least one sensed pair of the collection T whose second term is y. Then (1) the collection T is called a *transformation of M into N*, (2) if P is an element of M, $T(P)$ denotes the second term of the sensed pair of the collection T whose first term is P and $T(P)$ is called the *image* of P under the transformation T and P is said to be *transformed*, or *thrown*, into $T(P)$ by T, (3) if K is a subset of M, $T(K)$ denotes a set such that y belongs to it if and only if y is $T(P)$ for some element P of M, and $T(K)$ is called the *image of K under the transformation T*, and K is said to be transformed, or thrown, into $T(K)$ by T. If T_2 is a transformation of M into N_1 and T_1 is a transformation of N_1 into N_2, then by T_1T_2 is meant a transformation T of M into N_2 such that if P is a point of M, $T(P) = T_1(T_2(P))$.

If T is a transformation of M into N and there do not exist two sensed pairs of the set T with the same second term then (1) T is called a *reversible* transformation of M into N and (2) the notation T^{-1} is used to designate a collection such that z is an element of it if and only if for some sensed pair (x,y) of the collection T, z is the sensed pair (y,x). It is clear that if T is a reversible transformation of M into N and x is an element of M and $y = T(x)$ then $x = T^{-1}(y)$.

If M_1 is a point set in the space Σ_1, M_2 is a point set in the space Σ_2, and T is a transformation of M_1 into M_2 then (1) if P is a point of M_1, T is said to be *continuous at P relatively to Σ_1 and Σ_2* if and only if it is true that if ϵ is a domain in the Σ_2 sense containing $T(P)$ then there exists a domain δ in the Σ_1 sense containing P such that if X is a point of M_1 lying in δ, $T(X)$ lies in ϵ, (2) T is said to be *continuous relatively to Σ_1 and Σ_2* if it is continuous, at every point of M_1, relatively to Σ_1 and Σ_2, (3) T is said to be *reversibly continuous relatively to Σ_1 and Σ_2* if it is reversible and both T and T^{-1} are continuous relatively to Σ_1 and Σ_2.

If M_1 is a point set in the space Σ_1 and M_2 is a point set in the space Σ_2, and there exists a transformation of M_1 into M_2 which is reversibly continuous relatively to Σ_1 and Σ_2 then M_1 and M_2 are said to be topologically equivalent relatively to Σ_1 and Σ_2. In certain cases, variations of the phrase "relatively to Σ_1 and Σ_2" will be employed and this phrase may be entirely omitted in cases where it is clear what space is Σ_1 and what space is Σ_2. In particular, the statement that the point set M_1 is topologically equivalent to the set M_2 of real numbers signifies that M_1 is topologically equivalent to M_2 relatively to Σ_1 and Σ_2 where Σ_1 is some previously determined space and Σ_2 is a space in which every point is a number and in which the number x is a limit number of the set N of numbers if and only if it is true that for every positive number ϵ there exists a number of the set N distinct from x and differing from x by less than ϵ.

If T is a transformation of the arc, segment or open curve M into the arc, segment or open curve N, T is said to *preserve order* if and only if it is true that if the point Y is between the points X and Z in M then $T(Y)$ is between $T(X)$ and $T(Y)$ in N.

If T is a transformation of the simple closed curve M into the simple closed curve N, T is said to *preserve order* if and only if it is true that if the pair of points X and Z separates Y from W in M then the pair of points $T(X)$ and $T(Z)$ separates $T(Y)$ from $T(W)$ in N.

If T is a transformation of the arc, segment or open curve M into a set of numbers then T is said to preserve order if and only if it is true that if Y is between X and Z in M then the number $T(Y)$ is greater than one of the numbers $T(X)$ and $T(Z)$ and less than the other one. It is easy to see that in order that this condition be fulfilled it is necessary and sufficient that there exist two points A and B belonging to M such that if X precedes Y in the order from A to B then the number $T(X)$ is less than the number $T(Y)$.

A set N of real numbers is said to be *connected* if and only if it is true that if x and y are two numbers of the set N then every number between x and y also belongs to N.

THEOREM 1. *If T is a reversible transformation of an arc, segment or open curve into an arc, segment, open curve or a connected set of numbers then in order that T be reversibly continuous it is necessary and sufficient that it preserve order.*

Theorem 1 does not remain true if the word "connected" is omitted from its hypothesis.

THEOREM 2. *If a and b are real numbers every arc is topologically equivalent to the set of all real numbers from a to b.*

PROOF. Suppose $a = 0$ and $b = 1$. By Theorem 22 of Chapter I, there exists a simple countable sequence α of points of the arc AB, all distinct from each other and from A and B, such that if β denotes the point set whose elements are the points of this sequence, then every point of AB either belongs to β or is a limit point of it. If P_1 is the first point of the sequence α, denote the number $1/2$ by the symbol x_{P_1}. If P_2 is the first point of α between A and P_1, denote the number $1/3$ by x_{P_2}. If P_3 is the first point of α

between P_1 and B, denote 2/3 by x_{P_3}. If P_4 is the first point of α between A and P_2, denote 1/4 by x_{P_4}. If P_5 is the first point of α between P_3 and B denote 3/4 by x_{P_5}. If P_6 is the first point of α between A and P_4 denote 1/5 by x_{P_6}. If P_7 is the first point of α between P_2 and P_1 denote 2/5 by x_{P_7}. If P_8 is the first point of α between P_1 and P_3, denote 3/5 by x_{P_8}. If P_9 is the first point of α between P_5 and B, denote 4/5 by x_{P_9}. For each point P of α, let x_P denote the point which is associated with P by an obvious continuation of this process. If P is a point of AB not belonging to α, P is the sequential limit point of some subsequence of the sequence α and if $C_1, C_2, C_3, \cdots$ and $D_1, D_2, D_3, \cdots$ are two subsequences of α converging to P then $x_{C_1}, x_{C_2}, x_{C_3}, \cdots$ and $x_{D_1}, x_{D_2}, x_{D_3}, \cdots$ converge to the same number. Call this number x_P. The set of all sensed pairs (P,x_P), for all points P of AB, is a reversibly continuous transformation of the point set AB into the set of all real numbers from 0 to 1. A similar argument applies in case a and b are any two real numbers.

THEOREM 3. *If A is a point of a simple closed curve J and N is the set of all real numbers from 0 to 1 then there exists a continuous transformation T of N into the point set J such that $T(0) = T(1) = A$ but such that if x and y are any two numbers between 0 and 1 then the points $T(x)$ and $T(y)$ are distinct from each other and from A.*

Theorem 3 may be easily proved with the use of Theorem 2 of the present chapter and Theorem 107 of Chapter I.

THEOREM 4. *If O is a point of an open curve M, there exists a reversibly continuous transformation that throws M into the set of all real numbers and throws O into zero.*

Theorem 4 may be proved with the use of Theorem 2 and the fact that if OA and OB are the two rays of the open curve M that start from O, then there exist two infinite sequences of points $A_1, A_2, A_3, \cdots$ and $B_1, B_2, B_3, \cdots$ such that (1) for each n, A_n precedes A_{n+1} in the order from O to A on OA, and B_n precedes B_{n+1} in the order from O to B on OB, (2) if P is any point whatsoever of M, there exists a positive number m such that P lies between A_m and B_m.

THEOREM 5. *Every two arcs are topologically equivalent to each other and so are every two simple closed curves and every two open curves.*

Theorem 5 may be established with the use of Theorems 2, 3 and 4.

DEFINITION. The set of all sensed pairs of real numbers is called the number plane. If a, b and c are real numbers and a and b are not both 0, then the set of all sensed pairs (x,y) of real numbers such that $ax + by + c = 0$ is called a *straight line in the number plane*. If a, b and r are real numbers and r is positive, then the set of all sensed pairs (x,y) of real numbers such that $(x - a)^2 + (y - b)^2 = r^2$ is called a *number circle* or a *circle in the number plane*, and (a,b) is called its center and r is called its radius.

THEOREM 6. *Every space that satisfies Axioms* 0, 1, 2, 3, 4, 5_1, 5_2, 6, 7 *and* 8 *is topologically equivalent to the number plane in the sense that if S is the set of all points of a space* Σ *satisfying these axioms then there exists a reversible transformation T of S into the set of all sensed pairs of real numbers such that if, for each point P of S,* x_P *and* y_P *denote the first and second terms respectively of the sensed pair T(P), then the point A is a limit point of the point set M in* Σ *if and only if it is true that for every positive number* ϵ *there exists a point B of M distinct from A such that* $(x_A - x_B)^2 + (y_A - y_B)^2 < \epsilon$.

PROOF. By Theorem 14 of Chapter VI, there exist two sets W_1 and W_2 of open curves such that (1) through each point of S there is only one curve of the set W_1 and only one curve of the set W_2, (2) each curve of W_1 has only one point in common with each curve of the set W_2. Let O denote some definite point of S. Let OX and OY denote the curves of the set W_1 and W_2 respectively that contain O. By Theorem 4, there exist a reversibly continuous transformation T_1 of the point set OX into the set of all real numbers and a reversibly continuous transformation T_2 of OY into the set of all real numbers such that $T_1(O)$ = zero and $T_2(O)$ = zero. If P is any point of S, let x_P denote the number $T_1(P_1)$ where P_1 is the point in which OX intersects the curve of the set W_2 that contains P and let y_P denote the number $T_2(P_2)$ where P_2 is the point in which OY intersects the curve of the set W_1 that contains P. Let T denote the collection of all sensed pairs t such that, for some point P of S, the first term of t is P and the second term of t is the sensed pair of numbers (x_P,y_P). The transformation T satisfies all the requirements of the conclusion of Theorem 6.

THEOREM 7. *Under a reversibly continuous transformation, into a number plane, of a space satisfying these axioms, every number circle is the image of a simple closed curve and every straight line in the number plane is the image of an open curve.*

DEFINITION. A collection G of point sets is said to be a *topological collection* if and only if every point set which is topologically equivalent to a point set of the collection G is itself a member of the collection G.

THEOREM 8. *In a space satisfying Axioms* 0 *and* 1, *the collection of all closed and compact point sets is a topological collection. So is the collection of all connected point sets, the collection of all compact continuous curves, the collection of all arcs, the collection of all simple closed curves.*

However, even in the plane, the collection of all closed point sets is not a topological collection. Neither is the collection of all compact point sets, the collection of all open curves, the collection of all segments of arcs, the collection of all rays.

It is not true that, in every space satisfying Axioms 0, 1 and 2, the collection of all domains is a topological collection. Let AB denote a straight line interval in the plane and let Σ denote the compact subspace of the plane whose points are the points of AB. Let X and Y denote two points of AB distinct from A and B and let AY and XY denote intervals of AB with endpoints as indicated. The point set $XY - Y$ is the image of $AY - Y$ under some reversibly continuous transformation. But, in the space Σ, $AY - Y$ is a domain and $XY - Y$ is not. However, in every locally compact space satisfying Axioms 0 and 1–5, the collection of all domains is a topological collection.

In the plane every segment of an arc is topologically equivalent to an open curve. Furthermore there exists, in the plane, a point set M topologically equivalent to a segment of an arc but such that $\bar{M}$ is an irreducible continuum which is not an arc. Consider the following example.

EXAMPLE. In a Cartesian plane, let M denote the set of all points of the graph of $y = \sin(1/x)$ which lie between the graphs of $x = 0$ and $x = 1$ and let K denote the set of all points on the X-axis which lie between these same two vertical lines. Let T denote the transformation of K into M which throws each point of K into the point of M with the same abscissa. The transformation T is reversibly continuous. But, though K is a segment of an arc, $\bar{M}$ is not a continuous curve.

DEFINITION. The subsets H and K of the point set M are said to be *strongly topologically equivalent with respect to M* if and only if there exists a reversibly continuous transformation of M into itself that throws H into K.

It is not true that every two topologically equivalent compact continuous curves in the plane are strongly topologically equivalent with respect to the plane.

EXAMPLE. In a Cartesian plane E, let OA denote the interval of the X-axis whose extremities are the origin and the point (1,0), let AB denote the one whose extremities are (1,0) and (2,0) and let N denote the circle represented by the equation $x^2 + y^2 = 1$. Let H and K denote the point sets $N + OA$ and $N + AB$ respectively. These point sets are topologically equivalent but not strongly so with respect to E.

THEOREM 9. *If J and J' are two simple closed curves and α is a reversibly continuous transformation of J into J', there exists a reversibly continuous transformation ω of S into itself such that if P is a point of J then $\omega(P) = \alpha(P)$.*

PROOF. Let A, B, C and D denote four points such that $A + C$ separates B from D on J. Let AB, BC, CD and DA denote arcs of J, with endpoints as indicated, such that no one of them contains any one of the points A, B, C and D except its endpoints. Let A', B', C' and D' denote the points $\alpha(A)$, $\alpha(B)$, $\alpha(C)$ and $\alpha(D)$ respectively and let $A'B'$, $B'C'$, $C'D'$ and $D'A'$ denote the arcs $\alpha(AB)$, $\alpha(BC)$, $\alpha(CD)$ and $\alpha(DA)$ respectively. In view of results established in the course of arguments employed in the proof of previous theorems of this chapter, it is clear that there exist two reversibly continuous transformations T and T' of S into the set K of all sensed pairs of real numbers such that if, for each point P of S, x_P and y_P denote the first and second term respectively of the sensed pair $T(P)$, and x'_P and y'_P denote the first and second term respectively of the sensed pair $T'(P)$, then (1) $T(A) = T'(A') = (0,0)$, $T(B) = T'(B') = (1,0)$, $T(C) = T'(C') = (1,1)$ and $T(D) = T'(D') = (0,1)$ and (2) $x_P = 0$ if P belongs to AD, $x'_P = 0$ if P belongs to $A'D'$, $x_P = 1$ if P belongs to BC, $x'_P = 1$ if P belongs to $B'C'$, $y_P = 0$ if P belongs to AB, $y'_P = 0$ if P belongs to $A'B'$, $y_P = 1$ if P belongs to CD and $y'_P = 1$ if P belongs to $C'D'$. Let I denote the interior of J. Suppose P is a point of $J + I$. There exist on AB, BC, CD and DA respectively points E, F, G and H such that $x_E = x_G = x_P$ and $y_F = y_P = y_H$. Let E', F', G' and H' denote the points $\alpha(E)$, $\alpha(F)$, $\alpha(G)$ and $\alpha(H)$ respectively. Let I' denote the interior of J'. There exists an arc $F'H'$, lying entirely in $J' + I'$, such that $T'(F'H')$ lies on a straight line in the number plane. Similarly there exists an arc $E'G'$, lying in $I' + J'$, such that $T'(E'G')$ lies on a straight line in the number plane. The arcs $E'G'$ and $F'H'$ have only one point P' in common. Let ω_1 denote a transformation of $J + I$ into $J' + I'$ such that, for each point P of $J + I$, $\omega_1(P) = P'$. It may be shown that this transformation is reversibly continuous. If P is a point of J then $\omega_1(P) = \alpha(P)$.

Now let $I_1, I_2, I_3, \cdots$ denote the simple domains into which S is subdivided by the open curves of the sets U and V where U is the set of all open curves of the sequence $\cdots, k_{-3}, k_{-2}, k_{-1}, k_0, k_1, k_2, k_3, \cdots$ and V is the set of all those of the sequence $\cdots, t_{-3}, t_{-2}, t_{-1}, t_0, t_1, t_2, t_3, \cdots$ as described in the proof of Theorem 14 of Chapter VI, it being understood that I_1 is the interior of J and that AB and CD are intervals of curves of the set U and BC and DA are intervals of curves of the set V. For each n, let γ_n denote the boundary of I_n. Similarly there exist sets U' and V', satisfying the conditions (1), (2) and (3) which are stated as being satisfied by the sets U and V in the proof of that theorem, and satisfying the fourth condition that, of the subdivisions into which the open curves of these sets subdivide S, one is I', the interior of J', and the arcs $A'B'$ and $C'D'$ of J' are intervals of curves of the set U' and the arcs $B'C'$ and $D'A'$ are intervals of curves of the set V'. There exist a sequence $\omega_1, \omega_2, \omega_3, \cdots$ and a labelling $I'_1, I'_2, I'_3, \cdots$ of the subdivisions into which the curves of U' and V' divide S such that if, for each n, γ'_n denotes the boundary of I'_n then (1) ω_1 is as already indicated, (2) for each n, ω_n is a reversibly continuous transformation of $I_n + \gamma_n$ into $I'_n + \gamma'_n$ that throws I_n into I'_n and γ_n into γ'_n, (3) if the point O is common to the simple closed curves γ_i and γ_j then $\omega_i(O)$ is identical with $\omega_j(O)$. There exists a transformation ω of S into itself such that if P is a point of $I_n + \gamma_n$ then $\omega(P)$ is identical with $\omega_n(P)$. The transformation ω satisfies all the requirements of Theorem 9.

Suppose S is the set of all points of a space Σ satisfying Axioms 0, 1, 2, 3, 4, 5_1, 5_2, 6, 7 and 8. By Theorem 6, Σ is topologically equivalent to the number plane. Let T denote a definite reversibly continuous transformation of S into the set of all sensed pairs of real numbers. For each point P of S, let x_P and y_P respectively denote the first and the second term of the sensed pair $T(P)$ and call them the abscissa and ordinate respectively of the point P. If A and B are points of S, the non-negative square root of $(x_A - x_B)^2 + (y_A - y_B)^2$ will be called the distance between A and B. The point whose abscissa and ordinate are a and b respectively may sometimes be referred to as "the point (a,b)". It is to be understood that this phrase is *merely* an *abbreviation* for the phrase "the point whose image in the number plane is the sensed pair (a,b)". It is not to be interpreted to mean that the sensed pair (a,b) is a point of S. Such use of language seems rather unfortunate but it is customary and has the advantage of brevity and may perhaps be considered allowable where it is clearly understood to be merely an abbreviation. A subset M of S is said to be a straight line provided $T(M)$ is a straight line in the number plane, that is to say provided there exist a, b and c such that either a or b is distinct from O and such that M is the set of all points P such that $ax_P + by_P + c = 0$. A subset M of S is said to be a circle provided $T(M)$ is a circle in the number plane, that is to say provided there exist a point O and a positive number r such that M is the set of all points of S whose distance from O is equal to r. In an entirely analogous manner, other concepts of ordinary Euclidean plane geometry may be defined. With respect to the notions of distance, straight line and so forth thus introduced, S is indeed a Euclidean plane and in the statements and proofs of some of the following theorems free use will be made of the ordinary terminology of Euclidean geometry. By the *diameter* of a compact point set M is meant the greatest number which is not exceeded by the distance between any two points of M. A point set will be said to be linear if it is a subset of some straight line.

THEOREM 10. *If D is either S or a simple domain and the upper semi-continuous collection G of mutually exclusive compact continua fills up $\bar{D}$ and is a regular curve with respect to its elements, then G is acyclic with respect to its elements.*

PROOF. If x and y are two elements of G, there exists a finite set h_1, h_2, $h_3, \cdots, h_n$ of elements of G such that x is separated from y in $\bar{D}$ by the point set $h_1 + h_2 + h_3 + \cdots + h_n$. By Theorem 27 of Chapter IV, x is separated from y in $\bar{D}$ by a continuum T which is a subset of $h_1 + h_2 + h_3 + \cdots + h_n$. The continuum T is a subset of some one of the point sets $h_1, h_2, h_3, \cdots, h_n$. It follows that x and y are separated in G by some one element of G. Hence G is an acyclic continuous curve with respect to its elements.

THEOREM 11. *Suppose D is either S or a simple domain and F is a family of upper semi-continuous collections of mutually exclusive compact continua such* *that each collection of the family F fills up $\bar{D}$ and for every two collections G_1 and G_2 of the family F either every element of G_1 is a subset of some element of G_2 or every element of G_2 is a subset of some element of G_1. Suppose furthermore that every collection of the family F is a regular curve with respect to its elements. Let G denote the set of all continua g such that, for every element H of F, g is a subset of some continuum of H but no continuum of which g is a proper subset has this property. Then G is a dendron with respect to its elements and therefore it is not the collection of all degenerate subsets of $\bar{D}$.*

Theorem 11 may be proved with the help of Theorem 10 of the present chapter and Theorem 47 of Part 1 of Chapter V.

There exists, in the plane, an upper semi-continuous collection G of mutually exclusive compact and nondegenerate continua such that (1) no continuum of G separates S, (2) G is an arc with respect to its elements, (3) G^* contains no domain. Consider the following example.

Example. Let A, B, A', B', t_1, t_2, $t_3, \cdots$, K, A_1A_1', A_2A_2', $A_3A_3', \cdots$, B_1B_1', B_2B_2', $B_3B_3', \cdots$ and Q be the same as in Example 1 of Part 1 of Chapter V. For each n, let h_n denote the continuum $t_n + A_nA_n' + B_nB_n'$. Let G denote the collection whose elements are the continua h_1, h_2, $h_3, \cdots$ and the elements of Q.

There exists, in S, an upper semi-continuous collection G of mutually exclusive compact continua such that G is topologically equivalent to S but such that G^* contains no domain.

Example. There exists a sequence P_1, P_2, $P_3, \cdots$ whose elements are the rational points, that is to say the points whose coordinates are both rational. Let K_1 denote a circle with center at P_1 and radius less than 1. Let n_1 denote 1. Let n_2 denote the smallest number n such that P_n neither belongs to K_n nor is enclosed by it. Let K_2 denote a circle, of radius less than 1/2, with center at P_{n_2}, and neither containing nor enclosing any point of K_1. Let n_3 denote the smallest number n such that neither K_1 nor K_2 contains nor encloses P_n. Let K_3 denote a circle, of radius less than 1/3, with center at P_{n_3}, and neither containing nor enclosing any point of K_1 or of K_2. This process may be continued indefinitely. Thus there exist an infinite sequence of positive integers $n_1, n_2, n_3, \cdots$ and an infinite sequence of circles K_1, K_2, $K_3, \cdots$ such that if j is any positive number then (1) K_j is a circle of radius less than $1/j$ and with center at P_{n_j}, (2) n_{j+1} is the smallest number n such that P_n is neither contained nor enclosed by any one of the circles $K_1, K_2, K_3, \cdots$, (3) K_{j+1} neither contains nor encloses any point of any one of the circles K_1, K_2, $K_3, \cdots$, K_j. Let G denote the collection whose elements are the circles of the sequence K_1, K_2, $K_3, \cdots$ and the points of S that neither belong to nor are enclosed by any circle of this sequence. The upper semi-continuous collection G is topologically equivalent to S. But clearly G^* contains no domain.

THEOREM 12. *If, in Σ, A and B are distinct points of a closed point set M which contains no domain, then there exists a circle J, separating A from B, such that either J does not intersect M or $J \cdot M$ is totally disconnected.*

PROOF. Suppose, on the contrary, that every circle which separates A from B contains a nondegenerate continuum lying in M. Let AB denote the straight line interval from A to B, let A_1 and B_1 denote points on AB in the order AA_1B_1B and let A_1B_1 denote the straight line interval whose extremities are A_1 and B_1. For each point P of A_1B_1, let J_P denote the circle with center at A that passes through P. Let a denote a definite positive number such that, for some point P of A_1B_1, the point set $M \cdot J_P$ contains an arc of length greater than a. For each positive integer m greater than $1/a$, let N_m denote the set of all points P of A_1B_1 such that $M \cdot J_P$ contains an arc of length greater than or equal to $1/m$. Then the nondegenerate continuum A_1B_1 is the sum of the point sets of the sequence N_m, N_{m+1}, $N_{m+2}, \cdots$. But each of these point sets is closed. It follows, with the help of Theorem 57 of Chapter I, that there exist a number k, greater than or equal to m, and two points A_2 and B_2 lying on A_1B_1 in the order $AA_1A_2B_2B_1B$ such that N_k contains the subinterval A_2B_2 of the interval A_1B_1. Let α denote the circle with center at A that passes through B. There exists a finite set W of arcs of α such that (1) every point of α belongs to some arc of the set W and (2) no two arcs of the set W have in common any point that is not an endpoint of both of them, (3) the length of every arc of W is less than $1/(2k)$. For each arc w of the set W, let Z_w denote the set of all points P of A_2B_2, if there are any, such that $M \cdot J_P$ intersects every straight line interval from A to a point of w. Let $w_1, w_2, w_3, \cdots, w_n$ denote the set of all arcs w of the set W for which Z_w exists. The point sets $Z_{w_1}, Z_{w_2}, Z_{w_3}, \cdots, Z_{w_n}$ are all closed and their sum is A_2B_2. It follows that there exist a number i and two points A_3 and B_3, lying on A_2B_2 in the order $AA_1A_2A_3B_3B_2B_1B$, such that Z_{w_i} contains the interval A_3B_3 of A_2B_2. Let X_i and Y_i denote the endpoints of the arc w_i. Let AX_i and AY_i denote straight line intervals with endpoints as indicated. Let D_1 denote the simple domain bounded by $AX_i + AY_i + w_i$ and let D_2 denote the annular domain consisting of all points that lie entirely within J_{B_3} and entirely without J_{A_3}. Let D denote the common part of the domains D_1 and D_2. The point set D is a domain and it is a subset of M. Thus the supposition that Theorem 12 is false leads to a contradiction.

THEOREM 13 (MENGER). *In order that the continuum M contain no domain it is necessary and sufficient that for every two points A and B of M there exists a totally disconnected closed point set that separates A from B in M.*

Theorem 13 may be proved with the help of Theorem 12 of the present chapter and Theorem 27 of Chapter IV.

DEFINITION. The set of all sensed triples of real numbers (x,y,z) such that $x^2 + y^2 + z^2 = 1$ is called the *number sphere*. If (x_1,y_1,z_1) and (x_2,y_2,z_2) are two sensed triples belonging to the number sphere, then the positive square root of $(x_1 - x_2)^2 + (y_1 - y_2)^2 + (z_1 - z_2)^2$ is called the *distance* from (x_1,y_1,z_1) to (x_2,y_2,z_2). If (a,b,c) is a sensed triple belonging to the number sphere and r is a positive number less than 2, then the set of all sensed triples belonging to the number sphere and at a distance from (a,b,c) equal to r is called a *circle on the number sphere* and (a,b,c) is called its center and r is called its radius. A circle on the number sphere is called a *great circle* on the number sphere if its radius is $\sqrt{2}$. If (a,b,c) is a sensed triple belonging to the number sphere, the set of all sensed triples belonging to the number sphere and at a distance less than $\sqrt{2}$ from (a,b,c) is called a *hemisphere* of the number sphere and (a,b,c) is called its *center*.

Each circle on the number sphere has only two centers and if J is a great circle on the number sphere there are only two hemispheres of the number sphere each having one of the centers of J as its center.

THEOREM 14. *Every space that satisfies Axioms* 1, 2, 3, 4, 5_1, 5_2, 6, 7 *and* 8′ *is topologically equivalent to the number sphere in the sense that if Σ is a space satisfying these axioms then there exists a reversible transformation T of S, the set of all points of Σ, into the set of all sensed triples of the number sphere such that if, for each point P of S, x_P, y_P and z_P denote the first, second and third terms respectively of the sensed triple $T(P)$ then the point A is a limit point of the point set M in Σ if and only if it is true that for every positive number ϵ there exists a point B of M distinct from A such that $(x_A - x_B)^2 + (y_A - y_B)^2 + (z_A - z_B)^2 < \epsilon$.*

PROOF. Let J denote some definite simple closed curve in S and let J' denote a definite great circle on the number sphere. Let D and E denote the components of $S - J$ and let D' and E' denote the hemispheres of the number sphere whose centers are the centers of J'. With the help of Theorem 9, it may be shown that there exist a reversibly continuous transformation T_1 of $D + J$ into $D' + J'$ and a reversibly continuous transformation T_2 of $E + J$ into $E' + J'$ such that if P is a point of J then $T_1(P) = T_2(P)$. There exists a transformation T of S into the number sphere such that if P is a point of $D + J$, $T(P) = T_1(P)$ and if P is a point of $E + J$, $T(P) = T_2(P)$. The transformation T satisfies the requirements of Theorem 14.

Suppose Σ' is a definite space satisfying Axioms 1, 2, 3, 4, 5_1, 5_2, 6, 7 and 8′ and S' is the set of all points of Σ'. By Theorem 14, S' is topologically equivalent to the number sphere. Let T denote a definite reversibly continuous transformation of S' into the number sphere. If A and B are two points of S', the distance from $T(A)$ to $T(B)$ will be called the *distance from A to B*. A subset of S' will be called a *circle* if and only if its image under the transformation T is a circle on the number sphere and it will be called a *great circle* if and only if its image under this transformation is a great circle on the number sphere. Similarly, by a *hemisphere* is meant a subset of S' whose image, under the transformation T, is a hemisphere of the number sphere.

THEOREM 15. *If G is an upper semi-continuous collection of mutually exclusive continua filling up S', then the space whose elements are the continua of G is topologically equivalent to a cactoid.*

PROOF. If the elements of G are called "points" and every region with respect to G is called a "region" then, by Theorems 15 and 16 of Chapter V and Theorem 117 of Chapter I, Axioms 0, 1′ and 2 hold true. Clearly G is compact with respect to its elements. Suppose now that Q is a nondegenerate link of G. The set Q of elements contains a simple closed curve of elements of G and, by Theorem 109 of Chapter I, no such curve contains uncountably many cut elements of the continuum G. Thus there exists a continuum A belonging to the set Q and not separating G and such that not every continuum of the collection G distinct from A is separated from A by some continuum of the collection G. If no continuum of G separates G then, by Theorems 29 and 30 of Chapter III and Theorem 19 of Part 1 of Chapter V, G is a simple closed surface with respect to its elements. Suppose there exists at least one continuum of G that separates G. Let H denote the set of continua consisting of A together with all continua y of G such that there exists no continuum of G separating y from A. By Theorem 15 of Chapter II, if g is a continuum of G distinct from A and there exists a continuum of G that separates g from A then there exists a continuum X_g belonging to H and such that (a) X_g separates g from A, (b) no continuum of G separates A from X_g. The collection H is a simple closed surface with respect to its elements. It is identical with Q. For suppose, on the contrary, that there exists a simple closed curve J of elements of G that contains two elements h_1 and h_2 of H and an element g that does not belong to H. The continuum X_g separates g from one of the continua h_1 and h_2 and therefore there exists no simple closed curve of elements of G containing g, h_1 and h_2. Thus the supposition that H is not identical with Q leads to a contradiction. Hence every nondegenerate link of the continuum G of elements is a simple closed curve of elements. In other words, G is a cactoid with respect to its elements.

THEOREM 16. *If, in the space Σ', $M_1, M_2, M_3, \cdots$ is a sequence of continua, all proper subsets of S', and $P_1, P_2, P_3, \cdots$ is a sequence of distinct points then there exists, in Σ', a contracting sequence of mutually exclusive continua $N_1, N_2, N_3, \cdots$ such that* (a) *for each n, there exists a reversibly continuous transformation of S' into itself that throws M_n into N_n,* (b) *if the continua of the sequence $N_1, N_2, N_3, \cdots$ are regarded as elements and each point that neither belongs to a continuum of this sequence nor is separated from one continuum of* *this sequence by another one is regarded as an element then there exists a reversibly continuous transformation of S' into the set of elements so obtained which, for each n, throws P_n into N_n.*

DEFINITION. An *arc* of simple links of a compact continuous curve K is a collection t of simple links of K such that if M denotes the sum of all the elements of t then M is a continuum containing two elements A and B of t such that if X is an element of t, $(t - X)^*$ is connected if and only if X is A or B.

THEOREM 17. *If t is an arc of simple links of a cactoid K and M is the sum of all the elements of t then there exists, on the simple closed surface S', an upper semi-continuous collection G of mutually exclusive continua such that there is a reversibly continuous transformation T of M into G having the property that if X is any point of M then the components of $K - X$ and the components of $S' - T(X)$ are equally numerous and if $S' - T(X)$ has infinitely many components they form a contracting sequence.*

PROOF. Let r denote some definite arc in S'. Let H denote the point set consisting of all the cut points of M together with every point (if there is any) which is an end element of t. Let H' denote the set of all points X of H such that either (1) $K - X$ has more than two components or (2) X is an end element of t and $K - X$ has more than one component. By Theorem 33 of Part 1 of Chapter V, H' is countable. Let $P_1, P_2, P_3, \cdots$ denote a sequence whose terms are the elements of H'. There exists, on r, a closed point set $W(H)$ which is the image of H under a reversibly continuous transformation W such that if X, Y and Z are points of H then X separates Y from Z in M if and only if $W(X)$ separates $W(Y)$ from $W(Z)$ on r. There exists, on the simple closed surface S', a sequence of mutually exclusive continua $L_1, L_2, L_3, \cdots$ such that (1) this sequence is infinite if the sequence $P_1, P_2, P_3, \cdots$ is infinite and has the same number of terms as $P_1, P_2, P_3, \cdots$ if that sequence is finite, (2) the diameter of L_i is equal to or less than $1/i$, (3) L_i has, in common with r, only the point $W(P_i)$, (4) L_i is the sum of countably many simple closed curves forming a contracting sequence Q_i, (5) Q_i is an infinite sequence if $K - P_i$ has infinitely many components, and if the number of components of $K - P_i$ is finite and equal to n_i then the number of simple closed curves in Q_i is $n_i - 1$ or $n_i - 2$ according as P_i is or is not an end element of t, (6) no point of any curve of the sequence Q_i is separated in S' from any point of r by any other curve of this sequence, and no two points of r are separated from each other in S' by any curve of Q_i, (7) every two curves of Q_i have in common only the point $W(P_i)$.

If every continuum of the sequence $L_1, L_2, L_3, \cdots$ is regarded as an element and every point of S' which neither belongs to any continuum of this sequence nor is separated from any point of r by any continuum of this sequence is regarded as an element, the set U of elements so obtained is upper semi-continuous. It follows that there exists a reversibly continuous transformation W_1 of U into S' which throws the set of elements of U consisting of $L_1, L_2, L_3, \cdots$ and all points of r that belong to no continuum of this sequence into an arc z of points of S', z being a great semicircle or a proper subinterval of a great semicircle according as it is or is not true that both end elements of t are points. Let V denote a transformation of the set H such that $V(P_i)$ is $W_1(L_i)$ and, for each point X of the set $H - (P_1 + P_2 + P_3 + \cdots)$, $V(X)$ is $W_1W(X)$. If at least one of the end elements of t is a point, let F denote one such point and let O denote $V(F)$. If neither end element of t is a point, let O denote a point not lying on the arc z and not opposite to either endpoint of z but lying on a great semicircle of which z is a part. For each point X of the set H let $C(X)$ denote the point O or the circle with center at O that passes through $V(X)$ according as $V(X)$ is or is not identical with O, and let $U(X)$ denote the collection of all continua Y of U such that $W_1(Y)$ is a point of $C(X)$ and let $Q(X)$ denote the point set obtained by adding together all the continua of $U(X)$. It may be noted that for each X all but possibly one of the continua of $U(X)$ are degenerate. Let U_t denote the collection of all the point sets $Q(X)$ for all points X of H. Let N denote the closed point set obtained by adding together all the continua of the collection U_t. Let $S_1, S_2, S_3, \cdots$ denote the simple closed surfaces, if there are any, which are elements of t. If S_n is not an end element of t it contains two points X_n and Y_n which are cut points of M. If it is an end element of t then S_n contains one, and only one, point X_n which is a cut point of M. On the simple closed surface S', let D_n denote a complementary domain of N whose boundary contains every element of $U(X_n)$ which is a point, or every element of $U(X_n) + U(Y_n)$ which is a point, according as S_n is or is not an end element of t. Suppose S_n contains one or more cut points of K that are not cut points of M. By Theorems 106 and 22 of Chapter I, there are only countably many such points. Call them $P_{n1}, P_{n2}, P_{n3}, \cdots$. It may be shown that there exists a contracting sequence of mutually exclusive continua $E_{n1}, E_{n2}, E_{n3}, \cdots$ all lying in D_n, such that (a) the sequences $E_{n1}, E_{n2}, E_{n3}, \cdots$ and $P_{n1}, P_{n2}, P_{n3}, \cdots$ either have the same finite number of terms or are both infinite, (b) the number of components of $S' - E_{nj}$ equals the number of components of $K - P_{nj}$, (c) the point set $(N + E_{n1} + E_{n2} + E_{n3} + \cdots) - E_{nj}$ lies in one component of $S' - E_{nj}$, (d) if R denotes the collection whose elements are $Q(X_n)$, $Q(Y_n)$ in case Y_n exists, $E_{n1}, E_{n2}, E_{n3}, \cdots$ and the points of D_n that are not separated from N on S' by any E_{nj} then there exists a reversibly continuous transformation Z_n of S_n into the upper semi-continuous collection R such that $Z_n(X_n) = Q(X_n)$, for each P_{nj}, $Z_n(P_{nj}) = E_{nj}$ and, if Y_n exists, $Z_n(Y_n) = Q(Y_n)$. There exists a transformation of M such that (1) if the point X belongs to H then $T(X) = Q(X)$, (2) if X is a point which does not belong to H but which lies on the simple closed surface S_n then $T(X) = Z_n(X)$. The transformation T satisfies the conditions required of T in the statement of Theorem 17.

THEOREM 18. *If K is a cactoid there exists, on the simple closed surface S', an upper semi-continuous collection G of mutually exclusive continua such that the space whose elements are the continua of G is topologically equivalent to K and such that every nondegenerate continuum of G separates S'.*

PROOF. It may be shown that there exists a finite or countably infinite contracting sequence $t_1, t_2, t_3, \cdots$ of arcs of simple links of K such that (a) neither end element of t_1 separates K, (b) for each n greater than 1, one element of t_n is an element of t_m for some m less than n, (c) if i is distinct from j and t_i and t_j have an element in common that element is an end element either of t_i or of t_j, (d) if there exists a simple link of K that is not an element of any t_n, every such link is a point and the set of all such points is totally disconnected. For each n, let M_n denote the sum of all the elements of t_n. Let G_1' and T_1 respectively denote an upper semi-continuous collection of continua and a continuous transformation satisfying with reference to t_1 the conditions which, according to the statement of Theorem 17, are satisfied by G and T with reference to t, and satisfying the additional condition that if X is a cut point of K belonging to M_1 then every component of $S' - T_1(X)$ that contains no point of any element of G_1' is of diameter less than 1. Let E_1 denote the sum of all the continua of G_1' and, if X is a cut point of K belonging to M_1, let $D_{X1}, D_{X2}, D_{X3}, \cdots$ denote the components of $S' - T_1(X)$, if there are any, that contain no points of E_1. Let A_2 denote the end element of t_2 which belongs to t_1. If A_2 is a point, let X_2 denote A_2. If A_2 is a simple closed surface, let X_2 denote the point of this surface which is a cut point of t_2. In either case there exist a collection G_2' of mutually exclusive continua and a transformation T_2 satisfying with reference to t_2 the conditions which, according to the statement of Theorem 17, are satisfied by G and T with reference to t and such that if E_2 denotes the sum of all the continua of G_2' then (a) if X is a point of A_2, $T_2(X) = T_1(X)$, (b) if X is a point not belonging to A_2 but belonging to some element of t_2 then $T_2(X)$ lies in the domain $D_{X_2 1}$, and (c) if X is a cut point of M_2, not belonging to A_2, then every component of $S' - T_2(X)$ that contains no point of E_2 is of diameter less than 1/2. If X is a cut point of K belonging to M_2 and not to M_1, let $D_{X1}, D_{X2}, D_{X3}, \cdots$ denote the components of $S' - T_2(X)$, if there are any, that contain no points of E_2. Let A_3 denote the end element of t_3 which belongs to t_1 or to t_2. Let X_3 denote A_3 or the cut point of M_3 that lies on A_3 according as A_3 is a point or a surface and, in either case, let G_3 and T_3 denote a collection of continua and a transformation satisfying, with reference to t_3, the conditions which, according to the requirements of Theorem 11, are fulfilled by G and T with reference to t and such that if E_3 denotes the sum of all the continua of G_3 then (a) if X is a point of A_3, $T_3(X) = T_1(X)$ or $T_2(X)$ according as X_3 is or is not in M_1, (b) if X is a point not belonging to A_3 but belonging to some element of t_3 then $T_3(X)$ lies in the domain $D_{X_2 2}$, or the domain $D_{X_3 1}$, according as X_3 is or is not identical with X_2 and (c) if X is a cut point of M_3, not belonging to A_3, then every component of $S' - T_3(X)$ that contains no point of E_3 is of diameter less than $1/3$. This process may be continued. Thus, in case t_1, t_2, $t_3, \cdots$ is an infinite sequence, we have an infinite sequence G'_1, G'_2, $G'_3, \cdots$ of collections of continua and a sequence T_1, T_2, $T_3, \cdots$ of transformations such that, for each n greater than 1, G'_n and T_n satisfy with respect to t_n the conditions which, according to the requirements of Theorem 17, are satisfied by G and T with reference to t and such that if E_n denotes the sum of all the continua of G'_n and for each cut point X of K belonging to M_n but not to M_j for any j less than n, D_{X1}, D_{X2}, $D_{X3}, \cdots$ denote the components of $S' - T_n(X)$ that contain no points of E_n and m denotes the smallest integer j such that t_j contains an end element of t_n, and A_n denotes the end element of t_n that belongs to t_m, and X_n denotes A_n or the cut point of t_n that lies on A_n according as A_n is a point or a surface, then (a) if X is a point of A_n, $T_n(X) = T_m(X)$ and (b) if X is a point of $M_n - A_n$ then $T_n(X)$ lies in the first domain of the sequence $D_{X_n 1}$, $D_{X_n 2}$, $D_{X_n 3}, \cdots$ that does not contain $T_j(X)$ for any point X that belongs to M_j for some j less than n and (c) if X is a cut point of $M_n - A_n$, then every component of $S' - T_n(X)$ that contains no point of E_n is of diameter less than $1/n$.

If X is a point of M_n, let $T(X)$ denote the continuum $T_n(X)$. If X is a point of K not belonging to M_n for any n there exists a sequence of integers n_1, n_2, $n_3, \cdots$ such that if, for each j, P_{n_j} is a point belonging to M_{n_j} then X is the sequential limit point of the sequence P_{n_1}, P_{n_2}, $P_{n_3}, \cdots$. It may be seen that the sequence of continua $T(P_{n_1})$, $T(P_{n_2})$, $T(P_{n_3}), \cdots$ has a sequential limit point and that if P_{m_1}, P_{m_2}, $P_{m_3}, \cdots$ is another sequence of points such that, for each j, P_{m_j} belongs to M_{m_j} and such that X is the sequential limit point of P_{m_1}, P_{m_2}, $P_{m_3}, \cdots$ then $T(P_{m_1})$, $T(P_{m_2})$, $T(P_{m_3}), \cdots$ and $T(P_{n_1})$, $T(P_{n_2})$, $T(P_{n_3}), \cdots$ have the same sequential limit point $T(X)$. For every point X of K there has now been defined an element $T(X)$ of S'. The set of all such elements $T(X)$ is upper semi-continuous and the transformation T is reversibly continuous.

THEOREM 19. *If, on the simple closed surface S', H is a contracting sequence of mutually exclusive continua and G is the collection whose elements are the continua of the sequence H and the points of S' that belong to no continuum of H (in other words if G is a collection of mutually exclusive continua which fills up S' and the nondegenerate elements of G form a contracting sequence) then the space S'_G whose elements are the continua of G is topologically equivalent to a simple aspiculate cactoid.*

It is clear that the collection G is upper semi-continuous. It follows, by Theorem 15, that S'_G is topologically equivalent to a cactoid. It is easy to see that every such cactoid is a simple aspiculate one.

THEOREM 20. *If K is a simple aspiculate cactoid and S is a simple closed surface, there exists, on S, an upper semi-continuous collection G of mutually exclusive continua such that the space whose elements are the continua of G is topologically equivalent to K and such that the nondegenerate continua of G form a contracting sequence.*

The truth of Theorem 20 may be seen with the aid of results established in the proof of Theorem 18.

THEOREM 21. *If Axioms 0 and 1 hold true and S is completely separable then there exists, for each point X and point Y, a nonnegative real number $\delta(X,Y)$ such that (1) $\delta(X,Y) = \delta(Y,X)$, (2) $\delta(X,Y) = 0$ if and only if $X = Y$, (3) in order that the point P be a limit point of the point set M it is necessary and sufficient that for each positive number ϵ there exist a point X of M, distinct from P, such that $\delta(X,P) < \epsilon$, (4) there exists a function $f(e)$ $(e > 0)$, approaching 0 with e, such that if A, B and C are any three points such that $\delta(A,B) \leqq e$ and $\delta(B,C) \leqq e$, then $\delta(A,C) \leqq f(e)$; in other words, there exists a uniformly regular ecart in the sense of Fréchet.*

PROOF. Let $Q_1, Q_2, Q_3, \cdots$ denote a sequence satisfying, with respect to S, the conditions described, in the statement of Theorem 113 of Chapter I, as being satisfied with respect to M. Let Q'_1 denote Q_1. Let Q'_2 denote the collection of all regions g such that, for some m greater than 1, g belongs to Q_m and if x belongs to Q_m and intersects g then $x + g$ is a subset of some region of Q'_1. Let Q'_3 denote the collection of all regions g such that, for some m greater than 2, g belongs to Q_m and if x belongs to Q_m and intersects g then $x + g$ is a subset of some region of the collection Q'_2. This process may be continued. Thus there exists a sequence $Q'_1, Q'_2, Q'_3, \cdots$ such that (1) Q'_1 is identical with Q_1, (2) for each i, Q'_{i+1} is the collection of all regions g such that, for some m greater than i, g belongs to Q_m and if x belongs to Q_m and intersects g then $x + g$ is a subset of some region of the collection Q'_i. It may be shown that (1) for each n, Q'_n covers S, (2) if x and y are regions belonging to Q'_{n+1} and having a point in common then $x + y$ is a subset of some region of the collection Q'_n. For each pair of distinct points X and Y, let $\delta(X,Y)$ denote $1/m$, where m is the smallest positive integer k such that if $n > k$ then no region of Q'_n contains both X and Y. If X is Y, let $\delta(X,Y)$ denote 0. For the so defined $\delta(X,Y)$ all the conditions of Theorem 21 are satisfied. The function $2e$ may be used as $f(e)$.

THEOREM 22. *In a space satisfying Axioms 0, 1′, C, 2, 3, 4 and 5, if a compact continuum contains a web then it is a web.*

PROOF. Suppose that, in a space Σ satisfying these axioms, M is a compact continuum containing a web L. By Theorem 41 of Part 1 of Chapter V, there exist two upper semi-continuous collections H_1 and H_2 of mutually exclusive continua such that each of these collections fills up L and such that each of them is an arc with respect to its elements and every continuum of H_1 intersects every continuum of H_2. Suppose t is a component of the common part of a continuum x_1 of the collection H_1 and a continuum x_2 of the collection H_2. Let y_1 denote a continuum of H_1 distinct from x_1 and let y_2 denote a continuum of H_2 distinct from x_2. Since y_1 intersects y_2, $y_1 + y_2$ is a continuum. Clearly this continuum is a subset of $L - t$. Let N denote $L - (x_1 \cdot x_2)$. Suppose P is a point of N. There exists a continuum p containing P, lying in N and belonging either to H_1 or to H_2. If p belongs to H_1 it intersects y_2 and if it belongs to H_2 it intersects y_1. Hence in either case it intersects $y_1 + y_2$ and therefore $p + y_1 + y_2$ is a connected subset of N containing P. It follows that N is connected. By Theorem 52(a) of Chapter I, every component of $x_1 \cdot x_2$ contains a limit point of the connected point set N. Therefore $L - t$, the sum of N and all such components other than t, if there are any, is connected.

For each component t of the common part of a continuum of H_1 and a continuum of H_2, let h_t denote t plus the set of all points X of M, if there are any, such that t separates X from $L - t$. Let H denote the set of all such continua h_t. For each i less than 3, let H_i' denote the set of all sets h' such that, for some element h of H_i, h' is the set of all continua h_t for all t's which are subsets of h. Let S' denote the collection of all continua P' such that P' is either a continuum of the collection H or a point which neither belongs to any continuum of H nor is separated by any continuum of H from any other continuum of H. If the elements of S' are regarded as "points" and the subcollection Z of S' is regarded as a "region" if and only if Z^* is a domain with respect to $(S')^*$ then Axioms 0, 1′, C, 2, 3, 4 and 5 all hold true for the space Σ_1 so obtained and, *in this space*, H is a continuous curve with no cut point and there exist a simple domain I, a simple closed curve J, four points A, B, C and F and two collections G_1 and G_2 satisfying all the requirements of the hypothesis of Theorem 9 of Part 2 of Chapter V and such that, for each i less than 3, every continuum of the collection G_i is a subcontinuum of some continuum of the collection H_i'. In the argument given in Chapter V to prove that theorem, replace the sentence "Let G denote the collection of all point sets g such that g is the boundary of a continuum of the collection Q" by "In the space Σ_1, for each continuum q of the collection Q, let g_q denote the boundary of q plus the set of all points of q, if there are any, which either (a) belong to H or (b) are points of M which in Σ belong to no continuum of the collection H. Let G denote the collection of all such g_q's". It may be seen that, in the space Σ_1, the collection G so determined is an upper semi-continuous collection of mutually exclusive continua such that (a) if the elements of G are regarded as points and the subcollection Z of G is regarded as a region if and only if Z^* is a domain with respect to G^* then the space Σ_2 so obtained satisfies Axioms 0, 1′, C, 2, 3, 4 and 5, (b) in Σ_1 every nondegenerate continuum of the collection G is a subset of H plus the set of all elements of S' which are points of M and which in Σ belong to no continuum of the collection H, (c) if, in Σ_1, g is a continuum of the collection G and P is an element of S' which either belongs to H or is, in Σ, a point of M

which belongs to no continuum of the collection H then, in Σ_1, g does not separate P from any other continuum of the collection G, (d) if P is an element of S' which either belongs to H or is a point of M which, in Σ, belongs to no continuum of H then P either belongs to G or is, in the space Σ_1, an element of some continuum of the collection G, and (e) in the space Σ_2, there exists a connected domain D with a connected boundary β such that every degenerate element of G that belongs to D is, in Σ, a point of M or a continuum of the collection H. In the space Σ_2, let W denote the collection whose degenerate elements are the points of D and whose sole nondegenerate element is the continuum γ obtained by adding to β all points of Σ_2 which do not belong to D and which in Σ_1 are points of H or points which in Σ belong to M but to no continuum of the collection H. In Σ_2, W is an upper semi-continuous collection of mutually exclusive continua. There exists a reversibly continuous transformation throwing the continua of the collection W into the points of a sphere. But clearly a sphere is a web. It follows that W is a web with respect to its elements. Hence, in the space Σ_2, there exist two upper semi-continuous collections of mutually exclusive continua Z_1 and Z_2 such that (1) each of these collections fills up $D + \gamma$, (2) each of them is an arc with respect to its elements, (3) each element of Z_1 intersects each element of Z_2 and (4) one element of Z_1 contains γ. For each continuum z of the collection Z_i $(i = 1,2)$ let v_z denote the continuum, in the space Σ_1, obtained by adding together all the continua in that space which are points of z in the space Σ_2 and let t_z denote the continuum, in the space Σ, obtained by adding together all the continua, in the space Σ, which are points of v_z in the space Σ_1. Let T_i denote the collection of all such t_z's for all elements z of Z_i. *In the space* Σ, for each i, T_i is an upper semi-continuous collection of mutually exclusive continua filling up M, T_i is an arc with respect to its elements and every continuum of the collection T_1 intersects every one of the collection T_2. Hence M is a web.

It is not true that if a continuum in Euclidean space of three dimensions contains a web then it is a web.

Example. Let E denote a plane, let J denote a circle lying in E, let I denote the bounded component of $E - J$ and let K denote $I + J$. Let T denote a connected point set topologically equivalent to a ray and such that T and K are mutually exclusive but $\overline{T} = T + K$. The point set $\overline{T}$ is an irreducible continuum from K to some point not lying in E. This continuum is not a web. But it contains the web K.

Theorem 23. *In a space satisfying Axioms* 0, 1′, C, 2, 3, 4 *and* 5, *if M is a compact continuum, G is an upper semi-continuous collection of mutually exclusive continua filling up M, and H is another such collection and, for every two continua g and h belonging to G and H respectively, $g \cdot h$ exists and is totally disconnected then each of the collections G and H is a continuous curve with respect to its elements and M is a continuous curve.*

PROOF. Suppose M is not a continuous curve. With the help of Theorem 11 of Chapter II it may be shown that there exist a sequence $M_1, M_2, M_3, \cdots$ of mutually exclusive subcontinua of M and a simple domain I bounded by a simple closed curve J containing four points A, B, C and D such that $A + C$ separates B from D on J and such that if, for every two points X and Y of this set of four, XY denotes the arc of J from X to Y which contains no one of the remaining ones, then (1) no element of G or of H is a subset of I, (2) for each n, M_n is a component of $M \cdot \bar{I}$, and M_{n+1} separates AD from $M_n + BC$ in $\bar{I}$. Let C_1D_1 denote an arc lying wholly in I except for its endpoints C_1 and D_1 which lie between B and C and between A and D, respectively, on the arcs BC and AD. Let H' denote the set of all point sets x such that x is a component of the common part of $\bar{I}$ and some continuum of the collection H. For each n, M_n contains a point P_n of C_1D_1 and there exists a continuum h'_n belonging to H' and containing P_n and therefore lying in M_n. For each n, h'_n intersects either AB or CD. It follows that there exist two points A_1 and B_1 and an infinite ascending sequence of positive integers $n_1, n_2, n_3, \cdots$ such that (1) either A_1 is D and B_1 is C or A_1 is A and B_1 is B and (2) for every j, h'_{n_j} intersects A_1B_1. Let I_1 denote the component of $I - I \cdot (C_1D_1)$ whose boundary contains the arc A_1B_1. Let C_2D_2 denote an arc lying wholly in I_1 except for its endpoints C_2 and D_2 which lie between B_1 and C_1 and between A_1 and D_1, respectively, on the arcs BC and AD. Let G' denote the collection of all sets x such that x is a component of the common part of $\bar{I}_1$ and some continuum of the collection G. There exist points A_2 and B_2 and an infinite ascending subsequence $m_1, m_2, m_3, \cdots$ of the sequence $n_1, n_2, n_3, \cdots$ such that (1) either A_2 is D_1 and B_2 is C_1 or A_2 is A_1 and B_2 is B_1, (2) for each j there is a continuum g_{m_j} belonging to G' and intersecting both A_2B_2 and C_2D_2 and lying in M_{m_j}. Let I_2 denote the component of $I_1 - I_1 \cdot (C_2D_2)$ whose boundary contains A_2B_2 and C_2D_2. For each j there exist two continua g_j and h_j such that (1) each of them intersects both A_2B_2 and C_2D_2, (2) g_j is a component of $\bar{I}_2 \cdot g'_{m_j}$ and h_j is a component of $\bar{I}_2 \cdot h'_{m_j}$. The sequence $g_1, g_2, g_3, \cdots$ converges to a continuum L and $h_1, h_2, h_3, \cdots$ converges to the same continuum. Hence there exist two continua g and h, belonging to G and H, respectively, such that L is a subset of each of them. But, since it intersects both A_2B_2 and C_2D_2, L is non-degenerate. This involves a contradiction. Hence M is a continuous curve. Therefore G is a continuous curve with respect to its elements and so is H.

If, in the statement of Theorem 23, the words "In a space satisfying Axioms 0, 1′, C, 2, 3, 4 and 5" are replaced by "In Euclidean space of three dimensions", the resulting proposition is false.

EXAMPLE. In Euclidean three dimensional space let AB denote a straight line interval of length 1, let K denote a cube having AB as one of its edges and let F and L denote the two faces of K that contain AB. Let T denote the point set consisting of K plus its interior. For each n, let F_n denote the common part of T and a plane parallel to, and at a distance equal to $1/n$ from, the plane in which F lies. Let M denote the continuum $F + L + F_1 + F_2 + \cdots$. Let G denote the collection of all straight line intervals g, of length 1, lying in M such that g either coincides with AB or lies on a straight line parallel to the one containing AB. Let H denote the collection of all continua h such that h is the common part of M and some plane perpendicular to AB. The collections G and H are upper semi-continuous, each of them fills up M and, if g and h are continua belonging to G and H respectively, $g \cdot h$ is a single point. The collection H is an arc with respect to its elements. But G is not a continuous curve with respect to its elements and M is not a continuous curve.

THEOREM 24. *If every two points of the compact continuous curve M are separated from each other in M by each of uncountably many mutually exclusive subcontinua of M but M has no cut point then M is not separated by any countable point set.*

PROOF. Suppose the points A and B are separated from each other in M by a countable point set K. Then, as may be seen with the help of Theorems 154 of Chapter I and 84 of Chapter II, there exists a closed and nondegenerate subset H of K such that H separates A from B in M but no proper subset of H does so. Every point of H is a limit point of both M_A and M_B, the components of $M - H$ that contain A and B respectively. By hypothesis, H is nondegenerate. Let E and F denote two of its points. There exists a subcontinuum L of M separating E from F in M and containing no point of K. Since $E + F + M_A$ and $E + F + M_B$ are connected subsets of M, L intersects both M_A and M_B. Hence $L + M_A + M_B$ is connected. But it contains A and B and no point of H. This involves a contradiction.

THEOREM 25. *If every two points of the compact continuous curve M are separated from each other in M by each of uncountably many mutually exclusive subcontinua of M and M has no cut point then, if H and K are two mutually exclusive closed point sets, there do not exist in M infinitely many mutually exclusive arcs such that each of them intersects both H and K and is the sum of countably many continua t such that t is either a point or a subset of the boundary of some complementary domain of M.*

PROOF. Suppose there exist two mutually exclusive closed point sets H and K such that M contains infinitely many mutually exclusive arcs intersecting both of them and satisfying these conditions. Then there exists a sequence $\alpha_1, \alpha_2, \alpha_3, \cdots$ of these arcs converging to some nondegenerate continuum N. Let E and F denote two points of N. There exists a subcontinuum T of M separating E from F in M and such that, if T intersects α_n, every component of $T \cdot \alpha_n$ is a subset of some segment of α_n lying on the boundary of some complementary domain of M. Let ω denote some point not belonging to M and regard it as the point at infinity. There exist (1) two simple closed curves J_E and J_F lying without each other and enclosing E and F respectively and (2) a sequence of mutually exclusive arcs E_1F_1, $E_2F_2, \cdots$ such that, for each n, E_nF_n is, for some m, an interval of α_m lying wholly without each of the curves J_E and J_F, except that E_n and F_n belong to J_E and J_F respectively, and furthermore T intersects every arc of the sequence E_1F_1, $E_2F_2, \cdots$ and lies wholly without both J_E and J_F. There exist two integers i and j, less than or equal to 3, and arcs E_iE_j and F_iF_j lying on J_E and J_F respectively, such that T intersects both I, the interior, and E, the exterior, of J, the simple closed curve $E_iF_i + F_iF_j + E_jF_j + E_iE_j$. It follows that some component Q of $T \cdot J$, and therefore either of $T \cdot (E_iF_i)$ or of $T \cdot (E_jF_j)$, contains both a limit point of $T \cdot I$ and a limit point of $T \cdot E$. But Q is a subset of a segment of E_iF_i or of E_jF_j lying on the boundary of some complementary domain D of M and, by Theorem 46 of Chapter IV, D is a simple domain. The domain D contains a point of M. This involves a contradiction.

THEOREM 26. *In a space satisfying Axioms* 0, 1′, C, 2, 3, 4 *and* 5, *if every two points of the compact continuous curve M are separated from each other in M by each of uncountably many mutually exclusive subcontinua of M but M has no cut point then if K is a countable subset of M, $M - K$ is locally connected.*

PROOF. A point set will be said to be of class 1 if, and only if, it is the sum of countably many continua such that each of them is either a degenerate point set or a subset of the boundary of some complementary domain of M.

For each point P of M let M_P denote the point set consisting of P and all points X of M, if there are any, such that X and P are the extremities of an arc of class 1.

Suppose P is a point of M. Suppose A is a limit point of M_P not belonging to it. There exists a sequence P_1, P_2, $P_3, \cdots$ of distinct points of $M_P - P$ converging to A. For each n there exists an arc PP_n of class 1. Let n_1 denote 1, let B_{n_1} denote P and let n_2 denote the smallest positive n such that P_m does not belong to PP_{n_1} for any m greater than or equal to n. Let B_{n_2} denote the first point in the order from P_{n_2} to P that PP_{n_2} has in common with PP_{n_1} and let $B_{n_2}P_{n_2}$ denote the interval of PP_{n_2} whose endpoints are B_{n_2} and P_{n_2}. There exist an infinite sequence of positive integers n_1, n_2, $n_3, \cdots$, an infinite sequence of points B_{n_1}, B_{n_2}, $B_{n_3}, \cdots$ and an infinite sequence of arcs $B_{n_1}P_{n_1}$, $B_{n_2}P_{n_2}, \cdots$ such that (1) B_{n_1}, B_{n_2} are as described, (2) for each j, n_{j+1} is the smallest positive integer n such that P_m does not belong to PP_{n_i} for any m greater than or equal to n, and i less than or equal to j, (3) $B_{n_{j+1}}$ is the first point, in the order from $P_{n_{j+1}}$ to P on $PP_{n_{j-1}}$ that $PP_{n_{j+1}}$ has in common with the point set $B_{n_1}P_{n_1} + B_{n_2}P_{n_2} + \cdots + B_{n_j}P_{n_j}$ and (4) $B_{n_{j+1}}P_{n_{j+1}}$ is the interval of $PP_{n_{j+1}}$ whose endpoints are $B_{n_{j-1}}$ and $P_{n_{j+1}}$. No two of the arcs $B_{n_1}P_{n_1}$, $B_{n_2}P_{n_2}$, $B_{n_3}P_{n_3}, \cdots$ have more than one point in common. Suppose there exist two mutually exclusive closed point sets H and K such that, for infinitely many integers i, $B_{n_i}P_{n_i}$ intersects both H and K. Then there exist two mutually exclusive closed point sets H' and K' such that for each such i there exists an interval of $B_{n_i}P_{n_i}$ intersecting both H' and K' and having neither B_{n_i} nor P_{n_i} as an endpoint. This is contrary to Theorem 25. It follows that no arc of the sequence $B_{n_1}P_{n_1}$, $B_{n_2}P_{n_2}, \cdots$ intersects more than a finite number of the others. Hence there exists an infinite sequence $Q_1, Q_2, Q_3, \cdots$ such that (1) for each n, Q_n is a finite collection of one or more arcs of the sequence $B_{n_1}P_{n_1}$, $B_{n_2}P_{n_2}, \cdots$, (2) $B_{n_1}P_{n_1}$ is the only element of Q_1, (3) each arc of Q_{n+1} intersects some arc of Q_n, but intersects no arc of Q_{n-1} if $n > 1$, (4) each arc of the sequence $B_{n_1}P_{n_1}$, $B_{n_2}P_{n_2}, \cdots$ belongs to some collection of the sequence $Q_1, Q_2, Q_3, \cdots$. By Theorem 114 of Chapter I, there exists an infinite sequence $m_1, m_2, m_3, \cdots$ of distinct integers of the sequence $n_1, n_2, n_3, \cdots$ such that, for each j, $B_{m_j}P_{m_j}$ belongs to Q_j and intersects $B_{m_{j-1}}P_{m_{j-1}}$. If, for each j, α_j denotes the point B_{m_j} or the interval $B_{m_j}B_{m_{j+1}}$ according as B_{m_j} is or is not identical with $B_{m_{j+1}}$, then $A + \alpha_1 + \alpha_2 + \cdots$ is an arc from P to A. But it is of class 1. Hence A belongs to M_P and M_P is closed.

With the aid of Theorem 25 and the fact that every two points of M_P are the extremities of an arc of class 1 lying in M_P, by an argument having much in common with that given to prove Theorem 11 in Chapter II, it may be shown that if A is a point of M_P and R_1 is a region containing A there exists a region R_2 containing A such that every point of M_P belonging to R_2 lies together with A in an arc of class 1 lying in R_1. It follows that not only is M_P a continuous curve but every two points of a connected domain D with respect to M_P are the extremities of an arc of class 1 lying in D.

By Theorems 46 and 47 of Chapter IV, every boundary of a complementary domain of M is a simple closed curve and no two of them have more than one point in common. Suppose AXB is an arc of class 1 lying in M_P and suppose no boundary of a complementary domain of M contains a segment of AXB containing X. Suppose X does not separate A from B in M_P. Then, since M_P is a continuous curve, if D is the component of $M_P - X$ that contains A, D is a domain with respect to $M_P - X$ containing B and therefore it contains an arc AYB of class 1. The point set $AXB + AYB$ contains a simple closed curve J of class 1 containing a segment of AXB containing X. The curve J is not the boundary of any complementary domain of M. But it is the sum of two mutually exclusive point sets N_1 and N_2 such that N_1 is countable and N_2 is the sum of two or more segments each of which is a subset of some complementary domain of M. If I and E denote the complementary domains of J, the point sets $M \cdot I$ and $M \cdot E$ exist and N_1 separates them from each other in M. But this is contrary to Theorem 24.

If A and B are points of M_P and the boundary of no complementary domain of M contains both of them, there exists an arc ACB of class 1 lying in M_P and there exists a cut point X of ACB belonging to no segment of ACB which is a subset of the boundary of a complementary domain of M. Furthermore if a point X of M_P is common to J_1 and J_2, the boundaries of two different complementary domains of M, and A and B are points of J_1 and J_2 respectively, distinct from X, and AX and BX are arcs lying on J_1 and J_2 respectively then, since X is the only point common to J_1 and J_2, $AX + BX$ is an arc AXB of class 1 and the point X belongs to no segment of AXB which is a subset of any complementary domain of M. It follows that (1) if two points of M_P do not belong to the boundary of the same complementary domain of M they are separated from each other in M_P by some point and (2) if the boundaries of two complementary domains of M have a point of M_P in common that point separates M_P. Furthermore if the boundary of a complementary domain of M intersects M_P it is a subset of M_P. Hence every nondegenerate simple link of M_P is a simple closed curve which is the boundary of some complementary domain of M.

Let G denote the collection of all continua g such that, for some point P of M, g is M_P. The continua of the collection G are mutually exclusive and if H and K are two mutually exclusive closed point sets there do not exist infinitely many elements of G each intersecting both H and K. Furthermore no continuum of G separates any two continua of G from each other, G fills up M and the boundary of each complementary domain of M is a subset of some continuum of G. Hence G is upper semi-continuous and, with respect to its elements regarded as points, G is topologically equivalent to the surface of a sphere. With the help of this fact and the fact that the surface of a sphere remains connected and locally connected on the removal of a countable point set, it may be shown that if K is a countable subset of M and H is the set of all elements of G that intersect K then (1) $M - H^*$ is connected and (2) it is locally connected at every one of its points not belonging to a nondegenerate continuum of G. But every point of a continuum of H is a limit point of $M - H^*$. Hence $M - K$ is locally connected at every point of $M - K$ belonging to no nondegenerate continuum of G. Suppose now that P is a point of $M - K$ belonging to the nondegenerate continuum g of the collection G. If L denotes the point set $(M - g) + P$ and N denotes $L - K\cdot(M - g)$ then both L and N are locally connected at P and every point of g that does not belong to K is a limit point of N. Hence $M - K$ is locally connected at P. Thus the omission of any countable point set leaves M locally connected.

There exist both separable and nonseparable spaces which satisfy Axioms 0, 1, 2, 3, 4, 5_1, 5_2 and 6 but for which there exists no uniformly regular ecart. Consider the following examples.

Example A. In a three-dimensional Euclidean space E, let β denote a definite straight line. Let G denote the set of all "half-planes" g such that, for some plane M containing β, g is the sum of β and a component of $M - \beta$.

Let γ denote a definite half-plane of the collection G. There exists a reversible transformation T of the point set β into the set of all elements of G except γ. For each point P of β let W_P denote the set of all straight line rays which start from P and lie, except for P, wholly in $\gamma - \beta$. If P is a point of β and a is a ray of W_P, let $Z(a)$ denote the image of a under a rotation around β that throws the half-plane γ into the half-plane $T(P)$. If P is a point of β, a and b are rays of W_P and ϵ is a positive number, let $R_{Pab\epsilon}$ denote a set such that x is an element of it if and only if either (1) x is the open curve $y + Z(y)$ for some ray y of the set W_P such that $y - P$ is separated from $\beta - P$ by $a + b$ in the half-plane γ or (2) x is a point which is at a distance less than ϵ from P and which is separated from $\beta - P$ by $a + b$ in γ or from $\beta - P$ by $Z(a) + Z(b)$ in $T(P)$.

Let Σ_A denote a space such that (1) X is a point of Σ_A if and only if either X is a point of $E - \beta$ or, for some point P of β, X is the open curve $a + Z(a)$ for some ray a of W_P, (2) R is a region in Σ_A if and only if either, in some half-plane of the set G, R is the interior of some circle containing no point of β or R is $R_{Pab\epsilon}$ for some P, a, b and ϵ. This space satisfies all of the above mentioned axioms. But it is not separable.

Example B. In a Euclidean plane E, let β denote a definite straight line, let A and B denote two definite points of β and let N denote a "half-plane" which is the sum of β and a definite one of the two components of $E - \beta$. If P is a point of β let A_P and B_P denote points of β such that the straight line rays PA_P and PB_P have only P in common and the common part of the straight line rays PB_P and AB is a ray. Let C_P denote a point of N such that the straight line ray PC_P is perpendicular to β. Let W_P denote the set of all straight line rays which start from P and lie, except for P, in the interior of the right angle C_PPB_P. For each ray a of W_P, let $Z(a)$ denote a straight line ray lying, except for P, in the interior of the right angle C_PPA_P and such that the acute angle whose sides are $Z(a)$ and PC_P is congruent to the one whose sides are a and PC_P.

If P is a point of β, a and b are rays of W_P and ϵ is a positive number, let $R_{Pab\epsilon}$ denote a set such that x is an element of it if and only if either (1) x is the open curve $y + Z(y)$ for some ray y, of the set W_P, such that $y - P$ lies in the interior of the acute angle whose sides are a and b or (2) x is a point which is at a distance less than ϵ from P and which lies either in the interior of the acute angle whose sides are a and b or in the interior of the one whose sides are $Z(a)$ and $Z(b)$.

If P is a point of β, a is a ray of the set W_P and ϵ is a positive number, let $R_{Pa\epsilon}$ denote a set such that x is an element of it if and only if either (1) x is the ray PC_P or (2) x is the open curve $y + Z(y)$ for some ray y of the set W_P such that $y - P$ lies in the interior of the acute angle whose sides are a and PC_P or (3) x is a point which is at a distance less than ϵ from P and which, in the half-plane N, is separated from $\beta - P$ by the open curve $a + Z(a)$.

Let Σ_B denote a space such that (1) X is a point of Σ_B if and only if either X is a point of $N - \beta$ or, for some point P of β, X is PC_P or the open curve $a + Z(a)$ for some ray a of W_P, (2) R is a region in Σ_B if and only if either, in the half-plane N, R is the interior of some circle containing no point of β or R is $R_{Pab\epsilon}$ for some P, a, b and ϵ or R is $R_{Pa\epsilon}$ for some P, a and ϵ. The space Σ_B satisfies Axioms 0, 1, 2, 3, 4, 5_1, 5_2 and 6. Furthermore it is separable. But it is not completely separable.

Neither Σ_A nor Σ_B is normal.

APPENDIX

For many of the most fundamental notions, definitions and propositions of point set theory the reader is referred to the earlier writers, in particular to G. Cantor, P. DuBois Reymond, Bendixson, Bolzano and Harnack. Specific references to some of their articles may be found in the first three chapters of Hobson's *Theory of functions of a real variable* (Cambridge, 1907). Somewhat later, extensive contributions were made by A. Schoenflies and W. H. Young.

The term "inner limiting set" was introduced by W. H. Young. The notion of a *connected* point set, in the sense in which it is used in this book, was introduced by N. J. Lennes.

With regard to Theorem 59 of Chapter I and its history, the reader is referred to an article by R. G. Lubben titled *Concerning limiting sets in abstract spaces* (cf. Bibliography).

The notion of an *irreducible continuum* between two points was introduced by Zoretti. Properties of such continua have been investigated by numerous authors including Zoretti, Janiszewski, Hahn, Yoneyama and Kuratowski. The subject of indecomposable continua has been treated by Brouwer, Mazurkiewicz, Yoneyama, Knaster, Kuratowski and others. Brouwer established the existence of such point sets. Much progress has been made in the study of such continua since the publication of the first edition of this book. In the 1921 issue of Fundamenta Mathematicae, S. Mazurkiewicz raised the question whether there exists, in space of m dimensions, a nondegenerate continuum which is not an arc but which has the property of being topologically equivalent to *every* one of its nondegenerate subcontinua. In 1947, in his Doctoral Dissertation, E. E. Moise showed that there exists, in the plane, an *indecomposable* continuum having this property. In 1959, in his Dissertation, G. W. Henderson showed that there does not exist, in any metric space, a compact *decomposable* continuum having this property. Recently R. D. Anderson and G. Choquet have shown that there exists a nondegenerate plane continuum such that *no* two of its nondegenerate subcontinua are topologically equivalent to each other.

The point set M is said to be *topologically homogeneous* if, for every two points A and B of M there exists a reversibly continuous transformation of M into itself that throws A into B. In the 1920 volume (Volume I) of Fundamenta Mathematicae, Knaster and Kuratowski raised the question whether every bounded topologically homogeneous plane continuum is topologically equivalent to a circle. In 1948, R. H. Bing answered this question in the negative. Since then, Anderson, Bing, C. E. Burgess,

H. J. Cohen, F. B. Jones and Moise have all made contributions to the study of homogeneous continua.

R. D. Anderson has shown that there exists a continuous collection of compact homogeneous indecomposable continua filling up the plane such that every two of them are topologically equivalent to each other. By showing the existence of a continuous collection G of homogeneous indecomposable continua such that G is an arc with respect to its elements and all of its elements are topologically equivalent to each other, he has answered a question raised, in Fundamenta Mathematicae, by Knaster in 1935.

In 1930, W. A. Wilson showed that if H and K are two mutually exclusive subcontinua of the compact continuous curve M then M is the sum of two continua such that neither of them intersects both H and K.

In 1931, G. T. Whyburn showed that, under this hypothesis, there exist two mutually exclusive connected domains, with respect to M, D_H and D_K, containing H and K respectively, such that $M - (D_H + D_K)$ is the common boundary of D_H and D_K with respect to M.

In 1949, or near that time, Bing and Moise each showed that if M is a compact continuous curve in a metric space and ϵ is a positive number there exists a finite collection G of mutually exclusive connected domains with respect to M, each of diameter less than ϵ, such that $M - G^*$ is the boundary of G^* with respect to M.

Analytical definitions of *simple continuous arc*, *simple closed curve and continuous curve* were given by Jordan. Point set-theoretic definitions of the first two of these notions were given by Lennes. Other definitions have been given by Janiszewski, Sierpiński, the author and others.

The notion of *connectedness im kleinen* was introduced both by Hans Hahn and by Mazurkiewicz. The term is due to Hahn. See also the paper of Pia Nalli referred to in the Bibliography. It was shown, both by Hahn and by Mazurkiewicz, that in order that a bounded continuum be a continuous curve in the sense of Jordan it is necessary and sufficient that it be connected im kleinen.

As was shown by Peano, and later by Hilbert and E. H. Moore, the plane point set consisting of a square together with its interior is a continuous curve in the sense of Jordan. Menger and Urysohn have adopted a definition of *curve* that restricts it to continua that are one-dimensional in the sense of their dimension theory. Thus not every continuous curve in the sense of Jordan is a curve of any sort in the sense of Menger and Urysohn.

F. Burton Jones has introduced the notion of a point set being aposyndetic. The point set M is said to be *aposyndetic* at the point P if P belongs to M and for each point X of M distinct from P there exists a domain with respect to M which contains P and is a subset of a connected subset of $M - X$ which is closed relatively to M. A point set is said to be *aposyndetic* if it is aposyndetic at each of its points.

Janiszewski established some of the most fundamental and important propositions relating to conditions under which the sum of two compact continua separate the plane or separate two given points of the plane from each other. A later contribution was made by Miss Anna Mullikin in her Dissertation in which she duplicated some of the results of Janiszewski, with whose paper (written in Polish) she was not, at that time, acquainted.

Schoenflies proved many theorems relating to the relation of a continuous curve to its complement in the plane, including certain propositions concerning accessibility. The author has the feeling that Schoenflies has not received sufficient credit for all of the contributions which he made to point set theory. In particular, in a paper titled *On the most general plane closed point set through which it is possible to pass a simple continuous arc*, J. R. Kline and the present author made references to the contributions of various mathematicians to the study of the proposition that every closed, bounded and totally disconnected plane point set is a subset of some arc, but we omitted any reference to Schoenflies, not knowing of his work on the subject. Years later, the author found a reference indicating that Schoenflies not only had considered the proposition in question but had given a satisfactory *proof* of it in his book.

For propositions related to Theorems 44 and 50 of Chapter IV, with applications, the reader is referred to certain articles by the author and to R. G. Lubben's paper *Separation theorems with applications to questions concerning accessibility and plane continua.*

For a treatment of *sense* on simple closed curves in the plane reference is made to an article by J. R. Kline included in the Bibliography.

The notion *cyclic element* of a continuous curve was introduced by G. T. Whyburn who has made extensive contributions to the theory of such elements and to its applications.

A proposition corresponding to Theorem 59 of Chapter II was proved, for the case of a continuous curve in the plane, by G. T. Whyburn and later, by W. L. Ayres, for the case of a locally compact continuous curve lying in a metric space. In the proof in the first edition of this book, use was made of a method of argument employed by R. L. Wilder in the course of the proof of Theorem 1 of his thesis, supplemented by a line of argument similar to that employed by H. M. Gehman in the proof of Theorem 4 of his paper *Concerning acyclic continuous curves.* In the present edition, a modification of that argument has been made to fit a weaker hypothesis.

E. W. Chittenden showed that every space for which there is a uniformly regular ecart is metric.

With the use of contributions occurring in the literature, Theorem 20 of Chapter VII could be established as follows. Urysohn showed that every normal and completely separable space is metric. Tychonoff showed that every regular and completely separable Hausdorff space is normal. But every space satisfying Axioms 0 and 1 is a regular Hausdorff space. Hence every completely separable space satisfying Axioms 0 and 1 is metric.

Sometime in 1930, Leo Zippin showed that there exists a space satisfying Axioms 0 and 1 but containing a compact point set M such that $\bar{M} - M$ is not compact, in other words that it is not true that, in every space that satisfies these axioms, the derived set of every compact set is compact.

John H. Roberts has shown that every metric space that satisfies Axioms 0 and 1 is complete.

In her Dissertation, Mary Ellen Estill (now Rudin) made a study of various modifications of Axiom 1.

In his 1958 Dissertation (cf. Fund. Math. **48** (1959), 15–29), J. N. Younglove has shown that if Σ is a space satisfying Axioms 0 and 1 then there exists a complete metric subspace Σ' of Σ such that the set of all points of Σ' is dense in Σ.

DEFINITION. A point set is said to be *compactly* connected if every two points of M lie together in some compact continuum which is a subset of M.

By Theorem 1 of Chapter II, every connected space satisfying Axioms 1 and 2 is arcwise connected and therefore compactly connected. It is clear that certain of the propositions of that chapter hold true in every compactly connected space that satisfies Axioms 0 and 1.

BIBLIOGRAPHY

ADKISSON, V. W.

Cyclicly connected continuous curves whose complementary domain boundaries are homeomorphic, preserving branch points, Comptes Rendus des Séances de la Société des Sciences et des Lettres de Varsovie, Classe III, **23** (1930), 164–193.

On extending a continuous (1-1) correspondence of continuous curves on a sphere, ibid. **27** (1934), 547–549.

Finite groups associated with certain cyclic curves, ibid. **29** (1936), 23–26.

Plane peanian continua with unique maps on the sphere and in the plane, Trans. Amer. Math. Soc. **44** (1938), 58–67.

AITCHISON, BEATRICE

Concerning regular accessibility, Fund. Math. **20** (1933), 116–125.

On the mapping of locally connected continua into simple arcs, Comptes Rendus des Séances de la Société des Sciences et des Lettres de Varsovie, Classe III, **27** (1934), 130–146.

ALBERT, G. E.

A note on quasi-metric spaces, Bull. Amer. Math. Soc. **47** (1941), 479–482.

The structure of locally connected topological spaces (with Youngs, J. W. T.), ibid. **48** (1942), 627–630.

ALEXANDROFF, P.

Zur Theorie der topologischen Räume (with Urysohn, P.), Math. Ann. **92** (1924), 258–266.

Über die Struktur der bikompakten topologischen Räume, ibid., 267–274.

Über die Metrization der im kleinen kompakten topologischen Räume, ibid., 294–301.

Une condition nécessaire et suffisante pour qu'une classe $\mathfrak{L}$ soit une classe $\mathfrak{D}$ (with Urysohn, P.), C. R. Acad. Sci. Paris **177** (1923), 1274.

Sur les ensembles de la première classe et les espaces abstraits, ibid. **178** (1924), 185.

Sur la décomposition de l'espace par des ensembles fermés, ibid. **184** (1927), 425.

Über stetige Abbildungen kompakter Räume, Math. Ann. **96** (1926), 555–571.

Zum allgemeinen Dimensionsproblem, Nachr. Akad. Wiss. Göttingen. Math.-Phys. Kl. IIa. (1928), 1–20.

Darstellung der Grundzüge der Urysohnschen Dimensionstheorie, Math. Ann. **98** (1927), 31–63.

Über nulldimensionale Punktmengen (with Urysohn, P.), ibid. **98** (1927), 89–106.

Über die allgemeine Dimensionstheorie und ihre Beziehungen zur elementaren geometrischen Anschauung, ibid. **98** (1928), 617–636.

Über stetige Abbildungen kompakter Räume, Proc. Roy. Acad. Sci. Amsterdam **28** (1925), 997–999.

Notes supplémentaires au Mémoires sur les multiplicités cantoriennes, rédigés d'après les papiers posthumes de Paul Urysohn, Fund. Math. **8** (1926), 352–359.

Beweis des Satzes, dass jede abgeschlossene Menge positiver Dimension in einem lokal zusammenhängenden Kontinuum von derselben Dimension topologisch enthalten ist (with Tumarkin, L.), Fund. Math. **11** (1928), 141–144.

Mémoire sur les espaces topologiques compacts (with Urysohn, P.), Verh. Nederl. Akad. Wetensch. Afd. Naturk. Sect. I. 14, No. 1 (1929), 1–96.

Einfachste Grundbegriffe der Topologie, mit einem Geleitwort von David Hilbert, Verlag von Julius Springer, Berlin, 1932, 48 pp.

Dimensionstheorie. Ein Beitrag zur Geometrie der abgeschlossenen Mengen, Math. Ann. **106** (1932), 161–238.

Sur la notion de dimension des ensembles fermés, J. Math. Pures Appl. Serie 9. **11** (1932), 283–298.

ANDERSON, R. D.

Concerning upper semi-continuous collections of continua, Trans. Amer. Math. Soc. **67** (1949), 451–460.

On monotone interior mappings in the plane, ibid. **73** (1952), 211–222.

Continuous collections of continuous curves in the plane, Proc. Amer. Math. Soc. **3** (1952), 647–657.

Monotone interior dimension-raising mappings in the plane, Duke Math. J. **19** (1952), 359–366.

Convex functions and upper semi-continuous collections (with Klee, V. L., Jr.), ibid., 349–357.

Continuous collections of continuous curves, ibid. **21** (1954), 363–367.

A note on continuous collections of continuous curves filling up a continuous curve in the plane (with Hamstrom, Mary-Elizabeth), Proc. Amer. Math. Soc. **5** (1954), 748–752.

One dimensional continuous curves, Proc. Nat. Acad. Sci. **42** (1956), 760–762.

Some remarks on totally disconnected sections of monotone open mappings, Bull. Acad. Polon. Sci. Math. **4** (1956), 329–330.

Open mappings of compact continua, Proc. Nat. Acad. Sci. **42** (1956), 347–349.

On spaces filled up by continuous collections of atriodic continuous curves (with Hamstrom, Mary-Elizabeth), Proc. Amer. Math. Soc. **6** (1955), 766–769.

Atomic decomposition of continua, Duke Math. J. **23** (1956), 507–514.

The algebraic simplicity of certain groups of homeomorphisms, Amer. J. Math. **80** (1958), 955–962.

A characterization of the universal curve and a proof of its homogeneity, Ann. of Math. **67** (1958), 313–324.

One-dimensional continuous curves and a homogeneity theorem, ibid. **68** (1958), 1–16.

A plane continuum no two of whose nondegenerate subcontinua are homeomorphic, Proc. Amer. Math. Soc. **10** (1959), 347–353.

ANTOINE, L.

Sur l'homéomorphie de deux figures et de leurs voisinages, J. Math. Pures Appl. Serie 8. **4** (1921), 221–326.

Sur les ensembles parfaits discontinus, C. R. Acad. Sci. Paris. **173** (1921), 284–285.

Sur les voisinages de deux figures homéomorphes, Fund. Math. **5** (1924), 265–287.

ARMENTROUT, STEVE

Concerning a certain collection of spirals in the plane, Duke Math. J. **26** (1959), 243–250.

ARONZAJN, N.

Über die Bogenverknupfung in topologischen Raumen, Fund. Math. **15** (1930), 228–241.

Ein Urbildproblem, ibid. **17** (1931), 92–121.

AYRES, W. L.

A new characterization of plane continuous curves, Bull. Amer. Math. Soc. **33** (1927), 201–208.

Concerning continuous curves and correspondences, Ann. of Math. **28** (1927), 396–418.

Concerning the boundaries of domains of continuous curves, Bull. Amer. Math. Soc. **33** (1927), 565–571.

Note on a theorem concerning continuous curves, Ann. of Math. **28** (1927), 501–502.

On the structure of a plane continuous curve, Proc. Nat. Acad. Sci. **13** (1927), 749–754.

On the separation of points of a continuous curve by arcs and simple closed curves, ibid. **14** (1928), 201–206.

An elementary property of bounded domains, Bull. Amer. Math. Soc. **34** (1928), 200–204.

Concerning the arc-curves and basic sets of a continuous curve, Trans. Amer. Math. Soc. **30** (1928), 567–578.

Concerning continuous curves of certain types, Fund. Math. **11** (1928), 132–140.

On continuous curves in n dimensions (with Whyburn, G. T.), Bull. Amer. Math. Soc. **34** (1928), 349–360.

Continuous curves which are cyclicly connected, Bull. Acad. Polon. Sci. Math. (1928), 127–142.

Concerning subsets of a continuous curve, which can be connected through the complement of a continuous curve, Amer. J. Math. **50** (1928), 521–534.

On continuous curves having certain properties, Proc. Nat. Acad. Sci. **15** (1929), 91–94.

On simple closed curves and open curves, ibid. **15** (1929), 94–96.

Conditions under which every arc of a continuous curve is a subset of a maximal arc of the curve, Math. Ann. **101** (1929), 194–209.

Concerning the arc-curves and basic sets of a continuous curve, Second paper, Trans. Amer. Math. Soc. **31** (1929), 595–612.

Continuous curves in which every arc may be extended, Bull. Amer. Math. Soc. **35** (1929), 850–858.

On continua which are disconnected by the omission of any point and some related problems, Monatsh. Math. Phys. **36** (1929), 135–148.

Über Verallgemeinerungen des Jordanschen Kontinuums, ibid. **36** (1929), 301–304.

Concerning continuous curves in metric space, Amer. J. Math. **51** (1929), 577–594.

Continuous curves homeomorphic with the boundary of a plane domain, Fund. Math. **14** (1929), 92–95.

On continuous images of a compact metric space, Fund. Math. **14** (1929), 334–338.

On the density of the cut points and end points of a continuum, Bull. Amer. Math. Soc. **36** (1930), 659–667.

A new proof of a theorem of Zarankiewicz, Fund. Math. **16** (1930), 134–135.

Some generalizations of the Scherrer Fixed-Point Theorem, ibid., 332–336.

On the regular points of a continuum, Trans. Amer. Math. Soc. **33** (1931), 252–262.

On avoidable points of continua with an application to end points, Mathematische Zeitschrift **34** (1931), 161–178.

Note on a property of continuous arcs, Proceedings of the Cambridge Philosophical Society **27** (1931), 543–545.

A note on the definition of arc-sets, Bull. Amer. Math. Soc. **46** (1940), 794–796.
A new proof of the cyclic connectivity theorem, Bull. Amer. Math. Soc. **48** (1942), 627–630.

BALL, B. J.

Continuous and equicontinuous collections of arcs, Duke Math. J. **19** (1952), 423–433.
Some theorems concerning spirals in the plane, Amer. J. Math. **76** (1954), 66–80.
Countable paracompactness in linearly ordered spaces, Proc. Amer. Math. Soc. **5** (1954), 190–192.
A note on the separability of an ordered space, Canad. J. Math. **7** (1955), 548–551.
The sum of two solid horned spheres, Ann. of Math. **69** (1959), 253–257.

BARRETT, LIDA K.

Regular curves and regular points of finite order, Duke Math. J. **22** (1955), 295–304.

BASYE, R. E.

Concerning two internal properties of plane continua, Bull. Amer. Math. Soc. **41** (1935), 670–674.
Simply connected sets, Trans. Amer. Math. Soc. **38** (1935), 341–356.
Some separation properties of the plane, Amer. J. Math. **58** (1936), 323–328.

BENTON, T. C.

On continuous curves which are homogeneous except for a finite number of points, Fund. Math. **13** (1929), 151–177.
A definition of an unknotted simple closed curve, Bull. Amer. Math. Soc. **36** (1930), 406–408.
On continuous curves which are homogeneous except for a finite number of points (second part), Fund. Math. **15** (1930), 38–41.

BETZ, E. E.

Accessibility and separation by simple closed curves, Amer. J. Math. **63** (1941), 127–135.

BING, R. H.

Collections filling up a simple plane web, Bull. Amer. Math. Soc. **51** (1945), 674–679.
Generalizations of two theorems of Janiszewski, ibid., 954–960.
The Kline sphere characterization problem, ibid. **52** (1946), 644–653.
Concerning simple plane webs, Trans. Amer. Math. Soc. **60** (1946), 133–148.
Generalizations of two theorems of Janiszewski. II, Bull. Amer. Math. Soc. **52** (1946), 478–480.
Sets cutting the plane, Ann. of Math. **47** (1946), 476–479.
Extending a metric, Duke Math. J. **14** (1947), 511–519.
Skew sets, Amer. J. Math. **69** (1947), 493–498.
Solution of a problem of R. L. Wilder, Amer. J. Math. **70** (1948), 95–98.
A homogeneous indecomposable plane continuum, Duke Math. J. **15** (1948), 729–742.
Some characterizations of arcs and simple closed curves, Amer. J. Math. **70** (1948), 497–506.
Partitioning a set, Bull. Amer. Math. Soc. **55** (1949), 1101–1110.
Complementary domains of continuous curves, Fund. Math. **36** (1949), 303–318.

A convex metric for a locally connected continuum, Bull. Amer. Math. Soc. **55** (1949), 812–819.

Coverings with connected intersections (with Floyd, E. E.), Trans. Amer. Math. Soc. **69** (1950), 387–391.

Metrization of topological spaces, Canad. J. Math. **3** (1951), 175–186.

Snake-like continua, Duke Math. J. **18** (1951), 653–666.

Concerning hereditarily indecomposable continua, Pacific J. Math. **1** (1951), 43–51.

Higher dimensional hereditarily indecomposable continua, Trans. Amer. Math. Soc. **71** (1951), 267–273.

A characterization of 3-space by partitionings, Trans. Amer. Math. Soc. **70** (1951), 15–27.

A homeomorphism between the 3-sphere and the sum of two solid horned spheres, Ann. of Math. **56** (1952), 354–362.

Partitioning continuous curves, Bull. Amer. Math. Soc. **58** (1952), 536–556.

A connected countable Hausdorff space, Proc. Amer. Math. Soc. **4** (1953), 474.

A convex metric with unique segments, ibid., 167–174.

Locally tame sets are tame, Ann. of Math. **59** (1954), 145–158.

Partially continuous decompositions, Proc. Amer. Math. Soc. **6** (1955), 124–133.

Some monotone decompositions of a cube, Ann. of Math. **61** (1955), 279–288.

A simple closed curve that pierces no disk, J. Math. Pures Appl. **9** (1956), 337–343.

Upper semicontinuous decompositions of E^3, Ann. of Math. **65** (1957), 363–374.

Approximating surfaces with polyhedral ones, ibid., 456–483.

A decomposition of E^3 *into points and tame arcs such that the decomposition space is topologically different from* E^3, ibid., 484–500.

Necessary and sufficient conditions that a 3-manifold be S^3, Ann. of Math. **68** (1958), 17–37.

Another homogeneous plane continuum, Trans. Amer. Math. Soc. **90** (1959), 171–192.

Each homogeneous nondegenerate chainable continuum is a pseudo-arc, Proc. Amer. Math. Soc. **10** (1959), 345–346.

An alternative proof that 3-manifolds can be triangulated, Ann. of Math. **69** (1959), 37–65.

Borsuk, K.

Sur les rétractes, Fund. Math. **17** (1931), 152–170.

Sur l'hyperspace d'un continu (with Mazurkiewicz, S.), Comptes Rendus des Séances de la Société des Sciences et des Lettres de Varsovie, Classe 3, **24** (1931), 149–152.

Quelques théorèmes sur les ensembles unicohérents, Fund. Math. **17** (1931), 171–209.

Ein Satz über Unikohärenz, ibid. **20** (1933), 35–38.

Sur la décomposition des courbes regulières en dendrites, Fund. Math. **22** (1934), 287–291.

Über eine Bedingung die dem lokalen Zusammenhange äquivalent ist, Mathematica **7** (1933), 144–146.

Einige Sätze über stetige Streckenbilder, Fund. Math. **18** (1932), 198–213.

Sur les rétractes absolus indécomposables (with Mazurkiewicz, S.), C. R. Acad. Sci. Paris **199** (1934), 110–112.

Un théorème sur les prolongements des transformations continues, Fund. Math. **29** (1937), 161–166.

Sur les prolongements des transformations continues, Fund. Math. **28** (1937), 99–110.

An example of a simple arc whose projection in every plane has interior points, Fund. Math. **34** (1947), 272–277.

Sur un espace compact localement contractile qui n'est pas un rétract absolu de voisinage, Fund. Math. **35** (1948), 175–180.

On the topology of recracts, Ann. of Math. **48** (1947), 1082–1094.

On an irreducible 2-dimensional absolute retract, Fund. Math. **37** (1950), 137–160.

Families of compacta and some theorems on sweeping, ibid. **42** (1955), 240–258.

BROUWER, L. E. J.

Zur analysis situs, Math. Ann. **68** (1910), 422–434.

Beweis des Jordanschen Kurvensatzes, Math. Ann. **69** (1910), 169–175.

Beweis der invarianz der dimensionzahl, Math. Ann. **70** (1911), 161–165.

Beweis der invarianz des n-dimensionalen Gebeits, Math. Ann. **71** (1912), 305–313.

Über den naturlichen Dimensionsbegriff, J. Reine Angew. Math. **142** (1913), 146–152.

Über den naturlichen Dimensionsbegriff, Proc. Roy. Acad. Sci. Amsterdam **26** (1923), 795–800.

Sur les continus irreductibles de M. Zoretti, Ann. École Normale **27** (1910), 565–566.

Zum naturlichen Dimensionsbegriff, Mathematische Zeitschrift **21** (1924), 312–314.

Intuitionischer Beweis des Jordanschen Kurvensatzes, Proc. Roy. Acad. Sci. Amsterdam **28** (1925), 503–508.

Intuitionistiche Einführung des Dimensionsbegriffes, ibid. **29** (1926), 855–863.

BURGESS, C. E.

Continua and their complementary domains in the plane, Duke Math. J. **18** (1951), 901–917.

Continua which are the sum of a finite number of indecomposable continua, Proc. Amer. Math. Soc. **4** (1953), 234–239.

Continua and their complementary domains in the plane. II, ibid. **19** (1952), 223–230.

Some theorems on n-homogeneous continua, Proc. Amer. Math. Soc. **5** (1954), 136–143.

Certain types of homogeneous continua, ibid. **6** (1955), 348–350.

Collections and sequences of continua in the plane, Pacific J. Math. **5** (1955), 325–333.

Separation properties and n-decomposable continua, Duke Math. J. **24** (1956), 595–600.

A note on the separation of connected sets by finite sets, Proc. Amer. Math. Soc. **7** (1956), 1115–1116.

Continua and various types of homogeneity, Trans. Amer. Math. Soc. **88** (1958), 366–374.

Chainable continua and indecomposability, Pacific J. Math. **9** (1959), 653–659.

CARATHÉODORY, C.

Über die Begrenzung einfach Zusammenhängender Gebiete, Math. Ann. **73** (1913), 323–370.

CHITTENDEN, E. W.

The converse of the Heine-Borel theorem in a Riesz domain, Bull. Amer. Math. Soc. **21** (1915), 179–183 and **20** (1914), 461.

On the equivalence of ecart and voisinage, Trans. Amer. Math. Soc. **18** (1917), 161–166.

On the Heine-Borel theorem in the theory of abstract sets, Bull. Amer. Math. Soc. **25** (1918), 60–66.

On the theory of developments of an abstract class in relation to the calcul fonctionnel (with Pitcher, A. D.), Trans. Amer. Math. Soc. **20** (1919), 213–233.

On the relation between the Hilbert space and the calcul fonctionnel of Fréchet, Rend. Circ. Mat. Palermo **45** (1921), 1–6.

Note on the division of a plane by a point set, Bull. Amer. Math. Soc. **28** (1922), 310–312.

Nuclear and hyper-nuclear points in the theory of abstract sets, ibid. **30** (1924), 511–519.

On the metrization problem and related problems in the theory of sets, ibid. **33** (1927), 13–34.

On general topology and the relation of the properties of the class of all continuous functions to the properties of space, Trans. Amer. Math. Soc. **31** (1929), 290–321.

On the reducibility of families of subsets and related properties (with Robinson, S.), Amer. J. Math. **55** (1933), 197–206.

On the reduction of topological functions, Lectures in Topology, pp. 267–285, University of Michigan Press, Ann Arbor, Michigan, 1941.

CHOQUET, G.

Homéomorphies, C. R. Acad. Sci. Paris **210** (1940), 129–131.

Points invariant et structure des continus, ibid. **212** (1941), 376–379.

Structure des domaines plans et accessibilité, C. R. Acad. Sci. Paris **216** (1943), 279–280.

Topologie de la représentation conforme, C. R. Acad. Sci. Paris **216** (1943), 330–331.

Représentation conforme et topologie, ibid., 402–404.

Caractérisation de la sphère en géometrie infinitesimale directe, Revue Sci. (Rev. Rose Illus.) **81** (1943), 447–452.

Étude des espaces métrique par les propriétés de leurs sous-ensembles finites, Bull. Soc. Math. France **71** (1944), 112–192.

Prolongements d'homéomorphies. Ensembles topologiquement nommables. Caractérisation topologique individuelle des ensembles fermés totalement discontinus, C. R. Acad. Sci. Paris **219** (1944), 542–544.

Convergences, Ann. Univ. Grenoble. Sect. Sci. Math. Phys. (N.S.) **23** (1948), 57–112.

Ensembles boreliens et analytiques dans les espaces topologiques, C. R. Acad. Sci. Paris **232** (1951), 2174–2176.

A plane continuum no two of whose nondegenerate subcontinua are homeomorphic (with Anderson, R. D.), Proc. Amer. Math. Soc. **10** (1959), 347–353.

CLAYTOR, W. S.

Topological immersion of Peanian continua in a spherical surface, Ann. of Math. **35**, No. 4 (1934), 809–835.

Peanian continua not imbeddable in a spherical surface, Ann. of Math. **38** (1937), 631–646.

CSÁSZÁR, Á.

Sur les courbes atriodiques, Acta Math. Acad. Sci. Hungar. **9** (1958), 329–332.

Sur la notion d'espace topologique. I, II, III, Mat. Lapok **8** (1957), 211–231.

CLEVELAND, C. M.

Concerning points of a continuous curve that are not accessible from each other, Proc. Nat. Acad. Sci. **13** (1927), 275–276.

On the existence of acyclic curves satisfying certain conditions with respect to a given continuous curve, Trans. Amer. Math. Soc. **33** (1931), 958–978.

COHEN, H. J.

Some results concerning homogeneous plane continua, Duke Math. J. **18** (1951), 467–474.

Sur un problème de M. Dieudonné, C. R. Acad. Sci. Paris **234** (1952), 290–292.

COLMEZ, J.

Sur les espaces precompacts, C. R. Acad. Sci. Paris **234** (1952), 1019–1021.

CORSON, H. H.

The determination of paracompactness by uniformities, Amer. J. Math. **80** (1958), 185–190.

DENJOY, A.

Continu et discontinu, C. R. Acad. Sci. Paris **151** (1910), 138–140.
Sur les ensembles parfaits discontinus, C. R. Acad. Sci. Paris **149** (1909), 1048–1050.
Sur l'analysis situs du plan, C. R. Acad. Sci. Paris **153** (1911), 423–426 and 493–496.

DIEUDONNÉ, J.

Une généralisation des espaces compacts, J. Math. Pures Appl. **23** (1944), 65–76.
Une critère de normalité pour les espaces produits, Colloq. Math. **6** (1958), 29–32.

DORROH, J. L.

Concerning a set of metrical hypotheses for geometry, Ann. of Math. **29** (1928), 229–231.
On a problem by G. T. Whyburn (with Roberts, J. H.), Fund. Math. **13** (1929), 58–61.
Some metrical properties of descriptive planes, Amer. J. Math. **53** (1931), 401–421.

DOWKER, C. H.

On countably paracompact spaces, Canad. J. Math. **3** (1954), 219–224.

DYER, ELDON

Irreducibility of the sum of the elements of a continuous collection of continua, Duke Math. J. **20** (1953), 589–592.
Continuous collections of decomposable continua on a spherical surface, Proc. Amer. Math. Soc. **6** (1955), 331–360.
Certain transformations which lower dimension, Ann. of Math. **63** (1955), 15–19.
Regular mappings and dimension, Ann. of Math. **67** (1958), 119–149.
Regular mappings and the space of homeomorphisms on a 2-manifold (with Hamstrom, Mary-Elizabeth), Duke Math. J. **25** (1958), 521–531.

EILENBERG, S.

Transformations continus en circonférence et la topologie du plan, Fund. Math. **26** (1936), 61–112.
Sur les espaces multicoherent. I, Fund. Math. **27** (1936), 152–190.
Sur les transformations à petites tranches, Fund. Math. **30** (1938), 92–95.

ETTLINGER, M. G.

On irreducible continuous curves. Bull. Amer. Math. Soc. **49** (1943), 569–574.

FLOYD, E. E.

A nonhomogeneous minimal set, Bull. Amer. Math. Soc. **55** (1949), 957–960.
The extension of homeomorphisms, Duke Math. J. **16** (1949), 225–235.
Some characterizations of interior maps, Ann. of Math. **51** (1950), 571–575.

Coverings with connected intersections (with Bing, R. H.), Trans. Amer. Math. Soc. **69** (1950), 387–391.

A characterization theorem for monotone mappings (with Fort, M. K., Jr.), Proc. Amer. Math. Soc. **4** (1953), 828–830.

Real-valued mappings of spheres, Proc. Amer. Math. Soc. **6** (1955), 957–959.

FORT, M. K., JR.

A specialization of Zorn's lemma, Duke Math. J. **15** (1948), 763–765.

A unified theory of semi-continuity, ibid. **16** (1949), 237–246.

A note on equicontinuity, Bull. Amer. Math. Soc. **55** (1949), 1098–1100.

Essential and non essential fixed points, Amer. J. Math. **72** (1950), 316–322.

A characterization theorem for monotone mappings (cf. Floyd, E. E.).

Open topological disks in the plane, J. Indian Math. Soc. (N.S.) **18** (1954), 23–26.

FOX, R. H.

Extension of homeomorphisms into Euclidean and Hilbert parallelotopes, Duke Math. J. **8** (1941), 452–456.

A remarkable simple closed curve, Ann. of Math. **50** (1949), 264–265.

FRÉCHET, M.

La notion d'écart et le calcul fonctionnel, C. R. Acad. Sci. Paris **140** (1905), 772–774.

Sur quelques points du calcul fonctionnel, Rend. Circ. Mat. Palermo **22** (1906), 1–74.

Une définition du nombre de dimensions d'un ensemble abstrait, C. R. Acad. Sci. Paris **148** (1909), 1152–1154.

Les dimensions d'un ensemble abstrait, Math. Ann. **68** (1910), 145–168.

Les ensembles abstraits et le calcul fonctionnel, Rend. Circ. Mat. Palermo **30** (1910), 1–26.

Sur les classes V normales, Trans. Amer. Math. Soc. **14** (1913), 320–324.

Sur l'homéomorphie des ensembles dénombrables, Bull. Acad. Polon. Sci. Math. (1920), 107–108.

Sur les ensembles abstraits, Ann. École Normale **38** (1921), 341–385.

Sur la notion de nombre de dimensions, C. R. Acad. Sci. Paris **178** (1924), 1782–1785.

Sur la distance de deux ensembles, Bull. Calcutta Math. Soc. **15** (1924), 1–8.

Sur la distance de deux surfaces, Ann. Soc. Polon. Math. **2** (1924), 232–247.

Sur l'homéomorphie de deux ensembles et sur les ensembles complets, Bull. Sci. Math. **49** (1925), 100–103.

L'expression la plus générale d "distance" sur une droite, Amer. J. Math. **47** (1925), 1–10.

Sur l'espace métrique universal de Paul Urysohn, Bull. Sci. Math. **39** (1925), 297–301.

Sur une représentation paramétrique intrinsèque de la courbe continue la plus générale, J. Math. Pures Appl. **4** (1925), 281–297.

Les transformations ponctuelles abstraits, C. R. Acad. Sci. Paris **180** (1925), 1816.

Les espaces abstraits topologiquement affines, Acta Mathematica **47** (1926), 25–52.

Quelques propriétés des ensembles abstraits, Fund. Math. **10** (1927), 328–355; second memoir, ibid. **12** (1928), 298–310.

Sur une définition du nombre de dimensions, Rev. Acad. Ciencias Madrid **33** (1928), 564–587.

Démonstration de quelques propriétés des ensembles abstraits, Amer. J. Math. **50** (1928), 47–72.

Esquisse d'une théorie des ensembles abstraits, Sir Asutosh Mookerjee Silver Jubilee Volumes, Volume 2, Science, 233–394; Calcutta Baptist Mission Press, 1922.
Les espaces abstraits, Gauthier-Villars et Cie, Editeurs, Paris, 1928, 11 + 296 pp.

GEHMAN, H. M.

Concerning the subsets of a plane continuous curve, Ann. of Math. **27** (1925), 29–46.
On irredundant sets of postulates, Bull. Amer. Math. Soc. **32** (1926), 159–161.
On extending a continuous correspondence of two plane continuous curves to a correspondence of their planes, Trans. Amer. Math. Soc. **28** (1926), 252–265.
Some conditions under which a continuum is a continuous curve, Ann. of Math. **27** (1926), 381–384.
Concerning irreducibly connected sets and irreducible continua, Proc. Nat. Acad. Sci. **12** (1926), 544–547.
Some relations between a continuous curve and its subsets, Ann. of Math. **28** (1927), 103–111.
Irreducible continuous curves, Amer. J. Math. **49** (1927), 189–196.
Concerning acyclic continuous curves, Trans. Amer. Math. Soc. **29** (1927), 553–568.
Concerning end points of continuous curves and other continua, ibid. **30** (1928), 53–84.
Concerning certain types of non-cut points, with an application to continuous curves, Proc. Nat. Acad. Sci. **14** (1928), 431–433.
Concerning irreducible continua, ibid., 433–435.
On extending a continuous (1,1) *correspondence* (*second paper*), Trans. Amer. Math. Soc. **31** (1929), 241–252.
On extending a correspondence in the sense of Antoine, Amer. J. Math. **51** (1929), 385–396.
Centers of symmetry in analysis situs, Amer. J. Math. **52** (1930), 543–547.
A special type of upper semi-continuous collection, Proc. Nat. Acad. Sci. **16** (1930), 609–613.
Concerning sequences of homeomorphisms, ibid. **18** (1932), 460–465.
On extending a homeomorphism between two subsets of spheres, Bull. Amer. Math. Soc. **42** (1936), 79–81.

GRACE, E. E.

A note on linear spaces and unicoherence, J. Elisha Mitchell Sci. Soc. **70** (1954), 33–34.
Cut sets in totally nonaposyndetic continua, Proc. Amer. Math. Soc. **9** (1958), 98–104.
Totally nonconnected im kleinen continua, ibid., 818–821.

GRIFFITH, H. C.

A characterization of tame surfaces in three space, Ann. of Math. **69** (1959), 276–290.

DE GROOT, J.

Topological characterizations of all subsets of the real number system, Nederl. Akad. Wetensch. Proc. **50** (1947), 876–884 = Indagationes Math. **9**, 387–395.
Some special metrics in general topology, Colloq. Math. **6** (1958), 283–386.

HAHN, HANS

Über die allgemeinste ebene Punktmenge, die stetiges Bild einer Streke ist, Jber. Deutsch. Math. Verein. **23** (1914), 318–322.

Mengentheoretische Charakterisierung der stetigen Kurve, Sitzungsber. Akad. Wiss. Wien, Math.-Nat. Klasse **123** (1914), 1–57.

Über irreduzible Kontinua, ibid. **130** (1921), 217–250.

Über die Komponenten offenen Mengen, Fund. Math. **2** (1921), 189–192.

HALL, D. W.

On a decomposition of true cyclic elements, Trans. Amer. Math. Soc. **47** (1940), 305–321.

Arc- and tree-preserving transformations (with Whyburn, G. T.), ibid. **48** (1940), 63–71.

Conditions for the continuity of arc-preserving transformations (with Puckett, W. T., Jr.), Bull. Amer. Math. Soc. **47** (1941), 468–475.

Periodic types of transformations (with Kelley, J. L.), Duke Math. J. **8** (1941), 625–630.

Strongly arcwise connected spaces (with Puckett, W. T., Jr.), Amer. J. Math. **63** (1941), 554–562.

A partial solution of a problem of J. R. Kline, Duke Math. J. **9** (1942), 893–901.

A note on primitive skew curves, Bull. Amer. Math. Soc. **49** (1943), 935–936.

Elementary topology (with Spencer, G. L.), John Wiley and Sons, Inc., New York; Chapman and Hall, Ltd., London, 1955, xii + 303 pp.

HALLETT, G. H.

Concerning the definition of a simple continuous arc, Bull. Amer. Math. Soc. **25** (1919), 325–326.

HAMILTON, O. H.

Fixed points under transformations of continua which are not connected im kleinen, Trans. Amer. Math. Soc. **44** (1938), 18–24.

Concerning continua in a separable space which do not cross, Bull. Amer. Math. Soc. **45** (1939), 114–118.

Concerning the decomposition of continua, Lectures in Topology, pp. 297–298, University of Michigan Press, Ann Arbor, Michigan, 1941.

A fixed point theorem for upper semi-continuous transformations of n-cells for which the images of points are non-acyclic continua, Duke Math. J. **14** (1947), 689–693.

Fixed point theorems for interior transformations, Bull. Amer. Math. Soc. **54** (1948), 383–385.

A fixed point theorem for pseudo-arcs and certain other metric continua, Proc. Amer. Math. Soc. **2** (1951), 173–174.

A short proof of the Cartwright-Littlewood fixed point theorem, Canad. J. Math. **6** (1954), 522–524.

Fixed points for certain noncontinuous transformations, Proc. Amer. Math. Soc. **8** (1957), 750–756.

HAMSTROM, MARY-ELIZABETH

Concerning webs in the plane, Trans. Amer. Math. Soc. **74** (1953), 500–513.

Concerning certain types of webs, Proc. Amer. Math. Soc. **4** (1953), 974–978.

Concerning continuous collections of continuous curves, ibid., 240–243.

Concerning the imbedding of upper semi-continuous collections of continua in continuous collections of continua, Amer. J. Math. **76** (1954), 793–810.

A note on continuous collections of continuous curves filling up a continuous curve in the plane (cf. Anderson, R. D.).

On spaces filled up by continuous collections of atriodic continuous curves (cf. Anderson, R. D.).

Characterization of plane continuous curves which are filled up by continuous collections of continuous curves, Amer. J. Math. **77** (1955), 914–925.

Regular mappings and the space of homeomorphisms on a 2-manifold (cf. Dyer, Eldon).

HARROLD, O. G.

The non-existence of a certain type of continuous transformation, Duke Math. J. **5** (1939), 787–793.

Hereditary arc sums, ibid., 111–117.

A note on strongly irreducible maps of an interval, ibid. **6** (1940), 750–752.

The role of local separating points in certain problems of continuum structure, Lectures in Topology, pp. 237–253, University of Michigan Press, Ann Arbor, Michigan, 1941.

Continua of finite sections, Duke Math. J. **8** (1941), 682–688.

A mapping characterization of Peano space, Bull. Amer. Math. Soc. **48** (1942), 561–566.

A characterization of tame curves in three-space (with Griffith, H. C. and Posey, E. E.) Trans. Amer. Math. Soc. **79** (1955), 12–34.

Some consequences of the approximation theorem of Bing, Proc. Amer. Math. Soc. **8** (1957), 204–206.

Locally tame curves and surfaces in three-dimensional manifolds, Bull. Amer. Math. Soc. **63** (1957), 293–305.

A sufficient condition that a monotonic image of the three sphere be a topological sphere, Proc. Amer. Math. Soc. **9** (1958), 846–850.

HAUSDORFF, F.

Grundzüge der Mengenlehre, Leipzig, Veit & Co., 1914.

Die Mengen G_δ in vollständigen Räumen, Fund. Math. **6** (1924), 146–148.

Über innere abbildungen, Fund. Math. **23** (1934), 279–291.

HEDRICK, E. R.

On properties of a domain for which any derived set is closed, Trans. Amer. Math. Soc. **12** (1911), 285–294.

HENDERSON, G. W.

Proof that every compact decomposable continuum which is topologically equivalent to each of its nondegenerate subcontinua is an arc, Ann. of Math. **27** (1960), 421–428.

HILDEBRANDT, T. H.

A contribution to the foundations of Fréchet's calcul fonctionnel, Amer. J. Math. **34** (1912), 237.

The Borel Theorem and its generalizations, Bull. Amer. Math. Soc. **32** (1926), 423–474.

HUREWICZ, W.

Über eine Verallgemeinerung des Borelschen Theorems, Math. Ann. **24** (1925), 401–421.

Über oberhalbstetige Zerlegungen von Punktmengen in Kontinua, Fund. Math. **15** (1930), 57–60.

Über die henkelfreien Kontinua, Proc. Roy. Acad. Sci. Amsterdam **35** (1932), 1077–1078.

Ein Einbettungssatz über henkelfreie Kontinua (with Knaster, B.), Proc. Roy. Acad. Sci. Amsterdam **36** (1933), 557–559.

Ein Satz über stetige Abbildungen, Fund. Math. **23** (1934), 54–62.

Über Abbildungen von endlichdimensionalen Räumen Auf Teilmengen cartesisicher Räume, S.-B. Deutsch. Akad. Wiss. Berlin. Kl. Math. Phys. Tech. **24** (1933), 1–17.

Homotopie, Homologie und lokaler Zusammenhang, ibid. **25** (1935), 467–485.

Dimension theory (with Wallman, Henry), Princeton Mathematics Series, vol. 4, Princeton University Press, Princeton, New Jersey, 1941.

Isbell, J. R.

Homogeneous spaces, Duke Math. J. **20** (1953), 321–329.

Iseki, K.

On hypercompact spaces, Portugal. Math. **13** (1954), 149–152.

Some properties of hypernormal spaces, Proc. Japan Acad. **30** (1954), 937–939.

A theorem on paracompact spaces (with Miyanaga, Y.), ibid. **32** (1956), 396–398.

Ishikawa, F.

On countably paracompact spaces, Proc. Japan Acad. **31** (1955), 686–687.

Janiszewski, S.

Contribution à la géometrie des courbes planes générales, C. R. Acad. Sci. Paris **150** (1910), 606–609.

Sur la géometrie de lignes cantoriennes, ibid. **151** (1910), 198–201.

Sur les continus irreductibles entre deux points, ibid. **152** (1911), 752–755.

Sur les continus irreductibles entre deux points, J. École Polytech. 2e Serie **16** (1912), 79–170.

Über die Begriffe "Linie" und "Fläche", Proceedings of the Fifth International Congress of Mathematicians, Vol. 2, pp. 126–128, University Press, Cambridge (1913).

Démonstration d'une propriété des continus irreducibles entre deux points, Bull. Acad. Sci. Cracovie (1912), 906–914.

Sur les coupures du plan faites par les continus (in Polish), Prace Mat.-Fiz. **26** (1913), 11–63.

Jones, F. B.

A theorem concerning locally peripherally separable spaces, Bull. Amer. Math. Soc. **41** (1935), 437–439.

Concerning normal and completely normal spaces, ibid. **43** (1937), 671–677.

Concerning certain topologically flat spaces, Trans. Amer. Math. Soc. **42** (1937), 53–93.

Concerning R. L. Moore's Axiom 5, Bull. Amer. Math. Soc. **44** (1938), 689–692.

Concerning the boundary of a complementary domain of a continuous curve, ibid. **45** (1949), 428–435.

Concerning certain linear abstract spaces and simple continuous curves, ibid., 623–628.

Certain equivalences and subsets of a plane, Duke Math. J. **5** (1939), 133–145.

Almost cyclic elements and simple links of a continuous curve, Bull. Amer. Math. Soc. **46** (1940), 775–783.

Monotonic collections of peripherically separable connected domains, ibid. **47** (1941), 661–664.

Aposyndetic continua and certain boundary problems, Amer. J. Math. **63** (1941), 545–553.

Certain consequences of the Jordan curve theorem, ibid., 531–544.

Connected and disconnected plane sets and the functional equation $f(x) + f(y) = f(x + y)$, Bull. Amer. Math. Soc. **48** (1942), 115–120.

Measure and other properties of a Hamel basis, ibid., 472–481.

Concerning the separability of certain locally connected metric spaces, ibid. **52** (1946), 303–306.

A characterization of a semi-locally connected plane continuum, ibid. **53** (1947), 170–175.

Concerning non-aposyndetic continua, Amer. J. Math. **70** (1948), 403–413.

A note on homogeneous plane continua, Bull. Amer. Math. Soc. **55** (1949), 113–114.

Certain homogeneous unicoherent indecomposable continua, Proc. Amer. Math. Soc. **2** (1951), 855–859.

Concerning aposyndetic and non-aposyndetic continua, Bull. Amer. Math. Soc. **58** (1952), 137–151.

On the separation of the set of pairs of a set, J. Elisha Mitchell Sci. Soc. **68** (1952), 44–45.

On certain well-ordered monotone collections of sets, ibid. **69** (1953), 30–34.

On a property related to separability in metric spaces, ibid. **70** (1954), 30–33.

On a certain type of homogeneous plane continuum, Proc. Amer. Math. Soc. **6** (1955), 735–740.

On the existence of weak cut points in plane continua, ibid. **9** (1958), 530–532.

R. L. Moore's Axiom 1′ and metrization, ibid., 487.

Moore spaces and uniform spaces, ibid., 483–486.

Product spaces in n-manifolds (with Young, G. S.), ibid. **10** (1959), 307–308.

Another homogeneous plane continuum (cf. Bing, R. H.).

Kelley, J. L.

Fixed sets under homeomorphisms, Duke Math. J. **5** (1939), 535–537.

A metric connected with property S, Amer. J. Math. **61** (1939), 764–768.

A decomposition of compact continua and related theorems on fixed sets under continuous transformations, Proc. Nat. Acad. Sci. **26** (1940), 192–194.

Periodic types of transformations (cf. Hall, D. W.).

Hyperspaces of a continuum, Trans. Amer. Math. Soc. **52** (1942), 22–36.

Simple links and fixed sets under continuous mappings, Amer. J. Math. **69** (1947), 348–356.

Exact homomorphism sequences in homology theory (with Pitcher, Everett), Ann. of Math. **48** (1947), 682–709.

General topology, D. Van Nostrand Company, Inc., Toronto-New York-London, 1955, xiv + 298 pp.

On mappings of plane sets, Colloq. Math. **6** (1958), 153–154.

Kline, J. R.

Concerning the complement of a countable infinity of point sets of a certain type, Bull. Amer. Math. Soc. **23** (1917), 290–292.

The converse of the theorem concerning the division of a plane by an open curve, Trans. Amer. Math. Soc. **18** (1917), 177–184.

A definition of sense on closed curves in non-metrical analysis situs, Ann. of Math. **19** (1918), 185–200.

On the most general plane closed point set through which it is possible to pass a simple continuous arc (cf. Moore, R. L.).

Concerning sense on closed curves in non-metrical plane analysis situs, Ann. of Math. **21** (1919), 113–119.

On the passing of simple continuous arcs through plane point sets, Tôhoku Math. J. **18** (1920), 116–125.

A new proof of a theorem due to Schoenflies, Proc. Nat. Acad. Sci. **6** (1920), 529–531.

Concerning approachability of simple closed and open curves, Trans. Amer. Math. Soc. **21** (1920), 451–458.

A theorem concerning connected point sets, Fund. Math. **3** (1922), 238–239.

Closed connected point sets which are disconnected by the removal of a finite number of points, Proc. Nat. Acad. Sci. **9** (1923), 238–239.

Closed connected sets which remain connected upon the removal of certain connected subsets, Fund. Math. **5** (1924), 3–10.

Concerning the division of the plane by continua, Proc. Nat. Acad. Sci. **10** (1924), 176–177.

Concerning the complementary intervals of countable closed sets, Bull. Amer. Math. Soc. **31** (1925), 409–410.

Concerning the sum of two continua each irreducible between the same pair of points, Fund. Math. **7** (1925), 314–322.

A condition that every subcontinu of a continuous curve be a continuous curve, Fund. Math. **10** (1927), 298–301.

Separation theorems and their relations to recent developments in analysis situs, Bull. Amer. Math. Soc. **34** (1928), 155–192.

What is the Jordan curve theorem?, Amer. Math. Monthly **49** (1942), 281–286.

KLIPPLE, E. C.

Two-dimensional spaces in which there exist contiguous points, Trans. Amer. Math. Soc. **44** (1938), 250–276.

KNASTER, B.

Sur les ensembles connexes (with Kuratowski, C.), Fund. Math. **2** (1921), 206–255.

Un continu dont tout sous-continu est indécomposable, Fund. Math. **3** (1922), 247–286.

Sur les continus non bornés (with Kuratowski, C.), ibid. **4** (1923), 14–19.

Sur un problème de M. R. L. Wilder, ibid. **7** (1925), 191–197.

Quelques coupoupures du plan, ibid., 264–289.

A connected and connected im kleinen point set which contains no perfect subset (with Kuratowski, C.), Bull. Amer. Math. Soc. **33** (1927), 106–109.

Remark on a theorem of R. L. Moore (with Kuratowski, C.), Proc. Nat. Acad. Sci. **13** (1926), 647–649.

Sur les ensembles connexes irréductibles entre deux points, Fund. Math. **10** (1927), 277–297.

Un continu irréductible à décomposition continu en tranches, ibid. **25** (1935), 568–577.

Sur la caractérisation topologique de l'ensemble de bouts d'une courbe (with Reichbach, M.), Fund. Math. **40** (1953), 13–28.

KURATOWSKI, C.

Une définition topologique de la ligne de Jordan, Fund. Math. **1** (1920), 40–43.

Sur les continus indécomposables (with Janiszewski, S.), ibid., 210–222.

Le théorème de Borel-Lebesgue dans la théorie des ensembles abstraits (with Sierpiński, W.), ibid. **2** (1921), 172–178.

Une méthode d'élimination des nombres transfinis de raisonements mathématiques, ibid. **3** (1922), 76–108.

Sur les ensembles connexes (with Knaster, B.), ibid. **2** (1921), 206–255.

Quelques propriétés topologiques de la demi-droite, ibid. **3** (1922), 59–64.

Théorie des continua irréductibles entre deux points. I, ibid., 200–231.

Contribution à l'étude de continus de Jordan, ibid. **5** (1924), 112–122.

Sur les coupures irréductibles du plan, ibid. **6** (1924), 130–145.

On the accessibility of an arc from its complement in space of three dimensions, Bull. Amer. Math. Soc. **31** (1925), 32.

Sur les continus de Jordan et le théorème de M. Brouwer, Fund. Math. **8** (1926), 137–150.

Remark on a theorem of R. L. Moore, Proc. Nat. Acad. Sci. **13** (1926), 647–649.

Über geschlossene Kurven und unzerlegbare Kontinua, Math. Ann. **98** (1927), 399–405.

A connected and connected im kleinen point set which contains no perfect subset (with Knaster, B.), Bull. Amer. Math. Soc. **33** (1927), 106–109.

A theorem on connected point sets (with Zarankiewicz, C.), ibid., 571–575.

Sur les décompositions semi-continues d'espaces métriques compacts, Fund. Math. **11** (1928), 169–185.

Sur le structure frontières communes à deux régions, ibid. **12** (1928), 20–42.

Sur la séparation d'ensembles situés sur le plan, ibid., 214–239.

Généralisation d'un théorème de Janiszewski (with Straszewicz, S.), ibid., 152–157.

Un système d'axiomes pour la topologie de la surface de la sphère, Atti del Congresso Internazionale dei Matematici, Bologna, 1928, 239–241.

Théorème sur trois continus, Monatsch. Math. Phys. **36** (1929), 77–80.

Sur le problème des courbes gauches en topologie, Fund. Math. **15** (1930), 271–283.

Remarques sur la théorie axiomatique de la dimension (with Menger, K.), Montsch. Math. Phys. **37** (1930), 169–174.

Sur les éléments cycliques et leurs applications (with Whyburn, G. T.), Fund. Math. **16** (1930), 305–331.

Sur une généralisation de la notion d'homéomorphie, ibid. **22** (1934), 206–220.

La notion de connexité locale en topologie, Enseignement Math. **35** (1936), 229–240.

Quelques théorèmes sur les prolongements topologiques des espaces, Fund. Math. **30** (1937), 8–13.

Quelques généralisations des théorèmes sur les coupures du plan (with Zarankiewicz, C.), ibid. **36** (1949), 277–282.

Sur une propriété topologique fondamentale du plan euclidien, Atti del Quarto Congresso dell'Unione Matematica Italiano, Taormina, vol. II, 1951, pp. 361–362. Casa Editrice Perella, Roma, 1953.

Teoria mnogosci (*Theory of sets*) (with Mostowski, Andrzej), Monografie Matematyczne, vol. 27. Polskie Towarzystwo Matematyczne, Warsaw-Wroclaw, 1952, ix + 311 pp.

Sur un problème concernant les coupures des régions par des continus, Fund. Math. **39** (1952), 15–24.

Topologie II, 2ème éd. Monografie Matematyczne, vol. 21. Polskie Towarzystwo Matematyczne, Warszawa, 1952. viii + 443 pp.

Un théorème sur les espaces complets et ses applications à l'étude de la connexité locale, Bull. Acad. Polon. Sci. Cl. III **3** (1955), 75–80.

On a characterization of connected domains in locally connected spaces, Bull. Acad. Polon. Sci. Cl. III **4** (1958), 211–214.

Sur une méthode de métrisation complèts de certain espaces d'ensembles compacts, Fund. Math. **43** (1956), 114–138.

Sur quelques invariants topologiques dans l'espace euclidien, J. Math. Pures Appl. (9) **36** (1957), 191–200.

Topologie I, 4ème éd. Monografie Matematyczne, vol. 20. Panstwowe Wydawnictwo Naukowe, Warsaw, 1958. xiii + 494 pp.

LEBESGUE, H.

Leçons sur l'intégration et la recherche de fonctions primitives, Paris, Gauthier-Villars, 1904, 7 + 136 pp.

Sur le théorème de Schoenflies, Fund. Math. **6** (1924), 97–99.

Sur les correspondances entre les points de deux espaces, ibid. **2** (1921), 256–285.

LENNES, N. J.

Curves in non-metrical analysis situs with applications in the calculus of variations, Amer. J. Math. **33** (1911), 287–326 and Bull. Amer. Math. Soc. **12** (1906), 395–398.

LOHWATER, A. J.

The Moore-Young theorem in point set topology, Arkhimedes (1956), 22–24.

LUBBEN, R. G.

Concerning limiting sets in abstract spaces, Trans. Amer. Math. Soc. **30** (1928), 668–685.

The double elliptic case of the Lie-Riemann-Helmholtz-Hilbert problem of the foundations of geometry, Fund. Math. **11** (1928), 35–95.

Separation theorems with applications to questions concerning approachability and plane continua, Trans. Amer. Math. Soc. **31** (1929), 503–522.

Concerning limiting sets in abstract spaces. II, Trans. Amer. Math. Soc. **43** (1938), 482–493.

Separabilities of arbitrary orders and related properties, Bull. Amer. Math. Soc. **46** (1940), 913–919.

Concerning the decomposition and amalgamation of points, upper semi-continuous collections, and topological extensions, Trans. Amer. Math. Soc. **49** (1941), 410–466.

MANSFIELD, M. J.

Some generalizations of full normality, Trans. Amer. Math. Soc. **86** (1957), 489–505.

On countably paracompact normal spaces, Canad. J. Math. **9** (1957), 443–449.

MAZURKIEWICZ, S.

Sur les lignes de Jordan, Fund. Math. **1** (1920), 166–209. See references, in this paper, to earlier papers of Mazurkiewicz in Polish.

Sur l'existence d'un ensemble plan connexe ne contenant aucun sousensemble connexe borné, ibid. **2** (1921), 96–103.

Un théorème sur les lignes de Jordan, ibid., 119–130.

Sur les ensembles quasi-connexes, ibid., 201–205.

Extension du théorème du Phragmen-Brouwer aux ensembles non-bornés, ibid. **3** (1922), 20–25.

Remarque sur un théorème de M. Mullikin, Fund. Math. **6** (1924), 37–38.

Sur les continus homogènes, ibid. **5** (1924), 137–146.

Sur la décomposition d'un domaine en deux sous ensembles punktiformes, ibid. **3** (1922), 65–75.

Sur les continus plans non bornés, ibid. **5** (1924), 188–205.

Sur les composantes dimensionels d'un espace compact, Fund. Math. **19** (1932), 243–246.

Sur la décomposition du plan en courbes, Fund. Math. **21** (1933), 43–45.

McAuley, L. F.

On decomposition of continua into aposyndetic continua, Trans. Amer. Math. Soc. **81** (1956), 74–91.

A relation between perfect separability, completeness and normality in semi-metric spaces, Pacific J. Math. **6** (1956), 315–326.

Paracompactness and an example due to F. B. Jones, Proc. Amer. Math. Soc. **7** (1956), 1155–1156.

A note on naturally ordered sets in semi-metric spaces, ibid. **8** (1957), 384–386.

A note on complete collectionwise normality and paracompactness, ibid. **9** (1958), 796–799.

Menger, K.

Grundzuge einer Theorie der Kurven, Math. Ann. **95** (1925), 272–306.

Über regulare Baumkurven, Math. Ann. **96** (1926), 572–582.

Zur allgemeinen Kurventheorie, Fund. Math. **10** (1926), 97–115.

Bericht über die Dimensionstheorie, Jber. Deutsch. Math. Verein. **35** (1926), 113–150.

Das Hauptproblem über die dimensionelle Struktur der Räume, Proc. Roy. Acad. Sci. Amsterdam **30**, 138–144.

Remarques sur la théorie axiomatique de la dimension (with Kuratowski, C.), ibid. **37** (1930), 169–174.

Eine dimensionstheoretische Bemerkung von O. Schreier, ibid. **37** (1930), 7–12.

Bemerkungen zur zweiten Untersuchung über allgemeine Metrik, Proc. Roy. Acad. Sci. Amsterdam **30** (1927), 710–714.

Zusammenhangsstufen und Cantorische Mannigfaltigkeiten, ibid., 705–709.

Untersuchungen über allgemeine Metrik, Math. Ann. **100** (1928), 75–163.

Dimensionstheorie, Leipzig und Berlin, B. G. Teubner, 1928, 4 + 318 pp.

Bericht über die mengentheoretische Überdeckungsatze, Ergebnisse Math. Kolloquium, Wien **2** (1932), 23–27.

Remarks concerning the paper of W. L. Ayres on the regular points of a continuum, Trans. Amer. Math. Soc. **33** (1931), 663–667.

Über die Dimension von Punktmengen. III. *Zur Begrundung einer axiomatischen Theorie der Dimension*, Monatsch. Math. Phys. **36** (1929), 193–218.

Kurventheorie, Leipzig und Berlin, B. G. Teubner, 1932, 374 pp.

An abstract form of the covering theorems of topology, Ann. of Math. **39** (1938), 794–803.

On linear sets in metric spaces (with Milgram, A. N.), Reports of a Mathematical Colloquium **1** (1939), 16–17.

Topology without points, Rice Institute Pamphlet **27**, No. 1 (1940), 80–107.

Michael, E.

Another note on paracompact spaces, Proc. Amer. Math. Soc. **8** (1957), 822–828.

MILGRAM, A. N.

Decomposition and dimension of closed sets in R^n, Trans. Amer. Math. Soc. **43** (1938), 465–481.

A general existence theorem and some applications, Ann. of Math. **39** (1938), 804–810.

On linear sets in metric spaces (cf. Menger, K.).

On shortest paths through a set, Reports of a Mathematical Colloquium **2** (1940), 39–44.

MILLER, E. W.

On subsets of a continuous curve which lie on an arc of the continuous curve, Amer. J. Math. **54** (1932), 397–416.

On certain properties of Fréchet L-spaces, Fund. Math. **26** (1936), 116–119.

On a property of families of sets, Comptes Rendus des Séances de la Société des Sciences et des Lettres de Varsovie, Classe II, **30** (1937), 31–38.

Some theorems on continua, Bull. Amer. Math. Soc. **46** (1940), 150–157.

MILLER, HARLAN CROSS

On unicoherent continua, Trans. Amer. Math. Soc. **69** (1950), 179–194.

A theorem concerning closed and compact point sets which lie in connected domains, Bull. Amer. Math. Soc. **46** (1940), 848.

MOHAT, J. T.

Concerning spirals in the plane, Duke Math. J. **24** (1957), 249–264.

MOISE, E. E.

An indecomposable plane continuum which is homeomorphic to each of its nondegenerate subcontinua, Trans. Amer. Math. Soc. **63** (1948), 581–594.

A note on the pseudo-arc, ibid. **67** (1949), 57–58.

A theorem on monotone interior transformations, Bull. Amer. Math. Soc. **55** (1949), 810–811.

Grille decomposition and convexification theorems for compact metric locally connected continua, ibid., 1111–1121.

A note of correction, Proc. Amer. Math. Soc. **2** (1951), 838.

Remarks on the Claytor imbedding theorem, Duke Math. J. **19** (1952), 199–202.

A remark on $\mathscr{L}^*$-spaces, Michigan Math. J. **1** (1952), 79–80.

Affine structures in 3-manifolds. I. *Polyhedral approximations of solids*, Ann. of Math. **54** (1951), 506–533.

Affine structures in 3-manifolds. II. *Positional properties of 2-spheres*, ibid. **55** (1952), 172–176.

Affine structures in 3-manifolds. III. *Tubular neighborhoods of linear graphs*, ibid., 203–213.

Affine structures in 3-manifolds. IV. *Piecewise linear approximations of homeomorphisms*, ibid., 215–222.

Affine structures in 3-manifolds. V. *The triangulation theorem and Hauptvermutung*, ibid. **56** (1952), 96–114.

Affine structures in 3-manifolds. VI. *Compact spaces covered by two Euclidean neighborhoods*, ibid. **58** (1953), 107.

Affine structures in 3-manifolds. VII. *Disks which are pierced by intervals*, ibid., 403–408.

Affine structures in 3-manifolds. VIII. *Invariance of the knot-types; local tame imbedding*, ibid. **59** (1954), 159–170.

Almost locally polyhedral spheres (with Harrold, O. G.), ibid. **57** (1953), 575–578.

MONTGOMERY, DEANE

Sections of point sets, Trans. Amer. Math. Soc. **35** (1933), 915–928.

A metrical property of point set transformations, Bull. Amer. Math. Soc. **40** (1934), 620–624.

Properties of plane sets and functions of two variables, Amer. J. Math. **56** (1934), 569–586.

Non-separable metric spaces, Fund. Math. **25** (1935), 527–533.

MOORE, R. L.

On a set of postulates which suffice to define a number-plane, Trans. Amer. Math. Soc. **16** (1915), 27–32.

The linear continuum in terms of point and limit, Ann. of Math. **16** (1915), 123–133.

On the linear continuum, Bull. Amer. Math. Soc. **22** (1915), 117–122.

Concerning a non-metrical pseudo-Archimedean axiom, ibid. **22** (1916), 225–236.

On the foundations of plane analysis situs, Trans. Amer. Math. Soc. **17** (1916), 131–164.

On the foundations of plane analysis situs, Proc. Nat. Acad. Sci. **2** (1916), 270–272.

A theorem concerning continuous curves, Bull. Amer. Math. Soc. **23** (1917), 233–236.

A characterization of Jordan regions by properties having no reference to their boundaries, Proc. Nat. Acad. Sci. **4** (1918), 364–370.

Continuous sets that have no continuous sets of condensation, Bull. Amer. Math. Soc. **25** (1919), 174–176.

Concerning a set of postulates for plane analysis situs, Trans. Amer. Math. Soc. **20** (1919), 169–178.

On the most general plane closed point set through which it is possible to pass a simple continuous arc (with Kline, J. R.), Ann. of Math. **20** (1919), 218–223.

On the most general class L of Fréchet in which the Heine-Borel-Lebesgue theorem holds true, Proc. Nat. Acad. Sci. **5** (1919), 206–210.

On the Lie-Riemann-Helmholtz-Hilbert problem of the foundations of geometry, Amer. J. Math. **41** (1919), 299–319.

Concerning simple continuous curves, Trans. Amer. Math. Soc. **21** (1920), 333–347.

Concerning certain equicontinuous systems of curves, Trans. Amer. Math. Soc. **22** (1921), 41–55.

On the relation of a continuous curve to its complementary domains in space of three dimensions, Proc. Nat. Acad. Sci. **8** (1922), 33–38.

Concerning connectedness im kleinen and a related property, Fund. Math. **3** (1922), 232–237.

Concerning continuous curves in the plane, Mathematische Zeitschrift **15** (1922), 254–260.

On the generation of a simple surface by means of a set of equicontinuous curves, Fund. Math. **4** (1923), 106–117.

An uncountable closed and non-dense point set each of whose complementary intervals abuts on another one at each of its ends, Bull. Amer. Math. Soc. **29** (1923), 49–50.

Concerning the cut-points of continuous curves and of other closed and connected point sets, Proc. Nat. Acad. Sci. **9** (1923), 101–106.

Report on continuous curves from the viewpoint of analysis situs, Bull. Amer. Math. Soc. **29** (1923), 289–302.

An extension of the theorem that no countable point set is perfect, Proc. Nat. Acad. Sci. **10** (1924), 168–170.

Concerning the prime parts of certain continua which separate the plane, ibid., 170–175.

Concerning relatively uniform convergence, Bull. Amer. Math. Soc. **30** (1924), 504–505.

Concerning the sum of a countable number of mutually exclusive continua in the plane, Fund. Math. **6** (1924), 189–202.

Concerning the common boundary of two domains, ibid., 208–213.

Concerning upper semi-continuous collections of continua which do not separate a given continuum, Proc. Nat. Acad. Sci. **10** (1924), 356–360.

Concerning sets of segments which cover a point set in the Vitali sense, ibid. **10** (1924), 464–467.

Concerning the prime parts of a continuum, Mathematische Zeitschrift **22** (1925), 307–315.

A characterization of a continuous curve, Fund. Math. **7** (1925), 302–307.

Concerning the separation of point sets by curves, Proc. Nat. Acad. Sci. **11** (1925), 469–476.

Concerning upper semi-continuous collections of continua, Trans. Amer. Math. Soc. **27** (1925), 416–428.

Concerning the relation between separability and the proposition that every uncountable point set has a limit point, Fund. Math. **8** (1926), 189–192; *An acknowledgement*, ibid., 374–375.

Conditions under which one of two given closed linear sets may be thrown into the other by a continuous transformation of a plane into itself, Amer. J. Math. **48** (1926), 67–72.

Concerning indecomposable continua and continua which contain no subsets that separate the plane, Proc. Nat. Acad. Sci. **12** (1926), 359–363.

Covering theorems, Bull. Amer. Math. Soc. **32** (1926), 275–282.

A connected and regular point set which contains no arc, ibid., 331–332.

Concerning paths which do not separate a given continuous curve, Proc. Nat. Acad. Sci. **12** (1926), 745–753.

Some separation theorems, ibid. **13** (1927), 711–716.

Concerning triods in the plane and the junction points of plane continua, ibid. **14** (1928), 85–88.

On the separation of the plane by a continuum, Bull. Amer. Math. Soc. **34** (1928), 303–306.

A separation theorem, Fund. Math. **12** (1928), 295–297.

Concerning triodic continua in the plane, ibid. **13** (1929), 261–263.

Concerning upper semi-continuous collections, Monatsh. Math. Phys. **36** (1929), 81–88.

Foundations of point set theory, American Mathematical Society Colloquium Publications, vol. 13, 1932, 7 + 486 pp.

Concerning compact continua which contain no continuum that separates the plane, Proc. Nat. Acad. Sci. **20** (1934), 41–45.

A set of axioms for plane analysis situs, Fund. Math. **25** (1935), 13–28.

Foundations of a point set theory of spaces in which some points are continuous to others, The Rice Institute Pamphlet, XXIII No. 1 (1936), 1–41.

Upper semi-continuous collections of the second type, ibid., 42–57.

On the structure of continua, ibid., 58–74.

Concerning essential continua of condensation, Trans. Amer. Math. Soc. **42** (1937), 41–52.

Concerning accessibility, Proc. Nat. Acad. Sci. **25** (1939), 648–653.

Concerning the open subsets of a plane continuum, ibid. **26** (1940), 24–25.
Concerning separability, ibid. **28** (1942), 56–58.
Concerning intersecting continua, ibid., 544–550.
Concerning a continuum and its boundary, ibid., 550–555.
Concerning domains whose boundaries are compact, ibid., 555–561.
Concerning continua which have dendratomic subsets, ibid. **29** (1943), 384–389.
Concerning tangents to continua in the plane, ibid. **31** (1945), 67–70.
A characterization of a simple plane web, ibid. **32** (1946), 311–316.
Spirals in the plane, ibid. **39** (1953), 207–213.

MORITA, KIITI

Star-finite coverings and the star-finite property, Math. Japonicae **1** (1948), 60–68.

MORREY, C. B.

The topology of (path) surfaces, Amer. J. Math. **57** (1935), 17–50.

MULLIKIN, ANNA M.

Certain theorems relating to plane connected point sets, Trans. Amer. Math. Soc. **24** (1922), 144–162.

NAGATA, J.

On topological completeness, J. Math. Soc. Japan **2** (1950), 44–48.
On a necessary and sufficient condition of metrizability, J. Inst. Polytech. Osaka City Univ. Ser. A. Math. **1** (1950), 93–100.

NALLI, PIA

Sopra una definizione di dominio piano limitato da una curva continua, senza punti multipli, Rend. Circ. Mat. Palermo **32** (1911), 391–401.

NIKODYM, S.

Sur les points linéairement accessibles des ensembles plans, Fund. Math. **7** (1925), 250–258.
Sur les coupures du plan faites par les ensembles connexes et les continus, ibid., 14–22.
Sur une condition nécessaire et suffisante pour qu'un sous-continu d'un continu jordanien et plan soit lui-même jordanien, ibid. **12** (1928), 160–187.
Sur quelques propriétés des ensembles partout localement connexes, ibid., 240–243.

PEANO, G.

Sur une courbe, qui remplit toute une aire plan, Math. Ann. **36** (1890), 157–160.

PEARSON, B. J.

A connected point set, in the plane, which spirals down on each of its points, Duke Math. J. **25** (1958), 603–613.

REID, W. T.

A certain three dimensional continuum, Bull. Amer. Math. Soc. **41** (1935), 683–684.
A theorem on plane continua, ibid., 684–688.

RIESZ, F.

Sur les ensembles discontinus, C. R. Acad. Sci. Paris **141** (1905), 650–655.

Stetigkeitsbegriff und Mengenlehre, Atti del IV Congresso Internazionale dei Matematici, Roma, 1908, vol. 2, 18–24.

ROBERTS, J. H.

On a problem of G. T. Whyburn (with Dorroh, J. L.), Fund. Math. **13** (1929), 58–61.

On a problem of C. Kuratowski concerning upper semi-continuous collections, ibid. **14** (1929), 96–102.

On a problem of Menger concerning regular curves, ibid., 327–333.

Concerning non-dense plane continua, Trans. Amer. Math. Soc. **32** (1930), 6–30.

Concerning collections of continua not all bounded, Amer. J. Math. **52** (1930), 561–562.

A note concerning cactoids, Bull. Amer. Math. Soc. **36** (1930), 894–896.

Concerning atriodic continua, Monatsh. Math. Phys. **37** (1930), 223–250.

Concerning metric collections of continua, Amer. J. Math. **53** (1931), 422–426.

A non-dense plane continuum, Bull. Amer. Math. Soc. **37** (1931), 720–722.

A point set characterization of closed 2-dimensional manifolds, Fund. Math. **18** (1931), 39–46.

A property related to completeness, Bull. Amer. Math. Soc., **38** (1932), 835–838.

Concerning uniordered spaces, Proc. Nat. Acad. Sci. **18** (1932), 403–406.

On a problem of Knaster and Zarankiewicz, Bull. Amer. Math. Soc. **40** (1934), 281–283.

Collections filling a plane, Duke Math. J. **2** (1936), 10–19.

Monotone transformations of two-dimensional manifolds (with Steenrod, N. E.), Ann. of Math. **39** (1938), 851–862.

Note on topological mappings, Duke Math. J. **5** (1939), 428–430.

Two-to-one transformations, Duke Math. J. **6** (1940), 256–262.

A theorem on dimension, Duke Math. J. **8** (1941), 565–574.

A nonconvergent iterative process, Proc. Amer. Math. Soc. **4** (1953), 640–644.

ROBINSON, S.

Covering theorems in general topology, Amer. J. Math. **55** (1933), 421–436.

On the reducibility of families of subsets and related properties (cf. Chittenden, E. W.).

ROOT, R. E.

Limits in terms of point and order, Trans. Amer. Math. Soc. **15** (1914), 51–71.

ROSEN, R. H.

Fixed points for multi-valued functions on snake-like continua, Proc. Amer. Math. Soc. **10** (1959), 167–173.

ROSENTHAL, A.

Teilung der Ebene durch irreduzible Kontinua, Bayer. Akad. Wiss. Math.-Nat. Kl. S.-B. (1919) 91–109.

Über Peanoflächen und ihren Rand, Math. Z. **10** (1921), 102.

RUDIN, MARY ELLEN ESTILL

Concerning abstract spaces, Duke Math. J. **17** (1950), 317–327.

Separation in non-separable spaces, Duke Math. J. **18** (1951), 623–629.

A primitive dispersion set of the plane, ibid. **19** (1952), 323–328.

Concerning a problem of Souslin's, ibid., 629–639.

Countable paracompactness and Souslin's problem, Canad. J. Math. **7** (1955), 543–547.

A topological characterization of sets of real numbers, Pacific J. Math. **7** (1957), 1185–1186.

A note on certain function spaces (with Klee, V. L., Jr.), Arch. Math. **7** (1957), 469–470.

A connected subset of the plane, Fund. Math. **46** (1958), 15–24.

An unshellable triangulation of a tetrahedron, Bull. Amer. Math. Soc. **64** (1958), 90–91.

RUTT, N. E.

Concerning the cut points of a continuous curve when the arc curve, AB, contains exactly n independent arcs, Amer. J. Math. **51** (1929), 217–246.

On certain types of plane continua, Trans. Amer. Math. Soc. **33** (1931), 806–816.

Concurrence and uncountability, Bull. Amer. Math. Soc. **39** (1939), 295–302.

Prime ends and order, Ann. of Math. **34** (1933), 415–440.

Some theorems on triodic continua, Amer. J. Math. **56** (1934), 122–132.

SAKS, S.

Sur l'équivalence de deux théorèmes de la théorie des ensembles, Fund. Math. **2** (1921), 1–3.

SCHOENFLIES, A.

Über einen grundlegenden Satz der Analysis situs, Nachr. Akad. Wiss. Göttingen. Math.-Phys. Kl. IIa. (1902), 185–192.

Die Entwicklung der Lehre von den Punktmannigfaltigkeiten, II Teil, Leipzig, B. G. Teubner, 1908, 10 + 331 pp.

SCHWEIGERT, G. E.

The analysis of certain curves by means of derived local separating points, Amer. J. Math. **58** (1936), 329–335.

SHIMRAT, M.

Simply disconnected sets, Proc. London Math. Soc. **9** (1959), 177–188.

SIERPIŃSKI, W.

L'arc simple comme un ensemble de points dans l'espace à m-dimensions, Annali di Matematica, Serie III **26** (1916), 131–150.

Un théorème sur les ensembles fermés, Bull. Acad. Sci. Cracovie (1918), 49–51.

Une démonstration du théorème sur la structure des ensembles de points, Fund. Math. **1** (1920), 1–6.

Sur un ensemble punctiforme connexe, ibid., 7–10.

Sur une propriété topologique des ensembles dénombrables, denses en soi, ibid., 11–16.

Contribution à la topologie des ensembles dénombrables (with Mazurkiewicz, S.), ibid., 17–27.

Sur la décomposition des ensembles de points en parties homogènes, ibid., 28–34.

Sur une condition pour qu'un continu soit une courbe jordanienne, ibid., 44–60.

Sur l'équivalence de trois propriétés des ensembles abstraits, Fund. Math. **2** (1921), 179–188.

Sur les ensembles connexes et non connexes, ibid., 81–95.

Le théorème de Borel-Lebesgue dans la théorie des ensembles abstraits, ibid., 172–178.
Une remarque sur la notion de l'ordre, ibid., 199–200.
Sur une propriété des ensembles frontières, Fund. Math. **3** (1922), 7–13.
Sur quelques propriétés topologiques du plan, ibid. **4** (1923), 1–6.
Sur l'espace D_ω de M. Fréchet, ibid. **9** (1927), 189–192.
Sur l'invariance topologique des ensembles G_δ, ibid. **8** (1926), 135–136.
Introduction to general topology, translated by C. C. Krieger, The University of Toronto Press, Toronto, x + 238 pp.
Hypothèse du continu, Warszawa, Seminarjum Matematyczne Uniwersytetu Warszawskiego, Monagrafie Matematycne vol. 4, 1934.
Sur un espace métrique séparable universel, Atti. Accad. Sci. Torino Cl. Sci. Fis. Mat. Nat. **75** (1940), 575–577.
Sur la décomposition des espaces métriques en ensembles disjoints, Fund. Math. **36** (1949), 68–71.
Sur une propriété des ensembles plans fermés et bornés, Matematiche, Catania **6** (1951), 132–134.
On a certain definition of a complete space, Wiadom. Mat. **1** (1955/1956), 206–207.
Sur un problème de H. Steinhaus concernant les ensembles du points sur le plan, Fund. Math. **46** (1959), 191–194.

SLYE, J. M.

Flat spaces for which the Jordan curve theorem holds true, Duke Math. J. **22** (1955), 143–151.
Collections whose sums are two-manifolds, ibid. **24** (1957), 275–298.

SMIRNOV, YU.

A necessary and sufficient condition for metrizability of a topological space, Dokl. Akad. Nauk SSSR. (N.S.) **77** (1951), 197–200.

SMYTHE, W. R., JR.

A theorem on upper semi-continuous decomposition, Duke Math. J. **22** (1955), 485–495.

SORGENFREY, R. H.

Concerning triodic continua, Amer. J. Math. **66** (1944), 439–460.
Some theorems on co-terminal arcs, Bull. Amer. Math. Soc. **50** (1944), 257–259.
Concerning continua irreducible about n points, Amer. J. Math. **68** (1946), 667–671.
On the topological product of paracompact spaces, Bull. Amer. Math. Soc. **53** (1947), 631–632.

STEENROD, N. E.

Finite arc-sums, Fund. Math. **23** (1934), 38–53.
Characterizations of certain finite curve sums, Amer. J. Math. **56** (1934), 558–568.
Monotone transformations of two-dimensional manifolds (cf. Roberts, J. H.).

STONE, A. H.

Paracompactness and product spaces, Bull. Amer. Math. Soc. **54** (1949), 977–982.
Incidence relations in multicoherent spaces. I, Trans. Amer. Math. Soc. **66** (1949), 389–406.

Incidence relations in multicoherent spaces. II, Canad. J. Math. **2** (1950), 461–480.
Incidence relations in multicoherent spaces. III, Pacific J. Math. **2** (1952), 99–126.
On infinitely multicoherent spaces, Quart. J. Math. Oxford Ser. (2) **3** (1952), 298–306.
On coverings of two dimensional spaces, Proc. London Math. Soc. **3** (1953), 338–349.
Metrizability of decomposition spaces, Proc. Amer. Math. Soc. **7** (1956), 690–700.

STONE, M. H.

Applications of the theory of Boolean rings to general topology, Trans. Amer. Math. Soc. **41** (1937), 375–481.

STRASZEWICZ, S.

Verallgemeinerung des Jordanschen Kurvensatzes, Fund. Math. **4** (1923), 128–135.
Über die Zerschneidung der Ebene durch abgeschlossene Mengen, ibid. **7** (1925), 159–187.
Généralisation d'un théorème de Janiszewski (with Kuratowski, C.), ibid. **12** (1928), 152–157.

SWINGLE, P. M.

An unnecessary condition in two theorems of analysis situs, Bull. Amer. Math. Soc. **34** (1928), 607–618.
A certain type of continuous curve and related point sets, Trans. Amer. Math. Soc. **33**, (1929), 544–556.
Generalized indecomposable continua, Amer. J. Math. **52** (1930), 647–658.
Generalizations of biconnected sets, ibid. **53** (1931), 385–400.
Two types of connected sets, Bull. Amer. Math. Soc. **37** (1931), 254–258.
End-sets of continua irreducible between two points, Fund. Math. **17** (1931), 40–76.
Biconnected and related sets, Amer. J. Math. **54** (1932), 525–535.
A finitely-containing connected set, Bull. Amer. Math. Soc. **46** (1940), 178–181.
Indecomposable connexes, ibid. **47** (1941), 796–803.
The closure of types of connected sets, Proc. Amer. Math. Soc. **2** (1951), 178–185.
Local properties and sums of trajectories, Portugal. Math. **15** (1956), 89–103.
Sums of sets with indecomposable properties, ibid. **16** (1957), 129–144.
Higher dimensional indecomposable connected sets, Proc. Amer. Math. Soc. **8** (1957), 816–819.
Connected sets of Van Vleck, ibid. **9** (1958), 477–482.
Indecomposable trajectories (cf. Hunter, R. P.).

SZYMAŃSKI, P.

Sur les constituants d'ensembles situés sur des continus arbitraires, Fund. Math. **10** (1927), 363–374.
La somme de deux continus irréductibles, ibid. **11** (1928), 1–14.

TIETZE, H.

Über stetige Kurven, Jordansche Kurvenbogen und geschlossene Jordansche Kurven, Mathematische Zeitschrift **5** (1919), 284–291.
Über einfach zusammenhängende Flächen und ihre Deformationen in sich, Sitzungsber. Akad. Wiss. Wien, Math.-Nat. Klasse **122** (1913), 1653–1658.
Beitrage zur allgemeinen Topologie. I. *Axiome für verschiedene Fassungen des Umgebungsbegriffs,* Math. Ann. **88** (1923), 290–312.

Beitrage zur allgemeinen Topologie. II. *Über die Einführung uneigentlicher Elemente,* ibid. **91** (1924), 210–224.

Beitrage zur allgemeinen Topologie. III. *Über die Komponenten offener Mengen,* Monatsh. Math. Phys. **33** (1923), 15–17.

Über Analysis Situs, Hamburger Mathematische Einzelschriften 2 Heft, 1923, 32 S. Im Verlag des math. Seminars der Hamburgischen Universität.

TORHORST, MARIE

Über den Rand der einfach zusammenhängenden ebenen Gebiete, Mathematische Zeitschrift **9** (1921), 44–65.

TORRANCE, C. C.

Tangent lines and planes in topological spaces, Trans. Amer. Math. Soc. **41** (1937), 193–207.

TUKEY, J. W.

Convergence and uniformity in topology, Annals of Mathematics Studies, No. 2, Princeton University Press, Princeton, New Jersey, 1940, ix + 90 pp.

TYCHONOFF, A.

Über einen Metrisationssatz von P. Urysohn, Math. Ann. **95** (1926), 139–142.

URSELL, H. D. AND YOUNG, L. C.

Remarks on the theory of prime ends, Mem. Amer. Math. Soc., no. 3, 1951, 29 pp.

URYSOHN, P.

Über die metrisation der kompakten topologischen Räume, Math. Ann. **92** (1924), 275–293.

Zur Theorie der topologischen Räume (with Alexandroff, P.), ibid., 258–266.

Zum Metrisationsproblem, Math. Ann. **94** (1925), 309–315.

Über die Mächtigkeit der zusammenhängenden Mengen, ibid., 262–295.

Une condition nécessaire et suffisante pour qu'une classe ($\mathfrak{L}$) *soit une classe* (ω) (with Alexandroff, P.), C. R. Acad. Sci. Paris **177** (1923), 1274.

Les classes (ω) *séparable et l'espace Hilbertien,* ibid. **178** (1924), 65.

Mémoire sur les multiplicités Cantoriennes, Fund. Math. **7** (1925), 29–137 et **8** (1926), 225–351.

Sur un espace métrique universel, Bull. Sci. Math., 2e série **51** (1927), 1–38; C. R. Acad. Sci. Paris **180** (1925), 803.

Über im kleinen zusammenhängende Kontinua, Math. Ann. **98** (1927), 296–308.

Mémoire sur les multiplicités Cantoriennes. Deuxième partie: *Les lignes Cantoriennes,* Verh. Nederl. Akad. Wetensch. Afd. Natuurk. Sect. I. 13, No. 4 (1928), 1–172.

VAUGHAN, H. E.

On a class of metrics defining a metrisable space, Bull. Amer. Math. Soc. **44** (1938), 557–561.

On locally bicompact spaces, Fund. Math. **31** (1938), 15–21.

VEBLEN, O.

Theory of plane curves in non-metrical analysis situs, Trans. Amer. Math. Soc. **6** (1905), 107–112.

Definition in terms of order alone in the linear continuum and in well ordered sets, ibid., 165–171.

The Heine-Borel Theorem, Bull. Amer. Math. Soc. **10** (1903–1904), 436–439.

VERONESE, G.

Fondamenti di Geometria a più dimensioni e a più specie di unità retilinee esposti in forma elementare, Padova, Tipografia del seminario, 1891, XVIII + 628 pp.

Grundzüge der Geometrie von mehreren Dimensionen und mehreren Arten geradliniger Einheiten in elementarer Form entwickelt. Mit Genehmigung des Verfassers und nach einer neuen Bearbeitung des Originals übersetzt von A. Schepp, Leipzig, B. G. Teubner, 1894, XLVI + 710S.

VICKERY, C. W.

Spaces in which there exist uncountable convergent sequences of points, Tôhoku Math. J. **40** (1935), 1–26.

Spaces of uncountably many dimensions, Bull. Amer. Math. Soc. **45** (1939), 456–462.

Axioms for Moore space and metric spaces, Bull. Amer. Math. Soc. **46** (1940), 560–564.

VIETORIS, L.

Stetige Mengen, Monatsh. Math. Phys. **31** (1921), 173–204.

Über stetige Abbildungen einer Kugelfläche, Proc. Roy. Acad. Sci. Amsterdam **29** (1926), 443–453.

WALLACE, A. D.

Monotone coverings and monotone transformations, Duke Math. J. **6** (1940), 31–37.

On non-boundary sets, Bull. Amer. Math. Soc. **45** (1939), 420–422.

Some characterizations of interior transformations, Amer. J. Math. **61** (1939), 757–763.

The acyclic elements of a Peano space, Bull. Amer. Math. Soc. **47** (1941), 778–780.

WARDWELL, J. F.

Continuous transformations preserving all topological properties, Amer. J. Math. **58** (1936), 709–726.

WHYBURN, G. T.

Two-way continuous curves, Bull. Amer. Math. Soc. **32** (1926), 659–663.

Concerning certain types of continuous curves, Proc. Nat. Acad. Sci. **12** (1926), 761–767.

Cyclically connected continuous curves, ibid. **13** (1927), 31–38.

The most general closed point set over which continuous function may be defined by certain properties, Bull. Amer. Math. Soc. **33** (1927), 185–188.

Concerning continua in the plane, Trans. Amer. Math. Soc. **29** (1927), 369–400.

Some properties of continuous curves, Bull. Amer. Math. Soc. **33** (1927), 305–308.

Concerning point sets which can be made connected by the addition of a simple continuous arc, Trans. Amer. Math. Soc. **29** (1927), 746–754.

Concerning the disconnection of continua by the omission of pairs of their points, Fund. Math. **10** (1927), 180–185.

Concerning connected and regular point sets, Bull. Amer. Math. Soc. **33** (1927), 685–689.
Concerning the open subsets of a plane continuous curve, Proc. Nat. Acad. Sci. **13** (1927), 650–657.
Concerning the structure of a continuous curve, Amer. J. Math. **50** (1928), 167–194.
Concerning the cut points of continua, Trans. Amer. Math. Soc. **30** (1928), 597–609.
Concerning accessibility in the plane and regular accessibility in n dimensions, Bull. Amer. Math. Soc. **34** (1928), 504–510.
Concerning plane closed point sets which are accessible from certain subsets of their complements, Proc. Nat. Acad. Sci. **14** (1928), 657–666.
On a problem of W. L. Ayres, Fund. Math. **11** (1928), 296–301.
Concerning Menger regular curves, ibid. **12** (1928), 264–294.
On certain accessible points of plane continua, Monatsh. Math. Phys. **35** (1928), 289–304.
In continuous curves in n dimensions (with Ayres, W. L.), Bull. Amer. Math. Soc. **34** (1928), 349–360.
Concerning the complementary domains of continua, Ann. of Math. **29** (1928), 349–411.
Concerning collections of cuttings of connected point sets, Bull. Amer. Math. Soc. **31** (1929), 87–104.
On regular points of continua and regular curves of at most order n, Bull. Amer. Math. Soc. **35** (1929), 218–224.
Concerning irreducible cuttings of continua, Fund. Math. **13** (1929), 42–57.
Local separating points of continua, Monatsh. Math. Phys. **36** (1929), 305–314.
On simple closed curves, Bull. Acad. Polon. Sci. Math. (1929), 280–283.
Concerning points of continuous curves defined by certain im kleinen properties, Math. Ann. **102** (1929), 313–336.
Continuous curves and arc-sums, Fund. Math. **14** (1929), 103–106.
A generalized notion of accessibility, ibid., 311–326.
Cut points of connected sets and of continua, Trans. Amer. Math. Soc. **32** (1930), 147–154.
A continuum every subcontinuum of which separates the plane, Amer. J. Math. **52** (1930), 319–330.
The rationality of certain continuous curves, Bull. Amer. Math. Soc. **36** (1930), 522–524.
On the structure of connected and connected im kleinen point sets, Trans. Amer. Math. Soc. **32** (1930), 926–943.
On the set of all cut points of a continuous curve, Fund. Math. **15** (1930), 185–194.
Sur l'accessibilité des continus plans, Fund. Math. **15** (1930), 322–323.
Sur les ε-séparations irreductibles, Fund. Math. **16** (1930), 77–80.
Potentially regular point sets, ibid., 160–172.
Concerning continuous curves without local separating points, Amer. J. Math. **53** (1931), 163–166.
Concerning hereditarily locally connected continua, ibid., 374–384.
The cyclic and higher connectivity of locally connected spaces, ibid., 427–442.
Non-separated cuttings of connected point sets, Trans. Amer. Math. Soc. **33** (1931), 444–454.
On the cyclic connectivity theorem, Bull. Amer. Math. Soc. **37** (1931), 429–433.
Concerning continuous images of the interval, Amer. J. Math. **53** (1931), 670–674.
A junction property of locally connected sets, ibid., 753–756.
On the divisibility of locally connected spaces, Bull. Amer. Math. Soc. **37** (1931), 734–736.
Concerning addition of regular curves, Monatsh. Math. Phys. **38** (1931), 1–4.
Concerning the subsets of regular curves, ibid., 85–88.

Sur les éléments cycliques et leur applications (cf. Kuratowski, C.).

On the decomposability of closed sets into a countable number of simple sets of various types, Amer. J. Math. **54** (1932), 169–175.

A certain transformation on metric spaces, ibid., 367–376.

On the construction of simple arcs, ibid., 518–524.

A note on spaces having the S property, ibid., 536–538.

Concerning the proposition that every closed, compact, and totally disconnected set of points is a subset of an arc, Fund. Math. **18** (1932), 47–60.

Characterizations of certain curves by continuous functions defined upon them, Amer. J. Math. **55** (1933), 131–134.

On the existence of totally imperfect and punctiform connected subsets in a given continuum, ibid., 146–152.

Sets of locally separating points of a continuum, Bull. Amer. Math. Soc. **39** (1933), 97–100.

Decompositions of continua by means of local separating points, Amer. J. Math. **55** (1933), 437–457.

Concerning S-regions in locally connected continua, Fund. Math. **20** (1933), 131–139.

Cyclic elements of higher orders, Amer. J. Math. **56** (1934), 133–146.

Concerning maximal sets, Bull. Amer. Math. Soc. **40** (1934), 159–164.

Non-alternating transformations, Amer. J. Math. **56** (1934), 294–302.

Concerning continua of finite degree and local separating points, Amer. J. Math. **59** (1935), 11–16.

A decomposition theorem for closed sets, Bull. Amer. Math. Soc. **41** (1935), 95–96.

Concerning perfect sets, Duke Math. J. **1** (1935), 35–38.

Regular convergence and monotone transformations, Amer. J. Math. **57** (1935), 902–906.

On sequences and limiting sets, Fund. Math. **25** (1935), 408–426.

On the structure of continua, Bull. Amer. Math. Soc. **42** (1936), 49–73.

Arc-preserving transformations, Amer. J. Math. **58** (1936), 305–312.

On continua of condensation, ibid., 705–708.

Completely alternating transformations, Fund. Math. **27** (1936), 140–146.

Semi-closed sets and collections, Duke Math. J. **2** (1936), 685–690.

Concerning rationality bases for curves, Ergebnisse eines Mathematische Kolloquiums, Wien **7** (1936), 58–60.

Interior transformations on compact sets, Duke Math. J. **3** (1937), 370–381.

Interior transformations on surfaces, Amer. J. Math. **60** (1938), 477–490.

A theorem on interior transformations, Bull. Amer. Math. Soc. **44** (1938), 414–416.

Interior surface transformations, Duke Math. J. **4** (1938), 626–634.

Semi-locally connected sets, Amer. J. Math. **61** (1939), 733–749.

The existence of certain transformations, Duke Math. J. **5** (1939), 647–656.

On irreducibility of transformations, Amer. J. Math. **61** (1939), 820–822.

A relation between non-alternating and interior transformations, Bull. Amer. Math. Soc. **46** (1940), 320–321.

Analytic topology, American Mathematical Society Colloquium Publications, vol. 28, 1942, 10 + 278 pp.

Coherent and saturated collections, Trans. Amer. Math. Soc. **57** (1945), 287–298.

Uniqueness of the inverse of a transformation, Duke Math. J. **12** (1945), 317–323.

On n-arc connectedness, Trans. Amer. Math. Soc. **63** (1948), 452–456.

Continuous decompositions, Amer. J. Math. **71** (1949), 218–226.

Open mappings on locally compact spaces, Memoirs Amer. Math. Soc., No. 1, 1950, i + 24 pp.

Topological characteristics of the Sierpiński curve, Fund. Math. **45** (1953), 320–324.

Topological analysis, Bull. Amer. Math. Soc. **62** (1956), 204–218.

WILDER, R. L.

On the dispersion sets of connected point sets, Fund. Math. **6** (1924), 214–228.

A theorem on continua, ibid. **7** (1925), 311–313.

Concerning continuous curves, ibid., 340–377.

A property which characterizes continuous curves, Proc. Nat. Acad. Sci. **11** (1925), 725–728.

A theorem on connected point sets which are connected im kleinen, Bull. Amer. Math. Soc. **32** (1926), 338–340.

A connected and regular point set which has no sub-continuum, Trans. Amer. Math. Soc. **29** (1927), 332–340.

A point set which has no true quasi-components, and which becomes connected upon the addition of a single point, Bull. Amer. Math. Soc. **33** (1927), 423–427.

The non-existence of a certain type of regular point set, Bull. Amer. Math. Soc. **33** (1927), 439–446.

On connected and regular point sets, ibid. **34** (1928), 649–655.

Concerning R. L. Moore's axioms Σ_1 for plane analysis situs, ibid., 752–760.

A characterization of continuous curves by a property of their open subsets, Fund. Math. **11** (1928), 127–131.

On a certain type of connected set which cuts the plane, Proc. Internat. Math. Congress in Toronto **1** (1928), 423–437.

Concerning zero-dimensional sets in euclidean space, Trans. Amer. Math. Soc. **31** (1929), 354–359.

Characterizations of continuous curves that are perfectly continuous, Proc. Nat. Acad. Sci. **15** (1929), 614–621.

Concerning perfect continuous curves, ibid. **16** (1930), 223–240.

A converse of the Jordan-Brouwer separation theorem in three dimensions, Trans. Amer. Math. Soc. **32** (1930), 632–657.

Concerning simple continuous curves and related point sets, Amer. J. Math. **53** (1931), 38–55.

Extension of a theorem of Mazurkiewicz, Bull. Amer. Math. Soc. **37** (1931), 287–293.

A plane, arcwise connected and connected im kleinen point set which is not strongly connected im kleinen, ibid. **38** (1932), 531–532.

Point sets in three and higher dimensions and their investigation by means of a unified analysis situs, ibid., 649–692.

On the imbedding of subsets of a metric space in Jordan continua, Fund. Math. **19** (1932), 45–64.

On the linking of Jordan continua in E_n by $(n-2)$-cycles, Ann. of Math. **34** (1933), 441–449.

Concerning a problem of K. Borsuk, Fund. Math. **21** (1933), 156–167.

On the properties of domains and their boundaries in E_n, Math. Ann. **109** (1933), 273–306.

Concerning irreducible connected sets and irreducible regular connexes, Amer. J. Math. **56** (1934), 547–557.

Generalized closed manifolds in n-space, Ann. of Math. **35** (1934), 876–903.

On free subsets of E_n, Fund. Math. **25** (1935), 200–208.

On locally connected spaces, Duke Math. J. **1** (1935), 543–555.

A characterization of manifold boundaries in E_n dependent only on lower dimensional connectivities of the complement, Bull. Amer. Math. Soc. **42** (1936), 436–441.

The strong symmetrical cut-sets of closed euclidean n-space, Fund. Math. **27** (1936), 136–139.

The sphere in topology, American Mathematical Society Semicentennial Publications, vol. 2, 1938, 136–184.

Sets which satisfy certain avoidability conditions, Časopis pro Pěstování Matematiky a Fysiky **67** (1937–38), 185–198.

Property S_n, Amer. J. Math. **61** (1939), 823–832.

Decompositions of compact metric spaces, Amer. J. Math. **63** (1941), 691–697.

Uniform local connectedness, Lectures in Topology, pp. 29–41, University of Michigan Press, Ann Arbor, Michigan, 1941.

Topology of manifolds, American Mathematical Society Colloquium Publications, vol. 32, 1949, 5 + 402 pp.

Introduction to the foundations of mathematics, John Wiley and Sons, Inc., New York; Chapman and Hall, Ltd., London, 1952, xiv + 305 pp.

Some mapping theorems, with applications to non-locally connected spaces. Algebraic geometry and topology. A symposium in honor of S. Lefschetz, pp. 378–388, Princeton University Press, Princeton, New Jersey, 1957.

Monotone mappings of manifolds, Pacific J. Math. **7** (1957), 1519–1528.

WILLIAMS, R. F.

Local contraction and the size of a compact metric space, Duke Math. J. **26** (1959), 277–289.

The effect of maps upon the dimension of subsets of the domain spaces, Proc. Amer. Math. Soc. **8** (1957), 580–583.

WILSON, W. A.

On the oscillation of a continuum at a point, Trans. Amer. Math. Soc. **27** (1925), 429–440.

On the structure of a continuum, limited and irreducible between two points, Amer. J. Math. **48** (1926), 147–168.

Some properties of limited continua, irreducible between two points, Trans. Amer. Math. Soc. **28** (1926), 536–553.

On the separation of the plane by irreducible continua, Bull. Amer. Math. Soc. **33** (1927), 733–744.

On bounded regular frontiers in the plane, ibid. **34** (1928), 81–90.

A curious irreducible continuum, ibid. **32** (1926), 679–681.

On irreducible cross-cuts of plane simply connected regions, Amer. J. Math. **51** (1929), 19–30.

Certain problems relating to the cutting of a simply connected plane region by a continuum, Trans. Amer. Math. Soc. **31** (1929), 552–562.

A property of continua equivalent to local connectivity, Bull. Amer. Math. Soc. **36** (1930), 85–88.

On the Phragmen-Brouwer theorem, ibid., 111–114.

A property of continua similar to local connectivity, ibid. **37** (1931), 294–300.

A property of unbounded continua, with applications, Amer. J. Math. **52** (1930), 537–542.

On semi-metric spaces, ibid. **53** (1931), 361–373.

On quasi-metric spaces, ibid., 675–684.

On upper semi-continuous decompositions of compact continua, ibid. **54** (1932), 377–386.

On cyclic numbers of one-dimensional compact sets, Trans. Amer. Math. Soc. **34** (1932), 263–273.

On unicoherency about a simple closed curve, Amer. J. Math. **55** (1933), 135–145.

A separation theorem, Bull. Amer. Math. Soc. **39** (1933), 440–442.

On certain types of continuous transformations of metric spaces, Amer. J. Math. **57** (1935), 57–68.

On the imbedding of metric sets in euclidean space, ibid., 322–326.

WOODARD, D. W.

The characterization of the closed n-cell, Trans. Amer. Math. Soc. **42** (1937), 396–415.

YONEYAMA, K.

Theory of continuous set of points, Tôhoku Math. J. **12** (1917), 43–158.

YOUNG, G. S., JR.

On continua whose links are not intersecting, Bull. Amer. Math. Soc. **50** (1944), 920–925.

A generalization of Moore's theorem on simple triods, Amer. J. Math. **66** (1944), 439–460.

Spaces in which every arc has two sides, Ann. of Math. **46** (1945), 182–193.

Spaces congruent with bounded subsets of the line, Bull. Amer. Math. Soc. **52** (1946), 915–917.

The introduction of local connectivity by change of topology, Amer. J. Math. **68** (1946), 479–494.

A characterization of 2-manifolds, Duke Math. J. **14** (1947), 979–990.

On compact fiberings of the plane, Bull. Amer. Math. Soc. **53** (1947), 295–298.

On continuous curves irreducible about compact sets, Bull. Amer. Math. Soc. **55** (1949), 439–441.

A generalization of the Rutt-Roberts theorem, Proc. Amer. Math. Soc. **2** (1951), 586–588.

YOUNG, W. H.

Zur Lehre der nichtabgeschlossenen Mengen, Leipziger Berichte, 1903.

On sequences of sets of intervals containing a given set of points, Proc. London Math. Soc. Second Series, **1** (1903), 262–266.

Theory of sets of points, University Press, Cambridge, 1906, 12 + 316 pp.

YOUNGLOVE, J. N.

Concerning dense metric subspaces of certain non-metric spaces, Fund. Math. **48** (1959), 15–25.

YOUNGS, J. W. T.

K-cyclic elements, Amer. J. Math. **62** (1940), 449–456.

Arc-spaces, Duke Math. J. **62** (1940), 449–458.

The structure of locally connected topological spaces (cf. Albert, G. E.).

A note on separation axioms and their application in the theory of locally connected topological spaces, Bull. Amer. Math. Soc. **49** (1943), 383–385.

The topological theory of Fréchet surfaces, Ann. of Math. **45** (1944), 753–785.
Homeomorphic approximation to monotone mappings, Duke Math. J. **15** (1948), 87–94.
Extension of a homeomorphism, Bull. Amer. Math. Soc. **54** (1948), 805–808.

ZARANKIEWICZ, C.

Remarque sur un théorème de M. Kline, Fund. Math. **5** (1924), 11–13.
Sur les points de division dans les ensembles connexes, ibid. **9** (1927), 124–171.
Sur la structure d'un ensemble de points de division dans les continus de Jordan, Bull. Acad. Polon. Sci. Math. (1926), 361–371.
Sur les coupures faites par les continus, ibid. (1927), 193–218.
Über Endpunkte, ibid. (1928), 445–453.
A theorem on connected point sets (with Kuratowski, C.), Bull. Amer. Math. Soc. **33** (1927), 571–575.
Über eine topologische Eigenschaft der Ebene, Fund. Math. **11** (1928), 19–26.
Über die Zerschneidungspunkte der zusammenhängended Menge, ibid. **12** (1928), 121–125.
Über die lokale Verschneidung der Ebene, Monatsh. Math. Phys. **39** (1932), 371–376.
Über eine Eigenschaft des Konvergenzkontinuums, Sitzungsberichte der Berliner Mathematischen Gesellschaft **31** (1932), 43–45.
Sur le nombre des points de ramification dans des dendrites et dans des graphes, Comptes Rendus des Séances de la Société des Sciences et des Lettres de Varsovie, Class 3, **39** (1946), 18–24.
Relations symétriques dans l'ensemble fini, Colloq. Math. **1** (1947), 10–14.
Quelques généralisations des théorèmes sur les coupures du plan (cf. Kuratowski, C.).
On the category of the set of cut points of continua of a certain type, Czechoslovak Math. J. **1** (1951), 57–62.

ZERMELO, E.

Beweis, dass jede Menge wohlgeordnet werden kann, Math. Ann. **59** (1904), 514–516.
Neuer Beweis für die Möglichkeit der Wohlordnung, ibid. **65** (1908), 107–128.
Untersuchungen über die Grundlagen der Mengenlehre. I, ibid., 261.

ZIPPIN, LEO

On continuous curves and the Jordan curve theorem, Amer. J. Math. **52** (1930), 331–350.
A study of continuous curves and their relation to the Janiszewski-Mullikin theorem, Trans. Amer. Math. Soc. **31** (1929), 744–770.
On a problem of N. Aronszajn and an axiom of R. L. Moore, Bull. Amer. Math. Soc. **37** (1931), 276–280.
Generalization of a theorem due to C. M. Cleveland, Amer. J. Math. **54** (1932), 176–184.
The Moore-Kline problem, Trans. Amer. Math. Soc. **34** (1932), 705–721.
Independent arcs of a continuous curve, Ann. of Math. **34** (1933), 95–113.
A characterization of the closed 2-cell, Amer. J. Math. **55** (1933), 207–217.
On continuous curves irreducible about subsets, Fund. Math. **20** (1933), 197–205.
Semi-compact spaces, Amer. J. Math. **57** (1935), 327–341.
On a problem of Čech, Časopis pro pěstování Matematiky a Fysiky **65** (1936), 49–52.
On monotonic, complete covering systems, Fund. Math. **27** (1936).

ZORETTI, L.

Sur les fonctions analytiques uniformes, J. Math. Pures Appl. **1** (1905), 1–51.
Sur les ensembles discontinus, C. R. Acad. Sci. Paris **142** (1906), 763–764.
La notion de ligne, Ann. École Normale **26** (1909), 485–497.
Sur les ensembles de points, C. R. Acad. Sci. Paris **150** (1910), 162–164.
Leçons sur le prolongement analytique, Gauthier-Villars, Paris, 1911, 6 + 114 pp.
Sur la représentation analytique d'un continu irréductible, Bull. Soc. Math. France **39** (1911), 246–250.
Contribution à l'étude des lignes cantoriennes, Acta. Math. **36** (1912), 241–268.

GLOSSARY

(Numbers refer to pages)

Abutting, of one arc on another, 180.
Accessibility, of a point set from a point set, 85; from all sides, 266.
Acyclic continuous curve, 113.
Aposyndetic point set, 379.
Arc, 39 and 99; of a simple closed curve, 142; of simple links of a compact continuous curve, 364.
Arcatomic subset of a continuum, 288.
Arc-curve chain, 267.
Arc-wise connected point set, 121.
Aspiculate cactoid, 140.
Atriodic continuum, 218.
Betweenness, in a point set on an interval, 33.
Boundary of a point set, 1; outer boundary of a domain with respect to a point set, 176.
Bounded point set, 141.
Cactoid, 140; aspiculate, 140; simple, 140.
Cantor set, 62.
Closed point set, 1; relatively to a point set, 273.
Closing down of a sequence on a point, 4.
Closure of a point set, 1; notation for, 1.
Coherent collection, 76.
Collection, ix; upper semi-continuous, 273; continuous, 289; of class 1, 294; equicontinuous with respect to a given point set, 331; self-compact with respect to a given point set, 331; topological, 356; semi-contracting, 107; topologically contracting, 285; coherent, 46.
Common part of the sets of a collection, 1.
Compact point set, 1; locally compact, 21; perfectly compact, 2.
Compactly connected point set, 76.
Complement of a point set relative to (or with respect to) a point set containing that set, 26.
Complementary domain of a closed point set, 141.
Component of a point set, 11.
Composant of a point set, 57.
Connected point set, 11; strongly connected, 121; arc-wise connected, 121; locally connected, 89; connected im kleinen, 89; uniformly connected im kleinen, 233; maximal connected subset of a point set, 11.
Connected set of real numbers, 354.
Continuum, 14; of condensation of a point set, 58; essential continuum of condensation, 294; irreducible continuum from one point set to another one, 14; irreducible continuum about a point set, 14; indecomposable continuum, 58; triodic, 254.
Continuous curve, 89; simple, 99; acyclic, 113; cyclicly connected, 138.
Continuous collection of mutually exclusive closed point sets, 289.
Convergent sequence each term of which is a point set, 24.
Countable set, xi.
Covering of a set by a collection of sets, 1.
Crossing of an arc at a point, 182; at an interval, 183.
Cut point, 25.
Cyclic element of a continuous curve, 138.
Cyclicly connected continuous curve, 138.
Decomposable continuum, hereditarily, 293.
Dedekind-cut proposition, 42.
Degenerate set, 11.
Dendratomic subset of a continuum, 288.
Dendron, 113.
Density of a point set in a point set, 85.
Disk, simple, 152.
Domain, 7; relatively to (or with respect to) a point set, 7; of elements of an upper semi-continuous collection with respect to that collection, 275; simple, 152; simply connected, 217.
Emanation point of a triod of type 1, 291.

1; $M \cdot N$ for the common part of M and N, 1; S for a set such that P belongs to it if and only if P is a point, 1; ω for the point at infinity, 141.

Term of a sequence, x.

Topological collection, 356.

Topological equivalence, 353 and 354; strong, with respect to a point set, 357.

Topological homogeneity, 378.

Topological space, 74.

Topologically contracting collection of point sets, 285.

Totally disconnected point set, 23.

Transformation of a set into a set; image of a set under a transformation; reversible transformation; continuous; reversibly continuous; continuous relatively to a space, 353; order preserving, 354.

Triod, 218; simple, 218; of type 1, 291; emanation point of, 291; rays of, 291.

Triodic continuum, 254.

Unicoherent continuum, 222; hereditarily, 222.

Upper semi-continuous collection, 273.

Web, 297.

Webless continuum, 297.

Zermelo Axiom, ix.

Zermelo Proposition, ix.

FGHIJ-8987